Mobile Satellite Communications

M. Richharia

Mobile Satellite Communications
Principles and Trends

▲▼ Addison-Wesley

An imprint of **PEARSON EDUCATION**

Harlow, England · London · New York · Reading, Massachusetts · San Francisco · Toronto · Don Mills, Ontario · Sydney · Tokyo · Singapore · Hong Kong · Seoul · Taipei · Cape Town · Madrid · Mexico City · Amsterdam · Munich · Paris · Milan

PEARSON EDUCATION LIMITED

Head Office	London Office
Edinburgh Gate	128 Long Acre
Harlow CM20 2JE	London WC2E 9AN
Tel: +44 (0)1279 623623	Tel: +44 (0)20 7447 2000
Fax: +44 (0)1279 431059	Fax: +44 (0)20 7240 5771

Website: www.informit.uk.com
www.aw.com/cseng

First Published in Great Britain in 2001

© Pearson Education Limited 2001

The right of Madhavendra Richharia to be identified as Author of
this Work has been asserted by him in accordance with
the Copyright, Designs and Patents Act 1988.

ISBN 0 201 33142 X

British Library Cataloguing-in-Publication Data
A CIP catalogue record for this book can be obtained from the British Library.

Library of Congress Cataloging in Publication Data
Applied for.

The programs in this book have been included for their instructional value.
The publisher does not offer any warranties or representations in respect of their
fitness for a particular purpose, nor does the publisher accept any liability for any
loss or damage arising from their use.

Many of the designations used by manufacturers and sellers to distinguish their
products are claimed as trademarks. Pearson Education Limited has made every
attempt to supply trademark information about manufacturers and their products mentioned
in this book.

Every effort has been made to trace and obtain permission to use copyright material, but if any have been
inadvertently overlooked the publishers will be pleased to make the necessary arrangement at the first
opportunity.

10 9 8 7 6 5 4 3 2 1

Typeset by Pantek Arts Ltd, Maidstone, Kent.
Printed and bound in the United States of America.

The Publishers' policy is to use paper manufactured from sustainable forests.

Contents

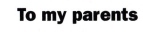
To my parents

Preface

For the proponents of satellite communications systems, the decade leading to the millennium will be remembered for the extraordinary growth and excitement caused by the breakthrough in mobile satellite system architecture with accompanying advances in ground and space technology. As the technology of the 1990s matures, the stage is set for us to reap benefits from the experience and to further expand the frontiers of space technologies as the revolution in telecommunications, computing and broadcasting ripples over to satellite communications.

In the last few years, mobile satellite services (MSS) imperceptibly entered the realm of personal communications as hundreds of thousands of terminals were introduced during the period. Thousands of terminals continue to be commissioned each month throughout the world across various MSS operators; however, though the growth is healthy, it is lower than the aggressive forecasts of the 1990s, which led to overoptimistic business plans.

Due to a rapid proliferation of systems and network architecture and the availability of a plethora of innovative applications, many telecommunication professionals and technical managers have lost sight of the mobile MSS perspective. A great amount of technical information related to MSS developments continues to flow into journals, magazines and conferences and lies embedded in expensive and lengthy specialist reference books, etc. It is therefore difficult for individuals and companies to obtain coherent up-to-date technical information.

This book, written to bridge the information gap, compiles current system concepts, architecture and trends in a structured and easily understandable style. The book is expected to serve as a reference source, as it sticks to technical concepts. Time-variant information and details which may distract a reader are given in an appendix. Mathematics is limited to essentials, and where possible equations are illustrated graphically; a comprehensive list of references has been included for the curious reader. Scientific principles have been amalgamated with business models for a balanced perspective of a commercial mobile satellite system. The subject treatment is unbiased and views expressed are the author's own; examples of commercial systems are chosen singularly on technical novelty and uniqueness.

The book, comprising 11 chapters and an appendix, presents an in-depth review of concepts, practices and trends of the gamut of mobile satellite systems. Topics include satellite constellations, propagation aspects peculiar to mobile communications including high frequency bands such as the K_a band targeted for next generation broadband systems, applicable modulation and coding techniques with emerging trends, design issues of hand-held communicators, including biological effects of radio frequency radiation, regenerative transponder technology with intersatellite links and multibeam antenna systems, business aspects of MSS highlighting the intertwined relationship between technical and business aspects of MSS; network optimization, vital elements of network operations, a summary of salient features of some innovative MSS system design, services with products similar to MSS such as navigation and direct broadcast sound systems and future system evolution covering topics such as asynchronous transfer mode (ATM), internet protocol (IP) technology, satellite component of universal mobile telecommunication systems (S-UMTS), K_a and higher band broadband emerging systems and growing convergence of FSS (fixed satellite service) and MSS for personal communications. Interesting data such as a list of recent proposals, interesting satellite footprints, launchers, etc., are included in an appendix.

The author wishes to express his gratitude to the reviewers of the book, Mr Jim Gifford, former editor of *Satellite Communications,* and Professor Barry Evans, Director of the Centre for Communication Research, University of Surrey, for their valuable comments. Thanks are also due to the editorial staff of Addison Wesley Longman, and in particular to Allison Birtwell for patiently guiding the book through the production process.

My son Anshuman deserves a very special word of thanks for creating a number of graphics, and his assistance in preparation of the typescript and an accompanying website during his vacations from Imperial College London.

My sincerest thanks are due to my wife Kalpana for her support and understanding as I endeavoured to grasp every spare moment to complete the book. Finally, it is a pleasure to remember the enthusiastic support of my daughter Meha throughout.

The reader can visit the website http://*www.SatellitesAndYou.com* for useful information and software pertaining to my books.

I hope the reader finds the book useful and enjoyable.

Madhavendra Richharia

Introduction 1

1.1 Scope and organization

This book has been written to promote an in-depth understanding of technical concepts bearing in mind the business aspects of mobile satellite communication systems, a branch of telecommunications rapidly becoming a part of our daily lives. Its relevance to the modern world is best demonstrated by a few illustrative real-life situations:

- Live satellite coverage from the summit camp brings a realism in people's homes never experienced before – a nation rejoices as the summit is conquered.
- A passenger ship cruising in high seas hits a rock and the crew, in a desperate attempt for survival, send a distress call; a rescue party arrives shortly.
- An old man's wizened face comes alive as he talks to his son from his thatched village home, tucked away where time stands still.
- An executive, preparing for a meeting on a transoceanic flight, contacts headquarters and receives vital documents within minutes.

Mobile satellite service or MSS systems provide communications to terminals which may be in motion, or moved at will anywhere within the service area. Terminals may be mounted on an aeroplane, a ship or carried by individuals; alternatively, the terminal may be a portable communicator set up at a convenient location. The vital elements are user mobility with minimal regulatory restrictions in the service area. Mobility is achieved by the use of a radio link for communication and incorporating network intelligence to manage mobility. Regulatory restrictions are minimized through appropriate spectrum selection and operating licences. Service areas of an MSS system can span a country, a region or indeed the world.

There was a steady and gradual growth in MSSs until early 1990, at which juncture demands began to accelerate as new services were introduced and public awareness heightened due to the success of terrestrial mobile systems. A further spurt in aggregate world-wide demand occurred when satellite phones were introduced in the third quarter of the decade, despite the business failure of some MSS operators. The failures were caused by differences in user expectations in terms of cost and quality and operator expectations in terms of market size and rate of penetration dented by advances in roaming arrangements of terrestrial operators. However, there is a growing public and commercial awareness of the role of satellite systems in supporting mobile and personal communications seamlessly, while capabilities of MSS systems continue to increase with shrinking terminal size, increasing throughput and integration with terrestrial mobile systems.

As happens in rapidly evolving technologies, collating cohesive information can be time consuming and difficult, as useful material lies buried and scattered in specialist reference books, journals, conference proceedings, trade magazines, etc. This book attempts to bridge the gap through a structured presentation of such knowledge to assist understanding of system architectures, their components, applications and trends. A comprehensive list of references is included for those interested.

The International Telecommunications Union (ITU), a United Nations body which regulates world-wide allocation of radio spectrum, has for spectrum planning categorized radio services according to their broad application – broadcast satellite service (BSS), fixed satellite service (FSS) or MSS, etc. Personal satellite communications services such as voice, facsimile and multimedia services require a radically different system design with a commercial approach quite new to the satellite community. The ITU has termed such services global mobile personal communications services (GMPCS). This book deals with system-level technical issues of MSS and GMPCS. We will refer to these services together as MSS, unless a specific distinction is essential.

This chapter introduces the topic, beginning with a review of the evolution of mobile communications. A subsequent section presents basic concepts of MSS architecture to familiarize the reader with the topic early in the book. A plethora of telecommunications products often leaves users uncertain regarding the most suitable solution. To enable users and prospective operators to make informed decisions, salient features of satellite and terrestrial systems are compared, followed by an overview of applications typical of an MSS. In recent years satellite navigation systems, direct sound broadcast systems and FSS products have entered the domain of personal communications. There is a certain degree of competition as well as synergy between various types of personal communication products. The principles and salient features of such systems are summarized, and finally, emerging trends reviewed.

Chapter 2 discusses an extensively reviewed topic of satellite communications – satellite constellations. After introducing the basics of orbital mechanics for the benefit of readers unfamiliar with this rather specialized topic, the chapter discusses characteristics of various types of constellations and compares them for a number of well-known recent constellation designs.

Chapters 3–6 introduce components and concepts applicable to mobile satellite communication systems. Some are applicable generally to satellite communication systems, but the treatment here is slanted towards an MSS perspective. Chapters 3 and 4 discuss various MSS air interfaces – spectrum issues, radio frequency (RF) propagation characteristics, modulation methods, coding schemes, link analysis and multiple access techniques. Chapter 5 addresses the main characteristics and technology of mobile terminals (MTs), which have profound implications for the success of MSS operators. Considerable public interest has arisen recently regarding RF radiation effects on humans; a section has been devoted to this topic where the current state of understanding is summarized. Gateways provide radio connection between a fixed network and the space segment; a section highlights MSS-specific features of gateways, which are otherwise identical to medium-sized earth stations of the FSS. Satellites are undoubtedly the most vital node of an MSS. With improvements in spacecraft device technology, there is a gradual shift from the traditional transparent transponder to the regenerative transponder because of advantages offered by the latter. Chapter 6 discusses the main features and emerging spacecraft and intersatellite link technologies.

This book views the MSS in a commercial perspective and in this context, business and technology are intricately entwined. Hence Chapter 7 changes the emphasis to MSS economics, illustrating its interrelationship with technology.

In Chapter 8 we explore MSS in a systems context by collating concepts discussed in previous chapters. Topics include system synthesis methodology with influences, constraints and trade-offs, networking issues and operational considerations such as radio resource management and quality of service.

A number of technically interesting systems have been proposed recently, each with some novel feature, and representative examples are discussed in Chapter 9 to illustrate how concepts have been translated into practice. The choice of system examples are made purely on the basis of their technical variety and merit, with little bearing on commercial performance or affiliations.

In Chapter 10 we discuss systems which offer services akin to mobile satellite systems but are not formally a part of them. In the past few years, the terminal size of the FSS has shrunk to an extent that they are portable, and hence the distinction between fixed and mobile services has become blurred in applications where mobility is either not essential or restricted. There is a significant interest in providing satellite sound broadcasts directly to small portable receivers mounted on cars or carried by individuals. Several applications, such as fleet management, are combining MSS communications with position fixes available through a global positioning system (GPS). Terrestrial mobile communications have now entered the third generation, and it is recognized that the third generation mobile systems will be generic with a satellite component, implying that there will be better appreciation of terrestrial mobile systems by the satellite community and vice versa. In such an environment, the treatment of mobile satellite communications cannot be taken in isolation. We will therefore also introduce salient features of terrestrial cellular systems in this chapter.

While first generation satellite phone services are barely established, plans for establishing second generation services are progressing in a number of international forums and the commercial world. The final chapter discusses various new concepts under investigation – integration of fixed and mobile networks, the concept of broadband satellite mobile systems, advanced space segment concepts and others.

An appendix summarizes the characteristics of several mobile satellite communication proposals.

1.2 Evolution of mobile telecommunications

For the purpose of this section, mobile communication systems are broadly categorized as terrestrial and satellite. In both cases, mobility is achieved by a RF link between the user and a relay station, which is connected to the fixed network. Until recently the evolution of these two systems progressed independently. Terrestrial systems were best suited for the urban environment, whereas satellites provided effective communication solutions for remote areas such as the high seas, air corridors and remote land masses. By the beginning of 1990, MSS technology had matured to an extent that system planners began to evaluate benefits and techniques of integrating these two technologies, leading to the introduction of partially integrated systems at the end of the decade and plans for full integration in the near future.

1.2.1 Terrestrial systems

The potential of mobile communications was recognized from the outset of radio dating back to the late 1800s. Earliest use of mobile radio was for maritime navigation and safety.

Before and during the Second World War, mobile systems were generally confined to military users. Vacuum tubes and heavy batteries in transceivers (i.e. a transmitter–receiver unit) made them bulky, restricting their use to specialized applications. By the 1950s, technology advancements enabled availability of man-pack very high frequency/frequency modulation (VHF/FM) radios, which extended the applicability of mobile systems to civil private mobile radio (PMR). The growth of mobile radio was slow until the 1970s, because the problem of extending the service range within the permitted frequency bands proved technically difficult. In general, mobile communication services have remained an expensive communication medium. But even if the cost of mobile services had reduced significantly, the factor that hampered growth of mobile services was spectrum scarcity. A concept called 'cellular radio', which offered a solution to support large numbers of users in a limited spectrum, was proposed at the Bell laboratories in the 1940s. However, enabling technologies to realize the concept would arrive much later, in the 1980s.

Mobile transceivers were permitted access to the public network of the United States by AT&T in 1946. The service was called the mobile telephone service (MTS), which operated in the 35 or 150 MHz band. The simple system was manual, where the user seized a vacant channel, requesting an operator for a connection to it.

An improved MTS (IMTS) which offered duplex operation, automatic dialling and switching was launched in 1964. The system capacity was initially limited to 11 channels in the 152–158 MHz band. Due to a rise in demand, the capacity was enhanced in 1969 by 12 channels in the 454–459 MHz band. Each service zone was served by a single base station, which used a channel only once and hence the demand could not be satisfied; typically, around 550 users could be served in a zone. Other service limitations included bulky transceivers which had to be vehicle mounted, and the need for high-capacity batteries.

Introduction of the cellular concept, mentioned earlier, provided the desired breakthrough in overcoming the spectrum capacity limitations. The first cellular system, known as the advanced mobile phone system (AMPS), was developed in the USA by AT&T and Motorola Inc. The analogue system was designed to operate in an 800 MHz band with a capacity of 666 paired channels.

However, the first cellular system was deployed in Japan in 1979, followed by the nordic mobile telephone (NMT) system introduced in 1981 in Denmark, Finland, Norway and Sweden, the total access communication system (TACS) in the UK and AMPS in 1983 in the USA. Yet other types of systems were developed and introduced in other countries.

Rapid developments were made in radio, very large-scale integration (VLSI) and computer technologies giving huge reductions in costs through economies of scale and consequent increases in personal communications. Users have since been inundated with a variety of cellular phones, systems and services. Figure 1.1 shows a snapshot of cellular phones' increasing penetration in some European countries in the first quarter of the year 2000; the rate of increase in penetration can be gauged by the projection that, in the UK, 60% of the population was expected to own a mobile phone by the end of the year 2000.

A consequence of the runaway commercial success of cellular phones was the proliferation of a variety of systems incompatible with each other, causing interoperability problems so that users could not operate their cellular phones outside a home territory. Concentrated efforts to harmonize evolution led to second generation digital standards such as the global system for mobiles (GSM) in Europe, which gave a cross-country roaming facility.

Despite high spectral efficiencies offered by cellular systems, capacity shortfall remained a pressing issue and efforts to enhance capacity continued. In the USA, carrier spacing of AMPS was reduced from 30 to 10 KHz, trebling the capacity. The second approach was to introduce digital voice compression with time division multiple access (TDMA), which resulted in a similar order of capacity enhancement. Spread spectrum modulation with code

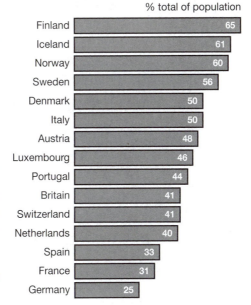

% total of population

Country	%
Finland	65
Iceland	61
Norway	60
Sweden	56
Denmark	50
Italy	50
Austria	48
Luxembourg	46
Portugal	44
Britain	41
Switzerland	41
Netherlands	40
Spain	33
France	31
Germany	25

Figure 1.1
Penetration of cellular phones in some European countries (1999)

Source: FT Mobile Communications

division multiple access (CDMA) was yet another approach. Its proponents claimed a 10–20 times capacity increase over AMPS. Recent personal communication systems (PCSs) are based around GSM technology and a spread spectrum technology called IS-136 or D-AMPS.

Third generation cellular systems based on common standards are being introduced in Europe and elsewhere to bring further homogeneity in mobile communications across the world, thereby offering benefits to users, manufacturers and operators; satellite systems are eventually expected to be a part of this all-pervading system. The system will offer wideband services with wider mobility.

1.2.2 Satellite systems

Table 1.1 summarizes some interesting milestones in the MSS evolution.

Table 1.1
Interesting events in the evolution of satellite systems

Year	Milestone	Responsible
1957	Launch of first satellite	Soviet Union
1965	Launch of first geostationary satellite	Comsat, USA
1976	First demonstration of intersatellite link	LES-8 satellite, USA
1977	Mobile experiments using ATS-6	NASA, USA
1978	Start of global positioning system (GPS)	USA
1979	Formation of Inmarsat	IMO
1980	IMO decides to use satellite communications for maritime safety	UN
1982	Start of GLONASS	Former Soviet Union
1982	First civilian mobile satellite system introduced	Inmarsat
1987–89	Architecture of non-GEO for mobile satellite communication investigated and proposed	University of Surrey
1990	First commercial satellite radio broadcast system filed	CD Radio Inc., USA
1990	First commercial non-geostationary hand-held system announced	Motorola/Iridium
1990–91	Introduction of commercial land and aeronautical mobile satellite services (MSS)	Inmarsat
1992	Introduction of GSM system	Europe
1992	Changes to mobile frequency allocation	WARC 1992
1993	Announcement of first commercial little-LEO satellite system (with secure finance)	Orbital Sciences Corporation – ORBCOM system, USA
1994	Announcement of first non-geostationary fixed satellite services (FSS) for personal communications	Teledesic Corporation, USA
1994–96	Announcement of several regional 'super-geostationary' satellite systems	Agrani (Indian consortium); APMT (China/Thailand); ACes; Thuraya, etc.

Table 1.1 continued

Year	Milestone	Responsible
1996	Introduction of paging service	Inmarsat
1997	Desktop-sized mobile terminals introduced	Inmarsat
1997	Introduction of first non-geostationary little-LEO satellite system	ORBCOM
1997	Frequency allocation for non-geostationary fixed system	WRC 1997
1997	Launch of first batch of non-geostationary satellite system for voice communications (big LEO)	Iridium
1997	Launch of first batch of non-geostationary satellite system for low bit rate data communications (little LEO)	ORBCOM
1997	Mobile experiments using ACTS	NASA
1997	Navigation system: geostationary overlay capability available	Inmarsat
1997–98	Start of world-wide spot beam operation for MSS	Inmarsat
1998	Introduction of first non-geostationary big-LEO satellite system	Iridium
1998	Introduction of dual-mode receivers (i.e. satellite and terrestrial)	Iridium
1998	SOLAS Treaty introduced	UN
1998	Introduction of processing satellites for MSS	Iridium
1999–2000	Serious financial difficulty experienced by new MSS systems; Iridium declares bankruptcy; ICO files for bankruptcy and is taken over by new investors; ICO name changed to New ICO.	Introduction of Globalstar
2000–05	Consolidation of new mobile satellite system operators; some become defunct due to losses	
2003–05	Introduction of wideband personal mobile communication systems	
2005–10	Convergence of fixed and mobile services for personal communications	

The earliest mobile communication experiments were conducted through the National Aeronautics and Space Administration's (NASA) Applications Test Satellites ATS-5 and 6. A number of early MSS proposals never took off due to the technical and financial risks. The potential of satellite systems to provide high-reliability communications to ships was recognized by the International Maritime Organisation (IMO). The prevailing communication systems used the high frequency (HF) band and proved unreliable under adverse propagation conditions, often requiring hours to establish communications from the high seas when weather conditions were unfavourable. Ships on the high seas would occasionally be lost without trace. Satellite communications radio links are reliable under most conditions. Recognizing the advantages offered by a satellite medium, the IMO initiated the formation of an organization, then called the International Maritime Satellite Organisation (Inmarsat), for the provision of safety and commercial public correspondence services to ships for peaceful purposes. Inmarsat was founded in 1979 and the maritime communications services became available in 1982 using satellites leased from a number of satellite operators. The success of maritime services led to the introduction of services in land and aeronautical environments and generally vast enhancements in MSS technology. The size of terminals reduced progressively with technology evolution, from heavy transportable terminals to desktop-sized personal telephones and briefcase-sized multimedia terminals; high-power satellites with spot beams were introduced, while throughput continued to increase. Next generation satellites are planned to use hundreds of spot beams and offer broadband services to a range of small personal communicators.

In the 1990s, a number of regional systems were introduced in the USA, Europe, Australia and Japan. For example, OmniTracs and EUTELTRACS provide low bit rate services in the USA and Europe, respectively, the American Mobile Satellite Corporation (AMSC) operates voice and data service in the USA and Canada and the Optus system provides voice and data service in Australia. Appendix 1 lists a number of regional systems deployed or planned in the 1990s.

Studies on non-geostationary orbits (GEOs) were first conducted in the 1960s, when it was difficult to launch satellites to the GEO, and therefore lower orbits were considered appropriate. Subsequently, GEO became a favoured choice, due to advantages arising from the static path geometry of geostationary satellites. Non-GEOs remained in use for specialized applications such as remote sensing, military reconnaissance, Earth resource survey, etc. The military were interested in the survival and robustness offered by low orbit satellite systems due to their inherently distributed architecture. Towards the end of the 1980s, a university group in the UK studied the possibility of deploying a low orbit satellite constellation for mobile communications with the conclusion that such systems were indeed feasible, compare favourably with a geostationary satellite system and could be implemented within a decade (Richharia *et al.*, 1989). Within a year, quite independently, Motorola Inc. announced plans of a low Earth orbit (LEO) satellite system to provide a personal voice communication service via hand-held phones (Nelson, 1998). The announcement triggered a feverish activity in the satellite industry and within the next three years a number of organizations and companies announced similar plans, most of them intending to use non-geostationary constellations. Notable among them were a medium Earth orbit (MEO) system, which was conceived and developed by Inmarsat but was to be implemented by a new privatized company called ICO

Global Systems Limited (referred in the remaining text as the ICO system; the company has now been taken over by a new investor and renamed New ICO – here, we will not make any distinction between the two), and a LEO system known as Globalstar. Appendix 1 summarizes the main technical features of a number of non-geostationary system proposals of the 1990s.

Another approach was to use low-risk technology and smaller spacecraft for low bit rate niche applications, such as messaging, etc., ensuring an early entry into the market, as proposed, for example, by companies such as Orbital Science System (ORBCOM).

In the mid 1990s, plans for hand-held satellite phone services using GEOs were announced. These systems planned to deploy hundreds of spot beams with powerful transmitters to compensate for the relatively higher altitudes of a GEO. Interestingly, only a few years ago geostationary systems were discarded due to the perceived complexity of payload; intense research and development in the ensuing four to five years had matured the technology. Close on its heel, a number of personal communication systems (PCSs) for fixed services were announced, converging high bit rate service offerings of the MSS with the lower end of FSS products. Most FSS systems intend to operate in the 20–30 GHz FSS band and a variety of orbits (see Chapter 11). Table 1.2 compares the salient technical features of various MSS non-geostationary satellite proposals for hand-held services with an equivalent regional geostationary satellite system (see also Appendix 1). Interestingly, all non-geostationary proposals listed in the table have experienced initial financial difficulties to the extent that the original Iridium system is now bankrupt. Table 1.3 presents a comparison of LEO, MEO and GEO MSS systems. Figure 1.2 compares the mass and power estimates of recent MSS proposals for personal communications; a few Inmarsat and INTEL-SAT satellites have been included as reference.

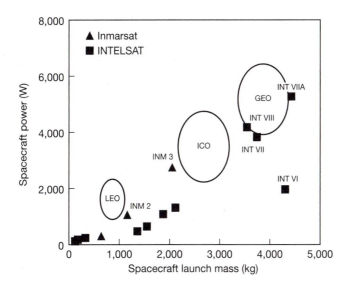

Figure 1.2

Mass and power consumption of recent MSS proposals compared. Inmarsat and INTELSAT satellites have been included for reference (Evans 1998) © 1998 IEEE

System parameter	System			
	Iridium	**Globalstar**	**ICO Global Communication (original system)**	**Typical regional system**
Service (voice bit rate)	Hand-held telephony/ data (4.8 kbps)	Hand-held telephony/ data (up to 9 kbps, depending on channel conditions)	Hand-held telephony/ data (4.8 kbps)	Hand-held telephony/ data (2.4–4.8 kbps)
Type of orbit	Low Earth polar orbit	Low Earth inclined orbit	Medium Earth orbit	Geostationary orbit
Number of spot beams/ satellite	48	16	163	100–300
Nominal capacity per satellite	1,100	2,400	4,500	16,000
Service area	Global	Global	Global	Regional
Service link frequency band (up/ down) in GHz	1.616–1.6265/ 1.616–1.6265	1.62–1.63/ 2.48–2.5	1.98–2.01/ 2.17–2.2	L(1.6)/L(1.5)
Feeder link frequency band (up/down) in GHz	30/20	5.1/6.9	5.2/6.9	14/12
Gateway antenna G/T (dB/K)	24.5	28.5	26.6	37.0
Multiple access	FDM/TDMA	FDM/CDMA	FDM/TDMA	FDM/TDMA
RF band- width per channel (KHz)	31.5	1,250	25.2	27
Modulation	DQPSK	SS/QPSK	QPSK	QPSK
User terminal RF power (W)	0.45	0.5	0.625	0.5
User terminal G/T (dB/K)	−23	−22	−23.8	−23.8
Service link Link margin (dB)	16.5	11	10	10

	Geostationary orbit	Medium Earth orbit	Low Earth orbit
Number of satellites (world-wide coverage)	3–4	10–12	40–300
Regional coverage	Well suited	Specific orbital design necessary (depends on region)	Specific orbital design necessary (depends on region)
Coverage limitations	Within ± 76° latitude	None	None
Approximate satellite lifetime	12–15	7–10	5–7
Order of system cost ($ billion)	2 (0.75 for regional system)	5	5–9
Operational cost and complexity	Low (matured technology)	High	High
Transmission delay (ms)	250	30–100	5–10
Inter and intrasystem spectrum coordination complexity	Least	Medium	High

Table 1.3
A broad comparison of system features using various types of orbit

Most satellite system designers, noting the benefits of integrating satellite systems with terrestrial systems, proposed system architecture which would combine satellite and terrestrial system networks to various extents. Dual-mode satellite/cellular hand-held telephones were conceived.

International organizations, developing standards and formalizing concepts for third generation terrestrial systems, began considering the role of satellite systems in future system architectures. It was well understood that cellular systems would remain concentrated in populated areas where they are economically viable, leaving gaps in vast, sparsely populated areas which could best be covered by satellite systems. Thus, future systems would incorporate interfaces and interworking arrangements to support interoperability between terrestrial and satellite systems.

In parallel, an unabated growth in *satellite navigation* technology for mobile and personal use continued throughout the 1990s. The cost of GPS receivers plummeted to levels affordable by the masses. A number of applications developed, combining navigation and communication capabilities. More recently, several non-geostationary MSS systems have included navigation as a value-added service. In Chapter 10, we will discuss the principles of satellite navigation systems.

Notable progress was made towards the introduction of a *satellite radio service*, a service aimed to broadcast directly to receivers carried by individuals or mounted on cars. Radio broadcast systems are expected to become available in the early part of this decade.

Technology penetration has recently accelerated due to the ease of advertising, accessibility of the general public to vast amounts of information and globalization of the economy. Figure 1.3(a) shows the relatively slow evolution of telephones when compared to the introduction of more recent technical innovations, such as video cassette recorders, and Figure 1.3(b) shows the evolution of personal aids over the past centuries and extrapolation to the future. Figure 1.4 portrays the estimated number of mobile subscribers by the end of the decade, which concurs with other similar forecasts. Section 1.11 discusses further the growth trend and section 7.7 covers the topic of traffic forecast methodology, taking into consideration factors such as population and affordability, the penetration rate of similar technology and constraints and extrapolates the market for such services up to the year 2010. In section 3.2.2 we discuss a spectrum forecasting methodology, taking into consideration technical parameters of RF carriers.

Figure 1.3
(a) Rate of market penetration for common electronic products (Stix, 1993) © 1993 *Scientific American*, Inc. All rights reserved. (b) Rate of market penetration for common electronic products (*Electronics and Communications Engineering Journal*, 1995)

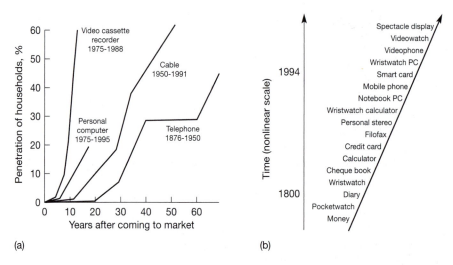

We may classify satellite system evolution on the basis of services into a number of phases as shown in Table 1.4.

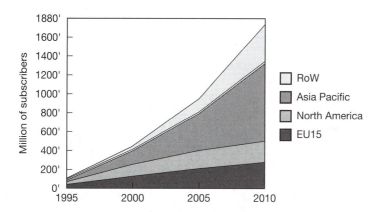

Figure 1.4
Predicted growth in
mobile subscriber
phone traffic (UMTS
Forum, 1999)

Phase	Year	Smallest terminal size/service range	Satellite	Orbit
I	1980–90	Large (1 m dish)/ analogue voice – 64 kbps data	Low power/ single beam	Geostationary
II	1990–98	Desktop size/ 2.4–64 kbps; pocket-sized pagers; tens of b/s	Medium power/ 1–8 beams	Geostationary
III (I generation hand-held and non-GEO system)	1998–2005	Hand-held-portable/ 10 bps–500 kbps	High power/ hundreds of beams/onboard processing	Geostationary/low Earth/medium Earth/elliptical/ hybrid
IV (II generation hand-held and non-GEO systems)	2005–10	Hand-held and portable/10 bps –2 Mbps [155 Mbps to fixed personal terminals]	Very high power/hundreds of beams/onboard processing	Ibid.

Table 1.4
Mobile satellite
communications
evolutionary trends

1.3 Satellite system architecture

The main components of a mobile satellite system are shown in Figure 1.5.

The system offers communication services to mobile users operating within a predefined service area. Users communicate with other mobiles or with fixed users through one of the visible satellites. Users in the fixed network are accessed through large fixed stations called gateways which carry large amounts of traffic, e.g. 100–200 Erlangs, whereas mobiles are small portable units capable of supporting 1–5 channels. Mobile terminals may be mounted on vehicles such as ships, aircraft, trucks or carried by individuals.

Figure 1.5

Main components of a
modern mobile satellite
system

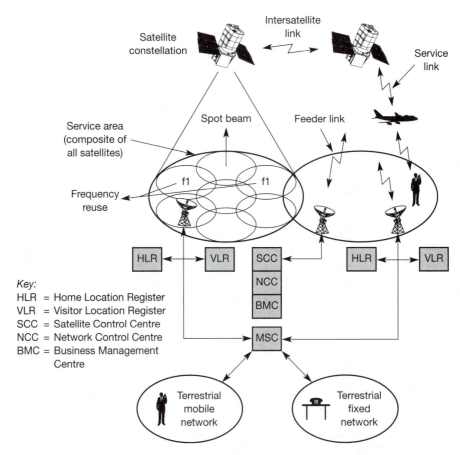

Key:
HLR = Home Location Register
VLR = Visitor Location Register
SCC = Satellite Control Centre
NCC = Network Control Centre
BMC = Business Management
Centre

Depending on the service area and application, the *space segment* may consist of one or more satellites. Telemetry and control ground stations, used for monitoring and controlling satellites, constitute a part of the space segment. To simplify the mobile terminals, complexity is shifted to the space segment and hence satellites tend to be large and complex – a 3–4 kW geostationary satellite with 5–10 footprints (or *spot beams*) is quite typical of second generation systems and 5 kW with 100–200 spot beams is typical of third generation systems. A geostationary satellite remains almost fixed with respect to the Earth and comprises of one or more *static* footprints. Therefore, fixed stations can operate with a single antenna with minimal tracking while network topology is simple.

The main problems in supporting mobile satellite communications are caused by difficult propagation environments and a small mobile terminal size. Figure 1.6(a) and (b) represents typical samples of signal level received on hand-held and car-mounted terminals, respectively (Jahn *et al.*, 1995).

As already mentioned, the first and second generation systems offered services to relatively large mobile terminals. Third generation systems, sometimes called 'super-geo systems' to distinguish them from second generation systems,

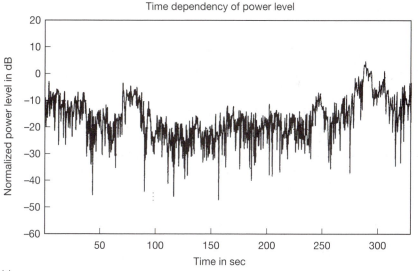

(a)

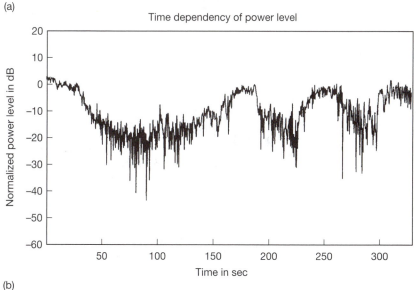

(b)

Figure 1.6
(a) Signal received on a hand-held terminal (Jahn *et al.*, 1995). (b) Signal received on a vehicle-mounted terminal (Jahn *et al.*, 1995)

provide voice or multimedia services to hand-held or desktop-size terminals. Furthermore, non-geostationary satellite systems have been introduced in the third generation time frame.

A network control station (NCS) manages traffic flow, broadcasts network information, manages call set-up/de-assignment and radio resources. On receiving a call request, the NCS searches and assigns a channel from its pool, returning it at the end of the call. Home location register (HLR) and visitor location register manage user mobility; network control centre (NCC) manages the network; while satellite control centre (SCC) manages and controls the satellite performance. Business management centre (BMC) is responsible for billing and other business functions.

The architecture of non-geostationary satellite systems is more complicated due to movement of satellite beams relative to a user. Therefore, a call could be routed through multiple beams and satellites, depending on satellite and spot beam visibility. This rerouting, known as *handover*, is of particular relevance to interactive services.

Routing of a call in the network can be achieved by a number of techniques. A call may be routed through intersatellite links or through one or more hops of satellite–ground links. Figure 1.7(a) and (b) illustrates architectures of non-geostationary satellite systems deploying satellite–ground hops and intersatellite links. The Iridium system deploys intersatellite links whereas the ICO system uses ground routing.

Figure 1.7
(a) Architecture of non-geostationary satellite systems using satellite–ground hops.
(b) Architecture of non-geostationary satellite systems using intersatellite links

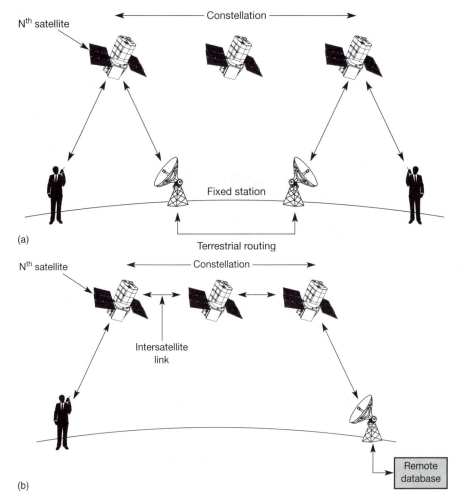

For store and forward systems, a discontinuous coverage is acceptable. A message is stored in the ground station or satellite buffer and delivered when the destination becomes visible. Figure 1.8(a) and (b) represents main entities of store and forward systems with satellite and ground-based buffers, respectively.

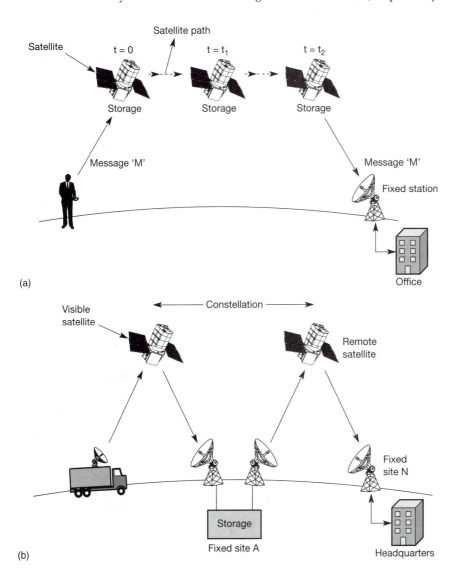

Figure 1.8
(a) Main entities of store and forward system with satellite buffer. (b) Main entities of store and forward system with ground station buffer

The architecture of a satellite system is influenced by service requirements and a number of technical considerations, some of which are summarized below. Chapters 2 and 8 address the subject in more detail.

Orbit

Orbits are categorized by altitude, inclination and eccentricity. The altitude determines the area covered by a satellite – higher altitude satellites cover a larger region. Inclination influences the minimum–maximum latitude covered on the Earth. The two extremes of orbital inclination are an equatorial orbit, which has an inclination of 0°, and a polar orbit, which has an inclination of 90°. A low Earth equatorial orbit would cover a belt around the equator, whereas a polar orbit would cover a belt around the pole orthogonal to the equator, thus covering the full Earth due to the Earth's west–east motion beneath. Eccentricity of an orbit determines the shape of the orbit. Satellites in a circular orbit provide an unbiased global coverage, whereas satellites in elliptical orbit favour coverage in specific areas by dwelling on them for a longer time. Examples of low-altitude polar constellations are the Iridium and Globalstar systems; the ICO system represents a medium Earth inclined orbit system; the Inmarsat system deploys a GEO; the Ellipso system uses a combination of circular and elliptical orbits, with the higher and mid-latitudes covered by the Sun-synchronous elliptical orbit and the equatorial region by a circular orbit (see Chapters 2 and 9).

Throughput

The throughput of a system is a measure of the capacity of the system. All MSS systems now use digital transmission and hence data rate is a common means of categorization as illustrated in Table 1.5 with example applications.

Table 1.5
Categorization of MSS by throughput with example applications

Category	Throughput	Typical applications	System example
Low bit rate	A few bps – 1 kbps	Paging, message transfer, X-400 services, SCADA, remote monitoring	EUTELTRACS (geostationary orbit); ORBCOM (low Earth orbit)
Medium bit rate	1 – 10 kbps	Voice, facsimile and data	Globalstar (low Earth orbit systems); ICO global communication systems (medium Earth orbit system); Inmarsat phone (geostationary system)
High bit rate	10 – 64 kbps	Voice, facsimile, data, some multimedia	Inmarsat B
Broadband	> 64 kbps	Multimedia	Personal Multimedia Service (Inmarsat); New ICO broadband system

Tolerable delay

Delay tolerated by a system is application dependent. An end-to-end delay of more than ~ 400 ms is quite disturbing in a conversation, but delays of minutes or hours are acceptable for e-mail delivery. The tolerable time delay influences several features of a mobile satellite communication system. A non-real-time system can tolerate a break in the communication link by data recovery techniques, whereas in a circuit-mode service, a continuous end-to-end connection must be maintained.

Non-real-time systems use packet switching where packets and messages are stored in buffers at intermediate nodes of a virtual link and transferred when conditions are favourable. Inmarsat-C and EUTELTRACS geostationary systems use such an architecture. The ORBCOM system represents a LEO store and forward system.

For a real-time geostationary system, ignoring RF link blockage due to obstructions, satellite visibility remains uninterrupted as users operate within static footprints. The boundaries of the spot beam coverage are fuzzy, extending several tens of kilometres which leads to a graceful degradation in signal quality for slow-moving vehicles. Thus, slow-moving mobiles need not handover a call to the next beam as the users spend a considerable period in this fuzzy zone where signals degrade gradually. However, handover is necessary for fast-moving mobiles such as aircraft. In non-geostationary satellite systems, satellite visibility is variable due to dynamic footprints. Therefore, a user is likely to communicate through different beams during a call, making handover essential for the network.

Connectivity

The network must be able to locate a called mobile and route signals between the parties as efficiently as possible and maintain the call.

In derivatives of the GSM, each mobile is registered in a database, called the home location register (HLR); if the mobile migrates outside the home territory, the mobile registers itself with the visiting system's visitor location register (VLR). The VLR conveys the location of each visitor mobile to the mobile's HLR, thereby refreshing the location status of the mobile. Whenever a call is addressed to a mobile, the mobile switching centre interrogates the mobile's HLR for the mobile's location and establishes the call through an appropriate route. Figure 1.7(a) shows a routing scheme in a non-geostationary system with satellite–ground hops, while Figure 1.7(b) represents an intersatellite link routing scheme. We have already introduced the concept of handover for maintaining RF link connectivity during a call, but Chapter 8 revisits connectivity issues in greater detail.

Physical environment

In a mobile satellite system, the physical environment around a mobile terminal exerts a crucial influence. The environment has two effects – it disturbs the stability of radio signals, setting a boundary on throughput, and limits the physical size of a terminal.

From a radio propagation viewpoint, maritime and aeronautical environments are more benign than land, where levels of fade and multipath fluctuations are significantly higher. Propagation conditions degrade progressively as the

Figure 1.9
(a) A personal mobile phone providing high-quality mobile office applications via the Inmarsat network. Inmarsat phones are small, light and a cost-effective mobile satellite communications unit, weighing about 2kg, resembling a laptop computer in size (courtesy: Thrane & Thrane. (b) A high-speed terminal with standard ISDN interface supporting Inmarsat Global Area Network/M4 service. Apart from the data facilities, the Capsat Messenger also offers reliable voice and fax services. It is easy to carry to even the most remote locations and easy to set up and use (courtesy: Thrane & Thrane)

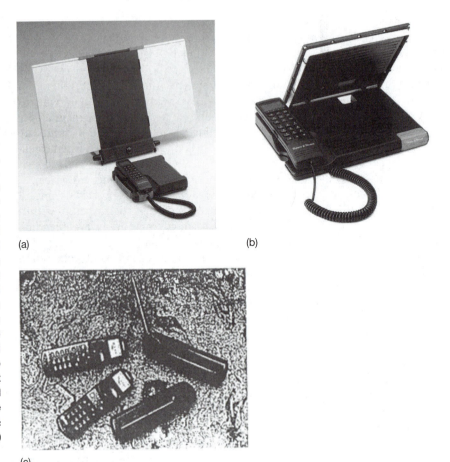

(a)

(b)

(c)

environment changes from highways to urban and generally gets worse as antenna size is reduced. Throughput and type of service are directly related to fading characteristics. Intermittent deep fades can break radio links, causing discontinuity to real-time services, while multipath influences modulation and coding schemes. Providing a higher link margin mitigates such impairments, but increases demands on satellite power. The Doppler Effect, introduced by relative motion between satellite and mobile, is yet another impairment and in this respect, aeronautical terminals exhibit the worst degradation.

The physical space available on a mobile limits the weight and size of the antenna and transceiver. A terminal mounted on a ship can be larger than a terminal on an aircraft; a vehicle-mounted terminal is smaller than either, whereas a hand-held terminal must be small enough to fit in a user's pocket or purse. Demands on spacecraft power and sensitivity increase as terminal size is reduced and throughput increased. Chapters 3 and 5 discuss these issues in greater detail. Figure 1.9(a)–(c) shows various types of mobile terminals.

Satellite access

In an MSS environment, thousands of users share satellite resources and therefore satellite access efficiency is paramount. Demand assigned (DA) single channel per carrier (SCPC) frequency or TDMA schemes, where a pool of channels is shared by all users on a per call or packet basis, offer an effective solution for circuit or packet-mode communication. The channel pool can be managed by either a central or distributed architecture. In a central architecture, a pool is managed centrally, whereas in a distributed architecture, separate pools are assigned to each participating fixed station for self-management.

A CDMA scheme offers an advantage in terms of interference and multipath mitigation, soft handover and capacity. A number of modern systems have preferred CDMA.

Data traffic tends to exhibit a variety of characteristics ranging from sporadic bursts to continuous streams and therefore accessing schemes are matched specifically to traffic characteristics. Common accessing schemes used for data communications include Aloha, slotted Aloha, Reservation Time Division Multiple Access, etc. (see Section 4.4).

In recent years, with the proliferation of data applications, a scheme known as asynchronous transfer mode (ATM), which allows a flexible combination of various types of data at the link layer, has been introduced. As ATM was originally proposed for fixed terrestrial systems, considerable effort was expended in adapting the scheme for wireless access including satellite systems. Internet protocol (IP) which operates at the application layer is now widely used. Several future broadband systems intend to use IP over ATM in conjunction with TDMA or CDMA as the transport mechanism (see Chapter 11).

Radio link characteristics

Techniques for improving radio link reliability include use of robust modulation and forward error correcting codes, fade countermeasures embedded in the system architecture, store and forward techniques to support communication in deep fades and adaptive power or code rate control.

Radio link considerations also include the selection of an appropriate frequency band and interference management. The selection of frequency band is governed by ITU's Radio Regulations in combination with engineering and commercial considerations. At present, most MSS systems operate in the L (~ 1.5 GHz) and S (2 GHz) bands and a few systems operate in the K_u band. The L and S bands are well suited for mobile communication because of relatively benign propagation characteristics and mature technology. In the future, it is likely that higher frequency bands will be used due to severe overcrowding in the 1–2 GHz bands coupled with a need for a larger bandwidth in support of higher throughput.

Due to heavy demands on the MSS spectrum, interference management is an important consideration in the planning and operation of mobile satellite systems. A certain level of interference is budgeted in link design to enable intra and inter-system frequency reuse. Techniques used to maximize frequency reuse include deployment of spot beams, selection of robust modulation/coding schemes and judicious frequency planning. Chapter 8 discusses techniques for maximizing spectrum efficiency and addresses the issue of interference management.

1.4 Business plans

A crucial element of modern mobile satellite ventures is a viable and credible business plan with a sound market analysis, investment strategy, financial returns, etc., for raising finances and revenue. In recent years, we have observed market needs rather than technology influencing mobile satellite system products and system architecture. Recently, prospective operators have tended initially to carry out extensive market research to select services, service areas and user expectations in terms of tolerance of transmission delay, signal fades, etc. System architecture is developed after acquiring a sound understanding of anticipated market requirements and user preference. Chapters 7 and 8 capture the system perspective essential in the development of business plans, demonstrating the intertwined relationship between technical and commercial aspects of MSS.

Invariably, all market forecasts of the 1990s projected a sharp growth in demands for satellite telephony, contrary to experiences of early entrants such as Iridium or Globalstar. Conclusions regarding the total number of systems required to support demands were less clear, with estimates varying between 2 and 4. Clearly, accurate forecasting is a vital element for the success of an expensive MSS venture. More recently, the emphasis is shifting towards mobile wideband data, shadowing the trend towards wider bandwidth in terrestrial mobile and fixed networks. Chapters 7 and 8 present methods of long-term and short-term forecasting.

1.5 Regulatory considerations

One of the first activities at the start of a venture is that of procuring an operating licence and frequency clearance from the regulatory authorities of the service area which may comprise a number of countries. This procedure may require considerable time due to a variety of reasons such as bureaucracy, regulatory policies and political issues. Other considerations include preparation of roaming arrangements with other operators, agreement of numbering schemes, etc. These and a range of associated topics are covered in Chapters 7 and 8.

1.6 Operational considerations

The operation and maintenance of an MSS system is quite challenging, involving a wide range of complex activities such as the launching and maintenance of satellites, network management, commissioning of new terminals, billing, marketing, constant review of business plans and long-term planning. The scale of effort increases as the size of the constellation and network grows.

The starting phase of a system is critical because of a need to set up technical operations, commissioning, billing, customer relations, publicity, etc., while maintaining cash flow and revenue projections. First generation non-geostationary systems are most susceptible to commercial risks due to uncertain markets, the introduction of untried system designs, the lack of operational experience and the pressure to introduce services on stringent deadlines.

Deployment of a new satellite constellation is time consuming, as it involves a number of launches spread over months. Moreover, satellites generally include a number of new technologies which make the system liable to a high number of early failures. The Iridium system, comprising 66 satellites, was deployed in about 12 months with about 10% satellite failures caused by a variety of technical problems. Globalstar lost 25% of its constellation in a single launch failure and ICO suffered an early setback due to the loss of its first satellite during launch.

Constellation maintenance requires regular monitoring of satellites, performing orbital manoeuvres and changes to spacecraft configuration, replacement of malfunctioning and ageing satellites, upgrading onboard software, etc. LEOs deploy the highest number of satellites with the lowest satellite lifetime.

Measures to provide services in the eventuality of a satellite failure include deployment of in-orbit and ground spares; and/or incorporating failure-resistant features in the network architecture. If a region is served by more than one satellite, users can communicate from the operational satellite in case of failure of any one of them, thus providing coverage redundancy as depicted in Figure 1.10 for a geostationary system.

Non-geostationary satellite systems have an inherent resistance to failure due to their distributed architecture and dynamic footprints. The consequence of a failed satellite in such a network is a coverage gap which propagates around the satellite's coverage belt. Thus, the outage is distributed and time shared by subscribers within the coverage belt. The gap can be filled by readjusting the position of other satellites and subsequently by the introduction of an in-orbit spare.

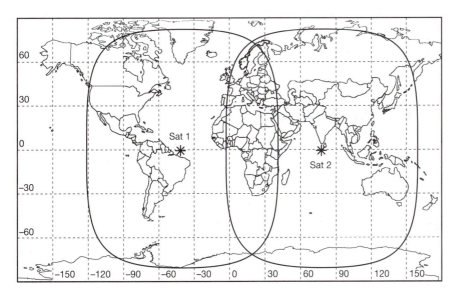

Figure 1.10
Coverage redundancy from two satellites

Network management involves monitoring satellite health, RF transmissions and traffic flow through the network, maintaining quality of service, traffic trend analysis, commissioning of terminals, etc. Satellite control centres monitor telemetered signals continuously and take corrective action when necessary. Typical actions include the firing of thrusters to maintain a satellite's orbit, reconfiguring a payload in case of failure of a subsystem or in response to an operational requirement, etc. An operational system is susceptible to a number of RF-associated problems such as inter and intrasystem interference, non-compliant transmissions from gateways, etc. Satellite transmissions are therefore monitored to maintain RF integrity of the network and assist in routine RF-related tasks.

The network operator has to ensure that the user is satisfied with the quality of service, which is usually measured as grade of service and signal bit errors in addition to matters related to commissioning, billing, after-sales service, etc. Sometimes failure of a critical subsystem, such as the network frequency management system, manifests itself indirectly as a loss in network traffic, and therefore monitoring of traffic flow through the network is essential. Forecasts used in the planning phase developed on theoretical assumptions and hypothesis can be refined in the operational phase by trending real data. Chapter 8 provides further details related to operational issues and network management.

The ITU has recognized personal mobile communications due to its potentially huge demands and unique technical requirements. The GMPCS is a specific category of MSS to support personal communications to individuals from hand-held personal communicators, much like the cellular telephone or pagers. A GMPCS can be categorized in various ways such as by orbit and service. Big LEO satellite systems deploy large, complex and powerful satellites in LEOs or MEOs capable of providing real-time communications, such as voice and facsimile to hand-held personal communicators; little LEO satellite systems deploy less complex and low-cost satellites operating in the lower part of the orbit designed to provide real-time or non-real-time low bit rate messaging services such as position reports and paging to pocket-sized terminals; 'super' GEO systems are modern GEO systems which deploy very large numbers of spot beams (>100) enabling services to use cellular telephone-sized terminals.

1.7 Mobile systems – a comparison

Satellite mobile systems are ideally suited for areas which are poorly served by terrestrial systems and they are generally complementary to terrestrial systems. In the future, when MSS call and terminal costs approach those of terrestrial systems, a certain degree of competition between their respective system operators is inevitable in some applications. However, together they will offer users a synergistic solution with a greatly extended service coverage.

The advantage of synergistic integration of terrestrial and satellite systems is well recognized. Most recent GMPCS systems support terrestrial services through dual-mode hand-held units with very little integration at network level. Table 1.6 compares the main features of terrestrial and satellite mobile services.

Satellite	Terrestrial
Wide area coverage is possible – typically thousands of kilometres	Relatively lower service area – typically hundreds of kilometres; some coverage breaks are possible
Roaming over a wide area is technically straightforward, as coverage is seamless; roaming is limited by licensing issues; usually a single operator owns the space segment with one or more service distributors	Roaming over wide areas encompassing several countries involves more than one operator/system requiring special system features and operational arrangements
Terminals are relatively large and heavy, but new systems offer hand-held services. Terminals deploying large antennas (0.3–1.0 m) are expected to be used for wideband services in the near future reducing to hand-held in a few years	Terminals are small with a wide variety
Cost of terminal ranges between $500 and >$25,000 (depending on application)	Terminal costs are typically < $200
Cost of call to end users ranges between $0.5 and $5 per minute (changes done regularly)	Call cost ranges from <$0.1 to $1.0 per minute (changes done regularly)
MSSs operate in aeronautical/land/maritime environments	Generally operate in land environments; limited coverage is possible in aeronautical and maritime environments
Service includes voice and data with a throughput up to ~ 450 kbps in next generation	Services include voice and data up to ~10 kbps; up to 2 Mb/s in third generation systems
Serve niche market – ships, aircraft, trucks, international travellers and businessmen, cellular extension, tourism	Serve individuals for social and business needs in populated areas; coverage limited to coastal areas in maritime environments; air coverage is available in Europe and the USA
Frequency is reused at distances of hundreds to thousands of kilometres	Frequency is reused at distances of 100 m–10 km
Handover between spot beams or satellites is not always essential for GEO but essential for non-GEO systems	Handover is frequent and necessary in all cellular systems
Suited for wide area coverage and thin routes (e.g. traffic density = < 0.1 Erlangs/sq. km); 30–70 channels per sq. million for existing non-geostationary satellite systems due to limited satellite power and large spot beam sizes	Suited for urban and suburban environment; uneconomic on thin routes (e.g. traffic density <1 Erlang/sq. km)

Table 1.6
A comparison of the main characteristics of mobile services through terrestrial and satellite media

1.8 Example applications

There are innumerable conventional and innovative applications of MSS. We have, therefore, categorized them here generically as follows:

1   Maritime environment
    - ship and cargo management;
    - distress;
    - social;
    - remote monitoring and control;
    - internet access;
    - fleet broadcast;
    - tourism and leisure;
    - journalism;
    - business;
    - others.
2   Land environment
    - business;
    - remote monitoring and control;
    - personal;
    - fleet management;
    - tourism and leisure;
    - journalism;
    - government and aid agencies;
    - internet access;
    - messaging;
    - others.
3   Aeronautical environment
    - cockpit communication;
    - passenger communications including facsimile/internet access;
    - automatic dependence surveillance;
    - others.

1.9 Practical limitations

In this section, we highlight some of the existing practical problems related to use of the second and third generation MSS and comment on their short- and long-term solution prospects. As illustrated in Table 1.7, many problems are likely to be mitigated as technology evolves.

In the 1990s call costs of satellite services dropped remarkably from ~ \$8–\$10 per minute to under \$1 per minute for voice; similarly, terminal cost and sizes plummeted from ~ \$25 000 to \$700 for voice communication. At the beginning of the 1990s, voice services were available from large terminals at a premium

Current limitations	Comments
Expensive in terms of infrastructure, call and terminal costs	Terminal and call costs continue to reduce rapidly
Terminals are large compared to terrestrial systems	A significant size reduction in recent years
User interface is complex	The limitation applies to specialist equipment such as ship-borne large terminals
A general lack of awareness of this technology	A growing awareness in recent years
Systems susceptible to local interference	Problem likely to remain as such; impact generally lower on spread spectrum systems
Routing arrangements can be complex and time consuming	
Concern from local operators about revenue loss due to possible bypass	Practical solutions feasible

Table 1.7
A summary of current limitations of MSS

cost, whereas by the turn of the millennium similar services were available from hand-held and laptop-sized terminals at a fraction of the cost. A trend towards integration of satellite and cellular systems is emerging as evidenced by first generation dual-mode hand-held telephones.

The first and second generation MSS products which are still in use tend to have complex human–machine interfaces, often requiring specialist training for proper operation. Some of the reported problems arise due to lack of proper understanding of the equipment; often workers in remote regions would be given satellite phones without proper instructions, leading to bad publicity. On most occasions, problems occur for simple reasons such as discharge of batteries and lack of provision of a standby.

Even in the late 1990s, there was generally a lack of awareness of the capabilities and cost effectiveness of satellite communications. For example, a survey of yacht users conducted by Inmarsat in the mid-1990s revealed only 10% of them fitting satellite terminals, with 70% stating price as the main reason for lack of fitting, even though with the budget typically stated, a satellite terminal could be readily afforded. There was a general lack of awareness about the benefits such as safety aspects of satellite communications. Older forms of communications such as HF were still in wide use. It is expected that with the introduction of hand-held services and high-profile advertisements, the public at large will rapidly become aware of the benefits.

Interference to terminals in specific parts of the world is possible. Radio frequencies used for satellite communications are shared with terrestrial systems in various parts of the world. Even when terrestrial allocations have a secondary status, in practice it is difficult to enforce termination of offending transmissions. Sometimes transmitters in the vicinity of a user, even if transmitting on a different frequency band, saturate the front end of the satellite receiver, causing unacceptable degradation to signal quality. Harmonics of powerful terrestrial transmissions such as radar or television can enter receivers, causing interference. The probability of interference increases in noisy radio environments such as near ports. These types of problems can be minimized by communicating via another visible satellite, tuning the equipment to another broadcast channel or communicating through another gateway. The operator can build a useful database of reported interference for better customer appreciation and possibly solutions to the problem.

Routing arrangements from the terrestrial segment to the space segment must be in place for the users to be connected efficiently. These arrangements are complex and time consuming for reasons of bureaucracy, politics, etc. The problem is compounded for an international operator, as the process has to be repeated, with each country having individual procedures and priorities.

With the introduction of personal satellite communication services, some concern has been raised regarding loss of business to local telecommunication operators because satellite systems can bypass them. Some concern regarding security aspects has also been raised by countries vulnerable to anti-social activities. Operators therefore make agreements with local authorities to minimize revenue losses to local operators and build features into the system design to switch off transmissions in specified geographical regions or to permit communications only via specific gateways.

1.10 Related satellite systems

Satellites provide a variety of mobile services, which are not MSSs *per se*, but complement or compete with MSS offerings. Some of the more interesting ones are:

1   navigation systems;
2   direct to individual radio broadcast systems;
3   personal communication system by FSS;
4   direct television broadcast to mobiles.

1   *Navigation systems*
    Navigation is used for estimating the position of a vehicle on sea, in air or space and on land to ensure that the chosen route is followed accurately, both in the short and long term. Short-term navigation is required for making instantaneous changes in direction, speed and acceleration to avoid an obstacle, and long-term navigation is used for making a general correction to a route.

In recent years receivers of satellite navigation systems, and in particular the GPS, have become a personal and mobile communications accessory due to a significant reduction in receiver costs, making them a part of regular gear carried by explorers, travellers, fleet managers, rally organizers, etc.

The GPS navigational system was introduced by then US military in 1978. A constellation of 18 GPS satellites can provide continuous worldwide two-dimensional coverage; increasing the constellation to 24 satellites can give three-dimensional position fixes world-wide. Satellites are in circular 63° incline orbit at an altitude of 20,200 km and an orbital period of 12 hours; accuracy of fixes for military users is 10m, degradingto about 100m for civilian users. Satellites transmit atomic clock-controlled timing signals together with their orbital parameters, which are used by receivers for a range estimation. The system operates on a similarprinciple as a Loran oe Decca ststem except that instead of hyperbola, hyperboloid of revolution are used for position estimation. A S receiver estimtes its location by measuring the range from three (or four) most favourably positioned satellites simultaneously and solving three (or four) simultaneous equations. A similar system using a different type of transmission signal was introduced by the former USSR at about the same time.

Satellite-aided navigation systems can be categorized on the basis of their operating principles: Doppler signature, range determination, single satellite and multiple satellite transmissions. Chapter 10 summarizes the principles of operation of the most commonly used satellite navigation systems.

2   *Direct to individual radio broadcast systems*

A number of studies have investigated the viability of satellite radio broadcast directly to individuals as they promise rapid wide-area coverage, high-quality sound and niche programme channels. The current disadvantages include large investment and expensive receiver technology (but technology is evolving very rapidly) and, on the programming front, possible lack of local coverage (e.g. city events, traffic reports, etc.).

In the early 1990s, the ITU assigned spectrum for direct sound broadcast in the S band. Since then, a number of commercial systems have been proposed and at least one system is close to service introduction.

Satellite radio systems require high-power satellite transmitters, robust modulation and coding schemes as shadowing and multipath countermeasures and most importantly affordable receivers. Chapter 10 addresses these issues in detail.

3   *Personal communication system by fixed satellite service*

The smallest terminals of FSS, commonly known as very small aperture terminals (VSATs), have been used in low-capacity routes since the 1970s. Advances in technology leading to the introduction of K_u (11–14 GHz) and K_a (20–30 KHz) band high-power satellites deploying spot beam antennas and VLSI have enabled a reduction in size and cost of VSATs to an extent that the distinction between VSAT and broadband MSS services is blurred at the upper end of MSS throughput (~ 64–500 kbps).

Several FSS systems, specifically targeting personal and private enterprise communications, are under development. The goal is to provide broadband services to very small terminals – a personal set could offer 2 Mb/s and a transportable set 155 Mb/s.

The formal definition, applicable radio regulations and frequency management methods between the services differ due to differences in application and terminals traditionally supported. At present, the majority of MSS spectrum allocations lie in the L and S bands, where propagation conditions are congenial for the service, whereas those of the FSS lie in a wide range from 3 to 30 GHz. Furthermore, MSS allocation tends to be exclusive for each operator within that area to permit users unrestricted movement within that area. In contrast, FSS allocations are shared with other terrestrial/satellite services, because terminals remain fixed at a specific geographical location. Each move ideally requires a further series of lengthy coordination meetings, making it impractical to move terminals freely. There is potentially an opportunity for unrestricted mobility within areas managed by a specific jurisdiction when the authority manages the spectrum accordingly.

Let us return to the issue of competition between these two services. Competition in core MSSs, such as communication from moving terminals, hand-held services, etc., is not feasible for reasons mentioned above. For fixed-site applications which allow use of directive antennas, both services are similar from a user's perspective. The main difference lies in economics, the regulatory advantage of MSS in terms of unrestricted connection to the public network and the ease of setting up MSS terminals at will. VSAT networks have the advantage in terms of high throughput and lower cost for large volume data transfer between fixed locations. Figure 1.11 shows a VSAT expected to be available in the near future; compare this to MSS terminals illustrated in Figure 1.9.

4   *Direct television broadcast to mobiles*

Television broadcasts from direct broadcast satellites can be received on moving mobiles, such as ships where a stabilized tracking antenna may be easily deployed. Conventional direct broadcast satellite systems require several MHz of bandwidth and therefore similar transmission schemes cannot be used in mobile satellite systems which have severely limited

Figure 1.11
A personal VSAT expected in the first half of the decade

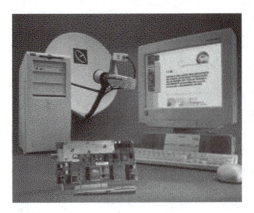

bandwidth. The challenge is to be able to transmit good quality television within bandwidths of the order of ~ 50 KHz. The key technologies to achieve this goal are video compression and highly efficient modulation and coding schemes. Television broadcasts to ships within a bandwidth of 400 KHz using the offset – Q phase shift keying (O-QPSK) modulation scheme at a modest quality were demonstrated in MSS systems.

1.11 Trends

1.11.1 General

The urge of people to remain in contact under all circumstances has been instrumental in the success of mobile communication technology. As people tend to expect from mobile communication systems services akin to those offered by fixed services, it is anticipated that trends in fixed services are likely to be a precursor to those in the mobile systems. Mobile wireless services give lower capacity than wired systems because of the difficulty in maintaining a high-quality mobile radio link, and at present, a limited available spectrum for these services.

In addition to continuing demands for voice communication, there is now an escalating demand for data traffic for both personal and business applications. Examples of such applications are internet access to large database, teleworkers transferring software, e-mails, image transfer, etc. Similar demands are already placed on the mobile communication sector due to the vigorous growth in portable computers. Thus, considerable effort is under way to provide high data rate mobile telecommunication services. Interestingly, both FSSs and MSSs are vying for the personal broadband service.

Convergence of computer technology with telecommunications has resulted in a considerable demand for data traffic. A number of new standards and technologies have been introduced in the fixed network to satiate the demand. Some of them apply to computer networking and others to generic data traffic. Examples of recent introduction are narrow and broadband integrated services data networks (ISDN and B-ISDN), ATM and IP transport, Frame relay technology, etc. ATM gives efficient use of resources when input contains a mix of data types such as audio, video, image data, etc. It is anticipated that satellite systems will provide an extension to the terrestrial ATM. Consequently, considerable research effort is directed towards adapting these technologies to a mobile satellite environment.

There is also a considerable interest in convergence of the fixed component of mobile networks with the fixed network due to a growing number of commonalities between them. It is recognized that in dense urban areas terrestrial systems are better suited, but satellite systems offer unique advantages in providing wide-area mobile communications to ships, aircraft and land mobiles. Thus, from a user's viewpoint, there are advantages in developing a unified satellite and terrestrial network. All first generation satellite hand-held systems offer a facility to switch from one or more of the existing cellular standards to another. Considerable research activity is in progress to unify terrestrial and mobile systems into a universal network.

Satellite systems have grown significantly but due to their niche services, their penetration is significantly lower. The services offered by first and second generation mobile satellite systems were targeted for specialized applications in aeronautical, maritime and land sectors on terminals which were heavy and expensive. However, third generation MSS systems, which we call first generation hand-held systems, provide a cellular system-like service at comparable costs and future multimedia terminals are anticipated to be about the size of desktop-size telephones. Figure 1.2 includes the increase in the effective isotropic radiated power (EIRP) of three generations of Inmarsat, which is representative of world-wide MSS growth trends in the 1980s and 1990s.

1.11.2 Market

Vast, sparsely populated areas throughout the world remain unserved by fixed or cellular systems, either because the service is uneconomic or due to a lack of infrastructure. Satellite systems are ideally suited to filling such coverage gaps. Experience of hand-held service providers, such as Iridium and Globalstar, indicates that the growth rate and perhaps the size of the market was vastly overestimated.

At present, personal terminal costs and the call charges are high, keeping the service beyond the reach of individuals in developing countries. Hence, individuals from affluent countries are likely to be key consumers. With the passage of time, economies of scale, improved technology, competition and operator incentives are expected to lower terminal and communication costs, paving the way for the services to be affordable in the developing world. This model follows the world-wide growth trend of the cellular service. However, in the case of satellite systems, the market is perhaps inverted, i.e. the real demand lies in the developing world for the reasons mentioned earlier.

An analysis of population and telephone density provides a simple approach in assessing the market potential. Table 1.8 represents an approximate global population by region and telephone lines per 100 inhabitants; the data indicates Asia as the most potential market. A more refined forecast model would account for affordability, existing infrastructure, etc., and is discussed in Chapter 7. Countries with large populations and modernization programmes such as India and China, the growing economies of a number of countries in the region, and heightened awareness of satellite communications in the Asia Pacific rim all indicate this region as a potential market. However, affordability is highest in North America and Europe.

Table 1.8
Approximate population and number of telephone lines in four continents (ITU, 1994 data)

	Approximate population (millions)	Telephone lines per 100 inhabitants
Africa	717	2.3
Americas	772	28.4
Asia	3417	5.1
Europe	772	33.5

A vigorous growth is expected in the personal communications market, though the rate of growth in the hand-held voice communication market has been markedly lower than anticipated, as evidenced by the bankruptcy of some well-known systems. It is believed that demand for broadband data services will escalate following the launch of terrestrial mobile broadband data services leading to the introduction of services up to 0.5 Mb/s for personal communications by 2005 and rates of 0.5 to several Mb/s in next generation mobile and personal FSSs.

1.11.3 System architecture

Until the late 1980s, most commercial mobile satellite systems used GEOs with bent-pipe transponders. In the 1980s, extensive studies were undertaken in the UK and by the European Space Agency (ESA) to investigate the feasibility of deploying intelligent satellites in highly eccentric elliptical orbits for mobile communications in the European region. Studies also investigated the feasibility of deploying LEO satellites for mobile satellite systems. One such study concluded that non-geostationary mobile satellite systems were feasible, offered advantages but required new technology which could be available by the end of the millennium (Richharia *et al.*, 1990).

The trend in space segment architecture diverged at the beginning of 1990 when a number of commercial non-geostationary satellite systems were proposed for hand-held voice and data communication services. The architecture proposed LEOs or MEOs, as hand-held services via geostationary satellites would require extremely complex spacecraft and suffer transmission delays. Their architecture varied widely in orbital choice, satellite complexity, transmission schemes, network routing and the market addressed.

By mid-1990 there was a re-emergence of geostationary systems based on powerful satellites deploying several hundreds of spot beams. A comparison of various types of systems and architectures is given in Chapters 2, 8, 9 and in the Appendix.

Third generation mobile systems have been standardized on the premise that satellites will form an integral part of the network. International forums which have been instrumental in the process include the ITU forum called the International Mobile Telecommunication – 2000 or IMT-2000 and the European Telecommunication Standardisation Institute (ETSI), a European forum where the third generation mobile systems are called universal mobile telecommunication systems or UMTS.

IMT-2000 proposes three data rates – 144 kbps for car and train travellers, 384 kbps for people on foot and 2 Mbps for stationary users. The system therefore consists of pico cells covering buildings and small areas, micro cells of a few hundred metres covering streets, macro cells which cover areas of ~1–10 km and overlay cells of ~100–500 km made up of satellite spot beams. Components of each system will be connected through standard interfaces, as conceptualized in Figure 1.12.

The services should be affordable and include:

- audio, video and data communication at high rates up to 2 Mbps;
- as far as possible, services available in the fixed network including the same range of multimedia services to the extent possible, considering the limitations of radio interface;
- value-added services such as navigation, vehicle location, road traffic information, etc.;
- international roaming using a variety of satellite terminals.

Figure 1.12
Concept of a universal
mobile communication
system (Richharia 1999)

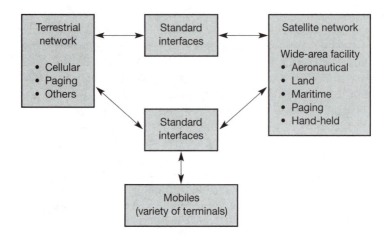

Embedded in the concept are broadband mobile systems in the range 2–155 Mbps. The earliest applications of these broadband services are expected to be specialized. Examples may be movable broadcast studios, mobile studio cameras for live broadcasting, emergency services, etc. (see Chapter 11).

While international standardization progresses, MSSs in the commercial world continue to evolve rapidly. Several systems were introduced in the late 1990s. The majority of them integrate with terrestrial systems to some extent, but without agreed standards integration with future universal systems can be rather limited. An interesting trend is the personal communications services being proposed by FSS in direct competition with the high throughput end of MSS.

1.11.4 Spectrum

The rapid evolution of MSSs, coupled with the introduction of broadband services, implies that demand for spectrum is on the rise. As mentioned above, most MSSs operate in L or S bands, the latter having been recommended for future universal systems. Consequently, these bands are facing congestion despite the introduction of spectrum enhancement techniques such as spectrally efficient modulation/coding and advanced spot beam technology, and there is a significant interest in higher frequency bands such as K_a and V. Chapter 3 provides further details on spectrum-related matters.

1.11.5 Technology

To realize new services and maximize spectral efficiency and revenue, several new technologies have been developed. A few representative examples follow:

- **Satellites with regenerative transponders and onboard computing:** Regenerative transponders can lower mobile terminal EIRP allowing mitigation of interference caused by emissions from terminals and a reduction in their size and cost. The ability of regenerative transponders to decouple up and down links enables optimization of multiple access separately and maximizes utilization of satellite resources.

- **Intersatellite links:** Intersatellite links are space links between satellites for routing signals in space rather than terrestrially.
- **Spot beam technology:** All modern mobile satellite systems deploy large numbers of spot beams. Typically, several hundred spot beams are used for geostationary mobile satellite systems.
- **Multiple-launch capability:** Large satellite constellations require a large number of reliable launches in quick succession and several launchers have been developed for multiple satellite delivery.
- **Mass-production technique:** Traditionally, several years were spent in manufacturing communication satellites, clearly an unacceptable scenario for manufacturing satellites which are part of large constellations. Satellite manufacturers have introduced mass-production techniques akin to those used in the automotive industry, reducing manufacturing time per satellite, in some cases, to a few weeks.
- **Software:** All aspects of satellite communications have benefited from the development in software and work stations. Applications include computer-aided design, remote Earth station operation, automated satellite control, constellation management, radio resource management, etc. Network control, data management, data flow, as well as a majority of applications are now heavily dependent on software.
- **Advanced receiver architecture and VLSI:** The challenge here is to produce low-cost, cellular-integrated satellite phones affordable to individuals. Phenomenal advances in VLSI, packaging, battery technology, etc. have already enabled the introduction of dual- and multimode receivers and a corresponding reduction in cost and weight of portable satellite transceivers.
- **Advancements** have been orchestrated in areas of modulation, coding, voice coding, compression, accessing technology, system architecture including the fixed network component, management of complex networks, etc.

Further reading

Chakraborty, D. (1989) 'Survivable communication concept via multiple low earth-orbiting satellites', *IEEE Transactions Aerospace and Electronics Systems*, 25 (6).

Clarke, A.C. (1945) 'Extra terrestrial relays', *Wireless World*, October.

Cosmas, J.; Evan, B.; Evci, C.; Herzig, W.; Persson, H.; Pettifor, J.; Polese, P.; Rheinschmitt, R.; Samukic, A. (1995) 'Overview of the mobile communications programme of RACE II', *Electronics and Communications Engineering Journal*, August, 155–67.

Dietrich, J.F. (1997) *The Globalstar satellite cellular communication system design and status*, proceedings of the Fifth International Mobile Satellite Conference, Pasadena, California, 16–18 June, co-sponsored by NASA/JPL and DOC/CRC; JPL Publication 97–11, Jet Propulsion Laboratory, Pasadena, California, 16 June.

Electronics and Communications Engineering Journal (1995) 'Technical demographics', December, 265–71.

Evans, J.V. (1998) 'Satellite systems for personal communications', *Proceedings of the IEEE,* vol. 86, no. 7, July 1325–41.

IEE (1984) IEE Colloquium on the Communication Engineering Research Satellite (CERS), April, 1984–39.

ITU (1990) *ITU-R Volume V*, Annex.

Jahn, A.; Buonomo, S.; Sforza, M.; Lutz, E. (1995) 'Narrow and wide-band channel characterisation for land mobile satellite systems: experimental results at L band', *International Mobile Satellite Conference,* Ottawa–IMSC 1995, The fourth International Mobile Satellite Conference Ottawa, co-sponsored by Communication Research Centre/Industry Canada and Jet Propulsion Laboratory/NASA, pp. 115–21.

Nelson, R.A. (1998) 'Iridium: from concept to reality', *Via Satellite*, **XIII**, (9), September, 62–70.

Richharia, M.; Hansel, P.; Bousquet, P.W.; O'Donnell, M. (1989) 'A feasibility study of a mobile communications network using a constellation of low earth orbit satellites', *IEEE Global Telecommunications Conference*, GLOBECOM '89, Dallas, 27–30.

Richharia, M. (1999) *Satellite Communications: Design Principles*, 2nd edn, MacMillan, UK.

Stix, G. (1993) 'Domesticating cyberspace', *Scientific American*, August.

Stuart, J.R. and Smith, D.J. (1989) 'Review of Archimedes study', *IEE Colloquium on Highly Elliptical Orbit Satellite Systems*, 24 May, 2/1–2/4.

Sutton, R. W.; Schroeder, E. H.; Thompson, A. D.; Wilson, S. G. (1973) 'Satellite-aircraft multipath and ranging experiment results at L band', *IEEE Trans. Comm*, **COM-21** 5, 639–47.

UMTS Forum (1999) 'The future mobile market', *Report*, (8), March.

Satellite constellations

2

2.1 Introduction

Satellite orbits have been one of the most extensively reviewed topics in the past five years due to a strong influence of orbit design on the space segment architecture of non-geostationary satellite systems. The trend away from GEOs began in the 1990s, due to a combination of events – availability of advanced technology and industry's willingness to support high-risk commercial ventures motivated by a potentially lucrative hand-held market. Low and medium Earth satellite systems, being closer to the Earth, would incur lower path loss and propagation delay, enabling the use of hand-held terminals while offering transmission delays approaching those of optical fibre and other terrestrial media. System designers were faced with the challenge of optimizing such networks and issues included the number of satellites, their orbital characteristics, efficient network routing schemes, cost, etc. System design takes into consideration the network in its entirety and is influenced by a variety of factors, such as service area, traffic distribution within it, acceptable space segment complexity, etc., though emphasis on each depends on the operator's preference. This is evident by the vast differences in orbital characteristics of various prevalent non-geostationary satellite constellations, ranging from LEOs and MEOs to elliptical, hybrid and GEOs. The topic of overall constellation optimization is covered in Chapter 8 to enable the reader to better appreciate the interrelationships.

In this chapter, the problem of constellation optimization is addressed solely from an orbital viewpoint, that is, taking into consideration the coverage area, space environment and orbital characteristics. To keep to the theme of the book, the subject is treated at a system level.

The chapter has been divided into two broad parts. The first part introduces the concepts of orbital mechanics and other related topics and the second addresses issues related to satellite constellations.

The first part provides a brief introduction to orbital mechanics which includes coordinate systems applicable to MSS, the definition of orbital parameters and various types of orbits. An understanding of the interrelationship between the most crucial parameters and their influence on orbit design is explained by a two-body model. The effects of perturbations caused by the Earth's non-uniform gravitational force and gravitational effects of the Sun and Moon are described and specific orbits related to these perturbations, namely *Sun-synchronous* and *Molniya* orbit, are discussed. Commonly used map projections for illustrating a satellite's coverage area are introduced with examples of coverage maps of GEO and non-GEO satellites. Subsequent sections discuss the solar eclipse, the Sun's interference, Doppler Effects and the growing problem of orbital debris.

The second part of the chapter discusses various aspects of satellite constellation including theoretical bounds of constellation size, and reviews well-known techniques used in the optimization of polar, inclined orbit and hybrid constellations. The chapter concludes with a comparison of orbit-related performances of constellations deployed or planned recently.

2.2 Satellite orbits

This section provides a brief overview of orbital mechanics fundamentals essential in developing an understanding of satellite constellations.

Kepler's Laws in conjunction with Newton's Laws quantify satellite motion around the Earth. Kepler's three Laws of Planetary Motion define the shape of orbit and the orbital period of satellites, while Newton's Laws of Motion and Gravitation explain the reason for this behaviour.

According to Kepler's Law, a satellite orbits the Earth on a well-defined path which can be quantified by a set of orbital parameters. These parameters are used to define the position of a satellite uniquely in a three-dimensional space at any given instant. A number of coordinate systems are used in mobile satellite systems. These are often called inertial coordinate systems and the space is referred to as an inertial space, as each is referenced to stars.

Figure 2.1
A geocentric-equatorial coordinate system

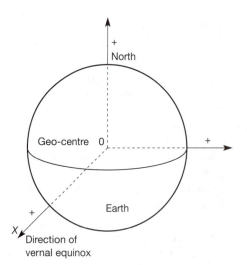

The coordinate system used for definition of orbital parameters is called the *geocentric-equatorial coordinate system* and is shown in Figure 2.1.

The Earth's centre constitutes the origin and the equator is the fundamental X–Y plane. The positive direction of the X-axis is towards the vernal equinox, which is the vector joining the geocentre to the direction of the Sun's position on the day of the spring equinox. Its equivalent spherical coordinate system is called the *right ascension-declination coordinate system* and is commonly used by astronomers for defining the position of heavenly bodies, such as the Sun and Moon. For satellite communications, a knowledge of the right ascension-declination system becomes necessary when the position of the Sun and Moon are of interest, such as when determining occurrence of the Sun's eclipse on a satellite. Another coordinate system named the *celestial horizon coordinate system* or *topo-centric coordinate system* is used for pointing the antenna towards a satellite. The axes in such a system are called elevation and azimuth. Elevation is the angle on the vertical circle from the local horizon to the position of the satellite, and azimuth is the angle of the satellite measured in an eastward direction from true north on the local horizontal plane as shown in Figure 2.2.

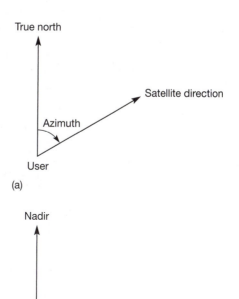

Figure 2.2

A celestial horizon coordinate system

The six orbital parameters, portrayed in Figure 2.3(a) and (b), are semi-major axis, eccentricity, inclination, right ascension of an ascending node, the argument of perigee and time elapsed from a reference point of the orbit.

1 *Semi-major axis* describes the size of the orbit as shown in Figure 2.3(b). The term semi-major axis is appropriate for an elliptical orbit. The semi-major axis reduces to a radius for a circular orbit.

2 *Eccentricity* represents the shape of an orbit. The eccentricity of a circular orbit is 1 and takes on a more elongated shape as its magnitude is increased.

3 *Inclination* describes the orientation of an orbit with respect to the equator, as shown in Figure 2.3(a). It is the angle between the plane of the orbit and the equatorial plane. Referring to Figure 2.3(a), it can be observed that the orbit crosses the equatorial plane at two points called the *ascending node*, where a satellite crosses the equator from the southern hemisphere to the northern, and the *descending node*, where a satellite crosses the equator when moving from the northern hemisphere to the southern.

4 *Right ascension* shows the orientation of an orbit with respect to the X-axis of the coordinate system. It is the angle between the X-axis and the *line of nodes* – a line joining ascending and descending nodes.

Figure 2.3
(a) A pictorial representation of orbital parameters; i = inclination; Ω = right ascension of an ascending node is the angle between the X-axis – the direction of the vernal equinox – and the ascending node; argument of perigee is the angle between the line of nodes and the perigee (Richharia, 1999). (b) Major parameters of an elliptical orbit (Richharia 1999)

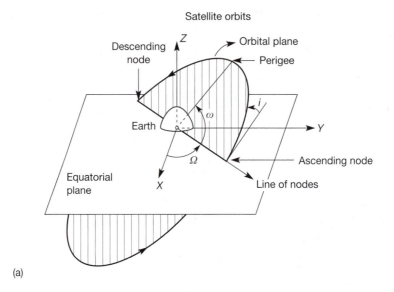

(a)

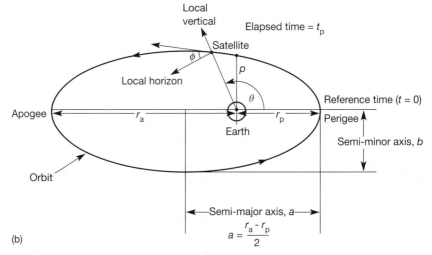

$$a = \frac{r_a - r_p}{2}$$

(b)

5   The *argument of perigee* describes the orientation of an ellipse with respect to the line of nodes.

6   Time t_p is the *time elapsed* since a satellite has passed a reference point in orbit called the epoch, usually its perigee.

Applying Newton's Law of Motion leads to the equation of motion of satellites in their orbit. A complete solution of the equation is quite complex, but a partial solution is adequate to illustrate the shape and size of an orbit. The simplified solution shows that the satellites follow a path of a conic section of the general form r = p / (1 + *e* cos (θ))   (2.1)

where r = the distance of any point on the trajectory from the geocentre, p = a geometrical constant – termed parameter of the conic, which determines the width of the conic at the focus, *e* = the eccentricity which determines the type of conic section and θ = the angle between r and the point on the conic nearest the focus.

The trajectory of a satellite depends on the final velocity vector and altitude at which the satellite is launched by the launch vehicle. Thus, for example, a 900 km circular orbit can be achieved by launching a satellite to an altitude of 900 km so as to achieve a velocity of 7.4 km/s, i.e. the velocity required at such an altitude, and tilting the satellite parallel to the Earth.

An orbit in which a satellite moves in the same direction as the Earth, i.e. west to east, is called a *direct* orbit and when a satellite moves in an opposite direction to the Earth's motion, the orbit is a *retrograde* orbit. Satellites may also be categorized according to their inclination, altitude and eccentricity. An orbit with an inclination of 0° is called an *equatorial* orbit; when the inclination is 90° the satellite is said to be in a *polar* orbit; orbits having an inclination between 0° and 90° are called *inclined* orbits. Orbits may be categorized according to their altitude as low Earth orbit, medium Earth orbit, also sometimes called *Intermediate Circular Orbit* (ICO), and geostationary Earth orbit. When eccentricity is 0, the orbit is a *circular* orbit and when eccentricity > 0, the orbit is an *elliptical* orbit. High-eccentricity orbits are sometimes called *highly elliptical* orbits.

For an elliptical orbit, the relationship between the period T of a satellite and the semi-major axis can be obtained by Kepler's Third Law

$$T^2 = 4\pi^2 a^3/\mu$$ (2.2)

For a circular orbit the equation reduces to

$$T^2 = 4\pi^2 (R+h)^3/\mu$$ (2.3)

where R = the radius of the Earth and h = the satellite altitude.

Figure 2.4(a) – (c) shows the relationship between altitude and period of orbit. Also marked on the figure are the approximate practical bounds of low and medium-altitude orbits. Figure 2.4(a) applies to altitudes up to the GEO, while Figure 2.4(b) and (c) pertains specifically to MEOs and LEOs. Similarly, Figure 2.5(a) – (c) demonstrates the relationship between altitude and orbital velocity for LEO, MEO and GEO satellites.

Figure 2.4
(a) Relationship between altitude and orbital period for altitudes up to the GEO (Graphics: AR).
(b) Relationship between altitude and orbital period for MEO (Graphics: AR).
(c) Relationship between altitude and orbital period for LEO (Graphics: AR).

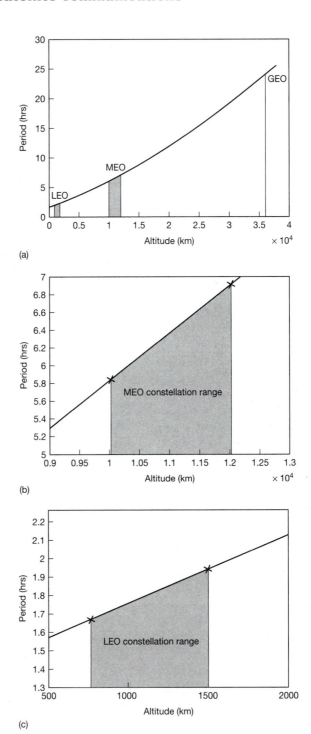

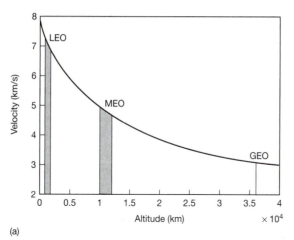

(a)

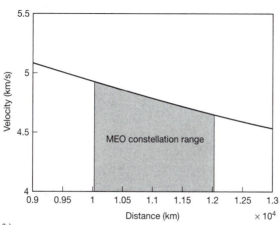

(b)

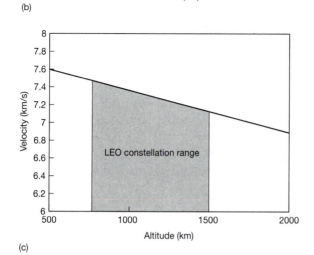

(c)

Figure 2.5
(a) Relationship between altitude and orbital velocity for altitudes up to the GEO (Graphics: AR). (b) Relationship between altitude and orbital velocity for MEO (Graphics: AR). (c) Relationship between altitude and orbital velocity for LEO (Graphics: AR)

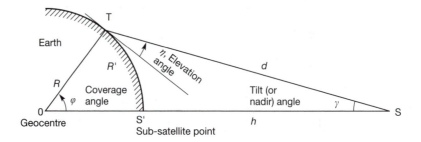

Figure 2.6
Orbital geometry of a
geostatuionary satellite

The region between low and medium altitudes consists of a strong radiation belt which can cause damage to satellite components (see section 2.2.2) and is not usable for commercial satellite systems. At low altitudes, satellite velocity decays due to friction from the Earth's atmosphere causing a loss in altitude until satellites re-enter the atmosphere and burn out due to frictional heat. For practical purposes, altitudes below ~180 km are not useful, except perhaps for military applications.

Figure 2.7
(a) Contour plot of
elevation as a function of
latitude and relative
longitude (Graphics: AR).
(b) Three-dimensional
plot elevation as a
function of latitude and
relative longitude
(Graphics: AR)

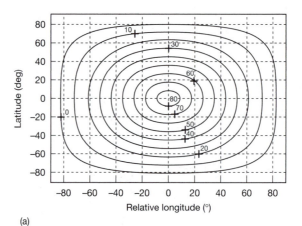

(a)

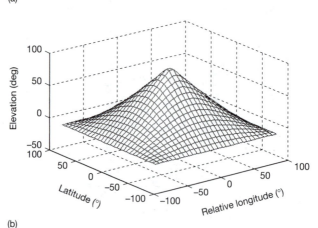

(b)

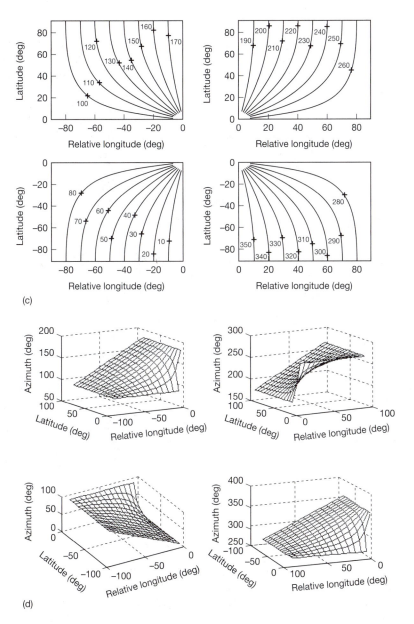

Figure 2.7
(c) Contour plot of azimuth angle as a function of latitude and relative longitude (Graphics: AR). (d) Three-dimensional plot of azimuth angle as a function of latitude and relative longitude (Graphics: AR)

At an altitude of 35,786 km, the orbital period becomes 23 h, 56 min and 4.1 s, which is the time the Earth takes to rotate 360° around its axis. This period is called a *sidereal* day. When the inclination of such an orbit is 0°, the relative motion of the satellite and Earth reduces to zero and therefore satellites in this orbit appear motionless to observers on the ground. This type of orbit is called geostationary, which has been used extensively for satellite communications.

To effect communication, it is necessary to determine position of the satellite at any instant. The position can be calculated by applying the orbital parameters of the satellite to a set of trigonometric equations. Knowledge of satellite geometry is also necessary for calculating radio link characteristics, such as satellite range and elevation angles.

For performing radio link analysis, simple geometric relationships between ground Earth station and satellite, assuming a spherical Earth, are often adequate. This is especially true for geostationary satellites, as shown in Figure 2.6.

Figure 2.7(a) and (b) shows the contour and three-dimensional plot of elevation as a function of latitude and longitude relative to the satellite subsatellite, while Figure 2.7(c) and (d) shows the contour and three-dimensional plot of azimuth as a function of latitude and longitude relative to the satellite subsatellite.

Figure 2.8

(a) A contour plot of geostationary satellite range as a function of latitude and relative longitude (Graphics: AR).
(b) A three-dimensional plot of geostationary satellite range as a function of latitude and relative longitude (Graphics: AR)

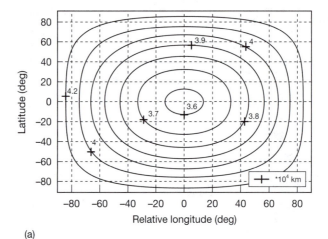

(a)

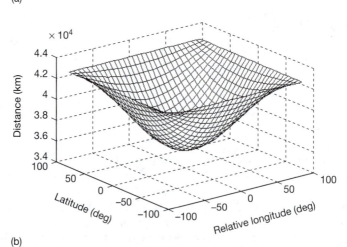

(b)

Figure 2.8(a) and (b) illustrates variations in range to a geostationary satellite with latitude and relative longitude as a contour plot and in a three-dimensional representation, respectively.

A number of external forces act on a satellite, causing its motion to deviate from the ideal. Hence, if a satellite is left uncorrected in its orbit, it drifts gradually – relative positions between satellites of a *phased* constellation change causing distortion of the coverage area, bearing in mind that in phased satellite constellations relative phase between satellites must be constant. The most significant perturbations are caused by:

- non-uniform gravitation of the Earth;
- gravitational effects of the Sun and Moon;
- atmospheric drag;
- solar radiation pressure.

Non-uniform gravitation of the Earth

The shape of the Earth is slightly ellipsoidal and the polar radius is ~ 21 km shorter than the equatorial orbit. A cross-section of the Earth would show a semi-major axis approximately along the line 165°E and 345°E (15°W) and a semi-minor axis approximately along the line joining 75°E and 255°E (105°W). Therefore, gravitational force is no longer directed at the geocentre, but towards the nearest ellipsoidal bulge.

Non-uniform gravitational fields cause the following effects:

- precession of perigee in orbital plane;
- precession of orbital plane around the Earth's north–south axis;
- perturbing force in a direction along the orbit.

The first two effects are most noticeable in LEO and MEO constellations, whereas the last mentioned mainly affects GEOs.

Precession of perigee

Precession of perigee is represented pictorially in Figure 2.9.

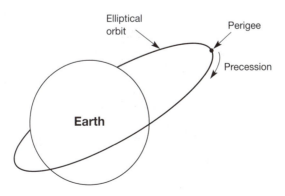

Figure 2.9
Precession of perigee

The rate of change of argument of perigee, ω, of an elliptical orbit is given as

$$\omega = 4.97 \, [R/a]^{3.5} \, (5 \cos^2 (i) - 1)/(1 - e^2)^2 \,°/\text{day} \tag{2.4}$$

where R = the mean equatorial radius (~ 6,378 km), a = the semi-major axis, i = the inclination and e = the eccentricity.

Figure 2.10(a) illustrates a variation in rate of change of argument of perigee versus semi-major axis for a number of inclination angles and Figure 2.10(b)–(d) shows a three-dimensional view of the variation for low, medium and geostationary satellites, respectively.

Figure 2.10
(a) Variation in rate of change of argument of perigee versus semi-major axis for a number of inclination angles (Graphics: AR). (b) A three-dimensional representation of variation in rate of change of argument of perigee for a LEO (Graphics: AR)

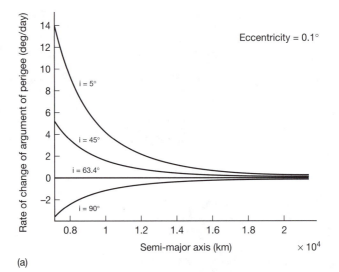

(a)

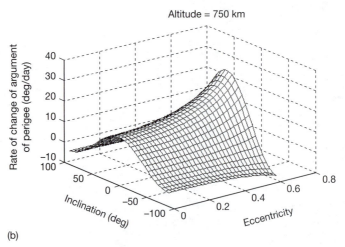

(b)

Altitude = 10000 km

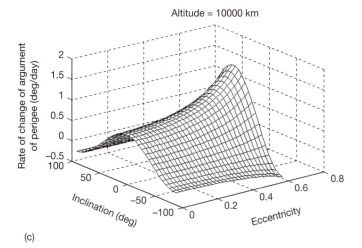

Figure 2.10
(c) A three-dimensional representation of variation in rate of change of argument of perigee for a MEO (Graphics: AR).
(d) A three-dimensional representation of variation in rate of change of argument of perigee for a geostationary orbit (Graphics: AR)

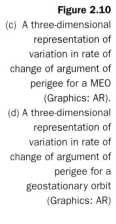

(c)

Altitude = 35786 km

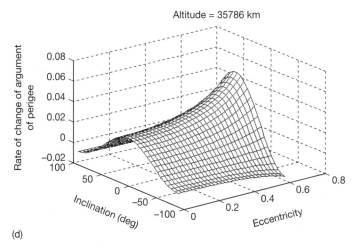

(d)

Note that at the inclination of 63.4°, ω = 0. This happens when the cosine term in the right-hand side of the equation equals zero. A satellite in such an orbit retains its perigee over the same region of the Earth. This characteristic can be usefully exploited for mobile satellite systems. Consider an eccentric orbit of 63.4° inclination with its apogee in the northern (or southern) hemisphere. From equation (2.4), note that the velocity of a satellite reduces with increasing altitude and hence also near the apogee of an elliptical orbit. By increasing the altitude sufficiently, the satellite can appear quasi-stationary at high elevation angles to an observer in the mid-to-high latitude of the northern (or southern) hemisphere. These features improve the radio link reliability of mobile satellite systems at mid/high latitudes (see Chapter 3). Figure 2.11 shows coverage of a satellite when the satellite is at the apogee of such an elliptical orbit.

Figure 2.11
Coverage of a satellite
from apogee of an
elliptical orbit
(Graphics: AR)

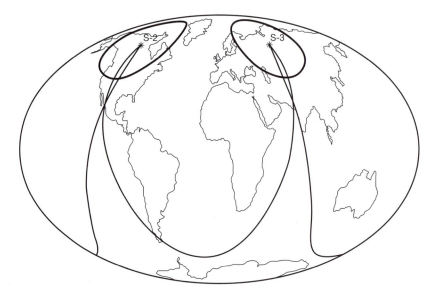

Figure 2.12
Precession of orbital
plane

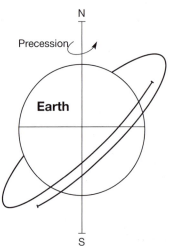

Examples of satellite systems designed for this class of orbit are Archimedese, a European design concept for mobile and broadcasting applications, and the Ellipso mobile system, a proposed US system for providing hand-held telephony service. This type of orbit, called the *Molniya* orbit after the Russian system, was used by the former USSR for providing satellite communications to high-latitude regions. *Tundra* orbits belonging to this class of orbit with an apogee of ~ 46,300 km and a perigee of ~ 25,250 km have also been proposed for mobile satellite communications. Note that these orbits cannot provide true world-wide coverage on their own and, hence, are combined with other types of orbit when world-wide coverage is desired.

Precession of orbital plane around the Earth's north–south axis
Figure 2.12 illustrates the precession of the orbital plane around the north–south axis, essentially precession of the ascending node.

The rate of precession Ω is given as

$$\Omega = 9.95 \, [R/a]^{3.5} \, (\cos (i)/(1-e^2)^2) \; °/day \tag{2.5}$$

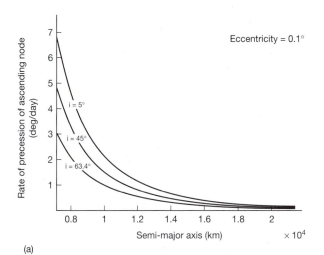

(a)

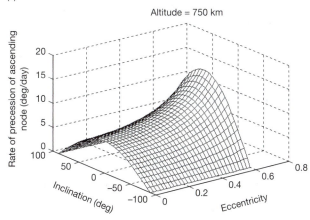

(b)

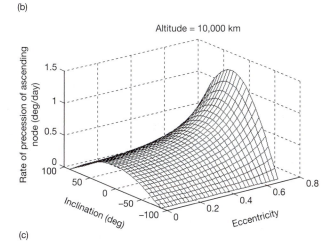

(c)

Figure 2.13

(a) Variation in rate of precession of ascending node versus semi-major axis for a number of inclinations (Graphics: AR). (b) A three-dimensional view illustrating rate of precession of ascending node for a LEO (Graphics: AR). (c) A three-dimensional view illustrating rate of precession of ascending node for a MEO (Graphics: AR)

Rotation is in a direction opposite to the motion of the satellite.

Figure 2.13(a) shows variation in rate of precession of ascending node versus semi-major axis for a number of inclinations and Figure 2.13(b) and (c) shows three-dimensional view of the variation for low and medium satellites respectively.

Precession of an orbital plane can, in fact, be utilized to give a useful class of orbit known as *Sun-synchronous* orbit.

A satellite's orbit is referenced to the Earth and as the Earth moves around the Sun, the angle between the Sun and a satellite's orbital plane changes. If the orbital parameters, a, e and i, are adjusted such that the rate of precession of the orbit equals that of the Earth around the Sun, i.e. 0.986° per day, then the Sun–Earth vector remains invariant. This is a *Sun-synchronous* orbit. Satellites in such an orbit always view the Earth under identical lighting, i.e. satellite rise time at a location remains unchanged, making it easy for users to locate satellites. Furthermore, satellite tracking requirements at Earth stations can be simplified because a satellite follows an identical arc on each pass. Further, if the orbital plane is made normal to the Sun, satellites never undergo eclipse. This simplifies satellite power supply requirements (see Chapter 6). Figure 2.14 illustrates an interesting perspective for the creation of a Sun-synchronous orbit using various combinations of a, e and i.

The non-uniformity in the gravitational field around the Earth also causes a force vector along the orbit which varies around the Earth. The resultant force cancels out in LEO or MEO, because a satellite's position relative to the ground keeps changing around the Earth. However, the resultant force for geostationary satellites does not change because the position of a satellite relative to the Earth remains unchanged and hence satellites undergo acceleration towards one of the two stable points located approximately on the minor axis of the Earth's ellipsoid. Thus satellites, when left uncorrected, drift towards the nearest stable position of ~ 79°E or ~ 252°E longitude and would oscillate around the stable point as they keep overshooting the point due to inertia.

Figure 2.14

Solution space for achieving Sun-synchronous orbit with various combinations of inclination, eccentricity and semi-major axis

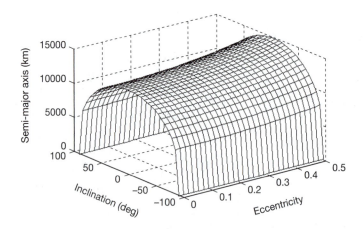

In practice, geostationary satellites must remain within closely bound limits around the designated location in compliance with Radio Regulations. Therefore, onboard thrusters are fired regularly to correct the drift and maintain the specified orbital location. This type of orbital manoeuvre is known as east–west station keeping.

Gravitational effects of the Sun and Moon

For satellites in LEOs and MEOs, the gravitational effects of the Sun and Moon are negligible in comparison to the Earth's gravitational effects. However, satellites in GEOs are perturbed quite notably by the gravitational forces of the Sun and Moon. The main effect of the gravitational pull by the Moon is a change in the inclination of the orbit between ~ 0.48°/year and ~ 0.67°/year with a period of about 18.6 years. The minimum occurred in 1997. The change is caused by variation in the inclination of the Moon itself. The yearly change in inclination due to the Sun is ~ 0.27°, which can be considered steady for practical purposes. The net effect of these two forces is to change the inclination of a geostationary satellite between 0.75° and 0.94° each year. As soon as the orbit of a geostationary satellite becomes inclined, the Earth begins to exert a force. The reader may remember that the Earth's non-uniform gravitational field causes a regression of the ascending node. This external force acts in a direction opposite to the force of the Sun and Moon. These forces cancel each other at an inclination of about 7.5°. Consequently, if the inclination of a geostationary satellite is left uncorrected, then the inclination of the satellite oscillates around a 7.5° inclination with a maximum inclination of 15° and a period of 53 years. In practice, inclination is corrected regularly by firing thrusters on satellites in an orbital manoeuvre called *north–south station keeping*.

Solar radiation pressure

When using large space structures it is necessary to consider the effect of solar radiation pressure. This is generally applicable to large geostationary satellites often deployed for MSSs. The net effect is an increase in the eccentricity together with disturbance along the north–south axis of the satellite and necessitates periodic corrections.

Atmospheric drag

Atmospheric drag is caused by friction to a satellite's body in the upper parts of the atmosphere. Therefore, satellites in LEO suffer the largest atmospheric drag. Below ~180 km, the friction is too high to sustain satellite motion, causing excessive heat on a satellite's surface and finally satellite burn out. From this perspective this altitude is considered as the lower limit of space. Atmospheric drag is directly related to the surface area and mass of a satellite and begins to become noticeable below ~ 750 km.

The orbital lifetime of a satellite is a complex function of the orbit, geometry and mass of the satellite and ionospheric conditions. Note the difference between the functional lifetime of a satellite and its orbital lifetime. The functional lifetime of a geostationary satellite may be of the order of 10–15 years, whereas its orbital lifetime may be thousands of years. On the contrary, a satellite in a 400 km orbit may have an orbital lifetime of a few months, whereas its functional lifetime is likely to be much longer.

Miscellaneous disturbances

A number of other types of forces act on a satellite, which must be compensated to maintain the desired orientation of satellite. These include effects of the Earth's magnetic field, meteorites, self-generated torque, etc.

2.2.1 Satellite coverage

Map projections

The region illuminated by a satellite is called the *coverage area or footprint*. Within this coverage, service may only be offered to a selected region called the *service area* of the satellite. The shape and size of the coverage area depends on the satellite's antenna radiation pattern (see Chapter 6). The most elementary antenna pattern consists of a bell-shaped circular pattern. The limit of useful coverage depends on the elevation of the satellite from the ground; the higher the elevation angle, the more reliable is a mobile satellite radio link but the lower is the coverage area. Minimum elevation angles of 5, 10 and 40° have been proposed. To make better use of satellite resources and reduce the size of the mobile terminal, it is usual to divide the coverage into smaller zones called spot beams. The number of spot beams range from a few to several hundred. To determine the coverage area of a satellite, its antenna pattern is projected to a map of the Earth as isocontours of antenna gain, or elevation angle, transmitted power or received power as necessary. A number of map projections are used, each suited to the application at hand.

Mercator projection is used commonly and is named after the Flemish cartographer Geraradus Mercator, who published a world map on such a projection for ship navigation in 1569. In this type of projection, the longitudes are represented by vertical lines spaced equally and latitudes by horizontal lines which are spaced closer near the equator than towards the pole. In such a map, coverage patterns get distorted. A number of plots with this projection appear in this book (for example, see Figure A.5 in Appendix A).

Satelli-centric or *hodocentric* projection is used when it is necessary to preserve the shape of the antenna pattern. This projection is a view of the Earth from a satellite. See Figure A.3 in Appendix A for an example.

Polar projection is used by radio amateurs as it is quite simple for plotting ground tracks. In *rectangular projection* the X-axis is represented as longitude and the Y-axis as latitude.

The coverage contour of a satellite in GEO is stationary and hence visualization and design of the service area is simple as the relative position of any location within the service area is invariant. In non-stationary orbit systems, the path geometry varies with time which requires a statistical description of visibility statistics from any given point. Section 2.3.8 summarizes visibility statistics of some non-geostationary satellite systems.

Consider the salient coverage features of a few satellite orbits used in an MSS.

Polar orbits provide unbiased coverage to all geographic regions – a single satellite can view the entire Earth over a period of time. Satellites in polar orbits have a large number of eclipses and require regular station keeping to maintain their orbital position. These orbits are well suited for applications, such as world-wide MSSs and global Earth resource surveys. Satellite constellations in polar orbit tend to favour polar regions because all the satellites in the constellation converge towards the pole.

Coverage provided by inclined orbits between 0 and 90° depends on the inclination of the orbit. Figure A.1 (see Appendix A) shows the ground trace of ICO satellite constellations which are in a 45° inclined LEO.

GEO is the most commonly used orbit in satellite communications because of its numerous advantages. Figure 9.3(a) (see Chapter 9) illustrates a 5° elevation contour of an Inmarsat satellite system, which deploys four geostationary satellites for its global mobile communication services. Some of the main features of GEOs are summarized in Table 2.1 (see section 2.2.7).

2.2.2 Space environment

Space environment affects the choice of *altitude* of an orbit as shown in Figure 2.15, which depicts the radiation environment around the Earth.

The Earth's magnetic fields deflect ions, which get trapped around the Earth's radiation belts called the Van Allen radiation belts, named after the American scientist who discovered them. From Figure 2.15, we note that the radiation intensity varies gradually with altitude, increasing to the highest intensity of 10,000 counts between ~ 2,000–4,000 km and ~ 13,000–25,000 km. The intensity also varies around the Earth, being highest above the equator and lowest above the poles. Regions in the south Atlantic between the middle region of south America and the lower tip of Africa exhibit much higher radiation levels than other regions at the same latitude. This effect, called the South Atlantic anomaly, affects satellites mostly in LEO.

High ion density causes adverse effects on electronic devices. Ionization causes an increase in background noise, increases the rate of degradation of solar cells and causes single-event failures which result in either temporary or permanent failure of electronic components. Single-event failures increase as the level of integration of electronic components in onboard electronic circuits increases.

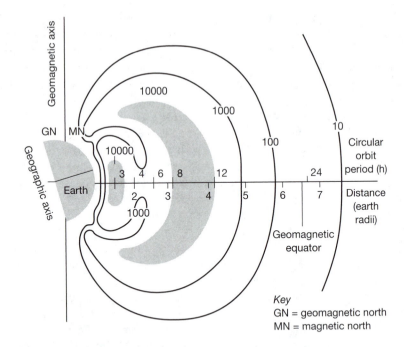

Taking atmospheric drag as the lower limit and the first Van Allen radiation belt as the upper limit, the approximate altitude of LEO is ~ 750 and ~1,500 km. The altitude of MEO usually lies in the range ~ 10,000 to ~ 12,000 km, the window between the first and second Van Allen belts.

2.2.3 Eclipse on satellites

Satellites derive their electrical power from sunlight. Just as we occasionally observe the solar eclipse on the Earth, so also do satellites in their respective orbits; the number and duration of eclipses depends on the altitude, inclination and eccentricity of the orbit. When the Sun is in an eclipse, satellites are powered by storage batteries which are subsequently charged. Regular charge/discharge cycles reduce the lifetime of batteries. Eclipses also cause load on a satellite's thermal control system as the system must react to rapid changes in temperature when satellites enter and leave an eclipse. As mentioned before, a Sun-synchronous orbit oriented at right angles to the Sun-orbital plane eliminates eclipse occurrence and therefore offers an advantage in this respect.

To understand the principle, consider the simple geometry of an Earth-induced solar eclipse on a geostationary satellite. Figure 2.16 shows the motion of the ecliptic plane (i.e. plane of the Earth around the Sun) with respect to the Earth's equatorial plane.

We note that at certain points the equatorial and ecliptic planes coincide and therefore the Earth lies in the same line as the Sun and the geostationary arc. These points correspond to the Spring and Autumn equinox, respectively.

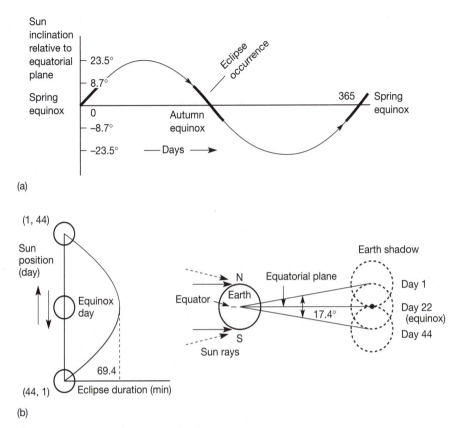

(a)

(b)

Figure 2.16
(a) Motion of the ecliptic plane with respect to the Earth's equatorial plane over a year (Richharia, 1999). (b) Eclipse duration in minutes on a geostationary satellite over the eclipse period; and the Earth's shadow on the geostationary arc on days 1, 22 and 44 of the eclipse (Richharia 1999)

The Earth's shadow subtends an angle of about 17.4° at the altitude of the GEO and the Sun is eclipsed on days when the shadow intercepts the GEO and during intervals when a satellite lies behind the Earth. Figure 2.16(b) shows the Earth's shadow on the geostationary arc progressively from the first day of the eclipse when the shadow begins to graze the GEO to the day when it moves just past the orbit. The duration of the eclipse increases progressively, maximizing on the day of equinox, and then subsides as the shadow moves away from the geostationary arc. On the day of the equinox, the Earth and the GEO are on the same line and hence the duration of the eclipse is highest, reaching around 69.4 minutes which is essentially the time it takes the satellite to traverse 17.4°. The Sun is shown moving relative to a fixed Earth and directions represent the motion of the satellite around the Spring and Autumn equinox.

Note that the eclipse occurs at the satellite's local midnight, i.e. when a satellite is directly behind the Earth. Therefore, by selecting the satellite to the west of the coverage region, the onset of the eclipse can be delayed past midnight when traffic carried by a satellite is usually low and hence storage battery

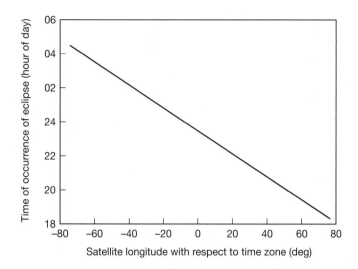

Figure 2.17

Time of occurrence of eclipse for various longitudes relative to satellite longitude for a GEO (Graphics: AR)

requirements are reduced. Figure 2.17 shows the time of occurrence of the eclipse for various longitudes relative to satellite longitude.

LEO satellites can undergo several thousand eclipses in a year; the number depends on the relationship between the Sun-Earth vector, altitude and the inclination of the orbital plane. Some eclipses could last for a considerable part of the orbital period. For example, a satellite in an equatorial orbit at an altitude of 780 km can remain in the Earth's shadow for 35% of the orbital period. Such eclipses can occur over 14 times per day during the equinox and the total eclipse period could last over 8 hours in a day. For a MEO under similar conditions, the maximum eclipse duration would be about 12.5% of the orbital period and the

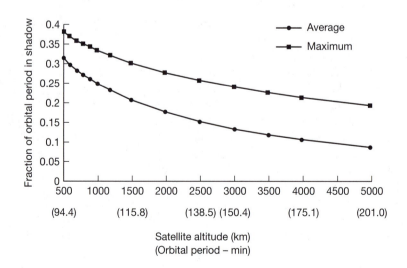

Figure 2.18

Maximum and average fraction of the orbital period that a satellite stays in the Earth's shadow as a function of orbital altitude (Graphics: AR)

total duration would be of the order of about 3 hours in a day, with a maximum of about 4 eclipses per day. Furthermore, the eclipse can occur at any time during the day. Figure 2.18 presents the maximum fraction and average fraction that low Earth satellites spend in the Earth's shadow (data from Gavish and Kalvenes, 1998).

The Sun can also be eclipsed by the Moon, when the Moon's shadow falls on a satellite. However, occurrence of the Moon eclipse is less predictable because of time-variant geometric relationships and may occur at any time on a geostationary satellite, unlike the Earth eclipse which is easily predictable and tends to occur at low traffic times.

2.2.4 The Sun's interference

The Sun is a strong radiator of electromagnetic radiation. Therefore, whenever the Sun appears behind a satellite, it receives strong radiation which can cause disruption to service. Generally, the effect is quite small in mobile terminals because receiver noise is significantly greater than the contribution from the Sun's noise. However, disruption in communication can occur in gateways which are far more sensitive. Solar interference occurs within ± 22 days around the equinox when the Sun's declination angle equals the angle between an Earth station and the equator. Figure 2.19 shows the maximum number of days when interference occurs and the duration of the Sun's transit on the day when the passage is the highest.

2.2.5 Doppler Effect

In a mobile communication system, the frequency shift caused by the Doppler Effect must be compensated to minimize the effect of frequency error on demodulation of signal. Whenever a moving object transmits a wave, the received frequency changes by an amount

$$\Delta f_d = \pm\, v_r f_t / c \qquad\qquad\qquad (2.6)$$

where v_r = the relative radial velocity between the observer and the transmitter, c = the velocity of light and f_t = the transmission frequency.

The sign of Doppler frequency shift is positive when the satellite is approaching the observer. Note that the Doppler shift is related to the radial velocity which can be predetermined and this feature can be utilized at the receiver for frequency compensation. For a LEO satellite, an observer typically sees an 'S-curve' with an increase in frequency as the satellite rises and a reduction as the satellite moves away. Figure 2.20 shows the Doppler curve of an L-band signal received at a point near Bhopal in central India for LEO and MEO satellite passes. In passing, we note that Doppler shift can be used to estimate the position of an observer, a principle used in some navigation systems.

Figure 2.19
(a) Maximum number of days when sun interference occurs for various antenna beamwidths (Graphics: AR).
(b) Duration of sun transit when sun passage is the highest versus half-power beamwidth of antenna (Graphics: AR)

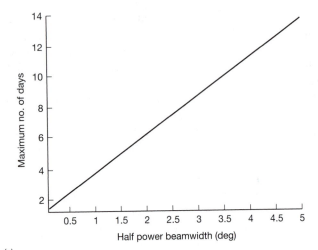

(a)

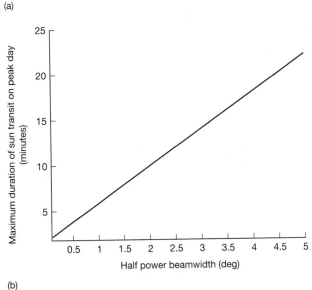

(b)

2.2.6 Orbital debris

There are two types of debris which may affect satellites – natural and man-made. Natural objects most likely to affect satellites are isolated meteorites or showers, such as the well-known Leonid showers which were in the news in 1999–2000, as their activities peaked. Satellite operators took precautions to minimize the impact of the showers on spacecraft electronics. There is a growing concern regarding the number of man-made objects which are cluttering space as cases of collision with orbiting satellites (e.g. the space shuttle) get reported. Hundreds of thousands of objects, from a few centimetres to several metres, are estimated to be orbiting the Earth and several hundred objects are added each year.

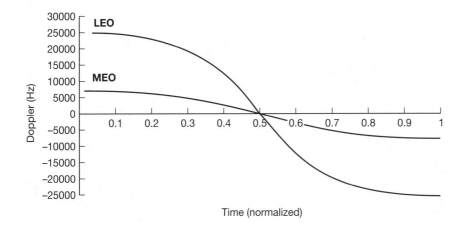

Figure 2.20
Doppler frequency shift
for a LEO (altitude =
1,400 km,
i = 52°, ascending node
= 60° E) and MEO
(altitude = 10,350 km;
inclination = 45°;
ascending node
= 60° E) satellite, as
viewed from a location
near Bhopal in central
India (77.5° E and 23° N)
(Graphics: AR)

It is recognized that proliferation of orbit debris must be minimized, if not eliminated, otherwise space will become unusable. As yet there is no radio regulation for controlling orbital debris, but there is an ITU recommendation to de-orbit a satellite from the GEO at the end of a satellite's lifetime. De-orbiting a satellite requires fuel which can be used to maintain the satellite in orbit, thereby earning revenue, and therefore operators may be reluctant to follow a recommendation unless it is mandatory. There is no regulation regarding tolerance in the orbital parameter of the non-geostationary satellite; to minimize the risk of collisions, regulations will become necessary if such systems proliferate.

There have been several analytical studies for investigating the evolution of current orbital debris in order to predict the probability of collision hazard in the future and thus facilitate formulation of suitable regulation. Studies conducted by Kessler and Burton (1978) demonstrated that a self-sustaining chain reaction of orbital collision may be initiated when the spatial density of orbital debris is high enough, causing severe restriction to the use of space. At present, atmospheric drag is the most practical way of removing debris from the Earth's orbit and hence it is expected that below ~ 750 km, where atmospheric drag effects become dominant, debris will decay and eventually get burnt in the Earth's atmosphere (Petro and Talent, 1989). However, above this altitude debris will continue to orbit the Earth for hundreds of years unless an artificial way is devised to remove it.

Debris evolution models have been developed, taking into consideration the debris removal characteristics of atmospheric drag below 750 km (McInnes, 1993) and modelling the current debris population by spline interpolation of observed spatial debris density, as shown in Figure 2.21 (Klinkrad and Jehn, 1992).

Debris evolution was approximated as a hydrodynamic process allowing the derivation of an analytical expression, which described collision evolution of debris under various scenarios. The most interesting scenario represented debris evolution with a slowly increasing rate of debris deposition to represent future debris deposits. Results demonstrated that catastrophic growth through collisions is likely to occur at an altitude of 1,000 km where the debris population is at its highest, though there is

Figure 2.21
Representation of
altitude profile of debris
population in the early
1990s, given as number
of objects per km³
(Klinkrad and Jehn,
1992)

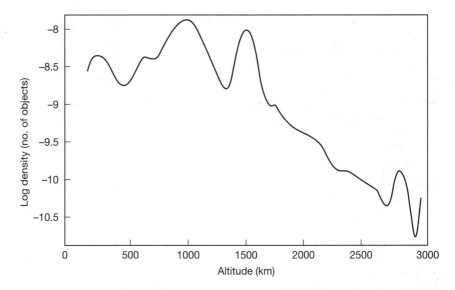

Figure 2.21 Representation of altitude profile of debris population in the early 1990s, given as number of objects per km³ (Klinkrad and Jehn, 1992)

a bias towards lower altitude due to the increased mean collision speed. For the case when the deposition rate is constant, the catastrophic growth occurs at an altitude of ~1,000 km after about 150 years (see Figure 2.22) and for the case when the rate of deposition and collision is increased by 1% per year due to the growing volume of satellites, the situation can occur at the same altitude by 2100 and if deposition and collision are increased by 10%, the catastrophic growth is brought forward to 2050 (see Figure 2.23). The dates may be postponed by depositing spent satellites at a local minima, such as within the high radiation zone of the Van Allen belt. The results provide an insight into the altitudes and approximate period at which problems may occur and its sensitivity to debris deposition. Another element of interest is the probability of occurrence of collision to an operational satellite. Clearly at each altitude, the collision probability will increase in the same proportion as the growth in orbital debris, rising exponentially in the orbital belts of catastrophic collision.

2.2.7 Summary of orbital characteristics

We have categorized satellites according to inclination (e.g. polar, equatorial, Molniya) eccentricity (i.e. circular or elliptical) and altitude (e.g. LEO, MEO or GEO). Equatorial, inclined or polar orbits favour equatorial, mid-latitude and polar regions, respectively. Eccentricity affects the dwell time of a satellite over specific regions. Satellites dwell longer over the apogee and it is therefore possible to take advantage of this prolonged dwell. Both path loss and propagation delay reduce as *altitude* is lowered, each is a useful attribute for provision of low-delay services to small terminals. Reducing the altitude reduces its coverage area, thereby increasing the number of satellites in a constellation and hence increasing the complexity of the network. Main features of LEO, MEO and GEO are summarized in Table 2.1 (Richharia, 1999).

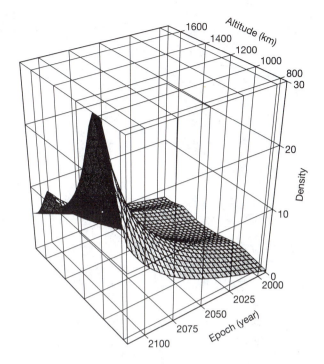

Figure 2.22
Growth in density of
orbital debris when
deposition and collision
growth are constant
(McInnes, 1993)

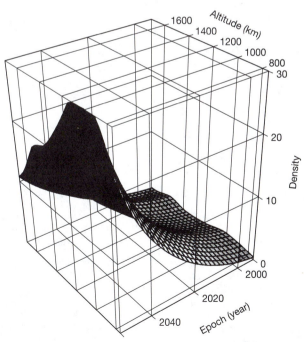

Figure 2.23
Growth in density of
orbital debris when
deposition and collision
growth are 10%
(McInnes, 1993)

Orbit	Advantages	Disadvantages
Geostationary orbit (GEO)	Well developed and proven technology; Signal strength stable due to constant ground–satellite range; Interference effects are easy to predict due to stable geometric relationship; Low Doppler; Advanced spot beam technology makes it possible to support hand-held narrow and broadband service; Coverage is available to most populated areas of world; Relatively low number of sun eclipse and hence battery charge/discharge cycle is low and battery lifetime is long; Time between launch and deployment/operation is relatively small – of the order of weeks; Only three satellites can provide near world-wide coverage; Well suited for regional coverage.	Coverage not available beyond ~ 76° latitude; Poor service link reliability at mid-to-high latitude; Large propagation delay – affects voice and time sensitive data protocols; Large path loss; At present, spectrum efficiency is low compared to low/medium orbit systems; High launch cost; Increasing threat of space debris; In-orbit backup satellite increases system cost disproportionately.
Highly elliptical orbit (HEO)	Reliable service links possible at mid-to-high latitudes; Lower launch cost; Distributed space segment architecture makes the system robust to satellite failures; 1:1 satellite is redundancy not required.	Inefficient for global coverage on its own; Propagation delay can be nearly as high as geostationary system; Doppler Effect notable; Hand-over is necessary (except for 'Loopus' orbit); Satellites near perigee may pass Van Allen radiation belt.

Table 2.1 Continued

Medium Earth orbit (MEO)	Can provide true global coverage;	Large number of satellites necessary (10–12);
	Lower path loss than GEO;	Receive signal strength is variable due to variability in range and elevation angle;
	Medium propagation delay (~ 30–100 ms from transmitter to receiver);	Doppler effects significant;
	Can offer efficient use of spectrum;	Complex network architecture: e.g., handover, intersatellite links, dynamic satellite resource management, routing etc.;
	Distributed architecture gives robustness; 1:1 satellite redundancy is not required.	Technology yet to mature;
		Relatively large number of eclipses increases charge/discharge cycle of batteries;
		Tend to increase orbital debris because of need of quite large number of satellite per system;
		Relatively long time required for constellation deployment;
		Space segment maintenance is more involved – higher number of satellite replacement than GEO but less than LEO.
Low Earth orbit	Can provide true global coverage;	Large number of satellites necessary;
	Lowest path loss;	Signal strength is variable due to variability in range and elevation angle;
	Lowest propagation delay (~ 5 to 10 ms from transmitter to receiver) comparable to optical fibre system time delay;	Doppler effects highest;
	Efficient use of spectrum;	Most complex space segment and network architecture: e.g., high rate of handover, intersatellite link, dynamic satellite resource management, complex routing;
	Possibility of including position determination as a value-added service.	Technology yet to mature;
		Very large number of eclipses with a large number of charge/discharge cycles;
		Rate of depositing orbital debris is the highest but it is easier to remove debris from LEO;
		Orbital maintenance is high – satellite replacement rate is more than in GEO or MEO.

2.3 Satellite constellations

In this section, the concepts developed in the preceding section are applied to the problem of optimization of satellite constellations from an orbital viewpoint. We will consider orbital altitude, space environment and geographical coverage as the optimization criteria in minimizing the number of satellites for *real-time* and *non-real-time* applications. Although orbital characteristics of constellations have a significant influence in the design of the overall mobile system architecture, a number of practical considerations influence the optimization of the space segment. These may be, for example, communication service, network issues, traffic distribution, technological risk, project schedules, etc. Such considerations have resulted in a wide range of constellation architecture, depending on the emphasis imposed by each operator. Optimization of constellations in their entirety is discussed in Chapter 8, after the reader has become familiar with the relevant concepts.

2.3.1 Considerations in constellation design

Considerable research has been directed to the problem of minimizing the number of satellites for continuous single or multiple visibility. The purpose has been varied – world-wide communications, communication capacity expansion, communications/distress services to polar regions, Earth resource survey, scientific data gathering, robust and survivable communications, military reconnaissance, navigation and mobile communications.

In a *random* constellation, all or some constellation parameters such as altitudes, inclination, interorbital plane separation, intersatellite phase are chosen at random whereas in a *phased* constellation, the phases have a well-defined relationship with one another. Random constellations, though simpler to maintain, are inefficient in terms of coverage property and tend to crowd the celestial sphere randomly. Such a constellation has been used in the past by the former Soviet Union for store and forward communications. Only phased constellations are considered here because they are economic, have a more reliable coverage property and do not clutter space. They can be categorized by inclination, eccentricity and altitude. We will summarize the main features of those designs which have been used in recent constellation designs (Richharia, 1999).

Constellations have been developed where orbital planes have a common intersection point, e.g. a polar constellation. Here, such constellations are called type 1 (Luders, 1961; Beste, 1978; Adams and Rider, 1987). Another type of design, which we call type 2 here, optimizes constellations in inclined orbits to distribute the satellites more uniformly on the celestial sphere (Walker, 1973; Mozhaev, 1972, 1973; Ballard, 1980). In both types of design, satellites are at a common altitude, i.e. their time period is the same.

Constellations can also be optimized by controlling eccentricity, which allows coverage to favour specific areas. Coverage from circular orbits is unbiased whereas eccentric orbits tend to be biased towards the field of view around the apogee. In all cases, altitude is an important consideration as the number of satellites and vital communications parameters depend on the altitude. It is also possible to deploy a hybrid constellation consisting of a combination of circular and elliptical orbits – an example of a hybrid constellation is the Ellipsat system (see Appendix A, section A.2).

With minor exceptions, to date Walker's constellations, which use an inclined orbit for single visibility, have been the most efficient up to a constellation size of 15 satellites, deployed at medium and high altitude. Examples of constellations based on this approach include Odyssey, Globalstar, MAGSS-14 (Benedicto *et al.*, 1992), Deligo (Meenan *et al.*, 1995), GIPSE (Sammut *et al.*, 1997), etc.

J. Draim, an American aerospace engineer, is reported to have developed a constellation which can provide continuous global coverage with only four satellites instead of five as demonstrated by Walker (Logsdon, 1995). The satellites are placed in a 31.5° inclined orbit with a period of 26.5 hours or greater and can see each other continuously; a US launch from Cape Canaveral would require little fuel for orbital plane change.

Polar constellations based on the work of Beste and Adams and Rider have been preferred to date for LEO. There are also some propriety constellation designs developed for specific applications.

Figure 2.6 shows the geometry of a single satellite, useful in understanding a constellation's optimization. Optimization is achieved by setting a minimum permissible elevation angle, η, thereby fixing the altitude and then adjusting orbital parameters to minimize the number of satellites. Another approach is to select a constellation pattern of known characteristics and minimize the largest value of great circle range R' for all observation points and instants of time by trial and error.

A useful constellation performance measure is its comparison with theoretical bounds which may be estimated in a number of ways; the Earth can be divided into a number of non-overlapping equilateral spherical triangles; this way the coverage is distributed uniformly around the world (Ballard, 1980). The bounds can also be approximated by dividing the Earth into a hexagon and enclosing circles around them (Beste, 1978):

$$N \sim 2.42/(1 - \cos\psi) \qquad\qquad (2.7)$$

where ψ = the great circle range for which the stationary bound is required and N = the number of stationary satellites.

The coverage efficiency of a constellation consisting of N satellites can be given as:

$$N\Omega/4\pi \qquad\qquad (2.8)$$

where $N\Omega$ = the total solid angle of the constellation, Ω = the solid angle bounded by a single satellite = $2\pi (1 - \cos\psi)$ Steradians and 4π = the solid angle of a sphere.

The best coverage efficiency of 1.0 transpires when two satellites are located diametrically opposite at a distance of infinity, theoretically the minimum number of satellites for world-wide coverage.

When comparing the uniformity of coverage of two constellations, the system designer is faced with the problem of quantifying the degree of uniformity of candidate schemes. The uniformity of coverage of constellations can be measured as the temporal uniformity index u_T and the spatial uniformity index u_S, which indicate how unbiased the coverage is over an orbital period and geographical area; an ideal constellation should give a value of 1 for each.

2.3.2 Polar constellations

Polar constellations have been studied by several researchers (Luders, 1961; Beste, 1978; Adams and Rider, 1987). As an illustration, consider the approach of Beste, who founded some of his development on Luder's earlier work.

Beste optimized single-visibility coverage for the complete world as well as for regions extending from the pole to any arbitrarily chosen latitude. Satellites are placed in orbital planes which intersect at a common point, such as over the pole. Plane separation and satellite spacing in each plane are adjusted to minimize the total number of satellites. Figure 2.24 shows the geometry used in optimization of a polar constellation for full Earth coverage.

Satellites in adjacent planes move in the same direction, shifted with respect to each other by half intraorbit satellite separations, i.e. half of $(2p/m)$ or p/m, where m is the number of planes. The arrangement staggers satellites at the equator giving a relatively uniform coverage. The separation between adjacent planes, where satellites move in the same direction, is $(\psi + \Delta)$. The relative geometry of such orbital planes remains constant because the satellites move in phase; however, there is one set of adjacent planes (see Figure 2.24) where satellites move in opposite directions. The relative geometry between adjacent planes is not constant and hence these planes are placed closer together at 2Δ, such that the equator is covered by the common area between adjacent satellites. The constellation is defined by the following parameter set:

Figure 2.24
Geometry for optimization of a polar constellation; the condition is $(n-1)Y + (n+1)D = p$; $Y =$ coverage angle; $2D =$ separation between planes where satellites move in opposite direction

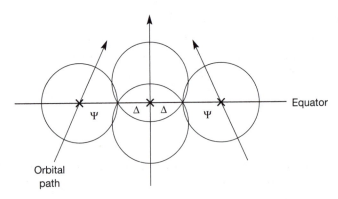

ψ = the coverage circle;
n = the number of orbital planes;
m = the satellites per plane;
$\psi + \Delta$ = the plane separation between adjacent planes containing satellites moving in the same direction;
2Δ = the separation between planes at the boundary where satellites move in opposite directions

where

$$\Delta = \cos^{-1}[\cos \psi/(\cos \pi/m)] \tag{2.9}$$

The optimization is based on minimizing the number of satellites at the equator, where the coverage requirement is most stringent due to the widest plane separation. The following condition must be satisfied for this region

$$(n - 1)\psi + (n + 1)\Delta = \pi \tag{2.10}$$

A trial and error is used to solve equation (2.10). The method can be extended for multiple visibility based on the same iterative search process so as to obtain multiple visibility at the worst point. The constellation geometry is similar to the single-coverage case.

Using the same definition of intra and interorbital spacing as used by Beste (Adams and Rider, 1987) provides optimum solutions using the following set of constellation definitions:

$$P = (2/3) k (\pi/\psi) \tag{2.11}$$
$$Q = (2/\sqrt{3}) j (\pi/\psi) \tag{2.12}$$
$$N = (4\sqrt{3}/9) n (\pi/\psi)^2 \tag{2.13}$$

where P = the number of orbital planes, Q = the number of satellites in each orbital plane, N = the total number of satellites, ψ = the coverage angle (see Figure 2.24), k = the multiple coverage factor in different planes, j = the multiple coverage factor in the same plane and n = the multiple coverage factor of the constellation.

For single visibility, k = j = n = 1.

2.3.3 Inclined orbit constellations

A family of constellations originally proposed by Walker (1973) provides a highly efficient regional or world-wide coverage using inclined orbits. The family provides unbiased world-wide coverage by deploying satellites in circular orbits each of the same period and inclination, distributed uniformly on the celestial sphere. Regional coverage can be provided by partial deployment of the constellation, according to the region of interest. Adjacent orbital planes are separated equally around an arbitrary reference plane which can be the Earth's equator. Within each orbit, neighbouring satellites are equally separated. The initial position of a satellite is proportional to the right ascension angle of the orbital plane – the right ascension angle being measured on the reference plane. Orbital perturbations are not included in optimization and hence in order to preserve constellation geometry with time, station keeping becomes necessary. A slight overdesign of the constellation can cancel station keeping errors. In the original work, Walker specified inclination as δ and hence called the coverage patterns 'delta patterns'. The orbital traces of satellites of a constellation on the celestial sphere resemble petals of a flower and therefore Ballard (1980), who formalized and extended Walker's work, called such constellations 'Rosette'. Identical results have also been independently obtained by Mozhaev (1972, 1973). Ballard's work is used here for definition.

The constellation is designated as (N, P, m) where N = the total number of satellites in the constellation, P = the number of orbital planes and m = the harmonic factor, which determines the initial distribution of satellites and the rate at which patterns precess over the celestial sphere. When m is an integer, the satellite constellation consists of 1 satellite per plane, when it is an unreduced integer S/Q, the constellation consists of a Q satellite in each orbital plane. For example, a (5, 5, 1) constellation consists of a total of 5 satellites in 5 planes with 1 satellite in each plane; a (12, 3, $\frac{1}{4}$) constellation consists of a total of 12 satellites in 3 planes with 4 satellites in each plane.

The constellation geometry is shown in Figure 2.25 where α_i = the right ascension angle of the ith orbit plane = $2\pi i/P$, i = the inclination angle of the ith orbit, the same for each orbital plane, γ_i = the initial phase angle of the ith satellite in its orbital plane at t = 0, measured from the point of right ascension = $m\alpha_i$

$$= mQ(2\pi i/N) \tag{2.14}$$

where m = (0 to N – 1)/Q, N = PQ, P and Q are integers; Q = the number of satellites per plane and i = the initial phase of the ith satellite in its orbital plane (i.e. t = 0). The phase at an arbitrary time, t = $2\pi t/T$, where T = orbital period.

The great circle range ψ, shown in Figure 2.24, is used as a measure of optimization. For a chosen set of constellation parameters, the highest value of ψ is minimized at all observation points for all instants of time. The worst observation point on Earth is the centre of a spherical triangle formed by joining the subsatellite points of three satellites. If the distance of the point to any of the subsatellite points is R_{ijk}, and a constellation provides usable coverage to $\psi <$ R_{ijk}, then the centre point of the spherical point is left uncovered, thereby failing the full coverage test. On the other hand, if $\psi \geq R_{ijk}$, the visibility at the worst point increases to three. A measure of goodness of coverage of a constellation is therefore the highest value of R_{ijk}, R_{ijk} (max) taken over all instants of time over the orbital time period T; a constellation with a lower value of R_{ijk}(max) is better suited from a communications viewpoint, because this would ensure that, in such a constellation, satellites are closer to an observer located at the worst point. A total of 2N–4 non-overlapping triangles are formed with N

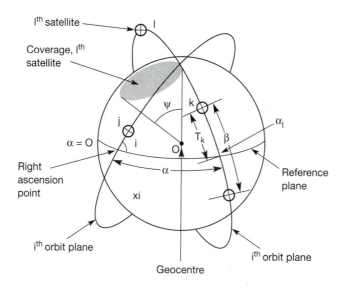

Ith satellite

Coverage, Ith satellite

$\alpha = 0$

Right ascension point

ith orbit plane

xi

Geocentre

Reference plane

ith orbit plane

Figure 2.25 Constellation geometry for Walker constellation; inclination angles, i, of each satellite in its orbital plane are equal; initial position of each satellite in its orbital plane is proportional to its right ascension angle; angular separation β between satellites in a plane are equal; adjacent plane separation α are equal. The constellation is defined as (N, P, m), where N = total number of satellites, P = number of orbital planes and m, the harmonic factor, determines the initial distribution of satellites and the rate at which the pattern precesses over the celestial sphere. Optimization for a set of constellation parameters is achieved by minimizing the highest value of coverage angle ψ at all observation points for all instants of time and inclination

satellites, all of which must be examined at each instant of time over the orbital period. Furthermore, the process must be repeated for all values of inclination to obtain the minimum value of $R_{ijk}(\max)$. A search of such a magnitude can best be performed by the use of a computer. The only effect of the Earth's rotation is that the location of R_{ijk} on the Earth varies with time and hence it is not necessary to include the Earth's rotation in the optimization.

Uniformity of coverage over an orbital period and geography is measured by temporal uniformity index, u_T, and spatial uniformity index, U_s. An ideal constellation should give a value of 1 for each. These uniformity indexes are defined as follows:

$$u_T = R_{ijk}(\max)_w / R_{ijk}(\max)_b \tag{2.15}$$

where $R_{ijk}(\max)_w$ and maximum $R_{ijk}(\max)_b$ are respective values of $R_{ijk}(\max)$ at the worst and the best phase angles

$$u_S = \text{the maximum of } [R_{ijk}(\max)/R_{ijk}(\min)] \text{ over all phase angles} \tag{2.16}$$

where $R_{ijk}(\max)$ and $R_{ijk}(\min)$ are the maximum and minimum value of R_{ijk}, respectively, at the same phase angle.

Ballard derived results of best single visibility using exhaustive searches for N = 5 to 15, which are summarized in Figure 2.26(a) and (b). The average temporal and spatial uniformity index of these constellations was calculated as 1.06 and 1.25, respectively. The minimum number of satellites required to guarantee single world-wide visibility is the constellation (5, 5, 1), i.e. one satellite in each of 5 planes at an altitude of 4.232 eru (~ 26,992 km). As expected, the total number of satellites N for world coverage reduces as the altitude is increased.

Figure 2.26
(a) Optimized constellation derived by various methods for altitudes up to 4,500 km. (b) Optimized constellation derived by various methods for altitudes between 5,000 and 25,000 km (Graphics: AR)

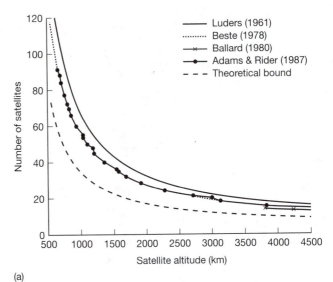

(a)

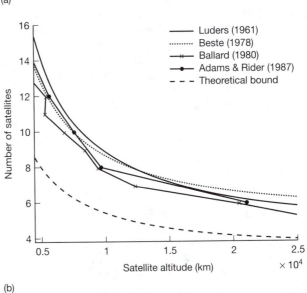

(b)

We can increase the reliability of the radio link by increasing the minimum look angle. This can be achieved by choosing a higher altitude at the expense of an increase in path loss.

From a practical viewpoint, there are some advantages in choosing the orbital time period to be the integer ratio of the Earth's rotation period; the ground track of such orbits follows fixed patterns on the ground. This makes algorithms for searching and tracking of satellites, and handover between satellites, simpler due to the relatively simple predictability of the satellite position from the ground. It is also possible to deploy a constellation in manageable stages, building up from the areas where coverage is most required, or for that matter, to deploy partial constellations for regional coverage. There are an infinite number of possibilities depending on the orientation between the constellation's reference longitude relative to the equator, and the constellation parameters (N, P, m).

Another interesting example of such a quasi-stationary constellation, known as 'Loopus' (geostationary loops in orbit occupied permanently by unstationary satellites), has been proposed by Donald (1984). The constellation parameters are chosen such that satellites follow the same track, forming small loops on the celestial sphere; changeover between satellites takes place at the loop intersection, i.e. just when a satellite is leaving the loop and another entering it. Satellites appear to remain within the small loop in the sky, much as a geostationary satellite. Loops can be created over high latitudes, enabling a high elevation angle service to a polar and high latitude region.

The discussion above dealt with single visibility of satellites; a system designer may consider multiple satellite visibility for achieving a path diversity advantage and/or to offer navigation capability. For the Rosette constellation, it has been shown that for world-wide V satellite visibility, at least 2V + 3 satellites are required. The definition of R_{ijk} in optimization in this case changes to the distance where V satellites are enclosed within the spherical triangle.

Figure 2.26 summarizes solutions derived by various methods including the theoretical bound as a reference. Figure 2.26(a) covers the LEOs, while Figure 2.26(b) covers the MEOs. Note that solutions by Luders, Beste and Adams and Rider apply well for LEOs, while Ballard's solution offers higher efficiency for MEOs.

2.3.4 Hybrid constellations

It is possible to combine various types of orbits for full Earth coverage. Such orbits may have a different orbital period. For example, it is possible to use circular orbits to cover the equatorial region and elliptical orbit to cover the higher latitude regions. Chapter 9 discusses an interesting hybrid constellation design so as to maximize coverage of populated areas of the world and in particular high latitude regions around Canada, proposed for the Ellipsat system. Other possibilities would be to use GEOs for covering the equatorial regions and inclined orbits for covering the polar regions. It could also be possible to use a combination of LEO and GEO constellations.

An example of this type of hybrid orbital design is the WEST broadband system. The system combines a 9 orbit MEO constellation called JOCOS comprising 9 satellites with a GEO (Pennoni and Bella, 1995). In this design, satellites are placed such that they follow the same subsatellite path. By dividing the Earth into three parts, 120° apart, and placing satellites in an 8 hour orbit, the design is capable of serving most of the inhabited regions.

2.3.5 Regional coverage

When regional coverage is required, it is necessary to ensure that all satellites in the constellation pass over the service area. Such a deployment, using delta patterns, has been discussed earlier. Partial deployment of such constellations can provide a continuous ground track of regions of interest. For example, equatorial regions may be covered by deploying satellites in an equatorial plane whose footprints follow each other.

Constellations using an elliptical orbit inclined at 63.4° have been used for covering high latitudes as satellites in such orbits dwell over high latitudes over a considerable time of the orbit. Such a system has been in use for a considerable time to cover high-latitude regions of Russia. More recently, such constellations have been proposed for mobile and broadcasting applications in Europe and for mobile communication in the USA and Canada (see Appendix A, section A.1).

2.3.6 Constellations for non-real-time systems

Constellation design for systems which do not require real-time coverage, e.g. messaging/paging systems, is less stringent because gaps in coverage can be tolerated, provided that the regions of interest have at least one satellite visible within $(t_a - t_d)$ seconds, where t_a is the specified end-to-end delay and t_d is the delay in message transfer, which is the sum of connection, processing and propagation delays. The ORBCOM system used this approach during the constellation deployment phase.

2.3.7 Use of spot beams

The optimization techniques discussed above ensure a minimum elevation angle within the coverage area. However, from link design considerations it becomes necessary to partition the coverage area into small segments covered by spot beams. Spot beams increase the satellite antenna gain and improve spectrum reusability; the penalty is the complexity in payload, an increase in the weight of a satellite and an increase in the number of handovers between beams, though they are low when the spot beams are Earth fixed (see Chapter 8).

2.3.8 Availability considerations for non-geostationary satellites

A fundamental requirement of an orbital design is to guarantee a minimum elevation angle visibility from all locations within a coverage area. A system should provide the desired visibility of one or more satellites from the coverage area, depending on the type of service, and other considerations such as whether path/satellite diversity is used. In a real-time service, communication is lost as soon as the radio link breaks and hence it is important to maintain at least one satellite within visibility. In a store and forward system, when the radio link breaks, the system stores information until such time as a satellite becomes available, when the stored data is transmitted in a burst.

The reliability of the service link depends on the radio path profile, or more generally, the skyline. The higher the elevation of a satellite, the less likely is the probability of path obstruction, but the number of satellites increases. Therefore propagation degradation caused by the environment reduces progressively in the following order: city, suburbs and open areas (see Chapter 3).

Link availability can be increased if communication can be established through separate paths, either from the same satellite (through multipath or reception from more than one spot beam) or through separate satellites. When path diversity is established through different satellites, there are other potential advantages. It is possible to reduce congestion on a satellite by diverting traffic through the less-congested satellite; signal quality can be improved by combining signals from each path. Table 2.2 shows an example of reduction in link margin for one, two and three satellite visibility from a measurement campaign during the design phase of the ICO system (Hart *et al.*, 1995). The diversity gain depends on elevation angles of visible satellites and azimuth separation of satellites. Generally, higher elevation satellites are likely to offer a clearer line of sight as shown in Figure 2.27, and wider azimuth separation reduces the probability of simultaneous fading from visible satellites; however, the extent of improvement is strongly influenced by the local environment (see Chapter 3). Figure 2.28 shows an estimated correlation in fading with azimuth in an urban environment (Hart *et al.*, 1995). For example, the figure shows a notable correlation in the azimuth range of ±30° from the boresight.

| Number of visible satellites | Link margin (dB) | |
	Rural environment	Urban environment
1	16.5	23.5
2	7	8
3	5	8

Table 2.2
Diversity versus link margin (Hart *et al.*, 1995)

Look angles of satellites in non-GEO are time and location dependent and therefore expressed statistically and generally obtained through computer simulation because of non-linear interaction of a number of system variables. Some useful statistics for a given location are summarized below. A comparison should be made for a large number of locations within the coverage area for better statistical representation and when considering link availability, geometric visibility cannot be taken in isolation as the radio link is severely affected by the local environment (see Chapter 3 for propagation effects):

- probability and cumulative angles of satellites visible from a location;
- average, minimum and maximum elevation and azimuth angle of the highest visible satellite;
- constellation availability for a constant elevation angle;
- maximum period when a satellite is unavailable;
- probability of a call of a given duration completing successfully;

Figure 2.27
Elevation angle
dependence of fade
margin for various
percent of locations
(ESA, 1988)

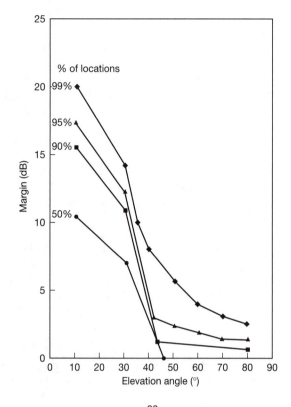

Figure 2.28
Azimuth angle
correlation of shadowing
in an urban environment
(Hart *et al.*, 1995)

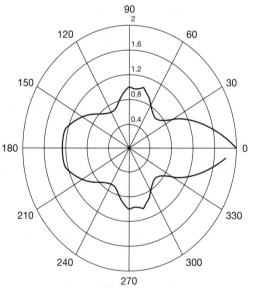

Note: 2 corresponds to correlation coefficient of +1
0 corresponds to correlation coefficient of −1

- sensitivity of the parameters identified above to incomplete constellation (e.g. caused by satellite failure or due to incomplete constellation at the start of operation);
- elevation angle and azimuth separation between two satellites – these parameters provide a measure of satellite availability and diversity improvements.

Consider, as an example, the performance analysis of a few LEO and MEO systems in a hypothetical urban environment with the following assumptions (Sandrin and Haschart, 1993):

- orbital characteristics specified in Table 2.3;
- no service when line of sight is absent, i.e. diffraction, shadowing, multipath were not considered;
- environment representative of a typical urban US environment;
- handover to the next satellite is perfect and based on a dynamic check of channel availability;
- call maintained as long as at least one satellite has an unobstructed view to the user;
- call occurrence at random times – user either initiating or receiving the call;
- 5° minimum look angle to the satellite.

Const-ellation	Number of satellites	Number of planes	Satellite/plane	Inclination (°)	Orbit altitude (km)	Orbital period (minutes)	Minimum elevation at 30° latitude	Approximate view time for 5° elevation (minutes)
LEO-35	35	5	7	90.0	1,584.9	117.9	10	21.0
LEO-54	54	9	6	55.0	1,800.0	122.7	20	26.5
ICO-12	12	3	4	50.7	10,355.0	359.0	20	144.8
ICO-15	15	3	5	53.5	11,622.2	400.6	30	164.5

Table 2.3
Main characteristics of orbits used for developing system statistics

A number of interesting conclusions were derived:

- Both ICO constellations provide a high probability of completing calls.
- LEO-54 provides a better probability of completion for short calls of 7 minutes or less in one case.
- LEO-54 provides a better probability of call completion than LEO-35.
- The system is very sensitive to the environment; a 45° rotation of city intersection causes a dramatic effect on probability.
- Satellites with higher minimum elevation angle have a better availability.
- The conditioned call initiation is expected to be beneficial only for incomplete constellations when there is a large gap in the visibility of satellites.

- The sky profile has a marked influence on performance even in a very specific environment; for example, the availability of satellites in a polar orbit on a street intersection with a clear north–south is better than if it were obstructed.

Comparison of visibility statistics is an effective way of comparing constellations and gives an insight into the criteria used by recent proposals; some chose to optimize coverage in specific regions; others chose an unbiased coverage, etc., illustrating that there is no unique solution to constellation optimization. The choice depends on the desired service area, i.e. regional, global, etc. service, i.e. real-time, store and forward, etc., and link reliability.

Visibility statistics of various constellations have been compared by various authors. Here we summarize the findings of Krewel and Maral (1998). Figure 2.29(a) – (e) demonstrates elevation angles exceeded for 90 and 100% of time

Figure 2.29
Visibility statistics in terms of the percentage of time that elevation is exceeded for 90% and 100% for (a) Iridium, (b) Globalstar, (c) ICO, (d) Ellipso, (e) ECCO (Krewel and Maral, 1998)
© John Wiley & Sons

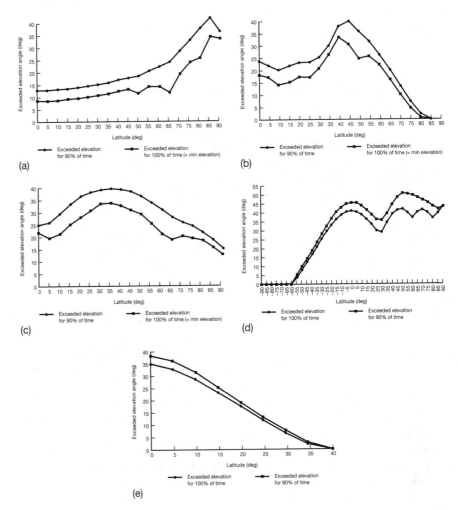

versus latitude for five types of constellations – Iridium (LEO/polar), Globalstar (LEO/inclined), ICO (Medium/inclined), ELLIPSO (hybrid – elliptical/circular) and ECCO (LEO/equatorial).

- **Iridium:** Has the best visibility performance above a latitude of ±65°; below this latitude the elevation angle varies between 8 and 15° (Figure 2.29(a)). Although the system has a good dual-satellite diversity performance beyond a latitude of around 50°, the system has not been designed to use diversity (see Chapter 9).
- **Globalstar:** Offers the best visibility in mid-latitude between 35 and 50° where elevation angle ranges between 20 and 32° (Figure 2.29(b)). The system offers good path diversity and has been designed to exploit this advantage.
- **ICO:** Visibility angle of 18° is exceeded up to a latitude of 80° with the highest elevation angle visibility at a latitude of 35° (Figure 2.29(c)). The system offers good path diversity with a feature to exploit diversity on demand.
- **ELLIPSO:** Has been optimized for the northern latitudes, with elevation angle > 27° at a latitude of 28.5° and 41° at a latitude of 50° (Figure 2.29(d)). The system offers the highest probability of dual and triple visibility in the latitude range of 20–40°.
- **ECCO:** Is an equatorial constellation offering the highest elevation angle at the equator and a minimum elevation angle of 8° within a latitude of ±30° (see Figure 2.29(e)). Satellite diversity is possible for 50% of the time.

Further reading

Adams, W.S.; Rider, L. (1987) 'Circular polar constellations providing continuous single or multiple coverage above a specified latitude', *J Astronautics Science*, **35** (2).

Ballard, A.H. (1980) 'Rosette constellations of earth satellites', *IEEE Trans. Aerospace and Electr. Systems* **AES–16**, (5), September, 656–73.

Beste, D.C. (1978) 'Design of satellite constellation for optimal continuous coverage', *IEEE Trans. Aerospace and Electr. Systems* **AES–14**, (3), May, 466–73.

Bischl, H.; Werner, M.; Lutz, E. (1996) 'Elevation-dependent channel model and satellite diversity for NGSO S-PCNs', *Vehicular Technology Conference (VTC)*, 1996.

Brunt, P. (1996) 'IRIDIUM® – overview and status', **Space Communications**, (14), 61–8.

Castiel, D. and Draim, J.E. (1995) 'The Ellipso™ mobile satellite system', *International Mobile Satellite Conference*, Ottawa, IMSC '95, The Fourth International Mobile Satellite Conference Ottawa cosponsored by Communication Research Centre/Industry Canada and Jet Propulsion Laboratory/NASA, 409–18.

Donald, P. (1984) 'Loopus opens a new dimension in satellite communications', *IJSC*, **2**, 242–50.

ESA (1988) *Archimidese Study Report.*

Galligan, K.P. (1989) 'Sound broadcasting applications of highly elliptical orbits', *IEE Colloquium on Highly Elliptical Orbit Satellite Systems* (1989/6), 24 May, Digest, 7/1–7/4.

Gavish, B; Kalvenes, J. (1998) 'The impact of satellite altitude on the performance of LEOS based communication systems', *Wireless Networks*, vol. 4, 199–213.

Hart, N.; Goerke, T.; Jahn, A. (1995) 'A discussion on mobile satellite systems and the myths of CDMA and diversity revealed', *International Mobile Satellite Conference*, Ottawa, IMSC '95, The Fourth International Mobile Satellite Conference Ottawa cosponsored by Communication Research Centre/Industry Canada and Jet Propulsion Laboratory/NASA, 469–75.

Haviland, R.P. and House, C.M. (1965) *Handbook of Satellites and Space Vehicles*, Van Nostrand–Reinhold, New York.

ICO Global Communications Ltd, *www.ico.com.*

International Journal of Satellite Communications (1994), **12**.

Kessler, D.J.; Burton, G.C. (1978) 'Collision frequency of artificial satellites: the creation of a debris belt', *Journal Geophys. Res.*, **83**, (A6), 2637–46.

Klinkrad, H. and Jehn, R. (1992) 'The space-debris environment of the Earth', *ESA Journal*, Vol. 16, no. 1, pp. 1–11.

Krewel, W. and Maral, G. (1998) 'Single and multiple satellite visibility statistics of first-generation non-geo constellations for personal communications', *International Journal of Satellite Communications*, vol. 16, 105–25.

Logsdon, T. (1995) *Mobile communication satellites*, McGraw-Hill International Editions, 160.

Luders, R.D. (1961) 'Satellite networks for continuous zonal coverage', *ARS J*, **31**, February, 179–84.

Maral, G.; Ridder, J.J.D.; Evan, B.G.; Richharia, M. (1991) 'Low earth orbit satellite systems for communications', *IJSC*, **9**, 209–25.

McInnes, C.R. (1993) 'An analytical model for the catastrophic production of orbital debris, *ESA Journal*, Vol. 17, pp. 293–305.

Meenan, C.; Sammut. A.; Tafazolli, R.; Evans, B.G. (1995) 'DELIGO: Repetitive ground track 100% dual-diversity LEO satellite constellation for S-PCN', *IEE Electronics Letters*, vol. 31, Part 17, August, 1407–8.

Mozhaev, G.V. (1972) 'The problem of continuous earth coverage and kinematically regular satellite networks, I', *Cosmic Res.* **10 (UDC 629.191)**, November–December 1972, translation in CSCRA7(Consultants Bureau, New York), vol. 10, no. 6, 729–882.

Mozhaev, G.V. (1973) 'The problem of continuous earth coverage and kinematically regular satellite networks, II', *Cosmic Res.*, 11 (UDC 629.191), January–February, 1973, translation in CSCRA7(Consultants Bureau, New York), vol. 11, no. 1, 1–152.

Pennoni, G. and Bella, L. (1995) 'JOCOS: A triply geosynchronous orbit for global communications. An application example', *IEE Conference*, publication no. 403/2, 646–52.

Petro, A.J. and Talent, D.L. (1989) 'Removal of orbital debris', *Progress in Astronautics and Aeronautics*, Vol. 121, pp. 169–182.

Richharia, M. (1999) *Satellite Communication Systems*, second edn, Macmillan Press Ltd, Basingstoke and London.

Richharia, M.; Hansel, P.; Bousquet, P.W. and O'Donnell, M. (1989) 'A feasibility study of a mobile communications network using a constellation of low earth orbit satellites', *IEEE Global Telecommunications Conference, GLOBE-COM '89*, Dallas, November 1989, 27–30.

Sammut, A.; Mertzanis, I.; Tafazolli, R.; Evans, B.G. (1997) 'GIPSE: A global integrated personal satellite multimedia environment', *4th European Conference on Satellite Communication* (ECSC-4), Rome, Italy, November 18–20, 11–15.

Sandrin, W.A. and Haschart, D.V. (1993) 'Availability prediction for hand-held communications system using non-geo-synchronous satellites', *COMSAT Tech Review*, **23**(2) Fall, 237–69.

Teledesic LLC, *www.teledesic.com.*

Walker, J.G. (1973) 'Continuous whole earth coverage by circular orbit satellites', *presented at the IEE Satellite Systems for Mobile Communications Conference*, Paper 95, March, 13–5.

Radio link   **3**

3.1 Introduction

This chapter provides an overview of concepts and techniques related to a satellite radio link. The satellite–mobile radio link is the most essential component of an MSS, as it gives mobility and concurrently sets bounds on throughput. Keeping to the theme of the book, topics are discussed at a system level to enable the reader to grasp interrelationships between system components.

The chapter introduces spectrum-related issues including interference estimation and management methods, international spectrum allocation procedures and trends. An understanding of radio wave propagation is of fundamental importance in the design and understanding of the mobile radio link, as propagation impairments set bounds to throughput. Beginning with a general introduction to radio wave propagation, the section reviews some of the prevalent propagation models applicable to land mobile, maritime and aeronautical links, followed by a review of radio link analysis highlighting factors which affect mobile radio links.

3.2 Spectrum issues

Radio waves are information carriers and a fundamental, albeit limited, natural resource of any radio communication system. Consequently, useful parts of the radio spectrum, particularly the MSS bands, are currently in great demand and precious. Commonly used MSS ranges are the L (1.5–1.6 GHz) and S (2.0–2.5 GHz) bands; however, higher frequency ranges such as the K_u band (~14–15 GHz) are also in use, and there is considerable interest in exploiting the K_a band (20/30 GHz) in the near future.

In an ideal world, an operator would select a frequency band which offered an optimum compromise in terms of propagation degradation, user equipment size and cost attractiveness, manufacture and launch of satellites, technology, etc. Such a freedom would result in a certain degree of chaos, as all the operators would quite probably converge to a few favourable bands, thereby causing interference to each other as well as to other radio services operating in the band. To avoid such a scenario, RF usage is regulated by the ITU. The ITU assigns spectra to all types of radio services and specifies procedures for their management on a global basis, taking into consideration the demands of each service by region and globally, radio wave propagation characteristics and technical status, while respecting the needs and concerns of each country. Within this framework, each country manages a spectrum through its own regulatory body such as the FCC in the USA or the DTI in the UK. Thus, an operator's freedom in the selection of spectra is constrained to specific ranges and within them those with technical and commercial advantages become contentious, leading to a need for recognized regulatory procedures for spectrum coordination between operators. Such procedures, developed by the ITU, require a new operator to notify the ITU with appropriate technical details, announcement of the intention by the ITU and coordination of the applicant with operators who may be affected by the new system.

The ITU has divided the world into three regions for the purpose of spectrum allocations, as shown in Figure 3.1.

These regions are broadly:

- *Region 1:* Europe, Africa, the Middle East and the European regions of the former USSR.
- *Region 2:* The Americas.
- *Region 3:* The remainder of Asia, plus Australasia.

Frequency allocations are listed in Article 8 of the *Radio Regulations* (ITU, most recent). A number of categories of allocations are provided. These are *primary, secondary, footnote, planned,* exclusive, shared etc. Exclusive primary frequency allocations are preferred for the MSS. In such allocations, a service has an exclusive right of operation, provided mutually satisfactory coordination between operators of the service has been achieved. Secondary allocations are of limited use for the MSS, because a service with such an allocation is not guaranteed interference protection and neither is it permitted to cause interference to services with primary status; in all, this has an adverse impact on user mobility. The problem can be mitigated by countermeasures such as use of robust modulation and accessing schemes, e.g. spread spectrum modulation with CDMA.

The majority of MSS allocations are in the L and S bands, with some secondary allocations in the K_u band. Due to spectrum overcrowding in these bands coupled with a demand for broadband applications, K_a and extra high frequency (EHF) bands have also begun to receive attention (see Chapter 11). Several World Radio Conferences (WRCs) addressed MSS-related issues, resulting in the merging of categories within the MSS bands, allocation of spectra for non-geostationary satellite systems, satellite components of third generation mobile systems (IMT-2000; see Chapter 11) and direct to person radio broadcast systems. Table 3.1 summarizes the main L and S band MSS link allocations, but this list is not exhaustive. The reader should refer to Article 8 of the most recent issue of the *Radio Regulations* for the latest allocations.

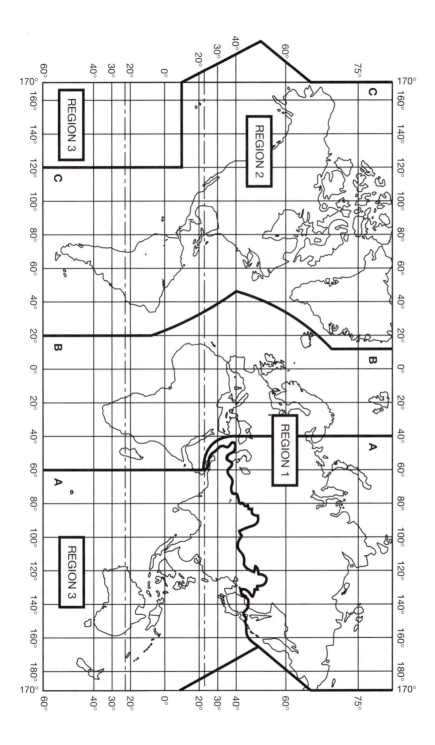

Figure 3.1
ITU regions of the world used for frequency allocation (Article 8 © ITU)

Table 3.1
Main service link
frequencies allocated
by the ITU

Uplink [MHz] (Earth–space)	Downlink [MHz] (space–Earth)	Comments
1,626.5 – 1,660.5 except 1,645.5 – 1,646.5	1,525 – 1,559 except 1,544 – 1,545	Generic MSS band; primary allocation; used by first/second generation MSS; in great demand now, previously the band was segmented into land, maritime and aeronautical bands.
1,645.5 – 1,646.5	1,544 – 1,545	Distress and safety band.
1,610 – 1,626.5	2,483.5 – 2,500	Primary allocations, allocated in WARC-92; expected to be used by big LEO systems, but allocation can be used by any type of MSS; in the USA the band is reserved for big-LEO MSS operators who cover the world (excluding polar regions) for 75% of the time and provide telephony service to the USA; a footnote permits use of 1,613.8– 1,626.5 MHz in space–Earth direction (used by Iridium).
2,670 – 2,690	2,500 – 2,520	Effective from 1 January 2005.
1,980 – 2,010	2,170 – 2,200	Satellite component of IMT–2000 world-wide.
1,970 – 2,010	2,160 – 2,200	Satellite component of IMT–2000 in the USA.

3.2.1 Spectrum sharing methods

As mentioned above, intersystem spectrum sharing must be compliant with a formal ITU procedure for which ITU has laid out specific guidelines, including standardized calculation methodology to avoid conflicting interpretation, while intrasystem interference is managed by the operator who will try to maximize the capacity following techniques best suited for its operational philosophy. For example, an operator may trade off interference against fade margin to achieve increased network capacity in favour of link reliability; essentially a trade-off between revenue and customer cooperation. At present, typical intrasystem inter-ference margins for DA SCPC multiple access schemes are 0.5–1.0 dB and intersystem margins are 6% of total noise, per interference entry. In the limiting case, such satellite systems become interference limited, similar to their terrestrial counterpart, rather than thermal noise limited, as are most satellite systems. We will see in Chapter 4 that this is particularly true for CDMA schemes. Degradation caused by an interfering carrier can be more severe if interference is coherent or intelligible. This can occur for example when its modulation charac-teristics are identical to those of the wanted carrier; gateways are more susceptible in this respect, because of their high sensitivity.

Interference calculations

Figure 3.2(a) and (b) shows general models for interference calculations.

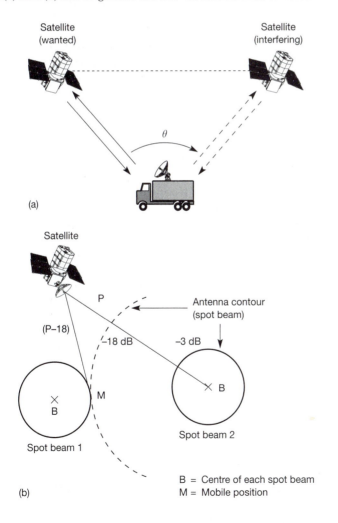

Figure 3.2

(a) A generic interference model of a mobile satellite communication system. θ is the offset angle on the ground between the wanted and the interfering satellite. (b) Mobile satellite communication interference model in a spot beam environment. When P dBW are transmitted towards the centre of spot beam 2, the power appears −18 dB below the peak at the point M at the edge of coverage of spot beam 1

Figure 3.2(a) shows a model for estimating intersatellite interference caused to a mobile or received on a satellite. Essentially, the offset angle θ between the wanted and interfering satellite determines the discrimination between the wanted and interfering signals; for a feeder link, the mobile station is replaced by a fixed Earth station.

Figure 3.2(b) illustrates a model for estimating the interference between spot beams of the same satellite; note that, in this case, isolation between the wanted and interfering carrier is provided by the spacecraft antenna. A carrier transmitted at an EIRP of P dBW towards the centre of coverage of spot beam 2 is received –18 dB lower at a location M at the edge of spot beam 1 resulting in a carrier to interference (C/I) noise ratio of 15 dB when the frequency is reused in spot beam 1 assuming M to be positioned on – 3 dB contour of spot beam 1. The interfering carrier power level in the forward direction is given as

$$I = I_e(\theta)\, P_i\, I_s(\phi)(lp)(\Delta c) \qquad\qquad (3.1)$$

where $I_e(\theta)$ = the ground station antenna gain towards interfering source, P_i = the satellite transmitter output of interfering carrier, $I_s(\phi)$ = the satellite antenna gain towards interfering source, lp = the path loss and Δc = the fraction of interfering carrier power captured within wanted receiver's bandwidth.

The main isolation is derived through discrimination of either the satellite antenna or the ground antenna or both and to some extent Δc, which is achieved by staggering the central frequency of the interferer, thereby reducing the interference power captured by the receiver demodulator.

Calculations may be summarized as follows:

1 Calculate θ.
2 Estimate ground station antenna gain towards the interfering satellite.
3 Estimate satellite antenna gain towards the ground station.
4 Calculate power (I) of each interfering carrier received within the predetection bandwidth of the wanted receiver (forward link) or satellite (return direction).
5 Calculate wanted carrier power (C).
6 Calculate C/I_t, where I_t is the sum of the interfering power. As an approximate rule, 20% of the total noise is budgeted for interference, which includes inter and intrasystem interference.

Carrier levels are estimated using standard link calculations (see section 3.4). Note that carrier levels at mobile receivers fluctuate randomly and hence C/I_t is a statistical quantity. Offset angles for geostationary satellites may be estimated using standard identities in the ITU recommendations. For non-geostationary satellite constellations, satellite dynamics causes randomness in C/I_t and therefore its statistics are conveniently estimated by computer simulation. To assist intersystem coordination, frequency planning and antenna manufacturers, satellite and mobile station antenna side-lobe patterns for a given application are specified generically as a mask which sets bounds on side-lobe performance (angular offset from boresight versus isolation). For hand-held and portable antennas, omni-directional or near omni-directional antennas are most cost effective and hence isolation must be derived from satellite antennas. This is one reason why MSS systems for hand-held services use a large number of spot beams to provide spectrum utilization efficiencies necessary for supporting these high-volume services.

Contrary to FSSs, where ground stations remain fixed, in MSS systems the mobile Earth stations are permitted to have unrestricted mobility within the service area. The traditional approach for intersystem coordination and interference analysis is to ensure that the (C/I) ratio remains within the limits when the mobile is situated at the point which causes or receives the highest levels of interference. For intrasystem interference management, this stringent specification can be relaxed due to the following considerations:

- The probability of simultaneous transmissions from worst locations is low, especially if the two locations happen to be in remote or inaccessible areas.
- MSS systems usually have voice-activated transmissions, which reduces average interfering power.
- MSS systems use power-controlled transmissions, which tend to reduce the average interfering power.
- The probability of occurrence of simultaneous fading when the link margin is exhausted in conjunction with the worst interference scenario is unlikely to be high and hence some fade margin may be available for accommodating increased interference degradation in a few specific regions.

It is therefore possible to derive C/I statistically as a function of percentage of locations and accept slight degradation to a small percentage of coverage.

The problem of interference calculations becomes quite complex in non-geostationary satellite systems, due to a dynamic path profile which causes variations in path loss, elevation angle and propagation conditions. Moreover, handover between satellites or beams causes discontinuity in interference levels, due to abrupt changes in path profile and transmission frequency. Interference in such scenarios is calculated through computer simulations and defined statistically, as mentioned above.

The simulation model for an interference analysis comprises the following components:

- **constellation definition** which essentially comprises orbital parameters of each satellite of the constellation with appropriate phase relationships between satellites;
- **ground projection of antenna pattern including the side-lobe performance** of each satellite. Figure 3.3 (Vatalaro *et al.*, 1995) illustrates an example of a spot beam pattern and its mask, assuming a Bessel function model of satellite antenna pattern. A Bessel function of the first kind and order 2 with a 20 dB edge taper was used for this diagram;

Figure 3.3

An example of a spot beam pattern and its mask (Vatalaro *et al.*, 1995) © 1995 IEEE

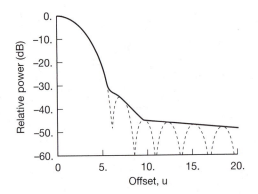

- **regions** of interest;
- **communication traffic model:** Traffic emanating from areas of interest gives a more realistic estimate of interference, as regions where mobiles are likely to be present in areas of interest require protection; for example, probability of a service targeted for land in oceans is low. Furthermore, traffic generation is strongly correlated to the time of day and hence level of interference is also time dependent; thus reuse between otherwise interfering areas may well be possible due to inherent staggering of busy hours of spot beams (see Chapter 8). Intersystem coordination is, however, done assuming a worst interference scenario to avoid unnecessary controversy;
- **mobile Earth station antenna pattern mask** including its side-lobe performance, which is usually provided by operators as an antenna radiation pattern envelope;
- **carrier parameters defined** in terms of EIRP, voice activity, modulation scheme and predetection bandwidth;
- **propagation model** characterizing fading and power control. Section 3.3 discusses a number of propagation models;
- **simulation granularity** in terms of time increments and number of Earth points, which together with run-time and simulation method sets bounds on accuracy of simulation results; a trade-off can be applied between accuracy and run-time for a given computer hardware;
- **statistical data processing:** Results obtained during a simulation must be presented statistically, such as cumulative distribution of C/I in terms of geographical location, to enable system-level performance evaluation;
- **graphical representation** of results is essential for comprehension and analysis, as trends in large amounts of data are easier to assimilate visually.

Figure 3.4 represents a pictorial view of a model which captures the essential units for interference calculations of a non-geostationary system.

Figure 3.5 shows an example of a simulation comparing four types of constellations similar to Iridium (LEO1), Globalstar (LEO2), Odyssey (MEO) and a four-satellite GEO constellation under some simplifying assumptions.

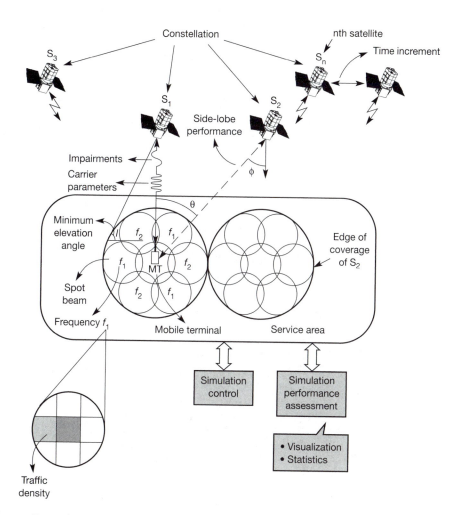

Figure 3.4

Essential entities of a simulation model for interference calculations

In each case, 49 spot beams were used comprising 7 clusters, each with 7 cells. A frequency pool was divided into 7 subsets and each subset was used in the corresponding cell of other sets (see Figure 3.6).

Each satellite uses the same set without attempting to minimize interference at the intersection of the orbital plane; the return link was evaluated with mobiles transmitting at the same power, i.e. without any power control. Simulation was carried on uniformly distributed grid points over the Earth within the latitude ±70°, assuming uniform traffic density. Performance was estimated for each grid point at each simulation instant and averaged, weighting appropriately for traffic density and service area (i.e. weight = 0 above ±70 and 1 below these latitude ranges). Finally, the results of each instant were averaged over the constellation period. The outage probability was defined as the probability of exceeding the C/I threshold. The high outage probability for

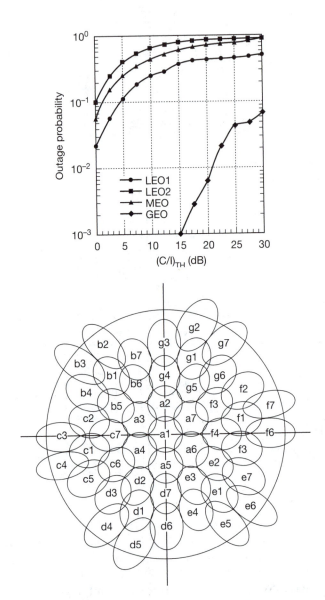

Figure 3.5

Example of a simulation
comparing threshold
carrier to interference
noise ratio of outage
probabiity of four types
of constellations similar
to Iridium (LEO1),
Globalstar (LEO2),
Odyssey (MEO) and a
four-satellite GEO
constellation under some
simplifying assumptions
(Vatalaro *et al.*, 1995) ©
1995 IEEE

Figure 3.6

Spot beam pattern used
in the simulation. Cluster
subdivisions are also
shown (Vatalaro *et al.*,
1995) © 1995 IEEE

non-geostationary systems was attributed to the simplistic frequency planning
arrangement used. The simulation was extended to study various strategies in
frequency planning to achieve better reliability (see Chapter 8). The model was
further refined to include Ricean fading.

The spectrum allocated by the ITU has to be shared equitably between operators. In fact, a regulatory authority may have to apportion spectrum within its
jurisdiction to more than one operator. Thus the spectrum sharing has to be
managed both at country and international levels, where more than one

jurisdiction is affected. The international MSS spectrum is managed by the region through regular meetings between operators, where each operator's requirements and usage are reviewed and mutually agreed interim sharing arrangements are coordinated. This way sharing arrangements are revised regularly on the basis of requirements. The process is expected to offer some control over spectrum hoarding, where an operator claims rights over the spectrum without actual usage.

The method by which a regulatory authority divides spectra between operators within its jurisdiction may vary. The regulatory authority may use its own licensing policy, such as first-come-first-served, an open bid, spectrum auction, etc. (see section 7.6). The authority has to negotiate the spectrum on behalf of the operators through the ITU when the demand is not covered by existing frequency allocation, as was the case, for example, when the licence for non-geostationary satellite systems was sought in the USA in the early 1990s.

3.2.2 Spectrum forecast methodology

Regulators require long-term spectrum forecasts for planning growth and decline of a service generically and operators target their forecast to specific markets for both short- and long-term planning. Spectrum forecasts are also necessary by each operator for operational planning and readjustment of sharing arrangements between operators. For operators with existing services, short-term forecasts can be based on trend analysis including the effects of known events. Chapter 8 outlines further details regarding such methodology. Such forecasts tend to deviate from reality in the longer term because of extraneous influences such as competition, technological obsolescence or breakthrough. There is usually no historic data at the start of service. In these circumstances, a model is developed from known data and influences. Chapter 7 discusses a methodology for developing forecasts in these latter situations.

Figure 3.7 illustrates a flow chart of a generic methodology, which may be used for deriving long-term MSS spectrum forecasts for a chosen service – a country, region or the world. The model can be modified and tailored on the basis of available or estimated data.

Spectrum requirements are categorized by the environment because of differences in traffic requirements and propagation environments. For each environment and service, the target population is estimated; traffic penetration within the population gives an estimate of market size, which is weighed for loss to competition, other service offerings such as VSAT or terrestrial mobile systems, etc. The extent of penetration is influenced by the existing infrastructure, affordability to the target population, their attitude towards acceptance of new technology, exposure/experience with similar technologies and social trends such as the way people communicate during work or during leisure, internet penetration, etc. It is then necessary to estimate traffic generated by each terminal in the busiest hour, the holding time of each call for circuit-mode services and the average message length for packet-mode services. The busiest

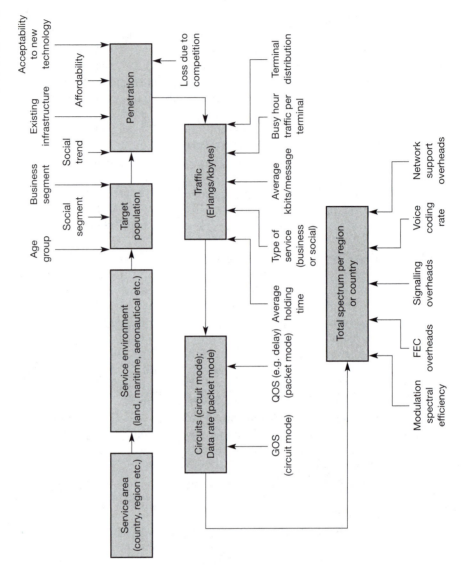

Figure 3.7
Methodology which may
be used for estimating
long-term spectrum
forecast in a given
service area

hour may be staggered if the service carries a mix of social and business traffic due to differences in their characteristics. The methodology is useful for early planning of a service, but its accuracy is highly sensitive to assumptions.

The total traffic carried for the circuit-mode traffic is

$$E = T_a * N \; Erlangs \qquad (3.2)$$

where T_a = average traffic per terminal and N = number of terminals.

The total traffic for the packet mode is

$$P = [M + C]N \; Kbytes \tag{3.3}$$

where M = average message length per user and C = coding overheads.

The total spectrum for each transport mechanism is estimated individually and summed, taking into consideration the modulation efficiency, coding overheads, grade of service (for circuit mode) and permissible delay with packet retransmissions (packet mode), including network overheads, such as signalling, network test and support channels.

3.3 Propagation characteristics

In an MSS system, feeder link propagation conditions are benign and easier to manage than in the service link. Feeder links use large fixed station, whereas mobile communication service links operate to small mobile Earth stations exhibiting variable path geometry. Generally, the propagation environment is more hostile in a land mobile environment than in maritime and aeronautical environments. The problem in the land mobile environment is exacerbated by a need to use small and hence wide beam-width antennas due to size and economic constraint. It is for this reason that land mobile communication systems have become viable only in the 1990s even though the market potential of this service had been known for some considerable time; it required almost two decades for the industry to develop the technology.

Propagation effects are frequency dependent. At present, feeder link frequencies range from 4 to 30 GHz, whereas service links are mostly in the L and S bands with considerable interest in the use of the K_a band and above.

Propagation characteristics are also affected by the orbital characteristics of the constellation. Additional variables to be considered for non-geostationary satellites are time-dependent range and elevation angle.

General propagation characteristics are summarized, which apply to both the feeder and service links, irrespective of the environment; followed by review of the propagation models applied specifically to land, maritime and aeronautical environments. The reader may observe that land mobile channels have been dealt with in more detail than others. This is due to the variability of land mobile communication environments, which has necessitated a large number of studies throughout the world resulting in a vast number of measurement campaigns as well as theoretical and empirical modelling methods.

3.3.1 General propagation characteristics

Signals travelling between a satellite and the ground are affected by the intervening medium causing changes to signal level, polarization and noise contamination. The effects are frequency-dependent and caused mainly by the troposphere and the ionosphere. The troposphere is the first few tens of kilometres of the atmosphere where clouds, rain and fogs form, and the ionosphere is the ionized region which extends between ~ 80 and 1,000 km around the Earth. The main sources of degradation in the troposphere are gaseous absorption in the atmosphere, absorption and scattering due to fog, cloud and rain, signal fluctuations due to atmospheric turbulence and depolarization due to rain. In the ionosphere, signals undergo changes in polarization and occasionally suffer rapid signal fluctuations. Furthermore, received signals become contaminated by noise from extra-terrestrial and man-made sources.

Tropospheric effects on radio-wave propagation are well documented, therefore only a brief summary is presented here. There is a considerable interest in the application of the K_a band to MSS and FSS (personal communications) links. Some results for the MSS K_a band link are presented in section 3.3.2.

The most noteworthy effects are:

- gaseous absorption;
- attenuation by hydrometers;
- scintillation;
- depolarization.

Gaseous absorption increases with frequency and peaks around 22.2 GHz due to water vapour absorption and close to 60 GHz due to oxygen (see ITU-R Rec 390-4). The absorption depends on temperature, pressure and humidity of the atmosphere as well as the elevation angle of the ground–satellite path. Absorption reduces with a reduction in the humidity elevation angle. For example, in the frequency range of 1–20 GHz gaseous absorption at the zenith varies approximately in the range 0.35–0.5 dB increasing to 0.35–5.7 dB at an elevation of 5° and 100% humidity. These values become critical when propagation margins are low, as in MSS links.

The CCIR (now the International Telecommunication Union's Radio Telecommunication Sector (ITU-R)) model for estimating gaseous absorption through the atmosphere is made up of the following four steps (ITU, 1990):

1 Determine frequency-dependent specific attenuation coefficients in dB/km for oxygen and water vapour.
2 Determine equivalent height for oxygen and water vapour (dependent on location latitude).
3 Adjust values to account for surface temperature.
4 Determine total gaseous attenuation.

Attenuation by hydrometers refers to attenuation caused by water particles existing in the atmosphere, such as fog, cloud, rain and ice, out of which rain produces the most significant attenuation by scattering and absorption. Over the past decades, considerable effort has been spent throughout the world for the development of rain attenuation prediction methods (e.g. ITU-R Rec 564-3; Ippolito, 1986; Crane, 1980). Prediction techniques have been developed for the FSS, which apply to the feeder links of the MSS and hence are well developed for 99.90 – 99.99% link reliability, where link margins are of the order of 3–15 dB at 11.7 GHz – 20 – > 30 dB at 20 GHz. For MSS links, margins of this magnitude are too demanding as, typically, MSS links operate with link margins of the order of 5–10 dB. Therefore, models for predicting margins at lower reliability links of the order of less than 99% are more appropriate for the K_a band. We note that tropospheric effects at the L-band are negligible compared to shadowing and multipath loss.

Attenuation A_{rain} of radio waves propagating through rain extending length L of the path is given as

$$A_{rain} = \int_o^L \alpha \tag{3.4}$$

where α = specific attenuation of rain (dB/km).

α can be estimated theoretically, but in practice, empirical statistical methods are used due to randomness associated with atmospheric conditions. The specific attenuation is estimated as

$$\alpha = aR^b \tag{3.5}$$

where a and b are frequency-dependent constants and R = the surface rain rate at the location of interest.

The most reliable estimates are obtained by measurements in geographical regions of interest. Measured data are presented as cumulative distribution of fade levels. Prediction models are based on these results and extended to other regions. Generally, prediction techniques use equations (3.4) and (3.5) as follows:

$$A_{rain} = aR^b \, L(R) \tag{3.6}$$

where L(R) = effective length of rain.

The difference in prediction method lies in the method used for estimating L(R). The most accurate method is to use a model which applies specifically to the region of interest, but this is not always possible due to inadequate databases. The ITU-R recommends a model which may be applied anywhere in the world (ITU-R Rec 618-2). The inputs to the model are point rainfall rate for the location of interest, height of the location above mean sea level, satellite elevation angle, latitude of the location and frequency of operation. Referring to Figure 3.8, steps involved in calculating excess attenuation due to rain by this method are given below:

Figure 3.8
Geometry used for
estimating attenuation
due to rain (Figure 1, Rec
618–2, ITU-R, 1992) ©
1992 ITU

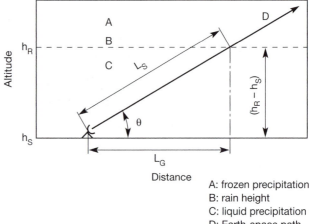

A: frozen precipitation
B: rain height
C: liquid precipitation
D: Earth-space path

1 Calculate the effective height (h_R) of the location (latitude, φ).
2 Calculate the radio path length, L_s.
3 Calculate the horizontal projection of the radio path, L_G.
4 Obtain 1 minute of integrated rain intensity, $R_{0.01}$, for 0.01% of an average year; if local data is not available then values given in the recommendation can be used. These values have been derived from world climate data.
5 Obtain the reduction factor, $r_{0.01}$, for $R_{0.01} \leqslant 100$ mm/h.
6 Calculate the specific attenuation, γ_R, using tabulated frequency-dependent coefficients.
7 Calculate the predicted attenuation for 0.01% of time using the equation

$$A0.01 = \gamma_R L_s r_{0.01} \text{ dB} \tag{3.7}$$

8 If necessary, obtain attenuation for other percentage of time using empirical relationship.

ITU-R Recommendation 840 provides a method for estimating attenuation, due to water clouds or fog, of known liquid content. It has been observed that attenuation due to clouds up to 30 GHz is relatively insignificant for fixed links, but at higher frequencies attenuation from clouds of high water content (e.g. cumulonimbus) becomes significant, for example, at 100 GHz the attenuation can be 4–5 dB, increasing up to 8 dB at 150 GHz.

When new bands are considered, reliable rain attenuation data are unavailable. This was the case when the K_u and K_a bands were initially introduced and is currently the case for bands above 30 GHz. One approach is to scale data from a frequency for which data are well characterized and a number of empirical scaling models have been developed (ITU-R Report 721).

To meet the specified link reliability throughout a year, it is essential to convert the annual p% attenuation statistics to the worst month of the year, as certain months are the wettest (e.g. monsoon season in the Indian subcontinent). Techniques for this type of scaling are well documented, for example, ITU-R Recommendation 581 provides a method of converting the worst-case statistics to annual statistics. However, it has been observed that there is a notable difference in year-to-year statistics; variations in excess of 20% rms are possible.

Site diversity offers a solution for mitigating effects of heavy attenuation occurring at the K_a band and above. It has been observed that intense rain cells typically have dimensions of a few kilometres. Thus, if sites are spaced several kilometres apart, it is likely that severe fading does not occur at both sites simultaneously. By selecting or combining signals from two sites, it is therefore possible to reduce demands on spacecraft power and improve link reliability. Such a scheme is designed in the feeder link of the Iridium and Teledesic systems.

Diversity performance is characterized by two measures – a diversity improvement factor and diversity gain. The diversity improvement factor is the ratio of single site time percentage for a given attenuation to that obtained after application of diversity for the same attenuation. Diversity gain is the difference in decibels between attenuation on a single site before and after application of diversity for the same time percent. Empirical methods for estimating both types of measures are available in the literature (ITU-R Rec 618-2). Empirical parameters depend on link characteristics which show the strongest dependence on distance between diversity stations with lower influence of elevation angle and frequency. Figure 3.9 represents improvements using diversity, shown as percentage of time with and without diversity for the same attenuation (ITU-R Rec 618-2).

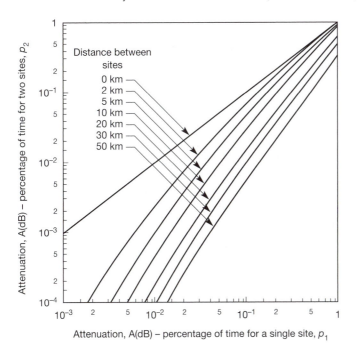

Figure 3.9

Improvements through two-site diversity (Figure 2, REC 618–2, ITU-R, 1992) © 1992 ITU

When a dual-polarized radio wave travels through rain or ice, power from one polarization gets coupled to the orthogonal component, due to the anisotropic behaviour of the medium causing degradation to the signal quality. Degradation is measured as cross-polar discrimination (XPD) or cross-polar isolation (XPI), given as

$$XPD = 20 \log |E_{11}|/|E_{12}| \tag{3.8}$$

where E_{11} = the received co-polarized electric field strength and E_{12} = the electric field strength coupled to orthogonal polarization.

$$XPI = 20 \log |E_{11}|/|E_{21}| \tag{3.9}$$

where E_{21} = the cross-polar component.

For practical purposes XPD and XPI are assumed equal.

It can be shown that XPD degrades with decrease in frequency and increase in copolar attenuation. It has also been observed that generally XPD of a vertically polarized wave is better than that for horizontal polarization and circularly polarized waves have ~ 10 dB lower XPD than horizontally polarized waves. Depolarization caused by ice occurs without accompanying co-polar attenuation; however, its magnitude is about 25 dB and therefore not very significant.

In the absence of measurements, prediction models such as recommended by ITU-R are useful. Empirical relationships show dependency on frequency, polarization, elevation angle and the canting angle of rain. Measurements at 19 and 28 GHz taken in the USA show XPD of 16–18 and 12–13 dB for 0.01% of the time.

Small-scale refractive index variations of troposphere (or ionosphere) cause signals to arrive via different paths causing rapid level variations, called scintillation. Tropospheric scintillation depends on the season, the local climate, frequency and elevation angle. Its magnitude increases with frequency and a reduction in elevation angle. Degradation caused by tropospheric scintillation becomes noticeable above 10 GHz and is significant in the K_a band and above. Scintillation can be accompanied with rain. The interdependency of rain fades and scintillation is not well understood and hence they are treated as statistically independent. Models for estimating the distribution of scintillation are available in the literature (e.g. ITU, 1982; Karasawa *et al.*, 1988). A technique for laboratory simulation of tropospheric scintillation has also been reported and used for evaluation of a tracking system. Typical values of scintillation in K_a band are of the order of 0.2–0.3 dB peak-to-peak level in winter, 1 dB in clear conditions of summer and 2–6 dB in some types of cloud while fade rates range from 0.5 to over 10 Hz.

The K_a band (20–30 GHz) has received considerable attention recently because the band is less crowded and high antenna gain is possible through small apertures, attributes attractive for wideband personal communication services (see

Chapter 10). Deep signal fades due to rain and scintillation pose the severest problems in this band. A large number of propagation measurements have been conducted since 1970, mostly applicable to fixed geostationary satellite systems. Due to continuous changes in elevation angle of non-geostationary satellites, these results are not directly applicable to systems deploying non-geostationary satellite constellations. In this section, considerations necessary for feeder links are considered; section 3.3.2 discusses issues related to the service link.

For radio link design of earth stations operating through non-geostationary satellites, it is necessary to combine attenuation due to path loss, rain, scintillation, gaseous absorption, shadowing and multipath as a function of elevation angle. Feeder links use large antennas and are sited to avoid blockage, so shadowing loss and multipath variations are minimal. Path loss, rain fade, tropospheric scintillation and gaseous absorption for the specific site and constellation are estimated as a function of elevation angle, using appropriate models such as discussed above; cumulative path loss distribution obtained, and standard link budget analysis used for sizing Earth station and satellite. Figure 3.10(a) illustrates the elevation angle variation from Bhopal, a city in central India (77.5° E and 23° N) for a LEO pass (orbital altitude = 1,400 km, inclination = 52°, ascending node = 60° E) while Figure 3.10(b) and (c) demonstrates respectively the path loss at 1.5 and 20 GHz for satellites in the same LEO, a MEO (altitude = 10,350 km; inclination = 45°; ascending node = 60°E) and a GEO from London, UK. Note that path loss decreases as the satellite rises until the satellite is closest to the ground station, increasing again as the satellite sets; this observation causes no surprise, as it essentially implies that the worst conditions occur at the lowest elevation angle.

Figure 3.11 illustrates variations in elevation angle, path loss and available link margin during a single pass of a LEO satellite link similar to that in the Iridium system, from White Sands in the USA (Ippolito and Russell, 1993).

Note that the link margin is a function of elevation angle, for example, when path loss varies from 69 dB to 59 dB, the margin increases from 5 dB to 15 dB in a system designed with a worst-case margin of 5 dB. It follows that link reliability depends on the minimum operational elevation angle; this rather obvious conclusion has considerable influence on the size and cost of a non-geostationary satellite constellation, as observed in Chapter 2. Figure 3.12 represents variation in 20 GHz gaseous attenuation and margin for rain effects for a single pass of the LEO satellite. Annualized outage probability for the link using the ITU-R global attenuation model is captured in Figure 3.13; note wide variations in outage probability for the pass; outage probability in the middle three minutes of the pass is 0.01% and increases to 1% at ±5.5 minutes in the 14.7 minute pass. A corollary to these observations is the need of a large dynamic range in Earth station receivers due to large variations of signal level.

Figure 3.10
(a) Elevation angle variation for LEO, MEO and GEO satellite pass over a location near Bhopal in central India (77.5° E and 23° N). [LEO: orbital altitude = 1,400 km, inclination = 52°, ascending node = 60° E; MEO: altitude = 10,350 km; inclination = 45°; ascending node = 60° E) (Graphics: AR). (b) A comparison of path loss at 1.5 GHz from satellites in LEO, MEO and GEO from London, UK (Graphics: AR). (c) A comparison of path loss at 20 GHz for LEO, MEO and GEO pass from London, UK (Graphics: AR)

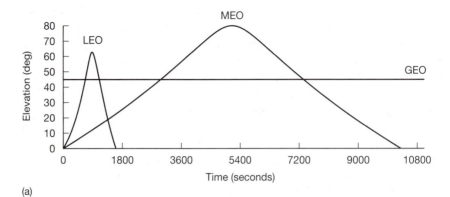

(a)

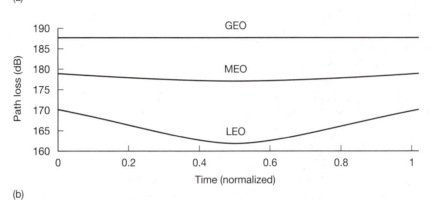

(b)

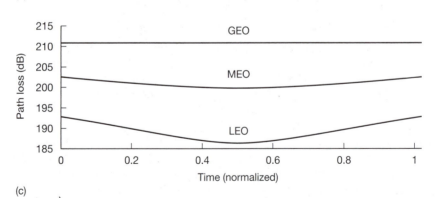

(c)

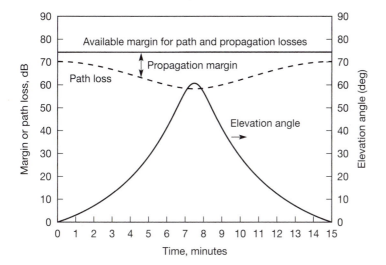

Figure 3.11
Elevation angle, path loss and available link margin for a 20 GHz link during a single pass of a LEO satellite from White Sands in the USA (ascending node = 100° W, altitude = 756.3 km) (Ippolito and Russel, 1993) © 1993 IEEE

Ionospheric effects

The ionosphere affects radio waves in a number of ways. It causes rotation of radio wave polarization, propagation delay, refraction, variations in angle of arrival, absorption, dispersion and scintillation. Table 3.2 provides the worst-case estimated values of these effects at 0.5, 1, 3 and 10 GHz.

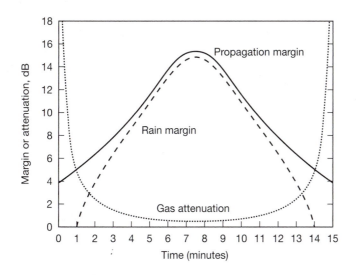

Figure 3.12
Variation in 20 GHz gaseous attenuation and margin available for rain effects for a single pass of a LEO satellite (Ippolito and Russel, 1993) © 1993 IEEE

Figure 3.13
Annual 20 GHz outage
probability for a single
pass of a LEO satellite
(ascending node 100° W,
altitude = 756.3 km)
(Ippolito and Russel,
1993) © 1993 IEEE

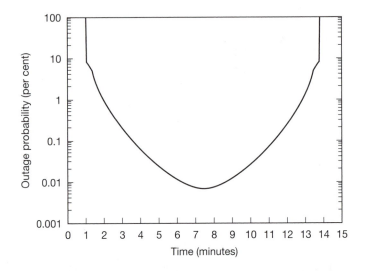

Figure 3.13
Annual 20 GHz outage probability for a single pass of a LEO satellite (ascending node 100° W, altitude = 756.3 km) (Ippolito and Russel, 1993) © 1993 IEEE

Table 3.2
Worst-case estimated values of rotation of radio wave polarization, propagation delay, refraction, variations in angle of arrival, absorption, dispersion and scintillation for 0.5, 1 and 3 and 10 GHz (ITU-R report 884-2)

Effect	Frequency (f) dependence	Frequency (GHz)			
		0.5	**1**	**3**	**10**
Faraday rotation (Deg)	$1/f^2$	432	108	12	1.1
Propagation delay (μs)	$1/f^2$	1	0.25	0.028	0.0025
Refraction	$1/f^2$	<2.4'	<0.6'	<4.2''	<0.36''
Variation in angle of arrival	$1/f^2$	48''	12''	1.32''	0.12''
Absorption (auroral and polar cap) (dB)	$\sim 1/f^2$	0.2	0.05	6×10^{-3}	5×10^{-4}
Absorption (mid-latitude) (dB)	$1/f^2$	<0.04	<0.01	<0.001	$<10^{-4}$
Dispersion (ps/Hz)	$1/f^2$	0.0032	0.0004	1.5×10^{-5}	4×10^{-7}
Scintillation (dB peak to peak) (Observed near equator, early local night time, at equinox under conditions of maximum sunspot number)	$1/f^3$	–	> 20	~10	~4

For mobile satellite communications operating in 1–2 GHz band, polarization rotation, also known as the Faraday Effect, and scintillation are dominant. The Faraday Effect is caused by the interaction of electromagnetic waves with Earth's magnetic field. Note that circular polarized waves are not affected by the Faraday Effect and therefore MSS links use circularly polarized waves, as this eliminates the need for polarization tracking, whereas feeder links can use linear (or circular) polarization as they can incorporate polarization tracking without significant cost impact. The effect is generally predictable and hence can be compensated by rotating the polarization of transmitted waves (or receiver antenna) in an opposite sense.

Peak-to-peak signal-level fluctuations of up to 20 dB are known to occur at ~ 1.5 GHz. Hence it is not surprising that scintillation can cause outage to MSS links at the L and S band, for several hours in a year, but fortunately the onset of scintillation events is late in the evening, when traffic volumes are quite low. Personal communication systems which often operate close to threshold are particularly susceptible. Note that navigation systems such as GPS and GLONASS also operate in the L band and therefore may also be susceptible.

	Region			
Parameter	**Equatorial**	**Mid-latitude**	**Auroral**	**Polar**
Scintillation dominant cause	Plasma bubbles	Occurrence of ionospheric spread, F	Related to auroral and geomagnetic activity	Related to auroral and geomagnetic activity
Severity	Largest intensity	Very quiet to moderate	Moderately active to very active	Intensity directly related to sunspot numbers
Diurnal behaviour	Night – Maximum Day – Minimum	Night – Maximum Day – sporadic	Night – Maximum	Night – Maximum
Seasonal behaviour	Longitude dependent; e.g. Ghana: Maximum – November/ March, Minimum – solstices Peru: Maximum – October to March Minimum – May to July	Maximum – spring/summer Minimum – winter	Function of longitude sector	Function of longitude sector
Effect of solar cycle	Activity/ intensity strongly dependent on number of sunspots	Limited data	Activity/ intensity strongly dependent on number of sunspots	Activity/ intensity strongly dependent on number of sunspots
Magnetic activity (measured in terms of K_p)	Depends on longitude (increasing or decreasing with the measure K_p)	Independent of K_p	Activity increases with K_p	Activity increases slightly with K_p

Table 3.3

Main features of scintillation (from ITU-R Rec 531–2)

Ionospheric irregularities occur due to certain solar, geomagnetic and upper atmospheric conditions. Hence ionospheric scintillation is affected by sunspot activity, position of the sun and hence time of day, latitude of location, season and magnetic activity. Table 3.3 summarizes the main features of scintillation.

The severity of scintillation is defined in a number of ways. A commonly used measure is known as the S4 scintillation index, defined as the standard deviation of received power divided by the mean value of the received signal power

$$S_4^2 = \sigma/\mu \tag{3.10}$$

The fading period is another useful system parameter. Its value is variable from event to event and ranges in the region of 1–10s in the GHz range. Longer periods lasting tens of seconds have also been observed at VHF and ultra high frequency (UHF).

Due to large regional variability, the best estimate of scintillation is obtained from measured local data. In the absence of measurements, an applicable model can be used (for example, see ITU-R Rec 531-2). Some useful conclusions of ITU-R recommendations are summarized as follows:

- Measurements taken at frequency f can be scaled to another frequency according to the dependence, $f^{-1.5}$.
- Instantaneous fluctuation of a scintillation event can be approximated by the Nakagami probability density function, which can be used for estimating cumulative fade distribution.
- Power spectral density of scintillation events varies widely due to variations in drift velocity of refractive index irregularities; power spectra density slopes of f^{-1}–f^{-6} have been reported; a value of f^{-3} can be used as an approximation in absence of real data.

Figure 3.14
Scintillation occurrence from observations at 1.6 GHz showing geomagnetic latitude, longitude and local time dependence of scintillation (Rec 531–2, ITU-R, RPI series, 1992) © 1992 ITU

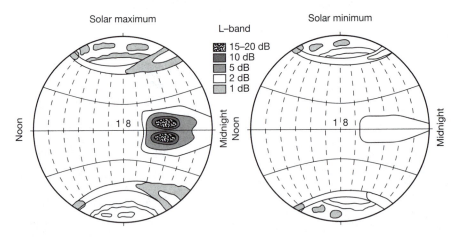

- Scintillation magnitude s_4^2 varies as $1/\cos(i)$ up to $i \sim 70°$ where i is the zenith angle and for lower values of i variation is between $(\cos(i))$ and $(\cos(i))^{1/2}$.
- Seasonal and longitudinal dependence of S_4 can be approximated as

$$S_4 \, \alpha \, \exp(-\beta/W) \tag{3.11}$$

  where β is a longitudinal and seasonal-dependent parameter and W is a location and day of year-dependent constant.
- Figure 3.14 (ITU-R Rec 531-2) illustrates geomagnetic latitude, longitude and local time dependence of scintillation at 1.6 GHz.
- The cumulative distribution of fading statistics for 1 dB peak-peak fluctuation as a function of monthly sunspot numbers, taken from 4 GHz measurements campaign of a number of locations is illustrated in Figure 3.15

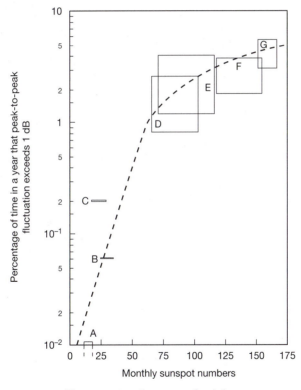

Figure 3.15
Cumulative distribution of fading statistics taken from a 4 GHz measurements campaign (Rec 531–2, ITU-R, RPI series Geneva 10/92, 1992) © 1992 ITU

The squares are the ranges of variations over a year
for different carriers

A: 1975–1976, Hong Kong and Bahrein, 15 carriers
B: 1974, Longovilo, 1 carrier
C: 1976–1977, Taipei, 2 carriers
D: 1970–1971, 12 stations, >50 carriers
E: 1977–1978, Hong Kong, 12 carriers
F: 1978–1979, Hong Kong, 10 carriers
G: 1979–1980, Hong Kong, 6 carriers

(ITU Rec 531-2). In an equatorial region (Hong Kong) 1 dB peak-to-peak variation can occur for up to 5% of the time in periods of high Sun activity.

- The probability of simultaneous occurrence of ionospheric and rain fading in equatorial regions is high, especially in years of high sunspot activity, resulting in differences to statistics. The occurrence probability of such events has to be considered in the design of high-reliability line-of-sight MSS radio links.

3.3.2 Land mobile channel

A service link provides mobility to a user, but at a cost! Motion of mobile terminals causes variability in signal path geometry, resulting in environment and velocity-dependent signal fluctuations at the receivers; additionally, reduction of antenna size (and hence terminal size) causes multiple-path signal reception because of wider antenna beam width. The rate at which the signal fluctuates due to shadowing and multipath depends on the user's velocity; coupled to this is a velocity-dependent Doppler component; satellite motion introduces a further Doppler component. Generally, propagation conditions progressively degrade in the following order: maritime, aeronautical and land.

In narrowband applications at data rates less than 100 kb/s, all the components of signals fade simultaneously; this is sometimes referred to as flat fading. However, when signals are sufficiently wideband, such as in high-fidelity sound broadcasting or wideband CDMA links, signals may suffer frequency selective fade, which affects only parts of the band causing degradation by intersymbol interference and thereby a reduction in throughput. Furthermore, when terminals operate near their threshold some otherwise relatively minor effects become noticeable.

Figure 3.16 depicts the general trend in throughput for various sizes and speeds of user terminal in a land environment. Throughput reduces as the size of user antenna reduces and the travelling speed increases. The upper limit of throughput of MSSs in the foreseeable future is of the order of 2 Mb/s.

Figure 3.17 illustrates a signal path geometry to characterize propagation characteristics in a land mobile channel.

Figure 3.16

Throughput vs. antenna size

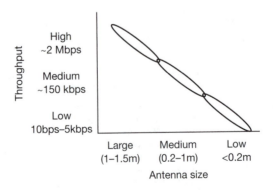

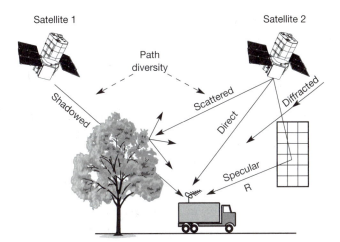

Figure 3.17
Geometry of an MSS communication channel showing all types of signal components – direct, shadowed, scattered, specular components and path diversity

Received signals, already affected by the ionosphere and troposphere, depending on the environment around the receiver, may comprise various components, as shown in Figure 3.18: a direct path; diffracted components, specular components caused by reflections from metallic or smooth objects such as the body of a car or a smooth sea surface; and diffused components caused by reflection and scattering of objects around a mobile. The magnitude of each component is strongly influenced by the environment around a mobile. In an urban environment, the direct path may be shadowed for a considerable period by buildings, etc. and therefore the diffracted or scattered components will dominate; in a maritime or an aeronautical environment, when a satellite is at a low elevation and the sea surface is rough, direct and diffused components are likely to prevail.

The resultant signal r(t) at a receiver can be represented as

$$r(t) = \sigma(x)a(t) + s(t) + d(t) \tag{3.12}$$

where $\sigma(x)$ is an environment, elevation and RF-dependent attenuation factor, $a(t)$ = magnitude of direct path, $s(t)$ = magnitude of specular components and $d(t)$ = the magnitude of diffused components.

The magnitude and behaviour of various components depend on a number of factors which include terminal antenna characteristics, speed of travel, user cooperation, elevation angle, type of environment, RF and signal bandwidth (narrowband versus wideband).

The following considerations are necessary to characterize propagation of MSS radio links:

1 tropospheric and ionospheric effects, which have already been discussed in section 3.3.1;

2   receiver antenna characteristics;
3   environment dependence;
4   elevation angle dependence;
5   frequency dependence;
6   path loss variation in non-geostationary satellite systems.

A number of approaches are used in characterizing land mobile channel propagation:

- **Empirical models** are based on applying regression fits to measured data. It is possible to derive simple algebraic equations relating fade level to experimental variables chosen by the experimenter for the application at hand. The fade level can be determined as a function of elevation angle, type of environment, antenna characteristics, frequency, season, foliage content, etc.
- **Statistical models** are based on applying numerical analysis to measured data to obtain statistical descriptions of signals such as probability distribution function and cumulative probability distribution, which are used to derive useful system parameters such as the fade margin for a given percentage of time, level crossing rate, power spectral density, etc.

    This type of model is sometimes also based on heuristic reasoning, based on understanding and simplification of propagation mechanism.
- **Deterministic models** are based on an accurate characterization of the environment and propagation mechanism, applying ray or electromagnetic wave theory together with models of receiver antenna pattern characteristics, vehicle velocity, elevation angle, etc. A received signal is given as the sum of individual wave components. This approach avoids the need for complex and expensive experiments in situations, such as characterization of signal in specific environments and at frequencies for which data are not available.

Environment dependence

Due to randomness associated with propagation, the most common modelling approach is application of statistical and empirical methods to measured or mathematically generated data or their combination. The most common technique has been to derive models through empirical fits to measured data. In situations where measured data are lacking, signals are generated through mathematical simulation. Examples of such instances are characterization of satellite signals within buildings or in urban areas for personal communications; mobile communications system studies at extra-high frequency (EHF); feasibility of applying diversity; characterization of unusual events such as specular reflection from the sea surface; site selection of fixed terminal in a dense urban area; wideband characterization of channels, etc.

    The first part of the section addresses the currently far more common statistical approach to propagation modelling with specific examples of prevalent models; this is followed by a review of prevalent propagation simulation techniques.

Referring to equation (3.12), each component in the equation varies randomly, and therefore may be characterized statistically. Depending on the physical interpretation of the underlying propagation mechanism, a number of variants of statistical models have been proposed.

Land mobile satellite channels are characterized by various combinations of Ricean, Rayleigh and log-normal probability distribution; the difference in models lies in the interpretation of the shadowing mechanism on direct and scattered paths, as illustrated in examples below.

- The probability distribution of the signal amplitude is considered Ricean, with the assumption that the component a(t) is constant; in-phase and quadrature components of the diffused signals are independent of each other and normally distributed with zero mean. In the absence of a direct component the signal is assumed Rayleigh distributed with a log-normal mean; both direct path and scattered signals are shadowed. Ricean distribution is characterized by the Ricean factor, defined as the ratio of direct signal to multipath ratio; and the Rayleigh function is characterized by a mean and standard deviation.
- In a variation on the approach above, only the direct path is assumed shadowed.
- Yet another approach is to assume scattered signals as a composite of clear and shadowed components.
- In the environment-dependent approach, the overall probability distribution is taken as a composite of a variety of probability distribution. A Ricean distribution is assumed for the fraction of time when a signal is unshadowed and a Rayleigh distribution with log-normally distributed mean for periods when the signal is shadowed. From a practical viewpoint, the inclusion of environment dependence provides a powerful technique for quantifying system performance in a mix of environments.
- A further refinement of the environment-dependent approach is to represent channel behaviour as a combination of a Markov process and a statistical model, where the state of a channel is described by a Markov process and the signal variations within each state are modelled by a statistical model appropriate for the environment. The Markov process allows the signal to assume one of the M states with a probability which only depends on its previous state. A number of authors have used this approach; their models differ in the number of states of M and the chosen probability distribution of signal within each state. We discuss this further in a following section.

Table 3.4 summarizes the characteristics of several prevalent narrowband models.

These statistical descriptions are strictly valid under simplifying mathematical assumptions, such as applicability of the central limit theorem; therefore statistics do not apply to every practical situation. When a specific environment is of interest, data stored from a measurement campaign can provide the most accurate characterization, although organizing a campaign is expensive and time consuming; in such cases, a deterministic modelling approach, such as described later in the section, can be an effective alternative.

Table 3.4
A summary of
characteristics of
several prevalent
models (Karaliopoulos
and Pavlidou, 1999)

Model	Complex channel process, $r = r_x + jr_y$
Loo (1985)	$r = Se^{j\phi o} + Re^{j\phi}$ S: log-normal R: Rayleigh distributed ϕ_o, ϕ: uniform
Rice-log-normal (Corazza and Vatalaro, 1994)	$r = RSe^{j\theta}$ S: log-normal R: Ricean distributed θ: uniform
Generalized Rice-log-normal (Vatalaro, 1995)	$r = RSe^{j\theta} + x_1 + jy_1$ S: log-normal R: Ricean distributed x_1, y_1: Gaussian
Hwang et al., (1997)	$r = A_cS_1e^{j\phi} + RS_2e^{j(\phi + \theta)}$ S_1, S_2: log-normal R: Rayleigh distributed A_c: constant
Poca-log-normal (Babalis and Capalis, 1998)	$r = RSe^{j\theta}$ S: log-normal R: Poca-log-normal distributed θ: uniform Note: Poca distribution is similar to Rayleigh distribution; here, the number of scattering components is assumed finite instead of infinite as in Rayleigh distribution.
Pätzold et al., (1998)	$r = Se^{j(2\pi f_pt + \theta_p)} + x_1 + jy_1$ S: log-normal x_1, y_1: Gaussian distributed f_p and θ_p : constant

Both primary and secondary statistics are useful in the analysis and design of satellite systems. Primary statistics refer to parameters which are independent of time, and secondary statistics relate to time. The primary statistical description, 'cumulative fade distribution', relates fade level and percentage of time that the level is exceeded and is used for estimating link margin for a given link reliability. Level crossing rate, average fade duration and Doppler frequency shift are useful secondary parameters for system design and evaluation. Their magnitude depends on the probability distribution of the signal envelope and the velocity of the mobile.

As mentioned, the probability distribution of signals received on a mobile receiver consists of an environment-dependent mix of probability distribution functions transitioning between each other as a Markov process. Two matrices are used in defining such a process – a 1×M state probability array whose ith element defines the probability of a system being in a state i and a M×M transition probability array whose element defines the probability of transitioning from one state (say, i) to another (say, j). Models have been proposed, ranging from two to four states. Probability distributions within each state depend on the environment and preference. The overall probability distribution function is given as the summation of $p_i w_i$ where p_i is the probability distribution function of signals in state i and w_i is the probability of the signal existing in state i. Figure 3.18(a) shows a two-state Markovian chain as proposed by Lutz *et al.*, (1993). The channel is assumed to alternate between a Ricean and Rayleigh/log-normal statistical state. The transition probabilities for various environments were determined through measurements.

A simulator of a land mobile channel based on the principle is presented in Figure 3.18(b) (Lutz *et al.*, 1986). Fading statistics are switched between Ricean

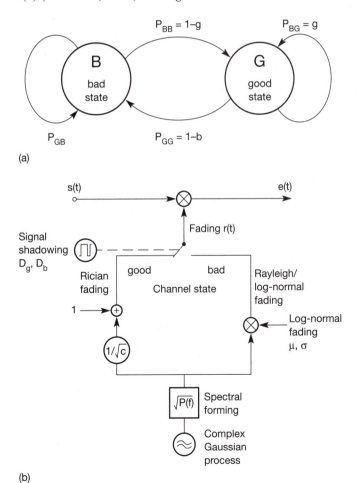

(a)

(b)

Figure 3.18

(a) A two-state Markovian model of a land mobile communication channel representing environment dependence; P_{BB} = probability of returning to a bad channel state; P_{BG} = probability of transitioning from a bad state to a good state; P_{GB} = probability of transitioning from a good state to a bad state; P_{GG} = probability of returning to a good state (Lutz *et al.*, 1996). (b) A narrowband propagation channel simulator based on the environment-dependent approach (Lutz *et al.*, 1991, © IEEE)

and Rayleigh distributions for mean periods of D_d and D_s, representing good and bad channel states respectively; for hand-held communicators, D_d and D_s can be measured in metres. The magnitude of D_d and D_s depends on environment and vehicle speed; and switching between the two states follows a two-state Markov model. The channel is shadowed on an average for $D_s/(D_d + D_s)$. Table 3.5 summarizes observations in highway and city environments of Europe for a number of satellite elevations for L-band transmissions (Lutz, 1998). Note the progressive improvement in channel conditions with elevation angle – increase in mean duration of the Ricean channel (good state), improvement in Rice factor and a reduction in fade levels.

Table 3.5
Characteristics of land mobile channels of the L band in Europe

Environment	City			Highway		
Satellite elevation	10	30	50	10	30	50
Mean duration (m): Ricean	7	25	50	100	1,000	10,000
Mean duration (m): Rayleigh	70	50	35	50	30	20
Rice factor (dB)	5	9	10	10	14	18
Time share: shadowing	0.9	0.7	0.4	0.3	0.2	0
Mean attenuation: shadowing	12	12	15	9	10	14
Standard deviation of received power: shadowing (dB)	4	4	4	4	4	4

Trees, utility poles and overpasses are known to affect propagation. Other physical objects and factors affecting propagation include terrain, orientation of the satellite with respect to direction of travel, and seasonal dependence.

Attenuation measurements caused by a variety of trees such as white pine, holly, burr oak and pin oak performed at 870 MHz for elevation angles between 20° and 45° show a mean value of 12 dB resulting in attenuation coefficients from 0.8 dB to 1.3 dB/m. A Fresnel diffraction theory was used to model the attenuation from a single tree by the authors (Vogel and Hong, 1988). Utility poles, often erected beside roads, cause chirp-like signal fluctuations, due to interference between direct and scattered signals. The authors present a theoretical model for estimating the electric field, treating the pole as a scatterer. Near motorway junctions or crossings of major roads, overpasses cause complete shadowing of signals. Measurements taken in Australia show typical values of shadowing of the order of >15 dB lasting for 0.5 – 1.0 s. occurring for 0.5 – 1% of the time.

Essentially, terrain features captured within the antenna pattern of the vehicle define the propagation characteristics. The effect of the local environment on fading characteristics is best illustrated with the help of measured data. Figure 3.19 illustrates cumulative distribution of U- and L-band fade in a clear environment without obstructions, typical of a desert (Vogel and Goldhirsh, 1990). It can be observed that a fade of 1.5 dB was exceeded for less than 1% of distance.

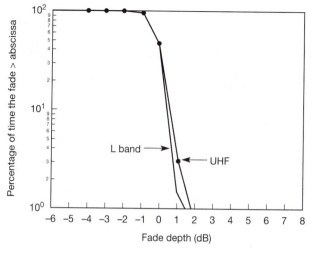

Figure 3.19
Cumulative distribution of U- and L-band fade in a clear environment without obstructions. (Vogel and Goldhirsh, 1990) © 1990 by American Institute of Aeronautics and Astronautics, Inc. Reprinted with permission.

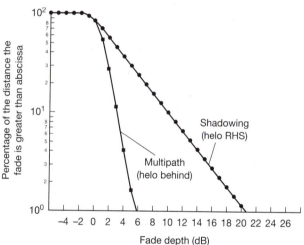

Figure 3.20
A comparison of L-band fading in two environments – one dominated by multipath and the other by shadowing. In one case the transmitting source mounted on a helicopter (helo) was behind and in the other to the right-hand side (RHS) of the data collection vehicle (Vogel and Goldhirsh, 1990) © 1990 by American Institute of Aeronautics and Astronautics, Inc. Reprinted with permission.

Figure 3.20 compares L-band fading of two environments – one dominated by multipath and the other by shadowing. It can be noted that for 1% level multipath fade is 6 dB, whereas for the shadowed environment the fade level is 20 dB. Similar results have been reported by several experimenters.

Propagation characteristics demonstrate a seasonal dependence in environments dominated by trees, due to variations in foliage and flowers throughout the year. Measurements indicate that the dominant attenuation is caused by branches of trees rather than foliage. Figure 3.21 (Butt *et al.*, 1992) reports cumulative fade distribution of the same environment for two seasons – spring 1992, when deciduous trees are without leaves, and summer 1991 when there is 100% blossom and leaves are rich in water content. As expected, the spring distribution gives the lower fade level, but the levels are only 3 and 1 dB worse for 1 and 10% levels respectively.

Figure 3.22 demonstrates that the direction of travel on a dual carriageway and its orientation with respect to the satellite affect the fade level. In each case, the reason is attributed to the path geometry of the environment.

An example of an empirical model for estimating L-band fade at a given probability level in the range 1 – 20% is the empirical roadside shadowing (ERS) model derived from 640 km measurements in various types of road in the USA in a predominantly tree environment with some utility roadside poles.

Figure 3.21
Cumulative fade distribution in the same environment for three seasons (Butt *et al.*, 1992)

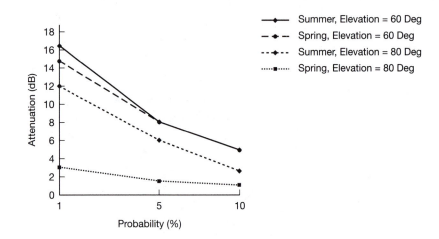

Figure 3.22
Effect of direction of travel in a dual carriageway and its orientation with respect to the satellite (Vogel and Goldhirsh, 1990) © 1990 by American Institute of Aeronautics and Astronautics, Inc. Reprinted with permission.

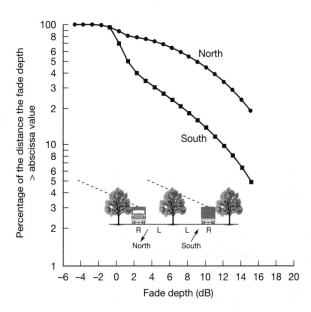

The fade depth F exceeded for P% of distance covered is represented as

$$F = -M \ln(P) + B \text{ dB} \qquad\qquad (3.13)$$

where 20% > P > 1%, M = 3.44 + 0.0975 θ – 0.002 θ^2, B = -0.443 θ + 34.76 and θ = elevation of satellite in degree.

Polarization of an electromagnetic wave changes on reflection, with complete polarization reversal when angle of incidence exceeds Brewster angle. Measurements show that at higher fade levels, XPI reduces to unacceptably low values, due to multipath scattering. Measurements taken in Australia on orthogonal polarization indicated that XPI reduced from 18 dB for 0 dB fade to 0 dB at 12 dB fade level. Further data are necessary to validate such a conclusion, because measurements on each polarization were taken at different times (Vogel and Goldhirsh, 1990). From a system perspective, this would imply that frequency reuse on opposite polarization can be difficult using conventional modulation schemes, except perhaps the spread spectrum scheme, and receiver architecture.

An example of fade duration statistics at fade threshold levels of 2–8 dB in 1 dB steps for a typical US rural road, consisting of trees and utility poles as main obstructions, is given in Figure 3.23. To make the results independent of the vehicle, they have been given in terms of distance travelled in metres.

The statistics followed a log-normal distribution. The non-fade duration for the same case followed an exponential variation (Vogel and Goldhirsh, 1990). The empirical relationship for fading was estimated as

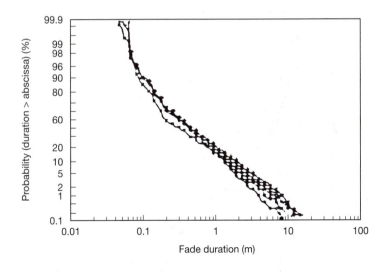

Figure 3.23
Fade duration statistics in the range 2–8 dB for a typical US rural road (Vogel and Goldhirsh, 1990) © 1990 by American Institute of Aeronautics and Astronautics, Inc. Reprinted with permission.

$$P (d_n > x) = bx^{-c} \tag{3.14}$$

where $P(d_n > x)$ = percentage probability of the non-fade distance d_n exceeding x; b and c are constants in the range 12–13 and 0.55–0.84, respectively.

Equation (3.15) defines the diversity improvement possible for the environment. Diversity improvement D_{im} is defined here as the ratio of the single antenna probability to the joint probability at a specified fade margin and antenna separation. A diversity improvement of 2, for example, implies that a single antenna system will fail at twice the number of locations for the chosen threshold. Figure 3.24 demonstrates the diversity improvement achievable when antennas are kept far enough apart.

$$D_{im} (d,F) = 1 + [0.2 \ln (d) + 0.23] F \tag{3.15}$$

where d = the antenna separation (metres) and F = the fade margin (dB).

ITU-R recommendation 681 provides five empirical prediction techniques for land mobile communications (ITU, 1992):

1 **The roadside fade estimation model,** applicable for a frequency of 1.5 GHz, takes path elevation angle and percentage of link availability as inputs to estimate fade depths.
2 **The attenuation frequency scaling model** extends the range of the roadside fade estimation model to the frequency range 0.8 – 2.7 GHz; fade depth

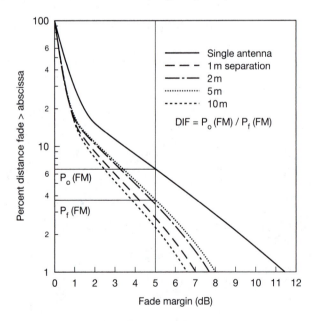

Figure 3.24
Illustation of diversity improvement (Vogel and Goldhirsh, 1990) © 1990 by American Institute of Aeronautics and Astronautics, Inc. Reprinted with permission.

at a given frequency and probability is converted to a fade depth at the same probability at the required frequency.

3   **The fade duration distribution model** provides fade duration statistics at 5 dB threshold for moderate to heavy shadowing; the model does not include elevation angle dependence.

4   **The non-fade duration distribution model** is useful for estimating non-fading statistics at a threshold of 5 dB.

5   **The multipath fading model** provides methods for predicting fading due to multipath in mountainous and tree-lined road environments, assuming a clear line of sight path.

The models are rather restricted in scope, due to lack of sufficient data and lack of sufficiently accurate environment definition, and generally apply to geostationary satellites. The models are refined regularly and hence the reader is encouraged to consider the most recent revision. It is expected that future enhancements to the model will include a more rigorous description of environments; an expanded range of models in terms of frequency, elevation angle, fade levels, percentage and orbits and the inclusion of diversity models.

We will now briefly review important second order statistics which influence signal and mobile terminal design – the level crossing rate and Doppler. The level crossing rate N_l is defined as (see Figure 3.25)

$$N_l = N_t/T_t \qquad (3.16)$$

where N_t = the total number of positive crossings and T_t = the total observation time.

Analytical expressions of level crossing rate have been derived in the literature (Loo, 1985).

Average fade duration T_a is the average duration of a fade at a depth r (Lee, 1986),

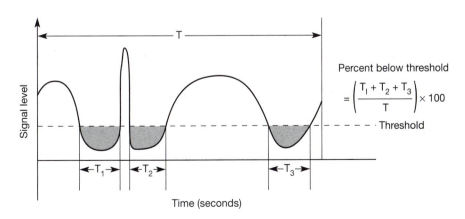

Figure 3.25
Representation of level
crossing rate

$$T_a = F\,(r)/\,N_l \tag{3.17}$$

where F (r) = cumulative distribution of fade for a fade of r and N_l = level crossing rate.

The Doppler spectrum of a mobile receiver using an omni-directional antenna in a two-dimensional space can be given as (Clarke, 1968),

$$S(f) = \cfrac{1}{\pi fm\,\sqrt{1-(\cfrac{f-f_c}{fm})^2}} \tag{3.18}$$

Where fm = maximum Doppler frequency.

A two-dimensional model can represent a terrestrial system, but is not adequate for a satellite system which is better represented by a three-dimensional system.

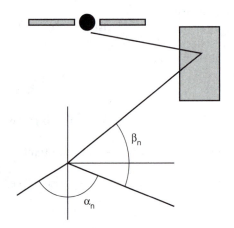

Using the geometry illustrated in Figure 3.26, Aulin derives signal auto-correlation which can then be readily Fourier transformed to give Doppler frequency in a three-dimensional space (Aulin, 1979).

Angles α_n, β_n, magnitude (c_n) and phase (ϕ) of wave components are random and statistically independent; ϕ_n and α_n are uniformly distributed between 0 and 2π; the angle β_n has a probability distribution of $p(\beta)$ which is zero outside $(-\beta_m, \beta_m)$ and $\beta_m \le \pi/2$. Figure 3.27 illustrates the power spectral density for two definitions of $p(\beta)$ and includes the results of a two-dimensional model for the case of β_m = 45 (Kanatas et al., 1995). Definitions of $p(\beta)$ are (Aulin, 1979) and (Parsons, 1992) respectively

$$p(\beta) = \cos \beta/2\sin(\beta),\ |\beta| \le \pi/2 \tag{3.19}$$

= 0 elsewhere.

$$p(\beta) = \pi/4\beta_m \cos ((\pi/2).\beta/\beta_m),\ |\beta| \le \pi/2 \tag{3.20}$$

= 0 elsewhere.

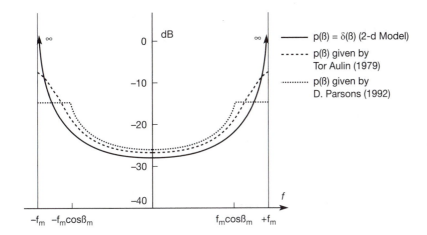

Figure 3.27
Theoretical Doppler
spectrum using a two-
dimensional model and
two cases of three-
dimensional model; f_m =
maximum Doppler
frequency (Kanatas *et
al.*, 1995)

The maximum Doppler frequency of the carrier is given as

$$f_m = u/\lambda \qquad\qquad (3.21)$$

where u = the velocity of the mobile and λ is the wavelength of the RF.

In practice, propagation data cannot be collected for all conditions and are often lacking when developing new concepts including use of untried frequency bands. In such cases, simulation methods involving mathematical modelling offer an economic and viable alternative.

A signal is received at a mobile via a number of paths, as shown in Figure 3.17, so it should be possible to construct the received signal at a mobile by vector summation of all major components. A number of authors have attempted such an approach to land mobile satellite channel modelling (e.g. Dooren *et al.*, 1993, Sforza *et al.*, 1993, Matsudo *et al.*, 1993). As an illustration, the approach adapted by Dooren *et al.*, is elaborated here.

The method, demonstrated for an urban environment, is based on Uniform Theory of Diffraction, which includes effects of non-perfect conductivity of obstacles and their surface roughnesses. The physical environment in the vicinity of mobiles is defined by approximating buildings as block-shaped obstacles and cylindrical shapes as a combination of polygonal cylinders or elliptic cylinders. Data for such a detailed definition of a location can be obtained from digital databases. The field strength at a mobile receiver is calculated by summation of contributions from line of sight waves, reflected waves, edge- or corner-diffracted waves, reflected-diffracted waves, diffracted-reflected waves and double-diffracted waves. Spatial filtering of the receiver antenna is included through either theoretical or measured antenna weight functions in amplitude, phase and polarization. A ray tracing procedure is applied at each observation point and each relevant parameter such as the complex received field strength, type of ray (line of sight, etc.), direction of propagation and the absolute path length are stored, from which system parameters, such as fading statistics, Doppler shift and time-delay spread, can be calculated by applying statistical methods. The propagation is defined as

$$E_o = E_i C \, A(s) e^{-jks} \tag{3.22}$$

where E_o and E_i are the outgoing and incident waves respectively; the coefficient C describes the physical interaction of a wave and an object and depends on the property of the material, direction of travel of incident/outgoing wave, wavelength and shape of the obstacle edges and surfaces; the factor $A(s)$ describes divergence of outgoing wave, s is the distance between the observation point and the point of interaction, and k is the wave number for free space.

The total received field at the output of the antenna is given as

$$E^{obs} = \varepsilon^{los} \, G^{los} \, E^{los} \, e^{los}_{pol} + \Sigma_l + \varepsilon^{refl}_l \, G^{refl}_l \, E^{refl}_l \, e^{refl}_{poi} + \\ \text{other types of rays} + \Sigma_m \, \varepsilon^{dd}_m \, G^{dd}_m \, E^{dd}_m \, e^{dd}_{pol} \tag{3.23}$$

where ε accounts for blockage by obstacles, the superscripts los, refl, dd respectively denote line of sight, reflected, double-diffracted components of electric field E or antenna voltage gain pattern G or polarization e_{pol}, as applicable; G applies to the angle between the applicable ray and antenna boresight and e_{pol} accounts for antenna polarization discrimination.

A simulation package developed on a similar principle, applicable in the range of 1–60 GHz, has been developed by (Sforza *et al.*, 1993). The user can emulate a full description of a communication channel for any environment through a user-friendly interface. The core of the package comprises the ray tracer and geometric theory of diffraction (GTD) solver with supporting functions performed by an Urban Area Modeler, an electromagnetic mesher and a post-processor unit, as illustrated in Figure 3.28.

The inputs include satellite position, urban area description, vehicle speed and path/number of electromagnetic interactions for inclusion; up to second-order electromagnetic interactions can be included. In addition to the GEO, a capability to simulate LEO, MEO and HEO has been included. The GTD solver can include the effects of leaf absorption through either user-entered weight functions or empirical relations. The urban layout can be characterized by features such as buildings, tunnels, overpasses, street lamps, phone booths, parked cars, etc. by using boxes, cylinders, spheres, etc. Trees are characterized by spherical leafage and a cylindrical trunk. The layout of cities can be introduced as digital

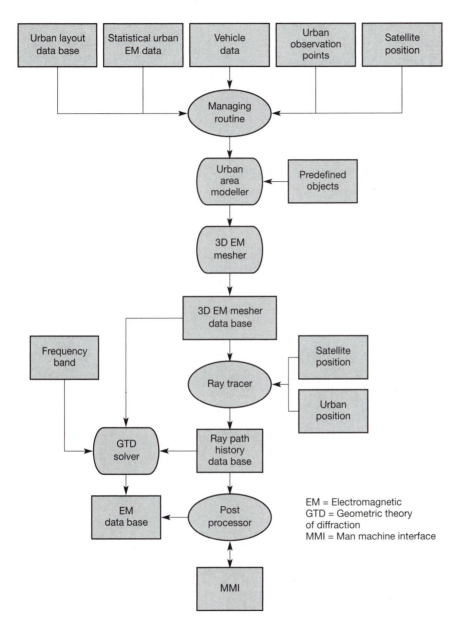

Figure 3.28
The main constituents of
a simulation package for
predicting the
propagation
characteristics of an
urban LMSS channel
(Sforza *et al.*, 1993)

terrain database through input files. The electromagnetic mesher interfaces the
urban layout modeller with the electromagnetic solver by verifying and reformatting input data, and introducing each object's electromagnetic property. The
post-processor unit extracts a range of parameters such as the time series of the
received field and antenna pattern weightings, and performs narrowband or

wideband statistics. Narrowband analysis includes probability and cumulative distribution functions, average fade duration, level crossing rates, fade/connection distributions and time shares of fades/connections. Wideband analysis includes Doppler and power delay profiles, delay spread and coherence bandwidth. Figure 3.29(a) shows a typical ray tracer output and Figure 3.29(b) illustrates a time series of the test case. A simulation package offering such features is useful to an LMS system designer for characterization of a specific location, without the need to perform expensive and time-consuming field measurements.

Vogel and Hong (1988) suggest a model based on simple physical and geometrical description of an environment developed to predict effects of elevation angle, distance of scatterer and bandwidth of a land mobile channel. Under simplifying assumptions of a single isotropic scatterer and a receiving antenna gain

Figure 3.29
(a) Ray model of a test case (Sforza *et al*, 1993). (b) Time series of signal strength for the test case (Sforza *et al.*, 1993)

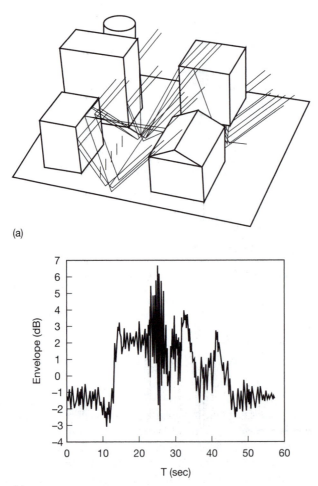

(a)

(b)

independent of azimuth, the model was used to duplicate measured data and predict the effects of elevation angle, scatter distance and bandwidth. Furthermore, the work demonstrated that modelling attenuation, caused by roadside trees as Fresnel diffraction, gave results of correct order of magnitude, despite violations of such an assumption.

Wideband

All components of narrowband transmissions undergo simultaneous variations in level with time and position. However, broadband signals suffer environment-dependent frequency selective fading which causes impairments to the received signal quality. The frequency dependence of fading is measured as *correlation* bandwidth, sometimes called *coherence* bandwidth, defined as the frequency

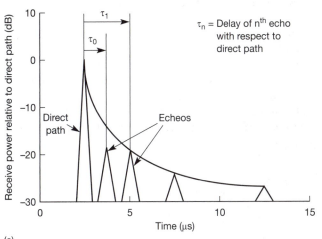

(a)

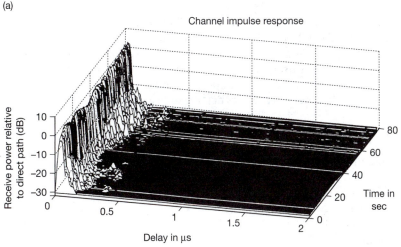

(b)

Figure 3.30

(a) Components for characterizing a wideband channel. (b) A representative power delay profile on a satellite channel in a suburban environment; 25 deg elevation; handheld receiver (Jahn et al., 1995–2)

separation, where correlation in frequency response is more than an arbitrary fraction K for which values of 0.5 or 0.9 have been used. A positive aspect of frequency selective fading is that frequency diversity can be designed into a system, as discussed later (see section 4.2.1, as an example).

Figure 3.30(a) shows a two-dimensional plot illustrating various components useful in characterization of a wideband channel, while Figure 3.30(b) presents a snapshot of power delay profile in a rural environment.

Typical measured values of delay spread, using an aircraft as the source, were observed to be less than 600 ns (coherence bandwidth = 0.265 MHz) on a mobile travelling on a European highway, and 100–500 ns (coherence bandwidth: 1.59–0.32 MHz) on hand-held receivers, neglecting echoes of less than –25 dB (Lutz, 1998). Measured delay spreads in two Japanese cities, Shinjuku Tokyo and Sapporo, were respectively 0.04s–0.18µs (coherence bandwidth = 3.97–0.88 MHz) and 0.05–0.19µs (coherence bandwidth = 3.18–0.837 MHz) (Ikegami *et al.*, 1993).

Considerable experimental work has been done in wideband characterization of terrestrial land mobile systems which indicates that the channel is approximated as quasi-wide sense stationary uncorrelated scattering (QWSSUS). The received signal of such channels consists of echoes received from independent point scatterers. As multipath mechanisms on slant path are identical, the same model can be applied to satellite channels. The complex-valued impulse response of QWSSUS channels can be represented as a tapped delay line with a complex valued impulse response

$$\sum_{k=0}^{k=\infty} a_k \delta(t - t_k) e^{j\theta k} \qquad (3.24)$$

The propagation medium can be characterized by a set of path amplitude (a_k), path arrival time (t_k) and phase (θ_k), the direct path being represented by amplitude a_0, delay t_0 and phase (θ_0). The components can be measured with a wideband probe which uses the correlation properties of a pseudo-random sequence (see section 4.2.2). The impulse response of the channel can be obtained from such a measured power delay profile as follows

$$p(t,\tau) = |h(t, \tau)|^2 \qquad (3.25)$$

Best-fit equations to measured data provide environment-specific characterization for system studies.

For an exponential power delay profile the coherence bandwidth is given as (correlation ≥ 0.5)

$$B_c = \frac{1}{2}(\pi)(\delta) \qquad (3.26)$$

The net effect of the multipath dispersion is an irreducible lower limit on the bit error rate (BER) of a digital transmission and a limit in minimum distortion in an FM system.

Elevation angle dependence

Shadowing of radio waves follows similar elevation angle dependence as sun rays, i.e. degree of shadowing reduces as elevation angle is increased. Degradation of signal quality at low elevation angle is of particular significance to high latitude regions where geostationary satellites appear at low elevation. This difficulty prompted investigations into other types of orbit which would provide higher elevation angle coverage at mid and high latitudes. In the European region, a highly elliptical orbit was considered to be a viable candidate.

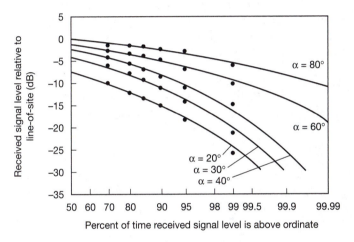

Figure 3.31
Comparison of measured (circles) and estimated fade (continuous line), as a function of elevation angle, α, in a European tree-shadowed environment (Corazza and Vatalaro, 1994) © 1994 IEEE

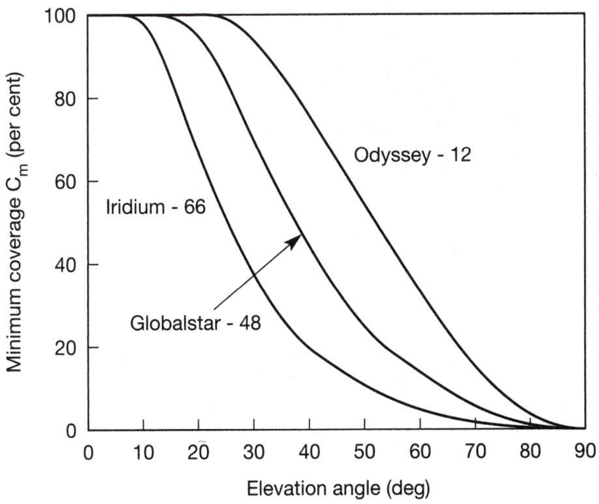

Figure 3.32
Percentage of worldwide coverage as a function of elevation angle for three constellations – Iridium, Globalstar and Odyssey (Corazza and Vatalaro, 1994) © 1994 IEEE

More recently there has been keen interest in characterization propagation channels operating with low and medium earth mobile satellite systems. ITU-R recommendation 681 provides a prediction technique showing elevation angle dependency of fade (ITU, 1992; see previous section).

In such systems, elevation angle changes continuously, even for a stationary observer, necessitating elevation angle dependence in channel characterization.

Corazza and Vatalaro (1994) propose a statistical model based on combination of Rice and log-normal probability distributions applicable to LEO and MEO in all types of environment. Shadowing is assumed to affect both direct and diffused components and its applicability is extended to all types of environment by altering model parameters obtained by empirical fitting to measured data over a wide range of elevation angles and environment. The model parameters applied to curve fitting are mean, standard deviation and Ricean factor of received signal probability distribution. Model parameters were derived by applying curve fitting to ESA's L-band database for a rural tree-shadowed environment, which provided empirical fits in the elevation angle range 20–80°. Measured and estimated cumulative distribution functions within the elevation angle range for the environment compared in Figure 3.31 demonstrate a close correlation.

The applicability of the model in system evaluation is demonstrated in Figures 3.32 and 3.33 which respectively report percentage coverage at various elevation angles for Iridium, Globalstar and Odyssey (which is similar to an ICO system) constellations and BER for Differential BPSK at various elevation angles.

Sforza *et al.* (1993) elaborated the empirical roadside shadowing (ERS) model or ERM (Goldhirsh and Vogel, 1992) using ESA's L- and S-band land mobile propagation database. The range of elevation angles extended from 20° to 80° and the percentage of optical shadowing ranged from 35 to 85% with roadside tree varieties being similar to the ERS model. The approach was to derive empirical expressions of fading probability by two different forms of curve fitting as follows:

Figure 3.33

Bit error probability as a function of energy per bit by noise power density (Eb/No) at various elevation angles, α (Corazza and Vatalaro, 1994) © 1994 IEEE

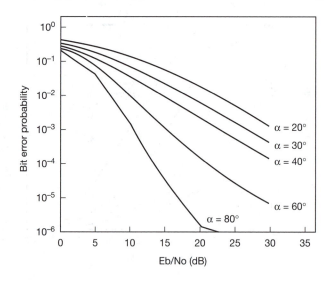

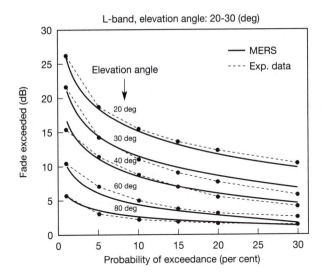

Figure 3.34
A comparison of the
MERS model with
experimental data
(Sforza *et al.*, 1993)

$$F(Pr, \theta) = -A(\theta) \ln(Pr) + B(\theta) \tag{3.27}$$
$$F(Pr, \theta) = \alpha (Pr) \theta^2 + \beta (Pr) \theta + \gamma(Pr) \tag{3.28}$$

where Pr = the % of distance where fade is exceeded, θ = the elevation angle and
$A(\theta)$ and $B(\theta)$ are elevation angle-dependent parameters and coefficients α, β
and γ depend on Pr.

Parametric curves of the modified ERS (MERS) model are plotted with experimental data in Figure 3.34; the results show an rms error lying within 0.5 dB.
The model was used for studying the diversity advantage of a multisatellite constellation with the conclusion that a MEO system can offer an improved link
reliability over a geostationary system.

Jahn *et al.* (1994) present results of a propagation measurement campaign conducted on behalf of ESA for narrowband and wideband characterization of LEO,
MEO and HEO constellations for hand-held and car-mounted terminals in a variety
of environments, and elevation angles in 0–80° range at 1.82 GHz, using a light aircraft. For the narrowband experiment, an unmodulated signal was transmitted,
while wideband measurements used 30 MHz wide pseudo-random transmissions.

The conclusions for the narrowband measurements may be summarized
as follows:

- For elevation angles between 10 and 30° shadowing is dominant, ranging
  20–30 dB.
- Car roof-mounted antennas provide better channels than hand-held antennas
  while stationary and when driving.
- Hand-held antenna signal performance is degraded because of shadowing
  from head and line-of-sight path, due to its lower height compared to car
  roof-mounted antenna.

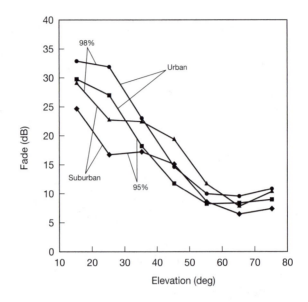

- The cumulative fade distribution of hand-held channels in urban and suburban environments for 95 and 98% are compared in Figure 3.35, demonstrating a need for around 6–10 dB link margins at elevation angles above about 50°, rising to over 15 dB below about 40°, which would make a system unfeasible with current technology without suitable countermeasures such as diversity.

Wide-band results can be summarized as follows:

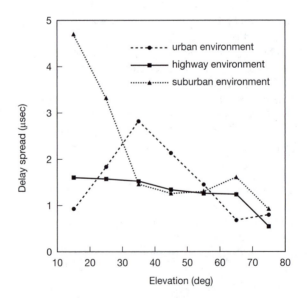

- Echoes appear with delays smaller than 600 ns, equivalent to a distance of the objects causing echo of 200 m, and are attenuated by 10–30 dB.
- Power of echoes with small delays decrease exponentially with delay.
- Figure 3.36 presents a delay spread for the hand-held terminal for urban, suburban and highway versus elevation angle. Delay spread decreases as elevation angle is increased. Generally, the delay spread lies in the range 0.5–2 μs.

K_a band propagation

Due to a need for large amounts of spectrum for next generation personal communication, there has been significant interest in utilizing the K_a band in the service link as the band offers large bandwidth and additionally offers the advantage of a lower sized directive antenna. Signals in this band undergo considerable degradation in the troposphere and are affected more adversely by the local environment than the L band, because diffraction advantage is less and penetration loss is more severe. This would, for example, result in notable shadowing loss from small branches and leaves which pose only minor problems in the L band.

Frequency	Probability (%)	Suburban (dB)		Wooded (dB)		Open (dB)	
		60°	80°	60°	80°	70°	80°
L band	1	16.5	12.0	18.5	8.0	2.8	2.0
	5	8.0	6.0	11.0	4.5	2.2	1.2
	10	5.0	2.5	7.5	3.0	1.8	0.9
S band	1	18.5	16.0	22.5	–	3.5	2.0
	5	9.0	8.5	13.0	–	2.6	1.6
	10	6.0	6.0	8.8	–	2.0	1.4
K_u band	1	27.5	26	28	24	4.0	2.5
	5	19.5	18.5	22.5	15.5	2.8	1.7
	10	13.0	13.0	18.5	10.5	2.3	1.4

Table 3.6
Fade variation as a function of frequency in various environments of southern England. [Polarization: circular; Frequencies: L band = 1,297.8 MHz, S band = 2,450 MHz in Summer and 2,320 MHz in Spring, K_u band = 10.368 MHz; Antenna beamwidth: L and S bands = 64° and K_u band = 80°]

At present, the K_a band propagation database for MSS is limited because the interest in this band for MSS is quite recent. It is currently believed that generally the statistical model applicable to the L band also applies to the K_a band. Therefore, empirical models have been developed by scaling data from L band to K_a band and measurement campaigns have been conducted for similar purposes. Table 3.6 (Butt *et al.*, 1992) lists examples of multiband (L, S and K_u band) fade data measured in southern England for 60°, 70° (for open environment) and 80° elevation simulated satellite path illustrating trends in attenuation with an increase in frequency and variation in the environment.

Consider a representative K_a band measurement campaign in Europe, using 18.7 GHz propagation transmissions of Italsat F1 (Murr *et al.*, 1995). Measurements were conducted in a number of European environments, open rural, tree-shadowed, suburban, urban and mixed constituting four types of environment – and in the elevation angle range 30–35°, maintaining an orientation of 0, 45 and 90° with respect to the satellite. The test gear comprised a vehicle-mounted high gain 2.4° beamwidth antenna system employing a dual-gyro stabilization system, Doppler correction, and a fast PLL which provided an overall measurement accuracy of better than 1 dB. Environments chosen included the Netherlands (open, suburban), France (tree-shadowed, open and mixed), Germany (urban, suburban and tree-shadowed) and Austria (mixed, suburban and tree-shadowed). To minimize ambiguity from tropospheric effects, measurements were conducted under clear sky conditions or under similar weather conditions. Statistical analyses included cumulative distribution function, probability distribution function, level crossing rate, cumulative connection time and time share of fade. As expected, statistics showed dependence on environment, orientation of mobile with respect to satellite, type of tree leaves, height and diameter of trees. Table 3.7 gives a summary of statistics for 80 , 90 and 95% probability levels.

Table 3.7
A summary of K_a band statistics in a European environment

Environment	Location	Cumulative probability, %	Power level with respect to line of sight		
			Azimuth 0°	Azimuth 45°	Azimuth 90°
Suburban					
	Netherlands	95	–3.3	–10	–24.7
	Germany		–1	—	–24.7
	Austria		—	–24.7	—
	Netherlands	90	–1.7	–5.7	–21
	Germany		-0.6	—	–19.3
	Austria		—	–15	—
	Netherlands	80	–1	–2	–14.6
	Germany		–0.3	—	–7
	Austria		—	-0.5	—
Tree shadowed		95			
	France		–12	–20	–20
	Germany		–1.7	—	–32.3
	France	90	–10	–17	–17
	Germany		–0.6	–34	–28.7
	France	80	–7.3	–13.9	–13.3
	Germany		–0.3	–29	–24.7
Urban					
	Germany	95	–1.3	—	—
	Germany	90	–1	–34.8	–33.3
	Germany	80	–0.3	-32	–29.3

Propagation measurement techniques

(a) Narrowband measurement

The main components of any mobile propagation measurement experiment are a transmission source, a mobile receiver, a representative environment and data analysis. A satellite source is preferred when experimental conditions are satisfied. A number of experiments have been conducted using L-band transmission from various geostationary satellites around the world. However, when measurements are required for typical conditions, such as characterization of elevation dependence, geostationary satellites are not the best choice, as elevation angle variations are very gradual. Platforms such as aircraft, helicopters and hot-air balloons have been considered as alternative sources. Similarly, there are not many wideband sources with the required signal characteristics for wideband measurements. With the launch of LEO and MEO satellites, the number of potential sources has now considerably increased for more accurate characterization of such channels.

Propagation receivers use standard coherent detection with a wide dynamic range such as used for propagation measurements of FSSs. The hardware is ruggedized to accommodate mechanical vibrations during vehicular motion. Antennas are chosen to be as close to the real system as is possible, due to the sensitivity of fading characteristics to antenna patterns. Data is gathered in a digital or analogue recorder and analyzed by statistical data processing software to give mean, standard deviation, cumulative distribution of fade, best-fit probability distribution or elevation dependence, etc. Care has to be exercised in setting a stable reference signal due to signal drift with time, and further to recognize the reference during data processing. One technique is to obtain a stable reference signal at the start of each run.

(b) Wideband measurement

Wideband measurement methods for satellite channels have been derived from terrestrial measurement systems. One of the most reliable transmission methods is the use of a PN sequence transmissions and correlation receiver. Each component, including the direct path, appears as a correlation peak at the receiver output. Taking the direct path as a reference, a signal path delay profile may appear as shown in Figure 3.30 (see above). The spatial resolution of the multipath depends on the chip rate of the pseudo-random sequence as (chip rate/velocity of light); for example, a 30 Mb/s chip rate gives a spatial resolution of 10 m.

3.3.3 Maritime channel

Satellite services for maritime applications were introduced first due to a genuine need for reliable communication on the high seas, aided by the relatively lower technology demand as maritime propagation conditions are quite benign and large terminals can be used aboard ships. The consolidated conclusions from investigations on maritime channels are that signal fades depend on elevation angle, sea conditions including wave height, slope and wind, receiver antenna characteristics (beamwidth, side lobe, axial ratio), height of antenna above the sea and the structure of the ship. Other influencing factors include antenna pointing accuracy and polarization mismatch.

The signal at the receiver comprises a summation of direct, specular and diffused components (see equation (3.12)). The magnitude of the specular component reduces with an increase in wave height and as satellite elevation angle and transmission frequency increase; diffuse components dominate under rough conditions; fade depth depends on amplitude and phase difference between the direct and indirect waves. The amplitude of the diffused component follows a Rayleigh distribution with a uniform phase distribution. The received signal amplitude has a Nakagami–Rice distribution (ITU-R report 1007).

To compare fading statistics in a consistent manner, the condition of the sea is characterized by wave height quantified by 'a sea-state number', as defined in Table 3.8. Sea-state numbers 0 and 1 represent calm seas with wave heights of < 0.15 m, sea-state number 4 represents a moderate sea with a wave height of 2–4 m and sea-state number 8 corresponds to an extremely high-sea condition with waves > 9.5 m. Statistics of the sea states are available for the whole world (Long, 1975); averages over all regions of the world are tabulated in Table 3.9.

Table 3.8
Relationship between sea-state number and significant wave height (ITU-R report 884-2, Hogben and Lumb, 1967)

% Occurrence	Height of waves	Sea condition
22	0–0.9	Calm
23	0.9–1.2	Smooth
20.5	1.2–2.1	Slight
15.5	2.1–3.6	Moderate
9.5	3.6–6	Rough
9.0	> 6	Very rough – high

Table 3.9
Average sea state for the world

Sea state number	State of sea	Significant wave height (m)
0 and 1	calm	<0.15
2	smooth	0.15-0.8
3	slight	0.8–2.0
4	moderate	2.0–4.0
5 and 6	rough	4.0–6.5
7	very rough	6.5–9.5
8	high	>9.5

The diffused component of the signal depends on the slope distribution of the sea-surface facets, the effects of which are RF dependent. Sea slopes range between 0.04 and 0.07 for rough sea conditions.

When specular reflection coefficient of the sea, rms height of the sea waves, RF, elevation angle and antenna gain pattern are known, it is theoretically possible to estimate the magnitude of specular components. Similarly, with the knowledge of average scattering cross-section per unit area of sea surface, scattering angles, receiver antenna pattern and height of antenna above sea level, the magnitude of diffused components can be estimated as demonstrated in Figures 3.37 and 3.38 (ITU-R 884-2). Figure 3.37 illustrates calculated values of direct to specular components of a 1.5 GHz circularly polarized wave received through an omni-directional antenna at elevation angles of 5, 10 and 15° as a function of wave height. Note the reduction in magnitude of the specular component as elevation angle is reduced and wave height increased. Figure 3.38 illustrates theoretical values of carrier to multipath ratio versus surface rms slope for a circularly polarized wave at a frequency of 1.5 GHz received through a 12 dB gain antenna at elevation angles of 5, 10 and 15°, assuming worst-case surface height. Observe a reduction in multipath noise as rms slope is reduced or elevation angle increased. The cumulative fade distribution can be determined for a given carrier to multipath ratio, or when statistics of coherent and diffused components are known (Beckmann and Spizzichino, 1963).

For practical purposes, the combined fading effects for a given frequency, antenna type and probability % may be approximated as

$$F_t = F_s(max) + F_d(\%) \tag{3.29}$$

where the $F_s(max)$ is the maximum fade depth due to specular reflection and $F_d(\%)$ is the diffused component of fade at the desired probability %.

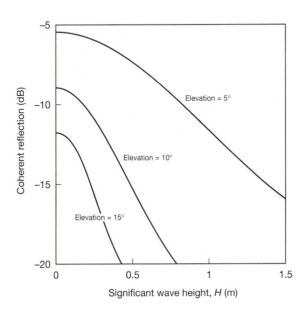

Figure 3.37
Magnitude of coherent reflection relative to direct signal as a function of significant wave height and elevation angle; frequency = 1.5 GHz, polarization: circular; antenna: omni-directional (ITU-R report 884-2) © ITU

Figure 3.38
Worst-case carrier to
multipath ratio as a
function of rms slope
and elevation angle;
frequency = 1.5 GHz;
circular polarization;
antenna gain = 12 dB
(ITU-R report 884-2)
© ITU

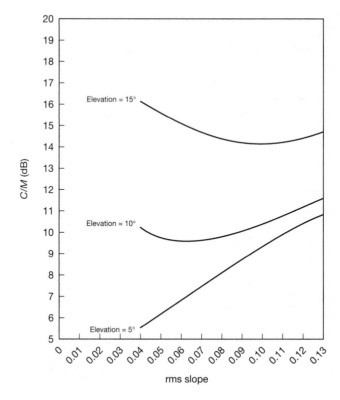

Figure 3.39 shows the calculated 99% fade depth for a 14 dB antenna at 1.5 GHz, as a function of wave height for both components. Fade caused by specular components Fs reduces with an increase in wave height until the magnitude becomes negligible beyond a wave height of 2 m, while fade caused by diffused

Figure 3.39
Calculated 99% fade
depth for a 14 dB
antenna at 1.5 GHz, as
a function of wave
height (ITU-R report
763-3) © ITU

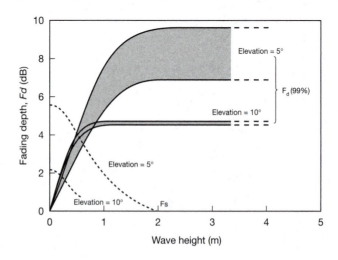

components F_d increases linearly until saturation occurs. The shaded portions indicate the range of slope variations at each wave height. Figure 3.40 depicts fade depth as a function of elevation angle for antenna gain of 8, 14 and 24 dB for wave heights ranging from 1.6 to 3.2 m.

Table 3.10 lists some measured results. A good general agreement is observed between theory and measurements where theoretical predictions were attempted.

Theoretical methods require solutions to complex mathematics and access to special databases so, where feasible, practising engineers prefer to use simple empirical methods. Figure 3.41 shows multipath power as a function of elevation angle for 1–3 m wave height corresponding to slight to moderate sea conditions, applicable in the 1–2 GHz frequency band calculated by using an empirical model (Karasawa and Shiokawa, 1988).

Most of the propagation work has focused on L-band systems, as this is the most used frequency band. Due to the imminent introduction of S- and K_a-band PCSs, research is in progress to extend L-band propagation models to higher frequency bands. Computed results from one model developed for the purpose indicate that the maximum fade level is unaffected by RF; however, for a specific sea condition, fade level reduces with an increase in frequency in the wave height range 1–3 m.

Circular polarization is widely used in mobile satellite systems. When such a wave is reflected at an elevation angle greater than the Brewster angle the polarization sense is reversed. At other angles the reflected wave become elliptically polarized. Theoretical studies indicate that circular polarization is superior

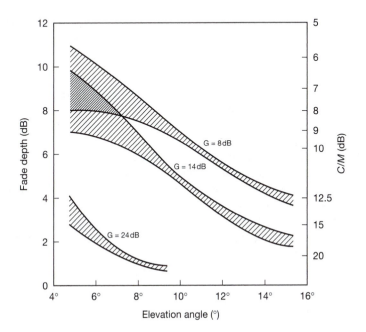

Figure 3.40

Fade depth as a function of elevation angle for various antenna gains (G) for 99% of time (ITU-R report 884-2) © ITU

Table 3.10
Measured fade depth
at the L band from a
number of campaigns,
as a function of
antenna gain and
elevation angle from
geostationary satellites
(ITU-R Reports 763
and 884)

Antenna gain (dB)	Elevation (degree)	Wave height (m) or (sea condition)	Fade depth (dB) not exceeded (99%, except where mentioned)	Reference
0	6	0.5	6.9-12.8	Higuchi and Shinohara (1988)
3	4	–	12	ITU-R report 763-3,
	19	–	10.5	Hagenaur et al., (1984)
5	4	–	14	ITU-R report 763-3,
	19	–	8	Hagenaur et al., (1984)
11	4	–	11	ITU-R report 763-3,
	19	–	6.5	Hagenaur et al., (1984)
15	5	1–4	7–10	ITU-R report 763-3, Karasawa et al., (1986)
20	5	3	4.8	ITU-R Report 763-2
24	5	1.6–3.2	1–5-3	Ohmori et al., 1985
	10	1.6–3.2	1	ITU-R Report 763
25	2	0–4	2 (95% time)	ITU-R Report
	8	(0–4)	1 (95% time)	884-2; ESA, 1977

to horizontal polarization in mitigating multipath noise pick-up. However, at elevation angles up to ~ 11°, vertical polarization can suppress multipath better.

The fading spectrum of multipath can similarly be determined using theoretical methods. Studies indicate that spectral bandwidth increases when any one of the following parameters is increased – wave height, velocity of a ship, ship roll and pitch movement and elevation angle. Figure 3.42 depicts the theoretical value of a –10 dB power spectral bandwidth of amplitude fluctuations of 1.5 GHz, as a function of elevation angle under typical maritime conditions – wave height of 1–5 m, ship speed 0–20 knots and roll of 0–30°.

Theoretical calculations based on a model proposed in the literature predict average values of fade duration and fade occurrence interval in elevation angle range of 5–10° as 0.05–0.4s and 5–40 s, respectively (Karasawa and Shiokawa, 1988). The probability distribution of these two parameters approximates an exponential distribution.

A further consideration is the effect of the ship's structure on propagation. Multipath fading occurs through reflections off ship structure, and shadowing

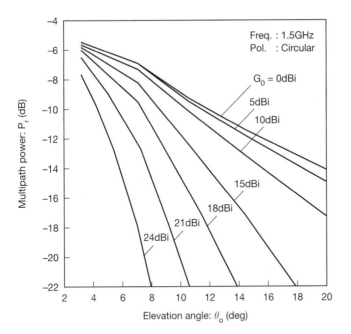

Figure 3.41
Multipath power as a
function of elevation
angle and antenna gain
for circular polarized
wave at 1.5 GHz under
slight to moderate sea
conditions (wave height
1–3 m) (ITU-R report
884-2) © ITU

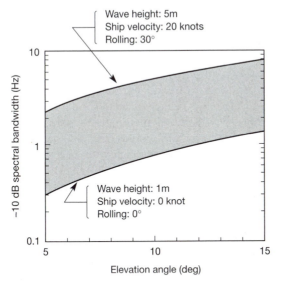

Figure 3.42
Elevation angle versus
−10 dB spectral
bandwidth of 1.5 GHz
multipath fading (ITU-R
report 884-2) © ITU

occurs when any part of a ship's structure blocks the signal. Both types of fading are
affected by the shape of the ship, antenna location, antenna pattern, elevation and
azimuth of satellite, axial ratio of ship and satellite antennas. The magnitude
of fade is best dealt with on a case-by-case basis using theoretical methods such as
discussed in the land mobile propagation section or through empirical models.
Simulation performed on a large ship model likely to give strong reflections showed

Figure 3.43
L-band blocking
attenuation due to a
column-type structure of
various diameters, D_o,
from an antenna of 20
dB gain (ITU-R
report 763-2) © ITU

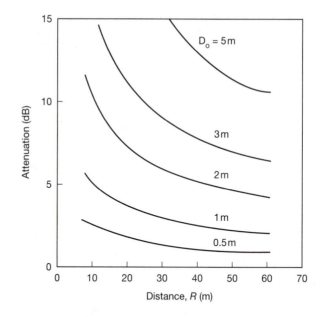

a fading loss of the order of 1–2 dB. Loss due to blocking depends on the distance of the blocking structure and its dimensions. As an example, Figure 3.43 shows attenuation due to a column structure of various diameters (D_o) spaced at various distances from an antenna of 20 dB gain (ITU-R report 763-2).

3.3.4 Aeronautical channel

Propagation effects of aeronautical channels differ from maritime and land mobile propagation because of the high velocity of aeroplanes, their distance from ground and influence of the aircraft body on antenna performance. Aeroplane manoeuvres can affect signal reception under conditions when antennas are shadowed. At present, most of the data exists for geostationary satellite systems, but several non-geostationary satellite systems intend to offer aeronautical services.

The troposphere lies in the first few tens of kilometres of the Earth–space path and its effect diminishes with altitude, so tropospheric degradations are altitude-dependent. To demonstrate this characteristic, Figure 3.44 (a) and (b) illustrates the attenuation due to a number of tropospheric causes at the Earth's surface and altitudes of 1 and 3 km for frequencies ranging from 1.5 GHz to 40 GHz and an elevation of 10°. It is evident that high-altitude flights are considerably less affected by attenuation. For example, in a 20 GHz link, the attenuation due to rain will reduce from 16.3 dB at the surface to a negligible amount at 4 km. Similarly, cloud attenuation will reduce from 3.48 dB at the Earth's surface to 0 dB at an altitude of 3 km.

Equation (3.12) also applies to aeronautical channels. The magnitude and characteristics of specular and diffused components depend on type and characteristics

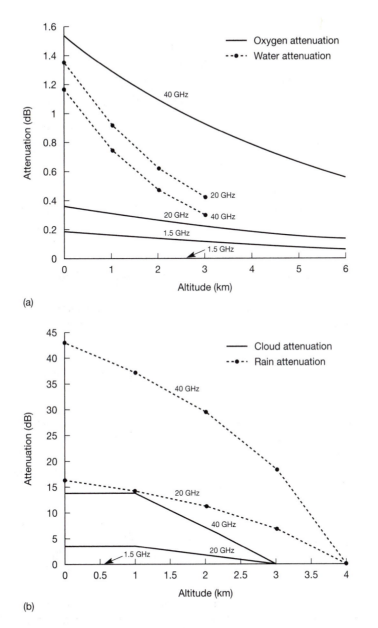

(a)

(b)

Figure 3.44
(a) Variation of attenuation due to Oxygen and water vapour (7.5 g/m3) as a function of altitude in 1.5–40 GHz range at an elevation angle of 10° (Data source: ITU-R report 1148) (Graphics: AR). (b) Variation in attenuation due to cloud (1g/m3) and rain (0.1% time, rain climate Zone K, as per ITU-R report 564) as a function of altitude in the frequency range 1.5–40 GHz, at an elevation angle of 10° (Data source: ITU-R report 1148) (Graphics: AR)

of the reflecting terrain, i.e. land forests, land urban, maritime under calm sea conditions, maritime with rough sea, etc. It has been observed that magnitude of multipath caused by the sea surface is considerably more than from the land surface and hence considerable research has been directed towards characterization of aeronautical channels over sea. Measurements conducted to date demonstrate that:

- composite received signals follow a Ricean distribution;
- multipath components exhibit a Rayleigh distribution;
- the magnitude of multipath components are elevation angle-dependent;
- multipath Doppler spectrum possesses a Gaussian distribution;
- fading is frequency selective, due to path delay associated with multipath components;
- the multipath signal is dominated by diffused components, i.e. specular components are likely to be negligible the majority of the time.

When the direct signal received at time t is $s_d(t)$, the reflected component can be represented as $s_r(t-\tau)$ where τ is the time delay of the reflected component whose magnitude depends on the geometry of the path. The magnitude of the reflected component depends on the reflection coefficient and geometry of the Earth below an aircraft. The reflection coefficient depends on the characteristics of the Earth's surface, i.e. whether it is sea or land; if the surface is sea, then the state of the sea, whether it is calm, rough, etc., and when land, whether the terrain comprises forest, built-up areas, desert, etc. Signal measurements on aircraft show that scattering from land is significantly lower than from the sea surface; furthermore, signals have a notable elevation angle dependence over sea but not over land; as already mentioned, most studies have therefore concentrated on propagation behaviour in overseas flights.

Theoretical multipath estimation techniques applicable to the maritime environment can also be applied to the aeronautical channels taking Earth's curvature effects into account. Carrier to multipath ratio is estimated using Fresnel's reflection coefficients for specular reflections and divergence factors (Beckmann and Spizzichino, 1963). Figure 3.45 (ITU-R report 1148) depicts estimated 99% fade levels received through an antenna of gain 7 dBi as a function of its altitude at 1.54 GHz for wave heights 1.5–3 m and elevation angle of 5°; the lower end represents the maritime environment and the upper end extending up to 10 km is an aeronautical environment. Carrier to multipath ratio for 0 and 3 dBi antenna for 1.5 GHz MARECS transmissions shows a good agreement with theory (see Figure 3.46). Interestingly, it is observed that fade is only about 2 dB lower at an altitude of 10 km than for a maritime environment. Some measurements are without multipath; this type of anomalous condition can arise when multipath signals are shielded by an aircraft's body.

Measured data illustrates that the power spectral density of multipath depends on aircraft speed, elevation angle and ascent/descent angle of the flight. Measurements on a jet aircraft travelling at a ground speed of 650 km/h broadside to the satellite and at a frequency of 1.6 GHz resulted in a multipath spectrum bandwidth in the range 40–80 Hz, at an elevation angle of 9° and 200 Hz at 31° (Sutton *et al.*, 1973). For level flight, –10 dB spectral bandwidth range was between 25 and 50 Hz at a 5 – 10° elevation and for ascending/descending flight trajectory the spectrum became skewed [ITU-R report 1148].

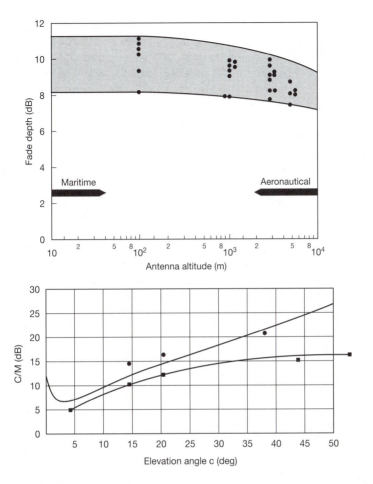

Figure 3.45
99% fade depth at
various altitudes for 1.5
GHz circularly polarized
waves received by a 7
dBi antenna (ITU-R report
1148) © ITU

Figure 3.46
Measured and calculated
carrier to multipath
ratio as a function of
elevation angles for 0
dBi antenna (.) and 3
dBi antenna (■) (ITU-R
report 1169, © ITU)

The aeronautical channel can be represented as wide-sense stationary uncorrelated scattering (WSSUS) channel (see section 3.3.2 – wideband) due to the randomly
changing nature of scatterers over the Earth's surface (Bello, 1973). The model can be
used to estimate a number of characteristics of the channel. The Doppler power spectrum of the reflected signal of such a channel is Gaussian and is given as

$$P(f) = \sqrt{2} / B_r \sqrt{\pi} \exp (-2f^2/B_r^2) \tag{3.30}$$

The rms Doppler spectrum is defined as twice the standard deviation of the
Doppler power spectrum

$$B_r = 4\alpha v \sin \varepsilon / \lambda$$

where α = the rms surface slope, v = the aircraft speed at constant elevation
angle, ε = the elevation angle and λ = the wavelength of the carrier.

The Gaussian shape, as well as values predicted by equation (3.30), agrees well with measurements (Sutton *et al.*, 1973). Figure 3.47 shows the estimated $\frac{1}{e}$ correlation bandwidth as a function of altitude up to 10 km, elevation angles of 5° and 10°, frequency of 1.54 GHz and a 10 dBi antenna (ITU-R report 1148). The correlation bandwidth is defined in this case as the frequency separation where the correlation coefficient has reduced to 1/e. Note a reduction in correlation bandwidth from several tens of MHz near the sea surface to 10–20 kHz at an altitude of 10 km.

Figure 3.48 shows the measured mean and standard deviation of 1.6 GHz fade duration received through a crossed dipole antenna over the Atlantic ocean and in parts of Europe. The aircraft altitude and ground speed were respectively 10 km and 700 km/s (Hagenaur *et al.*, 1987). The results of the campaign show that the mean fade duration at 5 dB fade level is 4 ms at 5° elevation, decreasing to 0.3 ms at 44°; the mean duration of connection level where the signal is above –5 dB threshold is 30 ms at 5°, increasing to 300 ms at 21°.

Table 3.11 lists a sample of results of a campaign using 1.6 GHz transmissions of an ATS-6 satellite. The aircraft at an altitude of 9.1 Km travelled at a ground speed of 740 km/h and used a two-element antenna array of 1 dB beamwidth of 20° in azimuth and 50° in elevation (ITU-R report 1148, Schroeder *et al.*, 1976).

Figure 3.47
Estimated ($\frac{1}{e}$) correlation bandwidth as a function of antenna altitude for an antenna gain of 10 dBi (ITU-R report 1148) © ITU

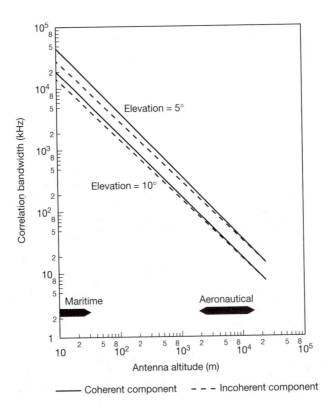

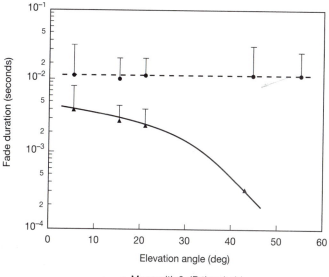

Figure 3.48
Fade duration mean and
standard deviation
statistics at 1.6 GHz at
aircraft altitude of 10 km
(Figure 4 Report 1169,
ITU-R, Hagenaur et al.,
1987); antenna gain =
3.5 dB; ƅ circular
polarization; data
collected over Atlantic
Ocean and Western
Europe © ITU

- Mean with 0 dB threshold
▲ Mean with −5 dB threshold
T Standard deviation added

Parameter	Measured range (bracketed values for land measurements)	Typical values at given elevation angles		
		8°	15°	30°
Two-sided 3 dB delay spread (µs)	0.25–1.8 (0.1–1.2)	0.6	0.8	0.8
3 dB one-sided correlation bandwidth (kHz)	70–380 (150–3,000)	160	200	200
3 dB Doppler spread (Hz)	4–190 (20–140)	5	70	140
3 dB decorrelation time (ms)	1.3 – 10 (1–10)	7.5	3.2	2.2

Table 3.11
A sample of measured
data over the ocean
from ATS-6
transmissions at 1.6
GHz (Schroeder *et al.*,
1976). Land
measurements did not
show any specific
elevation angle
dependency

System implication

Signal shadowing and multipath influence system design in a number of ways:

- signal blockage affects link reliability;
- multipath noise causes errors in digital transmission;
- frequency-selective fading causes intersymbol interference when signal bandwidth exceeds the coherence bandwidth;
- Doppler frequency shift has to be compensated.

The minimum elevation angle and the number of satellites simultaneously visible from user terminals set a bound on the radio link reliability of the system. The minimum elevation angle is directly related to the link margin offered by a system; the number of satellites simultaneously visible from a location and their relative separation provide an estimate of the diversity gain.

While constellations are designed to operate at as high an elevation angle as is economically feasible, in order to minimize propagation impairments, it is impossible to eliminate fading, therefore fade countermeasures are invariably applied. They include feedback power level control, static or adaptive channel coding, robust modulation methods and satellite diversity. In a feedback power control scheme, the transmitter power is adjusted to compensate for fade at a receiver through a feedback loop from the receiver to the transmitter. Channel coding is used in all MSS systems and a wide range of codes are in use. In an adaptive channel coding scheme the code rate is varied according to the state of channel, i.e. the code rate is reduced as the magnitude of fade increases, such that BER is maintained at the expense of signal throughput. An adaptive coding scheme was demonstrated in an ACTS experiment, whereby 10 dB of dynamic fade compensation could be introduced to a link. Chapter 4 discusses details of modulation and coding schemes. Figure 3.49 depicts the main features of power control and adaptive code rate control systems.

Note that for portable MSS terminals designed for use in fixed applications, system design considerations are similar to VSATs. A trade-off is applied between signal quality, system capacity and call cost. Figure 3.50 demonstrates the sensitivity of space segment capacity when operating a terminal with different fade margins for an MSS system with a total capability of 70 dBW spacecraft EIRP and an EIRP requirement of 30 dBW per channel without employing any fade countermeasures.

We have already addressed the applicability of path diversity to K_u and K_a band feeder links; deployment of separate antennas in the service link can achieve only a limited improvement, as shadowing events are strongly correlated; furthermore, the need for two antennas/receivers makes the user terminals more expensive and bulkier. For wideband channels some advantage can be achieved by combining various echoes. At present, a more effective approach is to combine signals from separate satellites.

Figure 3.49
Power level fade control
and dynamic code
control systems

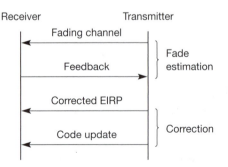

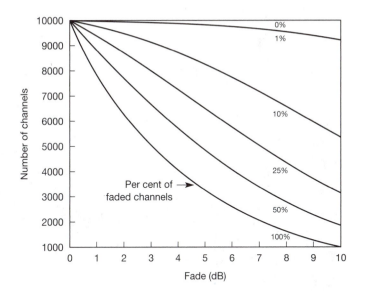

Figure 3.50
Variation in space
segment capacity as a
function of fade

At present, path diversity from separate satellites is rarely used for geostationary mobile satellite systems. Implementation becomes increasingly difficult for applications where receiver antennas are directional, as satellite separation is increased. Note the direct relationship between satellite separation and diversity improvement. Furthermore, deployment of two satellites is expensive; it wastes spacecraft resources as well as geostationary orbital slots.

Non-geostationary satellite systems have an inherent capability to exploit diversity, because the number of satellites is large and path diversity is a natural by-product. Constellations may be designed such that multiple visibility of satellites is a system feature, without incurring prohibitive costs. Therefore, a number of non-geostationary systems such as ICO and Globalstar have incorporated path diversity to improve link reliability. Figure 3.51 is a pictorial view of path diversity (also see Figure 3.17). Chapter 2 (section 2.3.8) also discusses the path diversity advantages, with examples of measurements, and Chapter 9 describes the application of diversity to specific constellations.

For system analysis and design it becomes necessary to characterize all elements of a system. In particular, propagation data was scarce until recently, mainly limited to investigations done for specific systems, and hence diversity advantages would be difficult to characterize. Therefore theoretical methods, involving simulation methods such as those mentioned earlier in the section, offered a cost-effective solution.

Another interesting approach proposed recently is first to characterize environments and derive a diversity advantage by superimposing constellation geometry. A method called photogrammetry proposes to use a combination of image processing of hemispherical fisheye lens photographs, constellation simulation and a propagation model to predict diversity and other propagation

Figure 3.51
A pictorial view of path
diversity

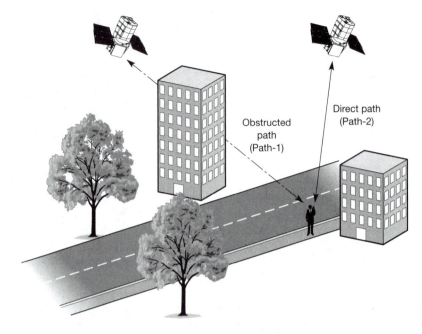

effects at locations of interest, thereby minimizing the need for a measurement campaign (Akturan and Vogel, 1997).

3.4 Radio link analysis

Radio link design is central in establishing essential system characteristics and in developing user terminal and satellite specifications within technological and economic constraints (see Chapter 8). The design is performed effectively with the aid of the transmission equation, which models the satellite link behaviour algebraically. It encapsulates transmitted power, path loss, propagation loss, interference effects, modulation, coding, multiple access, diversity improvements and signal quality requirements, allowing trade-offs to be done simply. The subject is well covered in the literature and therefore we will introduce essential and interesting aspects relevant for MSS. Table 3.12 lists some link equations, relates each parameter to the relevant link parameter and summarizes attributes of each component in a system context. Figures 3.50, 3.52 (a) and (b) demonstrates the utility of these equations applied to an MSS link. Satellite EIRP/channel, throughput per user and system capacity are shown for a variety of user terminals ranging from hand-held terminals to ship-borne terminals for a 70 dBW geostationary satellite and assuming a E_b/N_o of 5 dB to meet the quality objective. Figure 3.52(b) also marks the medium- and long-term capability of MSS.

Link equations	Component	Description	Comments
$R_x = P_s + G_s + G_d - 20 \log (4\pi D/\lambda) - L_l - L_i$ .... (3.31)	R_x	Received power level (dBW)	(1) Fundamental component which determines viability of a radio link (2) Determines signal quality (3) Requirement depends on service, modulation scheme coding type/rate, throughput and terminal sensitivity (4) Sizes, transmitted power and therefore traffic capacity (5) Used for estimating carrier to noise ratio (see equation (3.34)) (6) Unit is dBW For a given signal quality, the desired carrier to noise ratio is governed by modulation and coding scheme including implementation losses, data rate, service characteristics in terms of message delay and type of terminals and accessing scheme
	P_s	Transmitter power (dBW)	High satellite power is required to sustain MSS service links. Trend is to distribute power to a number of solid-state power amplifiers
	G_s	Transmitter antenna gain (dB)	High gain is necessary to sustain MSS service link and increase frequency reuse. Trend is to deploy large numbers of spot beams (maximum numbers reported ~ 300; work is in progress for thousands)
EIRP = $P_s + G_s$ ..(3.32)	EIRP	Effective isotropic radiate power (dBW)	A measure of effective power transmitted
	G_d	Gain of receiver antenna (dB)	
	$20 \log (4\pi D/\lambda)$ .. (3.33)	Free space path loss (dB)	λ = Wavelength; D = distance between transmitter and receiver
	L_l	Loss in the link due to fading, etc. (dB)	Loss can be caused by a number of factors; see text

Table 3.12
Link equations and explanation of each component in a system context

Table 3.12
Continued

Link equations	Component	Description	Comments
	L_i	Loss due to interference (dB)	A measure of extra power required to compensate for interference
$(C/N)_r = Ps + Gs - 20 \log (4\pi D/\lambda) - L_i - L_i + (G_d/T_r) - 10 \log (k) - 10 \log (B) \ldots$ (3.34)	$(C/N)_r$	Carrier to noise ratio	For a given signal quality, the desired carrier to noise ratio is governed by: modulation and coding scheme including implementation losses, data rate, service characteristics in terms of message delay and type of terminals and accessing scheme
	G_d/T_r	Receiver sensitivity (dB/K)	Fundamental in determining throughput and signal quality
	$10 \log (k)$	-228.6 dB/K	Boltzmann constant
	$10 \log (B)$	Bandwidth in dB	
$E_b/N_o = (C/N) + 10 \log B - 10 \log (R_t)$ .. (3.35)		Energy per bit by noise power spectral density; R_t = bit rate	Used in specifying quality of digital systems

Consider the model of a mobile satellite communications system illustrated in Figure 1.5. The radio link comprises a forward/return feeder link, satellite path and service link and intersatellite link(s) when applicable. By far the most important part of an MSS link is the service link, as it establishes fundamental bounds in terms of terminal size, throughput, traffic carrying capacity, etc. Depending on the emphasis, various types of trade-offs can be applied, as summarized below:

1. Forward service link
   - For a limited EIRP satellite, a trade-off can be exercised between the size of mobile terminal, throughput and spectrum; this situation was particularly applicable to first and second generation mobile satellite systems.
   - For a given user terminal size and throughput, a trade-off is possible between satellite power and bandwidth.
   - For a bandwidth limited system, given a terminal size and throughput, a trade-off can be applied between capacity, spot beam size and number, modulation/coding/multiple access scheme (see Chapter 4), etc.
2. Return service link
   - A trade-off can be applied between mobile EIRP, throughput and sensitivity of satellite and feeder stations.

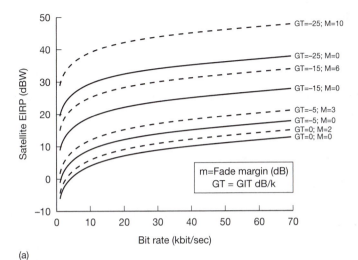

(a)

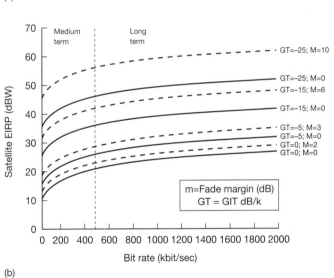

(b)

Figure 3.52
Satellite EIRP/channel for a range of user terminals from hand-held terminals to ship-borne terminals, throughput per user and system capacity for a 70 dBW geostationary satellite; E_b/N_o = 5 dB for bit rates up to (a) 70 kbps, (b) 2 Mbps respectively, showing the medium- and long-term capabilities of MSS (Graphics: AR)

- For a bandwidth limited system, a trade-off is possible between spot beam size, modulation and multiple access scheme.

Examples of practical considerations affecting trade-offs include (see also Chapter 8):

- Spacecraft service link EIRP limited by antenna and high-power amplifier technology.
- Mobile EIRP, which is limited by mobile battery capacity, HPA size/cost and radiation safety limits, the latter applying, in particular, for hand-held services.

- In the return direction, sensitivity of satellite receiver which influences mobile terminal size, EIRP and service capability.

A regenerative transponder de-couples feeder and service paths and therefore minimizes the interaction between them (see Chapter 6 for further details). The overall BER at a receiver for a regenerative transponder can be approximated as

$$BER_t \approx BERu + BER_d \qquad (3.31)$$

where BER_t = the total bit error rate, BER_u = the bit error rate of uplink and BER_d = the bit error rate of downlink.

In a transparent transponder, coupling between links is tighter, resulting in the overall BER becoming multiplicative, and therefore links have to be optimized in their entirety, respecting the limitations of the service links. The total E_b/N_o in this case is given as

$$(E_b/N_T) = [(E_b/N_u)^{-1} + (E_b/N_d)^{-1} + (E_b/N_I)^{-1} + (E_b/N_{iu})^{-1} + (E_b/N_{id})^{-1}]^{-1} \qquad (3.32)$$

where E_b/N_T, E_b/N_u, E_b/N_d, E_b/N_I, E_b/N_{iu}, E_b/N_{id} are energy per bit to noise power spectral density of total link, uplink, downlink, intermodulation, uplink interference and downlink interference respectively.

Note the large number of components which have to be balanced. The most vital component of an MSS is the service link due to its being inherently power limited, severe propagation impairments and susceptibility to inter- and intrasystem interference.

Due to invariant geometric relationships, link parameters such as path loss, look angles, and Doppler effect of a geostationary satellite system are nearly static for practical purposes. The main variables are propagation loss and interference. Note that mobiles move freely within the service area, and therefore receive and cause variable amounts of interference. The traditional method for interference management involves ensuring adequate interference margin at the worst-affected location, i.e. the location where the signal wanted is the weakest and the interfering signal is strongest. When considering intrasystem interference, an operator can tailor frequency reuse, such that dense traffic areas are better protected, or an interference margin is provided to a certain percentage of geographical areas (e.g 95%). This approach is no worse than the provisioning propagation margin based on statistics, the trade-off is instead applied to shadowing loss, multipath and interference. Table 3.13 (a) and (b) shows examples of link budgets for FDMA, TDMA and CDMA schemes for geostationary satellite systems.

	GSM (TDMA)		MSBN (CDMA)	
	Transportable	**Vehicular**	**Transportable**	**Vehicular**
K_u band uplink				
Fixed station EIRP (dBW)	57.8	57.8	57.8	57.8
Diameter (m)	1.8	1.8	1.8	1.8
Path loss (dB)	207.4	207.4	207.4	207.4
Other losses (dB)	4.5	4.5	4.5	4.5
Uplink C/N_0 (dB-Hz)	73.1	73.1	73.1	73.1
Voice activity factor				
(Fraction of time active)	0.4	0.4	0.4	0.4
Satellite				
Transponder type	**Transparent/ Globa 1 + 3 spots/15 MHz**	**Transparent/ Globa 1 + 3 spots/15 MHz**	**Transparent/ Globa 1 + 3 spots/15 MHz**	**Transparent/ Globa 1 + 3 spots/15 MHz**
Total L-band EIRP (dBW)	45	45	45	45
Capacity used	8 MHz; 8×1 MHz modules	8 MHz; 8×1 MHz modules	8 MHz; 8×1 MHz modules	8 MHz; 8×1 MHz modules
Satellite G/T (dB/K) in K_u band	−1.4	−1.4	−1.4	−1.4
L band downlink				
Satellite EIRP/carrier (dBW)	29.3	38.0	15.8	24.4
Path loss (dB)	188.3	188.3	188.3	188.3
Other losses (dB)	1.1	1.1	1.1	1.1
Shadowing margin (dB)	0	4.7	0	4.7
Mobile G/T (dB/K); Gain(dB)	−9.0; 16	−13.0; 12	−9.0; 16	−13.0; 12
Self noise loss (dB)	−	−	0.5	0.5
Downlink C/N_0 (dB-Hz)	59.5	59.5	45.5	45.4
Inter-system interference (dB-Hz)	−	−	59.0	63.6
Required E_b/N_0 (dB-Hz)	3.0	3.0	5.0	5.0
Bit rate (Kb/s)	270	270	6.76	6.76
Implementation margin (dB)	2.0	2.0	2.0	2.0
Total C/N_0 (dB-Hz)	59.3	59.3	45.3	45.3
Number of active simultaneous channels	592	80	832	115
Offered service	Speech	Speech	Speech	Speech

Table 3.13
(a) Comparative forward link budgets of TDMA and CDMA (Priscoli and Muratore, 1996)

	GSM (TDMA)		MSBN (CDMA)	
	Transportable	Vehicular	Transportable	Vehicular
L band uplink				
Peak mobile station				
EIRP/carrier (dBW)	34	30	22	18
Path loss (dB)	188.8	188.8	188.8	188.8
Other losses (dB)	1.1	1.1	1.1	1.1
Shadowing margin (dB)	0	4.7	0	4.7
Uplink C/N_0 (dB-Hz)	72.7	64.0	60.7	52.0
Voice activity factor				
(Fraction of time active)	0.4	0.4	0.4	0.4
Satellite				
Satellite L-band G/T (dB/K) in spot	0	0	0	0
Transponder type	**Transparent/ Globa 1 + 3 spots/15 MHz**	**Transparent/ Globa 1 + 3 spots/15 MHz**	**Transparent/ Globa 1 + 3 spots/15 MHz**	**Transparent/ Globa 1 + 3 spots/15 MHz**
Total K_u-band EIRP (dBW)	33.5	33.5	33.5	33.5
Capacity used	8 MHz; 8×1 MHz modules	8 MHz; 8×1 MHz modules	8 MHz; 8×1 MHz modules	8 MHz; 8×1 MHz modules
K_u *band downlink*				
Satellite EIRP/carrier (dBW)	20.8	22.4	11.9	14.0
Path loss (dB)	206.4	206.4	206.4	206.4
Other losses (dB)	3.5	3.5	3.5	3.5
FES G/T(dB/K); Gain(dB)	20, 46	20, 46	20, 46	20, 46
Self noise loss (dB)	–	–	0.5	0.5
Downlink C/N_0 (dB-Hz)	59.5	61.1	50.6	52.7
Intersystem interference (dB-Hz)	–	–	55.2	51.2
Required E_b/N_0 (dB/Hz)	3.0	3.0	5.0	5.0
Bit rate (Kb/s)	270	270	6.76	6.76
Implementation margin (dB)	2.0	2.0	2.0	2.0
Total C/N_0 (dB-Hz)	59.3	59.3	45.3	45.3
Number of active simultaneous channels	288	192	143	89
Offered service	Speech	Speech	Speech	Speech

Link analysis of non-geostationary constellations is more involved, due to continuous variations in path length, elevation angle and interference during a call. Figure 3.10 shows the temporal variation in path loss and elevation angle at a given location for LEO, MEO and GEO systems as a comparison.

The wide variation in path loss is further exacerbated by loss caused by spacecraft antenna pattern variation. The net effect is a need to incorporate a large dynamic range for the receiver coupled with dynamic power control of RF channel. Another consideration is the variability in propagation behaviour, which requires elevation-dependent modelling (see section 3.3.2). A further consideration related to propagation is to include diversity effects and methods of combining signals from diverse paths (see section 3.3.2 and Chapter 2, section 2.3.8). Estimating interference in non-geostationary satellite systems is rather involved, due to a need to consider the dynamics of the constellation as well as the frequency reuse scheme deployed in the network; in fact frequency planning and interference analysis are intertwined. Once propagation effects, diversity improvements and interference loss are established, link analysis can be applied through the appropriate link equation. Table 3.14(a) and (b) gives an example of the link budget for the Iridium LEO satellite system.

	Outermost cell	Innermost cell
Azimuth angle (Deg)	32.4	60.0
Elevation angle (Deg)	8.2	51.9
Slant range (Km)	2461.7	960.0
Satellite		
EIRP (dBW)	27.7	19.5
Propagation		
Space loss (dB)	164.5	156.3
Prop. losses (dBW)	15.7	15.7
Mobile terminal:		
Received power (dBW)	−152.5	−152.5
Antenna gain (dBi)	1.0	1.0
Received signal level (dBW)	−151.5	−151.5
G/T (dBi/K)	−23.0	−23.0
Required $E_b/(N_o+I_o)$ (dB)	5.8	5.8
E_b/I_o (dB)	18.0	18.0
Required E_b/N_o (dB)	6.1	6.1
Signal level required (dBW)	−151.5	−151.5
Link margin (dB)	0	0

Table 3.14
(a) Iridium link budget – forward link (Freeman, 1996)

Table 3.14
(b) Iridium link budget –
return link (Freeman,
1996)

	Outermost cell	Innermost cell
Azimuth angle (Deg)	32.4	60
Elevation angle (Deg)	8.2	51.9
Slant range (km)	2461.7	960.0
Mobile terminal		
EIRP (dBW)	6.0	6.0
Propagation		
Free space loss (dB)	164.5	156.3
Prop. losses (dB)	15.7	15.7
Satellite:		
Received power (dBW)	–174.2	–166.0
Effective coverage edge antenna gain (dBi)	23.9	16.4
Received signal level (dBW)	–150.3	–149.6
G/T (dB/K)	–3.1	-10.6
Required $E_b/(N_o+I_o)$ (dB)	5.8	5.8
E_b/I_o (dB)	18	18
Required E_b/N_o (dB)	6.1	6.1
Signal level required (dBW)	–148.5	–148.5
Link margin (dB)	–1.8	–1.1

Further reading

Akturan, R.; Vogel, W.J. (1997) 'Path diversity for LEO satellite-PCS in the urban environment', *IEEE Transactions on Antennas and Propagation*, **45**, (7), July, 1107–16.

Aulin, T. (1979) 'A modified model for the fading signal at a mobile radio channel', *IEEE Transactions on Vehicular Technology*, VT-28, (3), 182–203.

Babalis, P.G.; Capsalis, C.N. (1998) 'A new model for the characterisation of the combined shadowing and multipath effects in satellite systems', *ICT,. 98*, 25–8.

Beckmann, P.; Spizzichino, A. (1963) *The Scattering of Electromagnetic Waves From Rough Surfaces*, Pergamon Press, New York/Oxford.

Bello, P.A. (1973) 'Aeronautical channel characterisation', *IEEE Transactions on Communication*, **COM-21**, 548–63.

Butt, G.; Evans, B.G.; Richharia, M. (1992) 'Narrowband channel statistics from multiband measurements applicable to high elevation angle land-mobile satellite systems', multiband propagation experiment for narrowband characterisation of high elevation angle land mobile-satellite channels, *Electronics Letters*, **28** (15), 16 July, 1449–50.

Butt, G.; Parks, M.A.N.; Evans, B.G. (1995) 'Narrowband and wideband characterisation of satellite mobile/PCN channel', *Proceedings International Mobile Satellite Conference, IMSC '95*, The Fourth International Mobile Satellite Conference, Ottawa, co-sponsored by Communications Research Centre/Industry Canada and Jet Propulsion Laboratory/NASA 128–33.

Clarke, R.H. (1968) 'A statistical theory of mobile radio reception', *Bell System Technical Journal*, 957–1000.

Corazza, G.E.; Vatalaro, F. (1994) 'A statistical model for land mobile satellite channels and its application to nongeostationary orbit systems', *IEEE Transactions Vehicular Technology*, August, 738–41.

Cox, D.C.; Leek, R.P. (1972) 'Correlation bandwidth and delay spread multipath propagation statistics for 910 MHz urban mobile radio channels', *IEEE Transactions on Communication* **23** (11), Nov.

Crane, R.K. (1980) 'Prediction of attenuation by rain', *IEEE Transactions Communication Technology*, Vol COM-28, September, 1717–33.

Davarian, F. (1995) 'ITU recommendations regarding propagation effects on mobile satellite links', *Proceedings International Mobile Satellite Conference IMSC' 95*, fourth International Mobile Satellite Conference Ottawa co-sponsored by Communication Research Centre/Industry Canada and Jet Propulsion Laboratory/NASA, 99–104.

Dooren, G.A.J.; Herben, M.H.A.J., Brussaard, G.; Sforza, M.; Baptista Poiares, J.P. V. (1993) 'Electromagnetic field strength prediction in an urban environment: A useful tool for the planning of LMSS', IMSC 93, *Third International Mobile Satellite Conference*, Pasadena, California, 16–18 June, JPL Publication 93-009, 343–48.

ESA (1977) ESA/ESTEC Contract No 5323/83/NL/JS (in German) DFVLR D-8031 Oberpfaffenhofen, Federal Republic of Germany.

Fortuny, J.; Benedicto, J.; Sforza, M. (1991) 'Mobile satellite systems at Ku and Ka band', *Proceedings Second European Conference on Satellite Communications*, Liege, October, 55–62.

Freeman (1996) 'Reference manual for telecommunications engineering', John Wiley & Sons, second edition.

Gerace, G.; Smith, E. (1990) 'A comparison of code models', *IEEE Antennas and Propagation magazine*, October, 32–8.

Goldhirsh, J.; Vogel, W. (1992) 'Propagation effects for land mobile systems: Overview of experimental and modelling results', *NASA reference publication* 1274, Feb.

Hagenauer, J.; Dolainsky, F.; Edbauer, F.; Grabel, J.; Lutz, E.; Papke, W.; Plöchinger, E.; Schweikert, R. (1984) 'Multipath fading effects and data transmission for small ship earth stations (Standard C)', DFVLR Final report (in German) DFVLR D-8031 Oberpfaffenhofen, Federal Republic of Germany ESA/ESTEC Contract No 5323/82/NL/JS, 223 pages.

Hagenauer, J.; Neul, A.; Papke, W.; Dolainsky, F.; Edbauer, F. (1987) 'The aeronautical satellite channel', DFVLR Technical Report No NE-NT-T-97-17, May, DFVLR D-8031 Oberpfaffenhofen, Federal Republic of Germany.

Hagenaur, J. *et al.* (1986) 'The aeronautical satellite channel', Final report DFVLR 8031 Oberpfaffenhofen, Federal Republic of Germany.

Higuchi, T.; Shinohara, T. (1988) 'Experiment of Inmarsat Standard-C system', *Fourth International Conference on Satellite System for Mobile Communication and Navigation, IEE*, Conf. Pub no 294.

Hogben, N.; Lumb, F.E. (1967) 'Ocean wave statistics', *Her Majesty's Stationery Office*, London, UK.

Hwang, S.H.; Kim, K.J.; Ahn, J.Y.; Whang, K.C. (1997) 'A channel model for non-geostationary orbiting satellite systems', *Proc. IEEE Vehicular Technology Conference*, 41–5.

Ikegami, T.; Arakaki, Y.; Wakana, H.; Suzuki, R. (1993) 'Measurement of multipath delay profile in land mobile satellite channels', *Proceedings of the Third International Mobile Satellite Conference*, IMSC '93, Pasadena, California, JPL Publication 93-009, 16–18 June, 331–6.

Ippolito, I.J.; Russel, T.A. (1993) 'Propagation considerations for emerging satellite communications applications', *Proceedings of IEEE*, **81** (6), June, 923–9.

Ippolito, L. (1986) *Radiowave propagation in Satellite Communications*, Van Nostrand Reinhold.

ITU (1986) 'Attenuation by hydrometers in particular precipitation, and other atmospheric particles', *CCIR report 721-2*, Propagation in Non-ionised Media, Recommendations and Reports of the CCIR, Geneva, **5**.

ITU (1990) 'Propagation data and prediction methods required for earth–space telecommunication systems', *CCIR Report 564-3*, Propagation in Non-ionised Media, Recommendations and Reports of the CCIR, **V**.

ITU (1992) CCIR Recommendations, RPN Series, Propagation in Non-ionised Media, Geneva, October.

ITU (most recent) Radio Regulations.

ITU (1992) ITU-R Rec 531-2, RPI series, Geneva, pp 82.

Jahn, A.; Lutz, E. (1994) 'DLR channel measurement programme for low earth orbit satellite systems', *Proceedings International Conference on Universal Personal Communication ICUPC '94*, San Diego, September, pp. 423–9.

Jahn, A.; Buonomo, S.; Sforza, M.; Lutz, E. (1995–2) 'Wideband channel model for land mobile satellite system', *Ibid*, pp. 122–126.

Kanatas, A.G.; Kanderakis, E.C.; Constantinou, P. (1995) 'Narrowband land mobile satellite channel modelling', *IMSC '95, The fourth International Mobile Satellite Conference*, Ottawa, co-sponsored by Communication Research Centre/Industry Canada and Jet Propulsion Laboratory/NASA, 109–114.

Karaliopoulos, M.S.; Pavlidou, F.-N. (1999) 'Modelling the land mobile channel: a review', *Electronics & Communication Engineering Journal*, October, 235–48.

Karasawa, Y.; Matsudo, T. (1991) 'Characteristics of fading on low-elevation angle earth-space paths with concurrent rain attenuation and scintillation', *IEEE Trans Antennas and Propagation*, **39** (5), May, 657–61.

Karasawa, Y.; Shiokawa, T. (1984) 'Characteristics of L-band multipath fading due to sea surface reflection', *IEEE Transactions on Antenna and Propagation*, **AP-32 6**, 618–23.

Karasawa, Y.; Shiokawa, T. (1988) 'A simple prediction method for L band multipath fading in rough sea conditions', *IEEE Trans Communications*, **COM-36** (10), 1098–104.

Karasawa, Y.; Yasunaga, M.; Nomoto, S.; Seiokawa, T. (1986) 'On-board experiments on L-band multipath fading and its reduction by use of the polarisation shaping method', *Transaction IEICE* (Japan), **E69** (2), 124–31.

Lee, W. (1986) *Mobile communication design fundamentals*, Howard W Sams & Co.

Long, M.W. (1975) *Radar reflectivity of land and sea*, Lexington Books, Lexington MA USA.

Loo, C. (1985) 'A statistical model for the land mobile satellite link', *IEEE Transactions on Vehicular Technology*, **VT-34**, August, 122–7.

Lutz, E.; Papke, W.; Plöchinger, E. (1986) 'Land mobile satellite communications-channel model, modulation and error control', *Proc ESA Workshop on Land Mobile Services ESTEC*, 3–4 June, ESA, SP-259, September.

Lutz, E.; Cygan, D.; Dippold, M.; Dolainsky, F.; Papke, W. (1991) ' The land mobile satellite communication channel – recordings, statistics and channel model', *IEEE Transactions on Vehicular Technology*, VT–40, 375–86.

Lutz, E. (1996) 'A Markov model for correlated land mobile satellite channels', *International Journal of Satellite Communications*, 13, 333–39.

Lutz, E.(1998) 'Issues in satellite personal communication systems', *Wireless Networks*, 4, 109–24.

Maral, G.; Bousquet, M. (1987) *Satellite Communications Systems*, John Wiley & Sons.

Matsudo, T.; Minamisono, K.; Karasawa, Y.; Shiokawa, T. (1993) 'A prediction model of signal degradation in LMSS for urban areas', *International Mobile Satellite Conference*, IMSC 93, The third International Mobile Satellite Conference, Pasadena, California, 16–18 June, JPL Publication 93-009, pp. 355–60.

Murr. F.; Arbesser-Rastburg, B.; Buonomo, S. (1995) 'Land mobile satellite narrowband propagation campaign at Ka band', *IMSC '95*, The fourth International Mobile Satellite Conference, Ottawa, co-sponsored by Communication Research Centre/Industry Canada and Jet Propulsion Laboratory/NASA, pp. 134–8.

Ohmori, S.; Irimata, A.; Morikawa, H.; Kondo, K.; Hase, Y.; Miura, S. (1985) 'Characteristics of sea reflection fading in maritime satellite communications', *IEEE Trans Antenna and Propagation*, **AP-33 8**, 838–45.

Olsen, R.L.; Rogers, D.V.; Hodge, D.B. (1978) 'The aR^b relation in the calculation of rain attenuation', *IEEE Transactions Antennas and Propagation*, **AP-26**, March, 318–29.

Parsons, D. (1992) *The Mobile Radio Propagation Channel*, Pentech Press, London.

Patzold, M.; Killat, U.; Laue, F.; Li, Y. (1998) 'On the statistical properties of determinsitc simulation models for mobile fading channels', *IEEE Trans. Vehicular. Technology*, February, 254–69.

Priscoli, F.D.; Muratore, F. (1996) 'Radio and network comparisons for MSBN and GSM systems in a geostationary satellite environment', *International Journal of Satellite Communications*, **14**, 439–54.

Richharia, M. (1989) 'Review of propagation aspects of highly elliptical orbits', IEE Colloquium on Highly Elliptical Orbit Satellite Systems, 24 May, Digest No. 1989/86.

Richharia, M. (1999) *Satellite Communication Systems: Design Principles*, Macmillan Press Ltd, Basingstoke and London.

Richharia, M.; Evans, B.G.; Butt, G. (1990) 'A propagation experiment for modelling high elevation angle land mobile satellite channels', *International Mobile Satellite Conference*, Ottawa, pp. 231–7.

Richharia, M.; Pratt, T. (1985) 'Simulate tropospheric amplitude scintillation', *Microwaves and RF*, April , 79–82.

Sandrin, W.; Fang, D.J. (1986) 'Multipath fading characteristics of L band maritime mobile satellite links', *COMSAT Tech Rev*, **16** (2), 319–38.

Schroeder, E.H.; Thompson, A.D.; Sutton, R.W.; Wilson, S.G.; Kuo, C.J. (1976) 'Air traffic control experimentation and evaluation with NASA ATS-6 satellite', *The Boeing Company Report No FAA-RD-75-173*, **III NTIS**, Accession No. D6-44048, Vol V, NTIS Accession No. D6-44050, National Technical Information Service, Springfield, VA 22161, USA.

Sforza, M.; Bernardo, G.; Di; Cioni, R. (1993) 'Propagation model for the land mobile satellite channel in urban environments', *International Mobile Satellite Conference*, IMSC 93, The third International Mobile Satellite Conference, Pasadena, California, 16–18 June, JPL Publication 93-009, pp. 349–54.

Sforza, M.; Buonomo, S.; Baptisa, J.P.V.P. (1993) 'Global coverage mobile satellite systems: system availability versus channel propagation impairments', *International Mobile Satellite Conference*, IMSC 93, The third International Mobile Satellite Conference, Pasadena, California, 16–18 June, JPL Publication 93-009, pp. 361–6.

Solari, G.; Viola, R. (1992) 'M-HEO, the optimal satellite system for the most highly populated regions of the Northern Hemisphere', *ESA Bulletin no 70*, The Netherlands, May, pp. 81–8.

Sutton, R.W.; Schroeder, E.H.; Thompson, A.D.; Wilson, S.G. (1973) 'Satellite-aircraft multipath and ranging experiment results at L band', *IEEE Trans Comm*, **COM-21 5**, 639–47.

Vatalaro, F. (1995) 'Generalised Rice-lognormal channel for wireless communications', *Electron. Lett*, October, 1899–1900.

Vatalaro, F.; Mazzenga, F. (1998) 'Statistical channel modelling and performance evaluation in satellite personal communications', *International Journal of Satellite Communications*, 16, 249–55.

Vatalaro, F.; Corazza, E.G.; Caini, C.; Ferrarelli, C. (1995) 'Analysis of LEO, MEO and GEO global mobile satellite systems in the presence of interference and fading', *IEEE Journal on Selected Areas in Communications*, **13** (2), February, 291–300.

Vogel, W.J.; Goldhirsh, J. (1990) 'Propagation limits for land mobile satellite services', *AIAA-90-0847-CP*, pp. 564–74.

Vogel, W.J.; Hong, U. (1988) 'Measurement and modelling of land mobile satellite propagation at UHF and L band', *IEEE Transactions on Antennas and Propagation*, Vol AP-36, May, 707–19.

Vucetic, B.; Du, J. (1992) 'Channel modelling and simulation in satellite communication systems', *IEEE Journal on Selected Areas of Communications*, Vol 10, October, 1209–18.

Modulation, coding and multiple access

4

4.1 Introduction

This chapter gives an overview of a range of concepts and techniques related to modulation, coding and multiple access schemes. Topics are covered at a system level from an MSS perspective and treatment kept to a level essential to grasping the sensitivities of various components in a system context. Modulation and coding schemes applicable to MSS are widely used in other areas of radio communications and are hence well documented in the literature (e.g. Taub and Schilling, 1986), so the emphasis here is rather on their performance in an MSS environment. Some of the more recent schemes are addressed in more detail. Similarly, the treatment of multiple access schemes focuses on their applicability and performance to MSS.

4.2 Modulation

The first part of this section outlines the requirements of a modulation scheme for its applicability to an MSS system. The next section narrows the choice of modulation schemes to those best suited for MSS and in the final part, the performance of each is reviewed.

MSS requirements
Radio channel characteristics exert a marked influence on the performance of modulation schemes. In MSS systems, service link characteristics set a bound on system performance. The main sources of degradation affecting performance of modulation schemes in mobile communication links are:

- **signal fades** caused by environment and velocity dependent multipath, characterized by slow ($< \sim 1$ Hz) and rapid ($\sim$ tens of Hz) signal fluctuations;
- **signal fades** caused by tropospheric effects, applicable to K_a band links;
- **signal phase fluctuations** caused by multipath and Doppler Effects;
- **very low carrier to thermal noise ratio** due to small user terminals;

- **non-linearity** arising in various system components and in particular in mobile Earth stations due to class C amplifiers;
- **large Doppler frequency changes** associated with MEO and LEO, which have to be adequately compensated.

It is anticipated that a significant number of future MSS satellite transponders will be regenerative. Onboard applications will require lightweight, small-sized and power-efficient modems and high coding gain.

In coherent demodulation schemes, the carrier is recovered prior to demodulation. Carrier recovery circuits are susceptible to thermal noise and fading and therefore particular care has to be exercised in mobile Earth stations. Furthermore, carrier recovery becomes difficult as RF or data rate is increased, in thermal noise limited and in fading links. In this respect, non-coherent demodulation schemes are more robust.

Demodulation also requires symbol clock recovery for extracting the transmitted bit stream. The requirement becomes more stringent for quadrature modulation schemes such as parallel minimum shift keying (MSK), where it is essential to synchronize time, phase and amplitude of both I and Q channels. From this viewpoint, schemes which have less stringent synchronization requirements are preferred. Again, the problem worsens as frequency and symbol rate increase.

The choice of a modulation scheme is additionally affected by the multiple access scheme, which in turn influences spectrum utilization efficiency and is governed by service requirements, propagation considerations, etc. (see section 4.4). For example, the synchronization problem is quite stringent in TDMA schemes, due to the need to operate in burst mode, which imposes stringent timing requirements on carrier recovery and symbol/bit synchronization circuits. Additional complexity is introduced in demodulators by the need for frame synchronization and dynamic readjustments, due to satellite motion. For identical link and throughput requirements, carrier recovery and synchronization problems are less problematic for frequency division multiple access (FDMA). When CDMA is used, spread spectrum modulation remains the sole choice.

The spectral efficiency of the modulation scheme is vital in modern MSS because of the shortage of MSS spectrum, due to rising demands. Higher order modulation schemes are spectrally efficient but require higher transmission power, leading to classical power–bandwidth trade-off. Higher order modulation schemes, coupled with powerful codes, offer a possible solution.

Finally, it is essential that modulators/demodulators used in mobile terminals can be mass produced to give economies of scale.

Preferences

Within the set of constraints discussed above, choice is narrowed to a relatively few schemes. Due to the enormous advantages offered by digital systems, all emerging and evolving systems use digital transmissions and so digital modula-

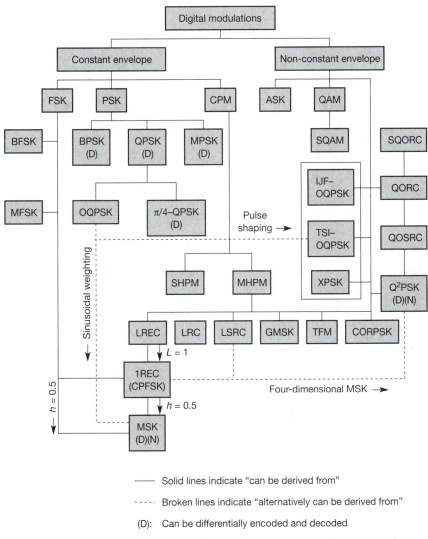

Figure 4.1
A tree diagram of digital modulation schemes (Xiong, 1994) © 1994 IEEE

———— Solid lines indicate "can be derived from"

- - - - Broken lines indicate "alternatively can be derived from"

(D): Can be differentially encoded and decoded

(N): Can be non-coherently detected

tion schemes are of main interest. Frequency modulation was used in first generation mobile satellite systems such as the Inmarsat-A service. Single-side band (SSB) systems were considered as a potential candidate due to their spectral efficiency. FM systems provide a more effective use of satellite power than the SSB systems, but are less spectrally efficient. We will not consider these analogue schemes any further here.

There are three basic types of digital modulation, depending on the particular parameter of the carrier frequency altered by the baseband digital stream – amplitude shift keying (ASK), frequency shift keying (FSK) and phase shift keying (PSK). As the names suggest, in ASK the amplitude of the carrier is altered, in FSK the carrier frequency is changed and in PSK the phase of carrier is varied in accordance with the incoming digital stream. Figure 4.1 (Xiong, 1994) illustrates a tree diagram of digital modulation schemes suitable for MSS.

Modulation schemes are divided into two broad categories – constant envelope and non-constant envelope modulation schemes. The relationships between various schemes are also illustrated. Note that some of the schemes have more than one parent, i.e. they can be derived from more than one technique. The schemes which can be differentially encoded are marked as D and those which can be detected non-coherently are marked as N. Differential encoding removes the phase ambiguity experienced in coherent detection in which a circuit is unable to distinguish between phase 0 and π radians. In general, constant envelope modulation schemes are preferred, as they offer a more robust performance. Of the constant envelope modulation schemes, FSK schemes are preferred for low bit rate transmissions, due to demodulator hardware simplicity; they have a lower spectral efficiency than other schemes and are therefore not suited for higher bit rate transmissions. PSK modulation schemes have a near constant envelope, but exhibit discontinuity in phase, whereas constant phase modulation (CPM) schemes have a constant envelope and a gradual change in phase, which results in better side-lobe performance than CPM. Traditionally non-constant envelope modulation schemes were not used for MSS, but due to the need for spectral efficiency, multilevel quadrature amplitude modulation (QAM) together with powerful convolution codes have recently been introduced for wideband MSS systems.

At present, the most common conventional digital modulation schemes are:

- binary phase shift keying (BPSK) and its variants such as aviation-BPSK (or symmetric BPSK);
- quadrature phase shift keying (QPSK) and its variants such as offset-QPSK (O-QPSK), aviation-QPSK, minimum shift keying schemes, Gaussian minimum shift keying (GMSK);
- multi-level frequency shift keying and its variants.

Modulation schemes introduced in recent systems include:

- spread spectrum modulation in conjunction with CDMA;
- coded orthogonal frequency division multiplexing (COFDM) modulation scheme;
- 16-QAM with turbo-coding.

PSK schemes

In broad terms, PSK modulation techniques may be divided into three categories according to the demodulation technique, as follows:

1 reference signals are transmitted with the main signal to aid carrier extraction;
2 the carrier is extracted from the received signals by appropriate signal processing;
3 carrier recovery is not required for demodulation.

The first two categories are coherent detection schemes and the last belongs to the non-coherent category. Coherent detection of PSK signals gives a better power efficiency than non-coherent detection schemes. For coherent detection, carrier synchronization (or recovery) is essential for extracting phase information; this can be problematic in the presence of multipath, low carrier-to-noise ratio and Doppler shifts. Furthermore, detection of the correct phase of the carrier is required to recover the transmitted symbol. Noise introduces phase error in the recovered carrier and causes timing jitters in the recovered bit stream. As an example, in a BPSK scheme, with a phase jitter of $\Delta\phi$ in carrier recovery and $\Delta\tau$ in bit synchronization timing, the bit error rate in Gaussian noise degrades to (see Table 4.1)

$$Pe = \tfrac{1}{2} \, \text{erfc} \, \sqrt{(E_s/\eta)} \, (\cos^2\phi)(1-2|\tau|/T)^2 \qquad (4.1)$$

Figure 4.2 shows the effect of phase and time jitter on the BER for a number of combinations where phase jitter and time jitter are respectively ±15° and $|\tau|/T$ of 0.025.

Differential detection schemes do not require carrier recovery and are therefore robust and capable of rapid synchronization, but at the expense of some

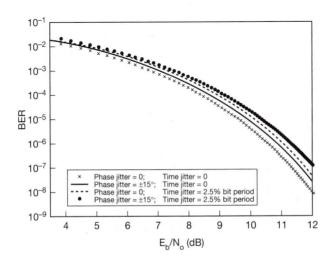

Figure 4.2

E_b/N_o vs BER with combinations of phase and time jitter

(Graphics: AR)

degradation in performance. Examples of such schemes are MSK and π/4 QPSK, differentially coded (D) BPSK, DQPSK, π/4-DQPSK, DMSK, etc. Two modes of transmissions are possible in the reference signal transmission method. The reference signal can be either an unmodulated pilot multiplexed with the main transmissions, or the signal may be multiplexed with information bits; numerous techniques have been proposed for the latter. A spectral null can be created at the centre RF, where a pilot is inserted. Symbols may be inserted once or periodically in each TDM frame, so that fading of data and symbol are correlated. The reference signal transmission scheme is wasteful of power and bandwidth, but carrier recovery is simpler. Examples of this technique are transparent tone in band (TTIB) and tone calibration (McGeehan and Bateman, 1984; Korn, 1989; Davrian, 1985, 1987; Caves, 1991; Simon, 1986). These schemes do not exhibit an irreducible error rate (explained later) and E_b/N_o in the presence of Rayleigh fading is high. Transmissions do not possess a constant envelope and therefore the scheme is sensitive to system non-linearity.

In the carrier recovery method, the phase of the carrier is derived through processing of the received signal. Well-established carrier recovery techniques include:

- the $(C)^M$ method, where C is the carrier signal and M denotes the number of symbols;
- the Costas loop;
- the decision feedback loop.

In the $(C)^M$ method, the carrier is recovered by passing the received carrier through a circuit which raises the carrier to the power of M, followed by a divide by M circuit. The detected bits are then synchronized to reproduce the transmitted signal.

The Costas loop, named after its inventor, consists of a two-phase lock loop (PLL) which uses a common loop filter and voltage control oscillator (VCO). VCO frequency is kept synchronized to the carrier frequency through a feedback loop comprising a multiplier, whose two inputs are outputs of an in-phase (I) comparator and a quadrature phase (Q) output. The multiplier output provides a corrective voltage to maintain the VCO synchronized to the carrier (see Figure 4.3).

Figure 4.3
VCO using Costas loop

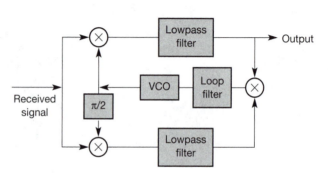

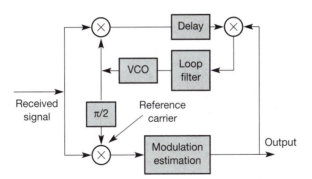

Figure 4.4

A decision feedback loop circuit

A decision feedback loop circuit recovers the carrier through a PLL using feedback from the demodulated signal. The recovered carrier is within the feedback loop, hence the name (see Figure 4.4).

Schemes which use non-coherent detection and constant envelope signals to minimize sensitivity to the amplitude and phase fluctuations are GMSK, differentially detected MSK (DMSK) and differentially detected O-QPSK (DOQPSK). In general, such schemes are affected significantly by fading and suffer irreducible BER, where BER cannot be reduced below a lower bound, irrespective of level of E_b/N_o.

Figure 4.5(a) and (b) show a simplified block schematic of a typical QPSK modulator and demodulator, respectively. The input digital stream is divided into I and Q components. Each stream is fed into a multiplier, the other input of which is an I or Q carrier. The I and Q components are then summed to accomplish a QPSK signal; at the same time, undesired components produced in multiplication are cancelled out. At the receiver, the signal is divided into two paths, each of which is fed into the I and Q channel multiplier, the other input of which consists of the recovered carrier. The output of each multiplier is low pass filtered and fed into an analogue to digital converter, synchronized to the incoming bit stream through the timing recovery circuit. BPSK modems operate in the same manner as QPSK but only with one arm.

Derivatives of QPSK are generated by altering characteristics of the I and Q channels in various ways so as to reduce the sensitivity of QPSK to non-linearity and reduce the amplitude of side lobes. For O-QPSK, the Q channel is delayed by half the symbol duration, which gives a smoother phase transition, thereby reducing the spectral side lobes of the transmitted signal. For MSK, I and Q channel timing is the same as O-QPSK, followed by co-sinusoid shaping for the I stream and sinusoid shaping for the Q stream. In O-QPSK detection, the I channel is delayed by half the symbol duration. The MSK detection scheme is similar to the O-QPSK scheme, with an additional circuit for shaping the digital stream as sinusoid and co-sinusoid compliant with transmissions. GMSK (Murota and Hirade, 1981), another derivative of O-QPSK, uses Gaussian pulse shapes.

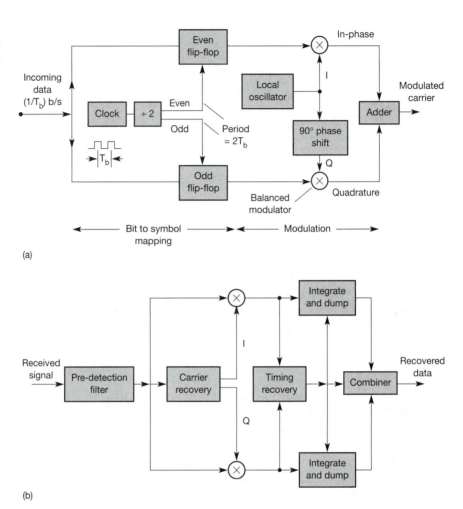

Yet another QPSK derivative, known as the $\pi/4$-QPSK scheme (Baker, 1962; Xiong, 1994) has been favoured recently for both terrestrial and mobile satellite communication. In this scheme, carrier phase transitions are limited to $\pm\pi/4$ and $\pm3\pi/4$ and therefore envelope variations caused by band limiting and subsequent spectral regrowth due to non-linear amplification are considerably reduced (see next paragraph). This scheme allows differential encoding and differential decoding, which has the advantage of removing the phase ambiguity inherent in coherent demodulation. By shaping the digital pulses as sinusoid and offsetting the transition in the I and Q channels, improved spectral restoration is achieved, which gives an improved performance when operating with saturated non-linear amplifiers. This type of modulation scheme is known as $\pi/4$-controlled transition PSK (CTPSK) (Feher, 1991).

CPM schemes use smooth changes in phase, which result in lower side lobes; these schemes are also power efficient (Sundberg, 1986) and are therefore a candidate for MSS. Depending on the pulse shape used to vary the phase, the modulation index (h) which determines the amplitude and size of symbol (M), a variety of CPM signals can be obtained. Examples of CPM schemes are continuous phase frequency shift keying (CPFSK), MSK, GMSK and tamed frequency modulation (TFM) (De Jaeger and Dekker, 1978). We have already mentioned MSK and GMSK.

In all well-designed schemes, a filter is placed at the modulator output to suppress side lobes to reduce the occupied bandwidth. Filter roll-off is a trade-off between occupied bandwidth and intersymbol interference. However, the

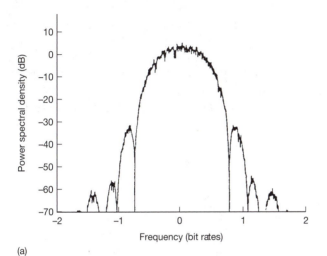

(a)

(b)

Figure 4.6

Spectrum of an O-QPSK signal (a) before and (b) after a transmit non-linearity (Lodge *et al.*, 1987) © 1987 IEEE

purpose of filtering may be defeated when spectral regrowth occurs elsewhere in the system, e.g. due to band limiting or by system non-linearities in the presence of amplitude fluctuations. Filtered PSK has some amplitude variation and therefore non-linear class-C amplifiers and filtering in mobile Earth stations cause spectral regrowth in the transmitted spectrum. O-QPSK has lower amplitude variations and consequently lower spectral regrowth. Figure 4.6 shows the spectrum of an O-QPSK signal before and after transmit non-linearity. The $\pi/4$-QPSK modulation scheme, another O-QPSK variant, offers good spectral efficiency, and low spectral regrowth when passed through non-linear amplifiers with a capability to use differential detection (Liu and Feher, 1991).

Modulation translates the baseband spectrum to an intermediate frequency (IF) which is up-converted to RF before transmission. Therefore several modulation schemes use a shaped baseband digital signal to minimize side lobes (see MSK and GMSK above).

Due to a shortage of spectrum in the L band, extensive research is in progress to improve spectral efficiency, while retaining robustness in the presence of channel impairments. A recent introduction is 16-QAM in conjunction with turbo-code (Feldman and Ramana, 1999). The scheme is discussed in some detail later in this section.

Figure 4.7
Constellation diagram of (a) binary-phase shift keying, (b) quadrature-phase shift keying, (c) 8-phase shift keying, (d) 16-quadrature amplitude keying (QAM)

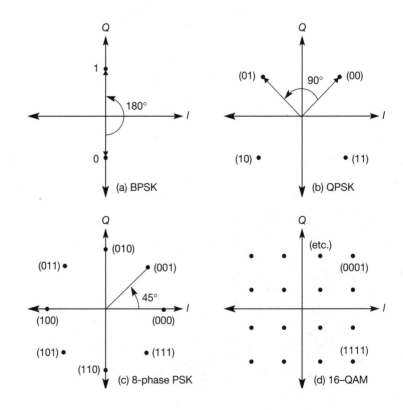

Multilevel frequency shift keying (M-FSK) and its derivatives, such as CPFSK, also offer good spectral and power efficiency (for a review of CPFSK, see Sundberg, 1986). In the M-FSK scheme, side lobes are generated by discrete changes in frequency at symbol transition; in CPFSK, such changes are smoothed to accomplish reduced side-lobe levels. MSK can be considered as a special case of binary CPFSK.

Performance comparison of conventional digital modulation schemes

A modulation scheme should provide the desired BER at acceptable energy per bit by noise power spectral density (E_b/N_o) and spectrum requirements, leading to the classical power–spectrum trade-off. In general, spectrally efficient modulation schemes are more sensitive to noise and hence require larger E_b/N_o. Figure 4.7 shows constellation diagrams of BPSK, QPSK, 8-PSK and 16-QAM schemes. As the distance between permissible states reduces, the signal becomes more susceptible to noise. Noise sources include modulator/demodulator circuit imperfections, adjacent channel interference, band-limiting effects of filters in transmit/receive sections, thermal noise, intermodulation noise, co-channel interference caused by frequency reuse, intersystem interference and multipath fading.

The power spectral densities of a few digital modulation schemes used in MSS, namely BPSK, MSK and QPSK, are compared in Figure 4.8.

A number of useful parameters are listed in Table 4.1. Parameters including various types of bandwidths; E_b/N_o for BER of 10^{-5}, implementation complexity and immunity to non-linearity are summarized in this table. Table 4.2 summarizes the power spectral density and BER versus energy per bit to noise density relationship (E_b/N_o) of various modulation schemes, and Table 4.3 lists examples of modulation and coding schemes used in MSS at present.

Power spectral density of commonly used digital modulation schemes

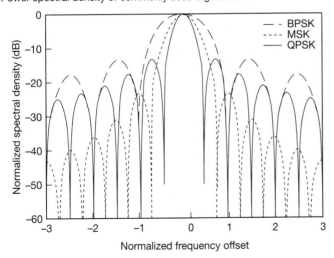

Figure 4.8

A comparison of normalized power spectral density of BPSK, MSK and QPSK (Graphics: AR)

Table 4.1
Comparison of common
modulation scheme
(Xiong, 1994; Amoroso,
1980)

Modul- ation	99% power bandwidth (Hz/bits/s)	Null–null bandwidth (Hz/bits/s)	Half-power bandwidth (Hz/bits/s)	Nyquist bandwidth PSK schemes ($R/\log_2 M$; R = bit rate; M = number of phases)	E_b/N_o dB-Hz (BER = 10^{-5})	Immunity to non-linearity	Implemen-tation complexity
BPSK	20.56	2.0	0.88	R	9.6	Poor	Very good
QPSK	10.28	1.0	0.44	R/2	9.6	Fair	Good
OQPSK and ($\pi/4$) PSK	10.28	1.0	0.44	R/2	9.6	Good	Fair
MSK	1.18	1.5	0.59	–	9.6	Very good	Poor

Table 4.2
Useful parameters of
common modulation
schemes

Modulation	Single-sided power spectral density centred at carrier frequency	Null–null bandwidth	Bit error rate in Gaussian noise	Comments
M-ary PSK	$P_s T_s/2$ [sin $\pi f T_s/\pi f T_s]^2$ P_s = carrier power, T_s = symbol duration	f_b/N	When number of symbols, M is large, erfc $[NE_b/\eta \sin^2 \pi/M]^{1/2}$ BPSK and QPSK, $1/2$ erfc$(E_b/\eta)^{1/2}$	M = 2^N, denotes number of symbols; N = number of bits/symbols; M = 2 for BPSK, M = 4 for QPSK; f_b = bit rate
QAM	Same as M-ary PSK	For 16 QAM; 2 erfc $(0.4E_b/\eta)^{1/2}$	2 erfc $[0.4E_b/\eta]^{1/2}$	$f_b/4$
M-FSK (M signals orthogonal)	At each discrete frequency; Impulse + BPSK PSD, i.e. $P_s T_b/2$ [sin $\pi f T_b/\pi f T_b]^2$; where T_b is the bit period		$[(M-1)/2]$ erfc $[NE_b/2\eta]^{1/2}$	Mf_b/N
MSK	$8P_s T_b/\pi^2[\cos 2\pi f T_b/1-(4f T_b)^2]^2$		$1/2$ erfc$(E_b/\eta)^{1/2}$	f_b/N

System	Transmission rate (Kb/s)	Modulation	Coding	RF bandwidth (KHz)
Inmarsat-A (SCPC)	Analogue	FM	Not applicable	25 (Voice) and 50 (Voice band data/data)
Inmarsat-Aero (SCPC)	21/10.5 (Evolved)	A-QPSK	$\frac{1}{2}$ rate convolution FEC; constraint length = 7; 8-level soft decision Viterbi decoder	17.5/10
Inmarsat-B Forward signalling	6	BPSK	$\frac{1}{2}$ rate convolution FEC; constraint length = 7; 8-level soft decision Viterbi decoder $\frac{1}{2}$ or $\frac{3}{4}$ rate convolution FEC; constraint length = 7;	10
Circuit-mode voice (SCPC)	24	O-QPSK	8-level soft decision Viterbi decoder	
Global area network (GAN)	128	16 QAM	Turbo-code	45
Inmarsat-C Forward link	1.2	BPSK	$\frac{1}{2}$ rate convolution coding with interleaving	10
Return link (Message)	1.2	BPSK		5
MOBILSAT	6.4	$\pi/4$ QPSK		7.5
MSAT	6.75	QPSK, 60% cosine roll-off	Rate $\frac{1}{2}$ and rate $\frac{3}{4}$ punctured code (K=7); Viterbi decoding	6.0

Table 4.3

Example of modulation schemes used in various MSS

Figure 4.9
Probability of bit error
versus energy per bit to
noise ratio for DOQPSK
with SEC over a Ricean
channel with various
values of k in the
presence of thermal
noise (Lodge *et al.*,
1987) © 1987 IEEE

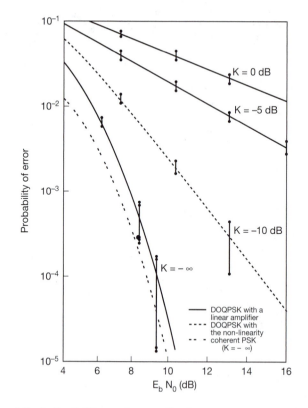

Because of their flexibility and cost effectiveness, computer simulations are widely used for comparing the performance of schemes. As an illustration, consider a simulation study conducted to compare candidate modulation schemes DMSK, DOQPSK and BPSK-TTIB for a land mobile satellite system (Lodge *et al.*, 1987). Here, DOQPSK with non-redundant single error correction out-performed other schemes in robustness to Ricean fading and amplitude compression by a mobile Earth station power amplifier. Figure 4.9 demonstrates the error probability versus energy per bit to noise ratio for the scheme with Ricean fading at various K factors in the presence of thermal noise. The K factor is defined as

$$K = 10 \log (P_m/P_d) \tag{4.2}$$

where P_m = the average multipath power and P_d = the direct path power.

Figures 4.10 and 4.11 show the model used in simulation, the spectrum of the Rayleigh process and the transfer characteristics of the mobile terminal.

Determination of a suitable fading margin in the presence of a multipath is vital for MSS link design. Analytical expressions suitable for numerical computations have been derived for estimating the fade margin required for BPSK and

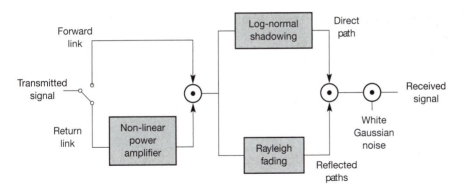

Figure 4.10
A model used for
performance comparison
of various modulation
schemes (Ball, 1982)
© 1982 IEEE

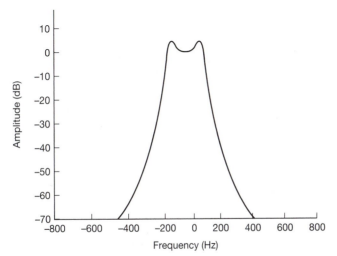

Figure 4.11
Spectrum of the Rayleigh
process used in
simulation (Ball, 1982)
© 1982 IEEE

QPSK at any specified BER and percentage coverage, assuming ideal carrier phase tracking (Davrian, 1985). Figure 4.12 shows numerically computed results of fade margin as a function of the K factor, defined by the author as the reciprocal of equation (4.2), i.e.

$$K = P_d/P_m \tag{4.3}$$

Figure 4.13 depicts the fade margin for 95% coverage as a function of K.

Modem implementation is now achieved through digital signal processing (DSP) chips which enable reconfiguring modulation and coding schemes by software. Thus it is possible to conceive of a multimode terminal where modulation may be altered according to the desired radio interface. This type of implementation also allows investigation and comparison of various combinations of modulation and coding schemes under similar conditions. For example, the performance of a pilot symbol-assisted carrier phase recovery scheme was compared with a differential PSK scheme for application to a helicopter-mounted mobile which suffers from deep fades synchronized to helicopter blade rotation (Cowley *et al.*, 1997). Flexible implementation of modems and codec is central to the concept of software radio.

Figure 4.12
Fade margin as
a function of K for
BPSK and QPSK BER =
10^{-2}, 10^{-3} and 10^{-4}
(Davrian, 1985)
© 1985 IEEE

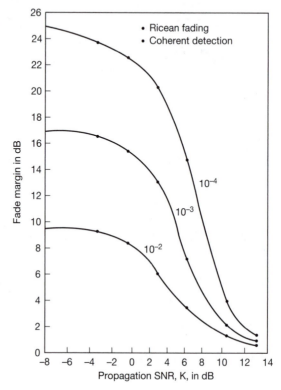

Figure 4.13
Fade margin for
95% coverage in a
Ricean channel as a
function of K
© IEEE

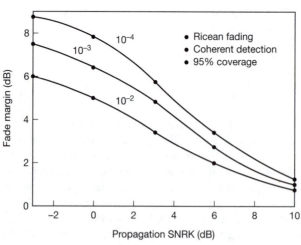

4.2.1 Coded orthogonal frequency division multiplexing (COFDM) modulation systems

Radio broadcasters have traditionally used analogue baseband signals with amplitude or FM, both of which are susceptible to weak signals in the multipath environment commonly encountered in moving vehicles. In the current market environment, a demand is being perceived for high-quality radio broadcasts, coupled with a larger variety of services. Clearly, such a service cannot be accomplished with existing transmission standards; consequently there is an interest in developing digital transmission standards applicable to such a need.

We have observed that traditional digital modulation schemes suffer in quality in a mobile environment, which worsens as transmission bit rate is increased. A multipath resistant transmission scheme called the COFDM modulation system has been proposed for terrestrial broadcasts (Pommier and Wu, 1986; Alard and Lassalle, 1988; Shelswell, 1995). The scheme can also be used for digital audio broadcasts from satellites.

As the name implies, COFDM uses coded data, which is frequency division multiplexed with carrier frequencies orthogonal to each other. Incoming data is segmented such that only a fraction of information is contained in each segment, each of which is transmitted on separate RF carriers. The reduction in bit rate per carrier increases the symbol period, with a consequent reduction in intersymbol interference, as shown in Figure 4.14.

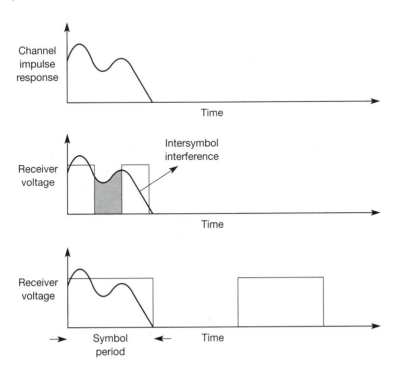

Figure 4.14

Reduction in intersymbol interference with an increase in the symbol period for a given channel impulse response. A smaller percentage of symbols are affected as the symbol period is increased

In theory, any suitable modulation scheme can be used to transmit the carrier. For example, π/4-DQPSK was used in the Eureka 147 project (a European collaborative project). Data was differentially encoded at the transmitter with the phase of the reference signal increased by π/4 for each symbol period, and differentially demodulated at the receiver.

If the band occupied by the multiplexed carriers is made greater than the coherence bandwidth of the propagation channel (see section 3.3), then the probability of all the carriers fading simultaneously is reduced. To obtain an optimal performance, it is essential that the characteristics of the propagation channels are well understood.

Further improvement to the signal quality is possible by interleaving and coding the incoming data stream. The choice of coding is determined by the propagation characteristics.

To maximize the use of spectrum, the frequencies are made linearly independent, i.e. orthogonal to each other. Orthogonal carriers permit a reduction in carrier separation and hence the occupied bandwidth. Nevertheless, COFDM is a wideband system and therefore several radio programmes must be combined to derive the maximum spectrum utilization. Mathematically, two signals' Ψ_p and Ψ_q, are orthogonal when

$$\int_a^b \psi_p(t)\,\psi_q^{*}(t)\,dt = K \qquad \text{for } p = q \tag{4.4}$$

$$= 0 \qquad \text{for } p \neq q$$

where the * indicates complex conjugate.

Figure 4.15
Conceptual building units
of a COFDM transmitter
(Shelswell, 1995)

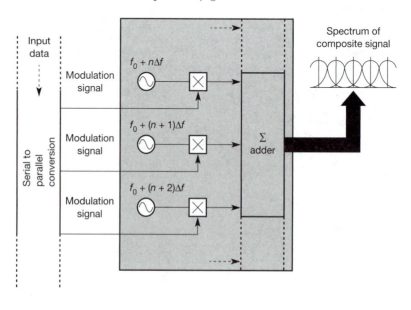

An example of an orthogonal series is sin (mx) for m = 1, 2 ... over a period $-\pi$ to $+\pi$. Applied to COFDM, the signals are orthogonal when carrier spacing is a multiple of $1/\tau$, where τ is the symbol period. The receiver consists of a bank of demodulators and a frequency translator which brings each carrier to the base-band, followed by integration over a symbol period. When carriers are orthogonal, integration reduces contributions from unwanted carriers to zero.

Figure 4.15 shows conceptual building units of a COFDM transmitter. Serial input data is converted into a parallel stream, each of which modulates a carrier from an orthogonal set which is finally summed to constitute a composite COFDM signal.

In practice, the boxed part is implemented in software by fast Fourier transform DSP chips. Mathematically, the composite signal can be expressed as

$$S_s(t) = \frac{1}{N} \sum_{n=0}^{N-1} A_n(t) e \, j \, [w_n t + \phi_n(t)] \tag{4.5}$$

where $A_n(t)$ and $\phi_n(t)$ are the amplitude and phase of the nth carrier and $\omega_n = \omega_0 + n\Delta\omega$, ω_0 is the angular frequency of carrier 1.

The digital equivalent of the signal can be represented as

$$S_s(kt) = \frac{1}{N} \sum_{n=0}^{N-1} A_n e^{j\phi n} e^{j(n\Delta\omega) \, k \, T} \tag{4.6}$$

For $\Delta f = 1/NT = \tau$, equation (4.6) becomes equivalent to an inverse Fourier transform where s(kT) is the time domain representation of the signal. A DSP chip performs an inverse transform on the incoming signal to give sampled time domain signals, which are converted to an analogue signal for transmission. To facilitate fast Fourier transform (FFT), the number of carriers N is made equal to 2^n, setting the power of the excess number of carrier numbers to 0 when the actual number of carriers is different.

A reverse process is applied at the receiver. The received carrier is synchronized, digitized and Fourier transformed to the frequency domain to provide individual carriers with the desired amplitude and phase. One of the practical problems is that of carrier synchronization. One solution is to use a coarse synchronization, followed by a precise synchronization. A coarse synchronization can be achieved by switching off all carriers regularly for a short duration; an amplitude detector can then be used to provide a synchronization pulse when carriers are switched on. Fine synchronization can be achieved by transmitting a reference signal which can be correlated at the receiver with a replica to achieve an accurate synchronization in time and frequency. Synchronization can be made robust at the expense of power efficiency by introducing guard intervals around each symbol, during which trivial or redundant data is transmitted. Guard intervals offer an additional advantage of reducing the effects of echo or co-channel interference, when the delay of interfering signals is small compared to the symbol period. A guard band of the order of 25% of the symbol period has been observed to be a reasonable compromise for terrestrial environments in the UK. The interference rejection property enables improvement to the spectrum efficiency of the network.

We have observed that the system provides frequency selective multipath rejection by frequency diversity which in effect is obtained by increasing the bandwidth of the signal. Thus it would appear that the system is less efficient than a single carrier system using the same base modulation scheme. For a single carrier COFDM, this observation may be true; however, when several programmes share the system, spectral advantages are distinct. For example, considering a bandwidth of 1.5 MHz with a base modulation scheme of QPSK and accounting for inefficiencies introduced by imperfect synchronization, coding and guard interval, a transmission rate of ~ 1 bit/Hz or about 1.5 Mbit/s is achievable. Thus the band can provide around six stereophonic broadcast channels. However, with a COFDM network, the same spectrum can be reused several times in a given geographical area to provide the same block of programmes; by contrast, if the programmes were transmitted over the conventional digital transmissions, spectrum requirements would be higher.

4.2.2 Spread spectrum modulation

Spread spectrum has been used extensively in military communications for a considerable time. Other applications include ranging and radar. A modulated signal is spread in spectrum using a pseudo-random code and retrieved at the receiver with a correlation receiver which uses the same code as the transmitted code. Spread spectrum modulation is used with a partner accessing technique called code division multiple access or CDMA, discussed later in the chapter. Use of distinct uncorrelated codes allows several users to utilize a channel with minimal effect on each other. Users can receive the signal only if the key to the transmitted pseudo-random code sequence is available and hence the system has inherent inbuilt security.

Recently the scheme has received considerable attention for application to mobile communications. In the first and second generation terrestrial systems developed in the late 1980s and early 1990s, schemes such as TDMA were superior in terms of technology. However, in the 1990s great strides in technology resulted in the application of spread spectrum to narrowband and, more recently, wideband CDMA terrestrial mobile standards. The trend has extended to MSS systems.

Classically, there are two methods for implementing spread spectrum modulation:

1　direct sequence spread spectrum;
2　frequency-hopped spread spectrum.

In each case, a pseudo-random sequence forms the basis of spread spectrum modulation systems. Figure 4.16 summarizes the main characteristics of pseudo-random sequences – auto-correlation function and power spectral density. The envelope of the power spectral density (PSD) has a magnitude of $(\sin x/x)^2$ comprising a line spectrum spaced at $(1/p)t_0$ where p is the number of bits in the pseudo-random sequence and t_0 is a period of one bit (or a chip). As the chip sequence is not truly random, there is a small component at the carrier frequency. The null of the PSD occurs at $\pm 1/t_0$ (code rate).

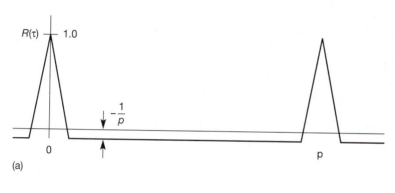

(a)

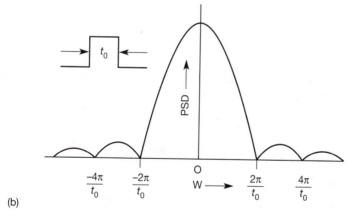

(b)

Figure 4.16

(a) Auto-correlation and (b) power spectral density (PSD) of pseudo-random sequence

Figure 4.17(a) and (b) illustrates the principle of the direct sequence spread spectrum scheme. The message is modulated using any standard scheme such as QPSK; the modulated signal is then spread by a spreading function which is up-converted, amplified and transmitted. At the receiver, the down-converted signal is correlated with a replica of the transmitted code. A correlation peak is obtained when codes match; the resultant signal is band pass filtered and

Figure 4.17
Principle of direct
sequence spread
spectrum scheme
(a) transmitter
(b) receiver

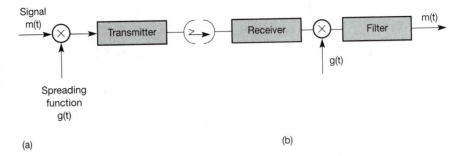

(a)   (b)

demodulated to obtain the data stream. It is possible to code the transmitted signal to improve performance.

In a frequency-hopped spread spectrum system, the transmit frequency is altered in a pseudo-random sequence. A pseudo-random chip generator changes the frequency of a synthesizer. The signal is down-converted by a synthesizer synchronized to the transmitted signal frequency and demodulated/decoded using an appropriate technique. The technique is illustrated in Figure 4.18(a) and (b).

The property of a code to discriminate interfering signals is determined by the processing gain of the scheme, defined as

$$G_p = B_c/B_m \tag{4.7}$$

Also,

$$G_p = R_c/R_m \tag{4.8}$$

where R_c = the chip rate, R_m = the message bit rate, B_c = the occupied channel bandwidth and B_m = the occupied message bandwidth.

Figure 4.18
Principle of frequency-
hopped spread spectrum
scheme (a) transmitter
(b) receiver

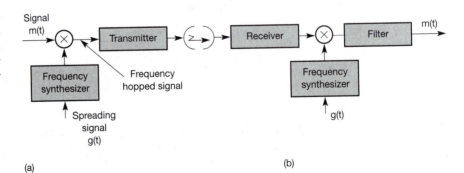

(a)   (b)

As an example, when a BPSK/spread spectrum system is interfered with by a single tone of mean power $P_j/2$, the interfering power is reduced by a factor $P_j/2(R_c/R_m)$. When the phase of the interferer is taken as a random variable, the error rate is given as

$$P_e = \tfrac{1}{2}\,\text{erfc}\,\sqrt{P_s/(P_j/2(R_c/R_m))} \tag{4.9}$$

where P_s is the power of the wanted carrier.

In addition to carrier recovery and bit synchronization, a spread spectrum demodulator must regenerate the code used for signal spreading. The process involves code acquisition, followed by tracking the code continuously.

4.3 Coding

MSS links operate with small terminals in a fading environment and limited satellite EIRP per channel amounting to a severely power-limited link (see a typical link budget in section 3.4). A robust spectrally efficient modulation scheme alone cannot accomplish the desired link reliability. Invariably, all MSS systems incorporate a channel coding scheme matched to the radio channel and application. In this section, the salient features of coding applicable to MSS are reviewed. The reader may check the literature for an in-depth appreciation of this interesting field (e.g. see Taub and Schilling, 1986). In essence, channel coding introduces redundant bits such that BER improves relative to uncoded information. There are two families of code used in MSS systems – block code and convolution code. It may be worth recapitulating some interesting features before proceeding:

- Coding increases the bandwidth of a signal. For the same transmitted power, therefore, E_b/N_o of the received signal degrades after coding.
- A coding advantage can be achieved only when the carrier-to-noise ratio is adequate for carrier recovery and bit synchronization.
- Low rate codes perform better than high rate codes when BER is high, whereas high rate codes give better performance for medium/low bit rate conditions.
- Under conditions where link continuity breaks or link quality is extremely poor, automatic repeat request (ARQ) schemes, in conjunction with coding, give an effective solution.

Block codes operate on groups of bits organized as blocks, i.e. information bits are assembled as blocks before coding (see Figure 4.19(a)). *Hamming distance* is used to measure the effectiveness of block codes. It is a measure of the minimum number of bits by which two coded words may differ. In practice, linear

Figure 4.19
(a) The concept of coding
code and code rate
(b) Main units of a
convolution coder

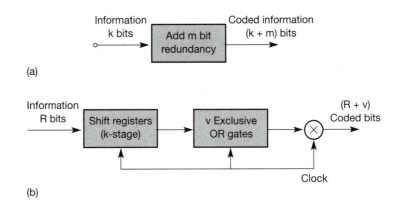

Figure 4.19 (a) The concept of coding code and code rate (b) Main units of a convolution coder

algebraic codes are commonly used due to simpler implementation. Cyclic codes provide algebraic structure amenable to decoding and are therefore commonly used. Some commonly used codes are Hamming, Bose, Chaudhari and Hocquenghem (BCH), Reed–Solomon (RS) code (a type of BCH code) and Golay code.

Convolution codes are formed by convolving information bits with the impulse response of a shift register encoder. Figure 4.19(b) depicts a block diagram of a convolution coder. Note that convolution codes use previous information bits in memory and continuously produce coded bits. The constraint length of a convolution code defines the number of information bits which influence the encoder output. Therefore the constraint length is decided by the number of shift registers, i.e. code memory v. The error correcting property of a convolution code depends on the constraint length; the error correcting property improves as code memory v is increased; however, decoding complexity increases. In a convolution code, the equivalent of Hamming distance is the *minimum distance*, defined as the minimum number of bits in a code which must be altered to obtain another valid code word. Figure 4.20 (Yasuda *et al.*, 1984) shows the coding gain achievable as a function of bandwidth expansion and code memory at a BER of 10^{-6}, using classical non-systematic convolution codes. Coding gain increases by ~ 0.5 dB for each memory addition, up to v of about 6. The coding complexity increases exponentially as $v\, 2^v$. The practical upper limit for v is about 7 for MSS applications.

Convolution codes can be decoded by sequential or Viterbi decoding. Both techniques can use either hard or soft decision code. Hard decision decoding uses code words assembled on a bit-by-bit basis, which are each independently derived using a 'hard' decision. In soft decision decoding, the decoder operates directly on unquantized analogue demodulator output and therefore knowledge of the bit state is not lost prior to decoding and for this reason it outperforms hard decision decoding, though at the expense of complexity. The soft decision decoding rule is to produce a sequence $\{a_r\}$ which is closest, in terms of minimum squared Euclidean distance, to a set of coded sequence $\{C\}$. The sequence set {C} is the specific set of code words which may be produced by the coder. The problem is defined more precisely as

$$|r_n - a_r|^2 = \text{Min } \Sigma |r_n - a_t|^2 \qquad (4.10)$$

where {a_r} lies within {C}, {a_t} is the transmitted sequence and {r_n} is the received sequence.

A Viterbi decoder is a near-optimum decoding algorithm when the code generation follows the rule of a finite state machine. From a practical consideration, a Viterbi decoder has a further advantage in that the same decoder chip can be used for decoding various coding rates by a technique known as *puncturing*. Some coded bits from a coded sequence are deleted according to a predetermined performance pattern. As an example, starting from a code rate of $\frac{1}{2}$, a $\frac{2}{3}$ rate code can be obtained by deleting the fourth bit; the matrix **P** for achieving the punctured code can be

$$\mathbf{P} = \begin{matrix} 11 \\ 10 \end{matrix} \qquad (4.11)$$

The limitation of conventional coding techniques becomes apparent in presence of error bursts lasting longer than the error correcting ability of the code. In MSS links, signal levels fluctuate widely. Fluctuations also occur due to scintillation or a combination of rain and/or cloud fades and scintillation. An effective countermeasure for correcting burst errors is *interleaving*. Information bits are dispersed over time such that consecutive bits are far apart. Provided that the separation, called *interleaving depth*, is greater than the duration of noise bursts, the technique overcomes error bursts.

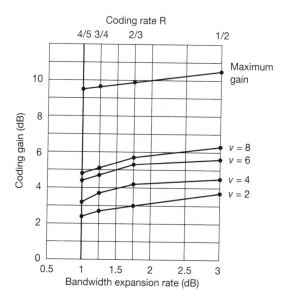

Figure 4.20
Coding gain achievable with three-bit quantization as a function of bandwidth expansion and code memory *n* at a bit error rate of 10^{-6} using classical non-systematic convolution codes with QPSK modulation (Yasuda *et al.*, 1984) © 1984 IEEE

The RS code is often used for error burst corrections. The RS code operates on groups of s bits called *symbols*. A block of RS code comprises k information symbols and r parity symbols. Hence, a code word comprises (k + r) symbols. Note that an RS code in conjunction with interleaving can considerably enhance the burst error correction capability. Moreover, decoding of RS codes is relatively simple and therefore RS codes are useful in MSS systems.

MSS links can suffer from random error as well as error bursts. It is possible to cascade two or more codes – one for correcting random errors and the other for correction of error bursts. This type of coding arrangement, called *concatenation*, can provide high coding gain with moderate complexity. Typically, in a concatenation scheme, a block code such as the RS code (outer code) is cascaded with a convolution code (inner code).

Recently a code called *turbo-code* comprising parallel concatenated convolution codes has drawn considerable attention and therefore we will discuss such schemes in some detail (Berrou *et al.*, 1993). When turbo-codes are decoded by an iterative process, they offer performance close to optimum (i.e. the Shannon limit). Furthermore, they offer high coding gain on fading channels. Efficient multilevel modulation schemes in conjunction with turbo-codes offer a potential solution to effect wideband communication at acceptable satellite EIRP and

Figure 4.21
(a) A basic rate $\frac{1}{3}$ turbo-coder (b) a turbo-code decoder (Berrou *et al.*, 1993) © 1993 IEEE

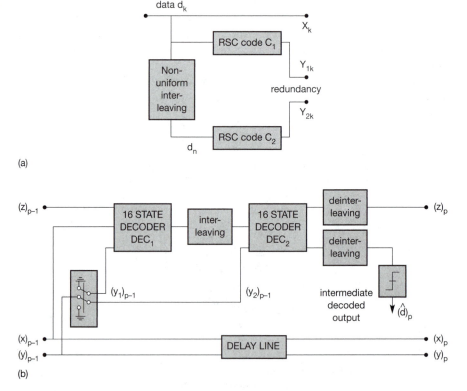

bandwidth. Recently a 16-QAM scheme with turbo-code has been introduced in the Inmarsat network to provide a 64 kbps circuit and packet-switched service via small mobile terminals (Trachtman and Hart, 1999).

Figure 4.21 shows a basic rate $\frac{1}{3}$ turbo-coder. A data stream d_k feeds directly into a recursive systematic convolution (RSC) coder C_1 and after interleaving into another RSC coder C_2, which is not necessarily identical to C_1. The transmitted bit stream comprises symbol X_k and redundancies Y_{1k} and Y_{2k} and is therefore a rate $\frac{1}{3}$ code, or it may be punctured to give higher code rate.

The turbo-code decoder comprises two cascaded elementary decoders – DEC_1 and DEC_2 (Berrou and Glavieux, 1996). The decoder DEC_1 is associated with the encoder C_1 (Figure 4.21(b)) and DEC_2 with the encoder C_2. The redundant bit y_k is demultiplexed in the DEMUX/INSERTION block and sent to DEC_1 when $y_k = Y_{1k}$ and to DEC_2 when $Y_k = Y_{2k}$. In case the redundant bit is not transmitted, the DEMUX/INSERTION block sets the corresponding input to analogue zero. The logarithmic of likelihood ratio $\Lambda_1(d_k)$, defined below, associated with each bit d_k is interleaved as in the transmitter and fed into DEC_2.

$$\Lambda_1(d_k) = \log \Pr[\{d_k = 1/\text{observation}\}/\{d_k = 0/\text{observation}\}] \qquad (4.12)$$

$\Pr[\{d_k = 1 \text{ or } 0/\text{observation}\}$ is a posteriori probability (APP) of bit d_k

The decoder DEC_1 uses a modified form of decoding scheme proposed by Bahl *et al.* (1974) and DEC_2 can use Viterbi decoding. The Bahl *et al.* algorithm is preferred to the Viterbi algorithm in DEC_1 because the Viterbi algorithm cannot provide APP of each decoded bit whereas Bahl *et al*'s algorithm does. However, the algorithm has to be modified for application to an RSC. The feedback loop gives additional redundant information Z_k, thereby improving the performance of the decoder. The name 'turbo' is given by analogy of this feedback scheme to the principle of a turbo engine. The decoder architecture is modular, comprising a number of decoder units pipelined as illustrated in Figure 4.21(b) (Berrou *et al.*, 1993). The input of the pth module at time k consisted of demodulator outputs $(X_k)_{p-1}$ and $(Y_k)_{p-1}$ fed through a delay line and the feedback information $(Z_k)_{p-1}$. Figure 4.22 shows E_b/N_o versus BER performance of a rate $\frac{1}{2}$ encoder with memory 4, generators G1 = 37, G2 = 21 and parallel concatenation R1 = 2/3, R2 = 2/3 using a Monte Carlo method; the interleaver is a 256 × 256 matrix; a modified Bahl *et al.* algorithm has been used. BER reduces as the number of iterations is increased. For 18 iterations, E_b/N_o is 0.7 dB, which is within 0.7 dB of the Shannon limit. The performance of the turbo-code improves as interleaver size is increased. Thus the code tends to increase delay as its performance improves. The turbo-code can give a several dB higher coding gain than the standard K = 7 convolution code commonly used in MSS with moderate interleaver size and shows better performance in fading channels. Coding gains of 10–11 dB have been reported, but this is at the expense of reduction in data rate or increase in bandwidth and signal delay. BER of 10^{-6} at E_b/N_o as low as –0.6 using a code rate of 1/15 has been reported. Implementation of the turbo-code is quite complicated, but use of DSPs has made implementation feasible.

Figure 4.22
E_b/N_o vs BER
performance of rate $\frac{1}{2}$
encoder with memory 4,
generators G1 = 37, G2
= 21 and parallel
concatenation R1 = 2/3,
R2 = 2/3 using a Monte
Carlo method
(Berrou *et al.*, 1996)
© 1996 IEEE

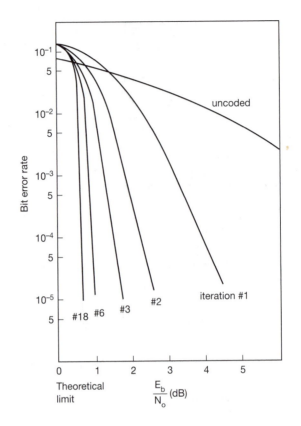

Significant research activity is in progress to refine and improve the concept (Divsalar and Pollara, 1997). Trellis-coded modulation (TCM) has been merged with turbo-code to obtain a high coding gain with improved spectral efficiency (see the next section for TCM). The concept of parallel concatenated code has been extended to block codes and interleaving property of turbo-codes has been demonstrated to provide an effective countermeasure for frequency selective fades encountered in wideband transmissions in aeronautical satellite systems (Akhter *et al.*, 1999). Turbo-code, in association with QAM constellations, has demonstrated a high gain in both Gaussian and Rayleigh channels and has outperformed 64-state TCM. Figure 4.23(a) and (b) (Le Goff *et al.*, 1994) compares BER of uncoded BPSK, 64-state TC-16-QAM and 16-QAM with rate $\frac{3}{4}$ turbo-code in Gaussian and Rayleigh channels, respectively.

In Table 4.4 the main results from this investigation for a Gaussian channel are summarized. Note the large coding gain and spectral efficiency possible with turbo-code.

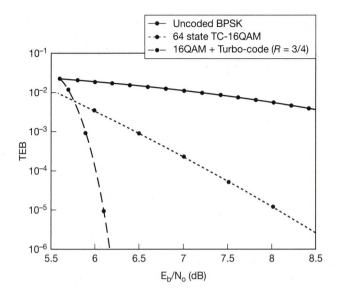

(a)

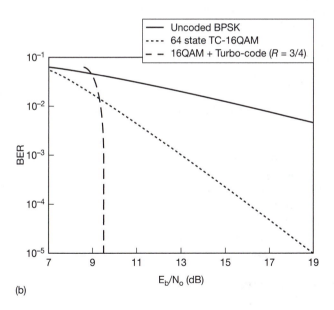

(b)

Figure 4.23
(a) BER of uncoded
BPSK, 64-state TC-16
QAM and 16-QAM with
rate $\frac{3}{4}$ turbo-code in a
Gaussian channel (b)
BER of uncoded BPSK,
64-state TC-16QAM
and 16 QAM with rate $\frac{3}{4}$
turbo-code in a Rayleigh
channel
(Le Goff *et al.*, 1994)
© 1994 IEEE

Table 4.4

Coding gain over
uncoded and 64-state
TCM offered by various
combinations of turbo-
code rate and
modulation schemes

Turbo-code rate	Modulation	Spectral efficiency	Coding gain (dB) at BER = $(10)^{-6}$ over uncoded modulation	Coding gain (dB) at 10^{-6} over 64-state TCM
$\frac{1}{2}$	16-QAM	2	6.0	2.4
2/3	8-PSK	2	5.5	1.9
$\frac{3}{4}$	16-QAM	3	7.8	2.6
2/3	64-QAM	4	5.8	2.2

When choosing a coding scheme, the error correcting capability of the code is matched to the channel and application characteristics. A vast number of codes have been devised over the years; each has a unique feature which offers an advantage under a specific set of conditions. For example, codes can be optimized to operate in low or high Gaussian noise environments, in Ricean or Rayleigh fading conditions or combinations of channel environments. In adaptive methods, code rates can be adapted to suit link conditions. Such a scheme was demonstrated under the Advanced Communications Technologies and Services (ACTS) experiment programme sponsored by NASA. Code concatenation is used when different sets of channel conditions co-exist.

Examples of block codes used in MSS include Hamming, BCH, RS and Golay. Covolution codes in common use have a constraint length of 7 and code rates of $\frac{1}{2}$ or $\frac{3}{4}$. Table 4.3 lists examples of codes used in mobile satellite systems.

Forward error correction (FEC) codes are used extensively in real-time applications. The use of time diversity for improving link reliability is also possible by a combination of interleaving or ARQ and FEC codes.

Considerable research is under way to determine suitable coding schemes for future broadband services such as proposed by SPACEWAY (Hughes Communication), WEST (Astrium), etc. These systems are meant for fixed personal multimedia-type services. Valadon et al. (1999) have studied a number of modulation/coding schemes for future K_a band systems for delivering data rates in the range 32–384 kbps. The schemes investigated included conventional convolution codes, TCM, several concatenated schemes and turbo-codes. Table 4.5 summarizes the candidates selected for further investigation in this particular study.

4.3.1 Trellis-coded modulation (TCM)

TCM combines coding and modulation techniques to provide coding gain without a loss in bandwidth or power efficiency (Ungerboeck, 1987, Divsalar and Simon, 1988). TCM schemes use multilevel modulation, whose states are coded with a finite state encoder so as to maximize the distance of signal sequences in Euclidean signal space. Signals are decoded using a soft-decision maximum-likelihood sequence decoder. Coding gains of over 6 dB are possible with more

Coding scheme	E_b/N_o	Spectrum efficiency
Convolution code (1/2, 7)	6.8	0.99
2/3-8PSK TCM (16 states) + Reed Solomon (71, 55, 8)	5.7	1.51
2/3-8PSK TCM (16 states) + Reed Solomon (63, 55, 8)	6.1	1.71
$\frac{3}{4}$-16QAM TCM (16 States) + Reed Solomon (63, 55, 8)	8.4	2.52
Convolution Code (3/4, 7) + Reed Solomon (237, 220, 8)	4.4	1.39
Convolution Code ($\frac{3}{4}$, 5) + Reed Solomon (237, 220, 8)	4.8	1.39
Turbo Code (rate $\frac{1}{2}$, M = 2, 4 iteration) + Reed Solomon (63, 55, 8)	3.2	0.87

Table 4.5

A comparison of candidate schemes for broadband personal communications

complex TCM schemes. The name of the modulation scheme stems from the fact that it can be represented by a trellis diagram similar to that used for the binary convolution code. In this case, trellis branches represent coded multi-level modulation signals instead of convolution code symbols. TCM schemes are widely used in voice-band channels for high data rate modems. Note that in conventional coding schemes redundancy is obtained at the expense of band-width expansion, but in TCM bandwidth is not expanded; coding advantage is obtained by increasing the number of modulation levels without increasing signal bandwidth or power. Take for example a rate $\frac{1}{2}$ the convolution code. Due to the addition of one redundant bit for each information bit, the transmission rate is twice, which doubles the signal bandwidth and gives a coding gain of the order of 5 dB. The TCM scheme can potentially achieve the same gain without compromising on power or bandwidth efficiency. In the case of convolution code, the Hamming distance is maximized to increase resistance to noise, whereas in TCM Euclidean distance is maximized.

Due to a combination of the convolution code with multiple-level coding, a rather complex demodulator is required. Use of block-coded modulation (BCM) simplifies demodulation (Imai and Hirakawa, 1977; Sayegh, 1986). Some of the well-known BCM decoding algorithms include Euclidean decoding and

Berlekamp–Massey decoding. It is also possible to use Viterbi decoding by representing BCM as a trellis (Li *et al.*, 1993).

The potential advantage offered by TCM has led to a number of experiments for application in mobile satellite communications. A 2/3 rate 16-state trellis-coded modem operating at a rate of 4.8 kbps was investigated for NASA's MSAT-X experimental programme (Divsalar and Simon, 1987). The TCM-coded symbols were further interleaved in 128-symbol blocks to improve performance in the presence of burst errors. A 100% raised cosine pulse shaping was used for matched filtering and Doppler shift estimation. In a Ricean fading channel with a Ricean factor of 10 dB, improvements of 1.6 and 3.1 dB in E_b/N_o (BER = 10^{-3}) over a conventional QPSK were observed with and without interleaving. Similar studies undertaken elsewhere have investigated the performance of concatenated RS code as the outer code and TCM inner code in mobile satellite channels (Francon and Bousquet, 1997).

4.3.2 Automatic repeat request

From the propagation behaviour of MSS channels, we know that there may be a break in the radio connection link due to obstructions in the signal path, sometimes lasting for several seconds. Under such drastic conditions, the FEC schemes become ineffective. ARQ schemes are used when high reliability in message transmission is required and when the associated delay is acceptable. Reliability in message delivery is further improved by storing the message and retrying transmissions at intervals until message delivery is acknowledged. In ARQ schemes, whenever an error is detected, the receiver sends a request for retransmission and the process is repeated until success is achieved. Note that in geostationary satellite systems a delay of at least ~ 1/2 s is inevitable and therefore delay intolerant services are not suitable for ARQ.

A number of schemes are possible. In the *stop and wait* ARQ scheme, a message is transmitted only when the correct receipt of the previous message is confirmed. In the *Go-back N* ARQ scheme, message transmission is continued until a request for transmission is received, when all the blocks beginning from the block where the error was detected are retransmitted. In the *Selective-request* ARQ scheme, only the corrupted block is retransmitted. The latter scheme would seem appropriate for satellite communications as it is efficient in terms of radio resources.

ARQ schemes are used in store and forward MSS systems such as the Inmarsat-C system and the EUTELTRACS system for message delivery (see Chapter 9).

4.4 Multiple access schemes

There are three basic multiple access schemes used in MSS for circuit-mode transmissions – FDMA, TDMA and CDMA. A variety of accessing schemes are used for packet-mode transmissions, which are typically characterized by traffic bursts. Such schemes may be classified as reservation and contention protocols.

In a reservation protocol, a channel is reserved for the duration of communication and, in contention protocols, users are granted access on a first-come-first-served basis. In practice, MSS systems combine a number of multiple accessing schemes to match conflicting requirements caused by different characteristics of service and practical considerations. For example, a request for channel assignment can be sent effectively through a short data burst, whereas voice communication requires a continuous end-to-end connection. Thus in this particular case, the requirement can be met by combining a data access protocol such as Aloha with a demand assigned scheme such as FDMA or CDMA.

In recent years, multiple access schemes have been extensively studied for MSS applications, due to their influence on critical system parameters such as satellite-radiated power, radio spectrum, adaptability to network expansion and mobile Earth station complexity. An access scheme should ideally provide radio spectrum efficiency, capability to manage varying traffic loads, i.e. diurnal, seasonal or event-driven, as well as general network growth, graceful degradation in case of unexpected traffic surge, flexibility in terms of managing a mix of traffic demands, resistance to multipath, shadowing, interference and technology capable of supporting low-cost terminals. Clearly, no single scheme can provide all these features. We will take a brief review of classical multiple access schemes

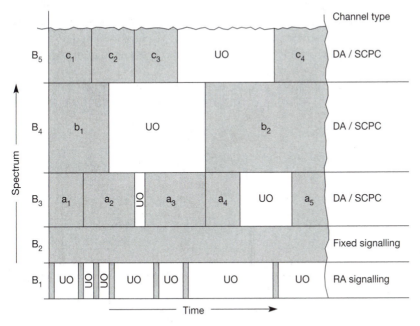

Figure 4.24

A circuit-mode frequency division demand assigned single channel per carrier multiple access for an MSS system schematic, portraying channels segmented by spectrum and channel occupancy with time

DA / SCPC = Demand assigned single channel per carrier
RA = Random access
B_n = Occupied bandwidth of channel type, n
a_m, b_m, c_m = m^{th} user
UO = Unoccupied channel

in the next few paragraphs to refresh our knowledge, followed by a comparison of accessing schemes in the present MSS context.

In the FDMA scheme the available bandwidth is partitioned into a number of segments. Each station is allowed to transmit in one or more segments according to communication needs. The concept is illustrated in Figure 4.24.

When spectrum blocks are preassigned to each Earth station of a network, the FDMA scheme is known as *fixed assigned* and if channels are allocated dynamically in response to requests, the scheme is a *demand assigned (DA)* FDMA. In the latter case, when each segment is a single channel wide, the FDMA scheme is known as a DA SCPC/FDMA scheme. DA schemes are better suited when traffic requirement per user is low, a prerequisite of MSS links. Fixed assignment is more suitable when communication demand is high or for network broadcasts to users. In an MSS environment, fixed assignments are used for communication between gateways, network broadcasts and for various types of signalling, such as call initiation.

Channel management in a DA SCPC/FDMA can be either centralized or distributed. In a centralized management scheme frequencies are maintained and managed centrally in a network control station (NCS); in the distributed management scheme, each participating fixed Earth station manages its individual pool of frequencies. The spectrum utilization efficiency of a centrally controlled system is superior, but it is prone to a single point failure. A back-up NCS can provide resilience. Distributed management offers a more resilient architecture, but it can be less spectrally efficient; as each station must have individual pools to meet its peak demand, many circuits remain idle after busy hours. Inefficiency increases in particular when traffic requirement per gateway is small and the busy hours of gateways are staggered, a situation which occurs when

Figure 4.25
Illustration of a time division multiple access scheme comprising a reference burst followed by a traffic burst. Each frame is repeated after a specified period and comprises the following fields: carrier bit recovery (CBR) for carrier and timing recovery, unique word (UW) for burst synchronization and control (C) for station indentification followed by traffic burst. The reference burst synchronizes the frame. Each participating Earth station transmits within its designated time slot

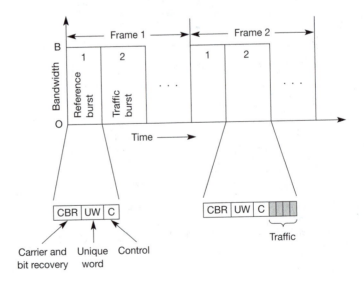

satellite coverage spans several time zones. In a hybrid scheme, each gateway has a few channels assigned on a permanent basis, while the remainder are assigned in real time from a central NCS in response to demand. This scheme allows a majority of channels to be shared, while retaining the distributed architecture.

In a TDMA scheme, users access the satellite in non-overlapping time bursts, as illustrated in Figure 4.25.

The time slots may either be fixed or demand assigned. As discussed above, for MSS, DA TDMA is better suited than fixed assigned TDMA. Again, either a central or a distributed control is possible. TDMA systems are complex, due to the need to maintain participating Earth stations in time synchronism. Complexities arise due to time difference in transmission between geographically dispersed Earth stations and a satellite's orbital motion, especially in non-geostationary systems. Moreover, synchronization of mobile terminals in fading conditions coupled with a low carrier-to-noise environment is technically challenging.

A variant of TDMA is time division duplex (TDD), which has been in use in terrestrial systems for a considerable period and was introduced in the Iridium system. In a TDD scheme, a time slot is used both for transmission and reception, in effect doubling the capacity. The time delay in each direction must be low so as to make the scheme effective. This is achievable in LEO systems where time delays are of the order of a few tens of milliseconds.

In CDMA, all users access the satellite without restriction within the full band. All users can co-exist simultaneously when spread spectrum modulation is used, since each user is assigned a unique code, which has low cross-correlation with codes used by others. Each user is able to extract transmissions destined for it by correlating the received signal with a replica of the code used at the transmitter. The correlation function peaks when codes match, whereas all other transmissions are rejected. With recent technological developments, the cost of implementing the CDMA system has reduced drastically. It is also possible to deploy a combination of FDMA and CDMA by segmenting the band appropriately.

The direct sequence CDMA scheme is affected by the so-called *near–far* problem, which occurs when interfering transmissions are received at a higher level than the transmissions wanted, resulting in excessive BER or even a loss of signal. The effect, however, is virtually absent in frequency-hopped schemes, as the probability of co-channel interference is low. The mechanism of self-interference in these two schemes is slightly different. In the direct sequence scheme, interference from all users is always present and noise-like, whereas in the frequency-hopped system interference may be coherent if the frequency of the interferer falls within the band of the carrier wanted. When the hopping rate is much smaller than the information rate, the interference is coherent, but intermittent; on the other hand, when the hopping rate is much larger, the interference becomes noise-like. One of the main problems in the frequency-hopped system is the need to maintain the phase noise within tolerance.

The characteristics of data traffic are generally different to speech and therefore dedicated channel assignments are inefficient. Data is typically characterized by bursts of high activity, followed by long pauses, which would be wasteful if a dedicated circuit were assigned. Note that some types of data traffic are continuous, for example transfer of large files. Another difference is that data traffic is less sensitive to message delivery time. Therefore more efficient use of satellite resource is possible if channels are assigned on a per message or per packet basis. A number of data accessing schemes have been developed, each matched to specific channel characteristics (e.g. Lam, 1979). We have already mentioned above that schemes range from random access on one end of the scale to reservation TDMA on the other. Random access schemes do not require any network coordination and are therefore quite simple; typical applications are registering a mobile to the network or requesting a channel assignment. The most commonly used random schemes in MSS are Aloha, Slotted Aloha and to a lesser extent Reservation Slotted Aloha. In the Aloha scheme, the user transmits a data packet whenever necessary; in Slotted Aloha the packet transmission is constrained within a time slot; this requires network time synchronization; in Reservation Aloha, a user continues transmission as long as necessary, after ceasing a time slot; at the end of transmission, the slot is again open to contention. Reservation TDMA is an example of a packet reservation scheme, useful when large amounts of data must be transferred. In this scheme, each user reserves a time before transmission identical to the conventional TDMA scheme. Table 4.6 summarizes some commonly used access protocols for various types of traffic used in MSS. The suitability of a protocol depends on a trade-off between permissible message delay, channel throughput, terminal cost and network complexity.

Table 4.6
Traffic characteristics
with a suitable protocol

Traffic characteristic	Suitable protocol
Bursty – short messages, e.g. request for a channel, a paging acknowledgement burst	Aloha, Slotted Aloha
Bursty – long messages, e.g. file transfer	Aloha for reservation; a reservation protocol for message transfer
Continuous, e.g. large file transfer, voice	Fixed assigned scheme (circuit mode)

In practice, hybrid schemes offer optimum performance. Typically frequency division is multiplexed with TDMA or CDMA; packet accessing schemes are mixed with circuit-mode DA schemes and fixed assignments are used for signalling. For the same reason, forward and return links may use different accessing schemes. For example, a fixed assigned channel carrying time division multiplexed data may be used in the forward link and a TDMA scheme in the return, as synchronization of a continuous TDM stream at the mobile is simple and synchronization of TDMA bursts is manageable at fixed Earth stations.

Similarly, a packet access scheme is far more efficient in the return direction for functions such as channel request or short messages. Table 4.7 gives examples of accessing schemes used in various MSS systems.

System	Channel function and accessing scheme
Inmarsat (Note: This is a representative sample and not an exhaustive list)	
Inmarsat-B	• Forward signalling, network broadcast, Telex, interstation signalling spot beam identifier: Fixed assigned TDM • SCPC Voice/data – Centrally controlled DA/FDMA • Return link – Request for channel assignment, call acknowledgement, ocean region registration: Aloha • Return link – low speed data, Telex: TDMA
Inmarsat-Aero	• Network broadcast, signalling and data in forward direction: Packet-mode, time division multiplexed fixed assigned transmission • Circuit-mode SCPC – DA/FDMA with distributed channel management • Request for channel assignment, signalling and data in return direction – Slotted Aloha • Data/messages in return direction – Reservation Time Division Multiple Access
Inmarsat C	• Forward signalling and messages: Fixed assigned TDM • Return signalling channel: Hybrid slotted Aloha with a provision for reserving some capacity • Return message channel: TDMA
Iridium	Multiple access – TDD
Globalstar	Multiple access – CDMA
ICO	Multiple access – TDMA

Table 4.7
Example of accessing schemes

4.4.1 Comparison of multiple access schemes

Numerous studies have been undertaken to compare the efficiency of multiple access schemes (Viterbi, 1985; Gilhousen *et al.*, 1990; Giubilei and Miracapillo, 1993; Giovanni *et al.*, 1995). Table 4.8 summarizes the main characteristics of the most commonly used schemes in MSS. It is evident that the selection of an accessing scheme involves trade-offs, as each scheme has its strengths and short-comings. Practical constraints may add further limitations. For example, an asynchronous CDMA may offer an advantage in terms of minimal synchroniza-tion requirements coupled with a self-regulating grade of service management. But it may not be possible to integrate the scheme with an operator's existing FDMA system, due to the inherent incompatibility and the operator's reluctance to modify the existing network.

Table 4.8
Main characteristics of common accessing methods

Access scheme	Characteristics
FDMA	Mature technology and low cost
	Network timing not necessary
	No restriction on type of modulation or baseband
	Susceptible to intermodulation noise introduced by system non-linearity such as Earth station or satellite high power amplifiers; weak carriers more susceptible
	Requires tight uplink power control to maintain equitable downlink transmitted power
TDMA	Efficient use of available satellite power and power control less critical when wideband TDMA carriers used (Note: this is not always the case in MSS)
	Well suited for digital systems
	Network timing is essential
	Fast synchronization of demodulator under fading and power limited conditions difficult
CDMA	Good resistance to interference, allowing closer spacing of satellites in geostationary orbits
	Soft handover possible
	Good resistance to frequency-selective fading and multipath
	Good signal security
	Graceful degradation in signal quality
	Grade of service degradation is gradual and can be perceived by a user without the specific need of a network congestion message
	More complex
	Technology is new
	A-CDMA requires power control to maximize capacity
	Wideband transparent transponders may get artificially loaded by pick-up of extraneous narrowband carriers
Packet access	Well suited for bursty traffic
	Flexible – can meet demands of sporadic or continuous traffic

A scrutiny of requirements reveals a close relationship between accessing and modulation schemes. This will become more evident later in the section when comparing their spectral efficiency. The following factors influence the efficiency of multiple accessing schemes:

- **RF interference:** A certain amount of interference is built into radio link design to allow spectrum reuse. In this context, schemes which offer higher interference rejection have a potentially higher capacity. Note the close relationship between modulation, coding and multiple access schemes.

- **Voice and data activity:** It is well known that, during a normal conversation, the average occupancy of transmission is about 40%. Similarly, it has been observed that circuit-mode data tends to have pauses (Feldman and Ramana, 1999). Therefore average interference power reduces when carrier suppression is used.

- **Variations in traffic mix:** MSSs tend to offer wide-area coverage which may cover areas with different types of communications, for example, data use may be prevalent in developed countries of the service area. Therefore a multiple access scheme offering traffic adaptability increases net system throughput.

- **Propagation effects:** The MSS propagation environment is characterized by multipath and shadowing. Accessing schemes which offer resistance to multipath can effect an increase in utilization of radio resources.

In practice, spectrum efficiency has to be considered in conjunction with other practical issues:

- **Maturity of technology** is vital for any commercial systems. Despite several potential advantages, CDMA could not be introduced in early MSSs due to insufficient technical development. For the same reason, TDMA on the forward service link was inconceivable in first generation MSS.

- For established operators, **continuity of service** is essential, which precludes the use of innovative/new technology for existing service offerings.

- The efficiency of multiple access is influenced by satellite **radio resource** availability, i.e. whether the link is spectrum or power limited.

To date, the highest revenue-earning MSS has been voice, and consequently, circuit-mode transmissions have been dominant. In the past few years, the gap between voice and data usage has been bridged very rapidly, mainly influenced by the internet phenomenon, and the trend is expected to continue. This has necessitated the introduction of more flexible higher layer transport schemes (see section 8.4), such as ATM, and more recently, IP technology. It is notable that technological limitations have been diminishing rapidly in recent years and hence the schemes which offer an optimal mix of spectrum efficiency, adaptability to traffic growth and capability to manage a different mix of traffic and support transport mechanisms will offer a distinct advantage. The non-trivial problem of backward compatibility has to be managed by operators through a carefully chosen evolution path suitable to them.

Comparison of spectral and power efficiency

The most common criteria for comparing multiple access schemes are spectral and power efficiency, which essentially translate to the revenue-earning potential of a system. Spectral efficiency is defined here as the number of bits transmitted per RF cycle.

The spectral efficiency η_s of a spread spectrum system is given as (Viterbi, 1985; Wolf, 1980)

$$\eta_s = [C/(N_0 W_s)]/(E_b/N_0')[1 + (C/N_0 W_s)(M - 1)/M] \tag{4.13}$$

where C = the total received carrier power from M Earth stations, M = the number of participating Earth stations, N_0 = the total thermal noise density, W_s = the RF bandwidth, $= 1/T_s$, where T_s = chip duration, E_b = the received energy per bit of information and $= 1/T_b$, where T_b = information bit duration, $N_o = N_o + I_o$, where I_o = interference power.

When interference from other systems exists, the total carrier power C in equation (4.13) is increased to $(1 + K)C$, where K represents interference from other systems.

Figure 4.26 shows spectrum efficiency as a function of E_b/N_o for BER of 10^{-5} and code rates using BPSK and values listed in Table 4.9 (Viterbi, 1985).

Table 4.9
Code rate vs E_b/N_o assuming convolution coding and soft Viterbi or sequential decoders

Code rate	E_b/N_o'
Uncoded	9.6
7/8	6.4
3/4	5.5
1/2	4.5
1/3	4.0
1/n (n → ∞)	3.4

Some conclusions which may be drawn from the figures are:

- CDMA systems are bandwidth limited rather than power limited.
- The limit of spectral efficiency approaches $(E_b/N_o')^{-1}$ for high (E_b/N_0) and therefore low code rates improve efficiency.
- The performance improvement is small for code rates less than 1/3.
- The reduction in code rate does not influence the bandwidth because of inherent spreading; the increase is absorbed by a corresponding reduction of chip rate.

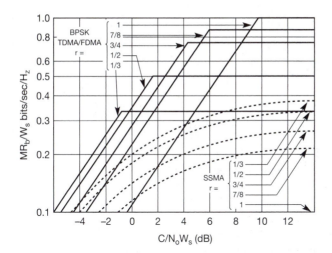

Figure 4.26

Spectral efficiency (Bits/sHz) versus $C/(N_oW_s)$ (N_o = single sided thermal noise density and W_s = occupied system bandwidth) for (a) CDMA (Viterbi, 1985) © 1985 IEEE

In practice, the following considerations apply (Gilhousen *et al.*, 1990).

- It is well known that **voice activity** in a typical telephone conversation is about 40% of call duration. By using voice-activated transmission therefore, the interference level is reduced by a factor of 2.5, increasing CDMA capacity by about the same factor.

- In a system deploying **spot beams**, the required isolation between beams where frequencies are reusable is significantly lower in the CDMA than in the FDMA scheme. When a very large number of beams are deployed, it is possible to reuse the same band in adjacent beams. This is possible because contributions to N_0' from users in adjacent spot beams can readily be absorbed in a spread spectrum system and, moreover, interference from spot beams further away becomes progressively lower.

- In FDMA or TDMA systems, **frequency reuse by polarization discrimination** is difficult due to the poor polarization discrimination possible in mobile antennas and polarization reversal caused in a multipath. The CDMA system offers a potential frequency reuse in an opposite polarization, as the interference from cross-polar pick-up is attenuated by cross-polar rejection of the user antenna. An increase in capacity of about 60% is possible with a modest antenna isolation of 6 dB.

Introducing the improvements made possible by the factors above, the total noise received can be modified as follows (Gilhousen, 1990):

$$N_0 + I_0 = N_0 + a\rho V(N-1)(E_bR_b)/W_s \tag{4.14}$$

where a = 1/number of spot beams (uniformly distributed traffic distribution), ρ = polarization isolation factor, V = voice activity factor, N= number of users E_b

and R_b are information energy per bit and bit rate, respectively and W_s = occupied RF bandwidth.

The efficiency η_s is given as

$$\eta_s = (C/N_0 W_s) \, / \, V \, [E_b/(N_0 + I_0)](1 + a\rho \, C/N_0 W_s) \tag{4.15}$$

Self-interference is one of the main causes of capacity limitation in asynchronous CDMA; transmissions from a user increase the noise to all users. A tight power control is necessary to keep system noise as low as possible, which does not actually eliminate the fundamental self-interference problem, but attempts to reduce it. An improved CDMA technique called band-limited quasi-synchronous CDMA (BLQS-CDMA) proposed by ESA researchers tackles the fundamental problem of self-noise itself (Gaudenzi et al., 1992). BLQS-CDMA offers all the advantages of CDMA, yet gives efficiencies comparable to orthogonal accessing schemes, such as TDMA and CDMA. The scheme requires chip clock and carrier frequency synchronization using the information embedded in the CDMA signal structure in the form of a master code. Another feature of BLQS-TDMA is Nyquist pulse shaping of signature chips, giving spectral compactness without compromising on detection performance, leading to simple digital implementation. The key to the improvement is the synchronization of the start epoch of the signature sequence to within $\pm$ 0.5 T_c, where T_c is the chip duration. Each transmitter has its individual signature sequence and a time reference with respect to the network reference. Preferentially phased gold codes (PPGC) provide optimal performance for QS-CDMA. By using different generator polynomials, it is possible to generate PPGC families with quasi-orthogonal cross-correlation properties within the same code family and pseudo-random correlation properties among signatures of different families. In a spot beam environment, the same family code can be used in each spot beam and different ones in other beams. Interference from the same beam will thereby be eliminated and interference from adjacent beams which should have some beam isolation will behave as self-interference of A-CDMA. The probability of error for a perfectly synchronized QS-CDMA $\rightarrow$ 0, whereas for A-CDMA the value $\rightarrow$ constant when code length $\rightarrow \infty$. Figure 4.27 (Gaudenzi et al., 1992) compares the theoretical BER for an ideal QS-CDMA with an A-CDMA for a 15-user system, showing the negligible self-noise experienced by QS-CDMA, which has been investigated and proposed for a number of mobile systems such as the European mobile system (Jongejans et al., 1993) and a K_a band mobile multimedia system using a regenerative transponder (Valadon et al., 1999).

Figure 4.28 compares the spectral efficiency, given in bits/chip, of these two CDMA techniques for a BPSK base modulation (Giubilei and Miracapillo, 1997). The efficiency of QS-CDMA is r bits/chip, where r = coding rate.

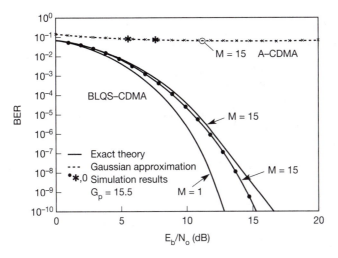

Figure 4.27
Comparison of
theoretical BER of an
ideal QS-CDMA with an
A-CDMA for a 15-user
system (Gaudenzi et al.,
1992) © 1992 IEEE

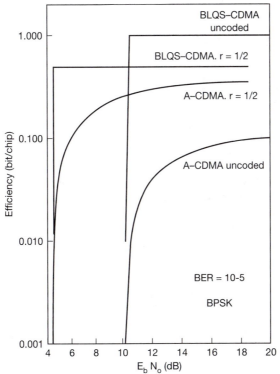

Figure 4.28
A comparison of spectral
efficiency (bits/chip)
versus E_b/N_0 of QS-
CDMA and A-CDMA
schemes (Giubilei and
Miracapillo, 1997)

In a thermal noise-dominated link, equation (4.15) also applies to FDMA and TDMA. For these schemes, C is the total EIRP of the satellite and B_t the total RF bandwidth.

(a) Power limited case

$$\eta_p = (C/N_oB_t)/(E_b/N_0), \text{ when } MR_b/B_t < G_1r \log_2 (m) \qquad (4.16)$$

(b) Bandwidth limited case

$$\eta_b = G_1r \log_2 (m), \text{ when } MR_b/B_t \geq G_1r \log_2 (m) \qquad (4.17)$$
R_b = user information rate.

where G_1 = guard band loss for FDMA and guard time/preamble loss for TDMA, r = code rate and m = PSK constellation dimension, R_b = user information rate.

Again, extending the formulation to a spot beam MSS environment and including voice activity advantage, the efficiency for each case is given as

(a) Bandwidth limited case

$$\eta_p = \alpha (C/N_oB_t)/V(E_b/N_0), \text{ when } MR_b/B_t < G_1r \log_2 (m) \qquad (4.18)$$

(b) Power limited case

$$\eta_b = \alpha G_1r \log_2 (m), \text{ when } MR_b/B_t > G_1r \log_2 (m) \qquad (4.19)$$

where α = frequency reuse factor,
v= voice activity factor.

Figure 4.29 (Gilhousen *et al.*, 1990) compares the spectral efficiencies of a CDMA and FDMA for a nine-beam system using the following link parameters:

1   CDMA: $E_b/(N_0 + I_0)$ = 2.5 dB; V = 0.35;
2   FDMA: E_b/N_0 = 8.4 dB; m = 8; r = 2/3; α = 2.

Note the increase in capacity of CDMA with the increase in $C/(N_0W_s)$, while the FDMA efficiency is limited to 2 bits/Hz. This is in contrast to Figure 4.26, where CDMA efficiency is lower than TDMA or CDMA.

Figure 4.29
A comparison of spectral efficiency of a CDMA (plot a) and FDMA (plot b) for a 9-beam system considering voice activity, antenna discrimination and polarization reuse. Spectral efficiency (Bits/sHz) is plotted versus $C/(N_oW_s)$ (N_o = single-sided thermal noise density and W_s = occupied system bandwidth). Parameters for: plot (a) – code rate $1/3$, constraint length = 9, and E_b/N_0 = 2.5 dB; and for plot (b) trellis code, rate = $2/3$, constraint length = 5, 8-DPSK and E_b/N_0 = 8.4 dB (Gilhousen *et al.*, 1990) © 1990 IEEE

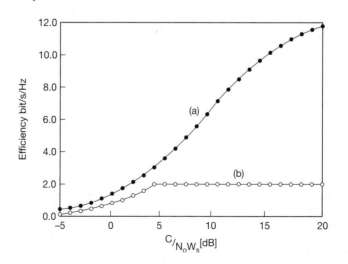

For a multiple satellite/multiple spot beam system, interference contribution is built into the total noise budget of the system. When considering non-geostationary satellite systems, intrasystem interference becomes time-dependent due to constellation dynamics, causing interference levels to become time variant and probabilistic due to signal fluctuations caused by fading, power control, etc. The multiple access scheme efficiency of such systems is also influenced by frequency planning or code design. Chapter 8 discusses various trade-offs when considering a system in totality.

Selection of a multiple access scheme for MSS requires a number of practical considerations and has been the subject of intense debate recently. The selection criteria differ according to the emphasis placed by the system designers. For example, the TDMA scheme was chosen in preference to the CDMA scheme for the ICO system after a number of studies conducted by Inmarsat, whereas Globalstar designers preferred the CDMA system.

Consider, for example, the rationale for selection of TDMA for ICO (Hart *et al.*, 1995). Note that some of the conclusions may be contradictory to the material presented above; such contradictions have been the subject of several interesting debates:

- Wideband measurements showed that frequency selective fades occur beyond 5 MHz and hence the fade resistance offered by CDMA could not be realized within the narrowband available for ICO operation.
- On the basis of an earlier investigation, TDMA was considered more efficient (Meidan, 1994).
- TDMA can benefit from satellite diversity and soft handover, much as CDMA.
- Satellite links are power limited and self-interference in CDMA links reduces link margins as the system approaches full capacity; the use of orthogonal CDMA increases capacity, but achieving signal orthogonality is difficult when satellite diversity is used, due to differential path delay.
- In TDMA, use of diplexer can be avoided if transmit and receive bursts are arranged to occur in different time slots, whereas diplexers cannot be avoided in CDMA, due to the need for continuous transmission and reception. Diplexers tend to add losses in the front end, thereby reducing a receiver's sensitivity.
- The inaccuracy introduced in power control loop due to propagation delay is likely to affect significantly the capacity of an asynchronous CDMA.
- Large interference into a CDMA signal can cause outage to all users when such interference occurs.
- Due to wideband characteristics of CDMA, the probability of interference outage in CDMA is higher than in narrowband TDMA, where interference can be counteracted by reassigning an interference-free channel.
- TDMA can better manage non-uniform traffic distribution in a spot beam environment, as it allows the peak capacity in a beam to be increased and switched between spots.

- TDMA disadvantage vis-à-vis CDMA in terms of return link power can be reduced by using non-linear amplifiers at the mobile whereas CDMA requires linear amplifiers.
- Spectrum sharing between CDMA systems can be problematic, as wideband transmissions from one system will affect the capacity of the other, whereas in TDMA the band segmentation traditionally used has minimal impact on the capacity of other operators.

In contrast, a study conducted in Italy comparing CDMA and TDMA concluded that neither was distinctly superior to the other (Priscoli and Muratore, 1996). The authors took two specific systems, the mobile satellite business network (MSBN) and a satellite extension of the global system for mobile communications (GSM), and compared radio and network aspects of the network. MSBN is a CDMA system developed by ESA for general applications whereas GSM uses the FDM/TDMA technique. ESA's L-band land mobile (LLM) satellite transponder, a multispot, geostationary satellite, was used as the payload. Comparison criteria included flexibility and number of available channels, EIRP, propagation environment and intersystem interference. It was observed that MSBN is more flexible than GSM; GSM generally has an advantage with regard to capacity when considering transportable mobile, but the reverse is true for vehicular mobiles. When considering the transportable terminal, the return link was the limiting factor with regard to the capacity for both technologies. For the GSM system satellite, EIRP was the limiting factor, whereas for the MSBN, self-noise and downlink thermal noise were the limiting factors. The CDMA system proved to be more resistant to shadowing loss in terms of system capacity for an average mobile power of 5 W; increasing the shadowing margin from 3 to 7.2 dB reduced the TDMA capacity by ~ 67%, from 192 to 64, whereas the reduction in capacity was ~ 40%, from 221 to 135, for the CDMA scheme. The study assumed that the power control and frame synchronization were perfect; neither was hardware complexity considered as an evaluation parameter. The forward and return link budgets and the maximum number of simultaneous channels are shown in Table 4.10.

The views and conclusions of system designers who have chosen CDMA will differ. Some of the advantages claimed by CDMA proponents include potential use of low power emissions from MES, the possibility of reducing call drop-outs by use of soft handover, the capability to use multipath signals to improve path diversity, etc. (see Table 4.8). Examples of operational systems are the EUTELTRACS/OmniTracs and Globalstar systems. For example, contrary to the opinions above, the Globalstar CDMA system uses techniques such as interleaving, open and close loop power control design, and diversity management to retain the advantages offered by the terrestrial CDMA system. The system uses a combination of FDM/CDM and orthogonal signal multiple access techniques. Chapter 9 provides details of Globalstar system design features including the CDMA waveform.

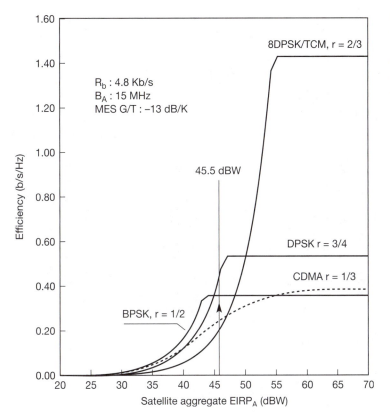

Figure 4.30
Access scheme trade-off analysis for ESA's L-band multibeam payload aboard ARTEMIS (Giubilei and Miracapillo, 1997)

The multiple access choice depends on the emphasis placed by the system designers on applicable constraints. The accessing scheme is only one element of a satellite system; other considerations include service requirements, spectrum availability, interference management technique/specifications, spacecraft characteristics, such as transponder type and antenna complexity, user terminal requirements, constellation geometry, etc. (see Chapter 8). For example, Figure 4.30 shows the results of access scheme trade-off analysis done for ESA's L-band multibeam payload aboard ARTEMIS (Giubilei and Miracapillo, 1997), showing the efficiency for CDMA and FDMA with a number of modulation schemes and code rates listed in Table 4.10.

Trellis-coded modulation parameters were taken from proposals in NASA's MSAT programme for land mobile communications. The trade-off was conducted within the specifications of the payload, i.e., transponder type, satellite EIRP, G/T, cross-polarization and adjacent beam isolation and other system parameters listed in Table 4.11.

	BPSK	8-DPSK	BPSK	CDMA
Code rate	$\frac{1}{2}$; constraint length (K) = 7	2/3, TCM 16 state	$\frac{3}{4}$; K = 5	$\frac{1}{4}$; K = 7
Signal constellation size	2	8	2	2
Channel bandwidth (KHz)	13.5	3.4	9	1000
Satellite EIRP/channel	16.7	21.7	18.2	16.4
Bit error rate	10^{-5}	10^{-5}	10^{-5}	10^{-5}
E_b/N_0 (dBHz)	3	8	4.5	–
$E_b/(N_0 +$self interference)	–	–	–	2.7
Ricean factor	10	10	10	10
Margin (dB)	4	4	4	4

Parameter	Value
Information rate per user	4.8 kb/s
Voice activity factor	0.4
Filter roll-off factor	0.4
Available EIRP	45.5 dBW
MES EIRP	–13 dB/K
Satellite transponder type	Transparent

For the case under consideration, FDMA/BPSK offers the best choice. 8 PSK FDMA becomes effective at higher levels. The performance of the CDMA scheme is pessimistic, as the voice activity advantage and polarization reuse advantages are not included. The triangular point shows the operating point of the payload which supports 445 simultaneous users using a BPSK/FDMA scheme.

Further reading

Akhter, M.S.; Rice, M.; Rice, F. (1999) 'Turbo code performance over aeronautical channel for high rate mobile satellite communications', Sixth International Mobile Satellite Conference, June 1999, 287–91.

Alard, M. and Lassalle, R. (1988) 'Principles of modulation and channel coding for digital broadcasting for mobile receivers', *EBU collected papers on concepts for sound broadcasting into the 21st century*, August, 47–9.

Amoroso, F. (1980) 'The bandwidth of digital data signals', *IEEE Communications Magazine*, 18 (6), Nov., 13–24.

Bahl, L.R., Cocke, J.; Jelinek, F.; Raviv, J. (1974) 'Optimal decoding of linear codes for minimising symbol error rate', *IEEE Transactions on Information Theory*, IT-20, 248–87, March.

Baker, P.A. (1962) 'Phase modulation data sets for serial transmission at 2000 and 2400 bits per second', Part 1, *AIEE Trans Communication and Electronics*, July.

Ball, J.R. (1982) 'A real time simulator for mobile radios', *The Radio Science and Electronics Engineer*, 52, 475–8.

Berrou, C.; Glavieux, A.; Thitimajshima, P. (1993) 'Near Shannon limit error-correcting code: Turbo Codes', *Proceedings IEEE International Conference on Communications*, Geneva, Switzerland, 1046–70, May.

Berrou, C. and Glavieux, A. (1996) 'Near optimum error correcting coding and decoding: Turbo-Codes', *IEEE Trans. on Comm.*, 44 (10) October, 1261–71.

Caves, J.K. (1991) 'An analysis of pilot symbol assisted modulation for Rayleigh fading channels', *IEEE Trans on Vehicular Technology*, VT-40, Nov., 686–93.

Cowley, W.G.; Lavenant, M.P.; Zhang, W. (1997) 'A mobile satellite modem for helicopter applications', International Mobile Satellite Conference, 479–84.

Davrian, F. (1985) 'High performance digital communications in mobile channels', *Proc IEEE Vehicular Technology*, 114–18.

Davrian, F. (1987) 'Mobile digital communications via tone calibration', *IEEE Trans. on Vehicular Technology*, VT-36 (2), May, 55–62.

De Jaeger, F. and Dekker, C.B. (1978) 'Tamed FM, a novel method to achieve spectrum economy in digital transmission', *IEEE Trans. Comm.*, COM.26 (5), 534–42.

Divsalar, D. and Pollara, F. (1997) 'Turbo Trellis coded modulation with iterative decoding for mobile satellite communications', *Proceedings of the Fifth International Mobile Satellite Conference*, Pasadena, California, June 16–18, co-sponsored by NASA/JPL and DOC/CRC; JPL Publication 97–11, Jet Propulsion Laboratory, Pasadena, June 16, 333–9.

Divsalar, D. and Simon, M.K. (1987) 'Trellis-coded modulation for 4800–9600 bits/s over a fading mobile satellite channel', *IEEE Journal of Selected Areas in Communications*, SAC-5 (2), February, 162–75.

Divsalar, D. and Simon, M.K. (1988) 'Multiple trellis coded modulation (MTCM)', *IEEE Transactions on Communications*, 36 (4), April, 410–19.

Feher, K. (1991) 'Baseband pulse shaping techniques for non-linearly amplified $\pi/4$-QPSK and QAM systems', *Proceedings of Onboard Processing and Switching Conference*, NASA, Lewis Research Center, Cleveland, OH, Nov.

Feldman, H. and Ramana, D.V. (1999) 'An introduction to Inmarsat's new mobile multimedia service', *IMSC '99, Sixth International Mobile Satellite Conference*, Ottawa, co-sponsored by Communications Research Centre and the Jet Propulsion Laboratory, 226–9.

Francon, M.G. and Bousquet, M. (1997) 'Performance analysis of concatenated Reed-Solomon Trellis coded modulation over mobile satellite channel', *Proceedings of the Fifth International Mobile Satellite Conference*, Pasadena, California, June 16–18, co-sponsored by NASA/JPL and DOC/CRC; JPL Publication 97–11, Jet Propulsion Laboratory, Pasadena, June 16, 303–8.

Gaudenzi, R. De; Elia, C.; Viola, R. (1992) 'Bandlimited quasi synchronous CDMA: A novel satellite access technique for mobile and personal communication systems', *IEEE Journal on Selected Areas of Communication*, 10 (2), February, 328–43.

Gilhousen, K.S.; Jacobs, I. M.; Padovani, R. and Weaver, L.A. (1990) 'Increased capacity using CDMA for mobile satellite communication', *IEEE Journal on Selected Areas of Communications*, 8 (4), May, 503–14.

Giovanni, E.C.; Ferrarelli, C.; Vatalaro, F. (1995) 'Analysis of multiple access techniques in multi-satellite and multi-spot mobile satellite systems', *International Mobile Satellite Conference*, Ottawa, 482–7.

Giubilei, R. and Miracapillo, L. (1991) 'The L-band land mobile (LLM) system: access schemes trade-off', *The Fourth International Mobile Satellite Conference*, Ottawa, co-sponsored by Communications Research Centre/Industry Canada and Jet Propulsion Laboratory/NASA.

Hart, N.; Goerke, T.; Jahn, A. (1995) 'A discussion on mobile satellite systems and myths of CDMA and diversity revealed', *Fifth International Mobile Satellite Conference*, Ottawa, June, 469–75.

Imai, H. and Hirakawa, S. (1977) 'A new multilevel coding method using error-correcting codes', *IEEE Transactions on Information Theory*, IT-23 (3), May, 371–7.

Jongejans, A.; Rogard, R.; Mistretta, I.; Ananasso, F. (1993) 'The European mobile system (EMS)', *Proceedings Third International Mobile Satellite Conference*, Pasadena CA, June, 305–10.

Korn, I. (1989) 'Coherent detection of M-ary phase-shift keying in the satellite mobile channel with tone calibration', *IEEE Transactions on Communications*, COM-37 (10), Oct., 997–1002.

Lam, S. (1979) 'Satellite packet communication – multiple access protocols and performance', *IEEE Transactions on Communications*, COM-27 (10), October, 1456–66.

Le Goff, S.; Glavieux, A.; Berrou, C. (1994) 'Turbo-codes and high spectral efficiency modulation', *Proceedings of IEEE ICC '94*, 1–5 May, New Orleans, 645–9.

Li, H.B.; Iwanami, Y.; Ikeda, T. (1993) 'Performance of a multidimensional BCM scheme with fading estimation based on time correlation', *IEEE ICC '93*, Geneva, May, 443–7.

Liu, C.L. and Feher, K. (1991) '$\pi/4$ QPSK modems for satellite sound/data broadcast systems', *IEEE Trans. on Broadcasting*, 37 (1), March, 1–8.

Lodge, J.H.; Moher, M.L.; Crozier, S.N. (1987) 'A comparison of data modulation techniques for land mobile satellite channels', *IEEE Trans.* VT-36 (1), February, 28–35.

McGeehan, J.P. and Bateman, A.J. (1984) 'Phase-locked transparent tone-in-band (TTIB): A new spectrum configuration particularly suited to the transmission of data over SSB mobile radio networks', *IEEE Trans. Comm.*, COM-32, January, 81–7.

Meidan, R. (1994) 'To spread or not to spread, this is the question', *Proceedings International Vehicular Technology Conference*, VTC '94, Stockholm, 56–9.

Murota, K. and Hirade, K. (1981) 'GMSK modulation for digital mobile radio telephony', *IEEE Trans. on Communications*, COM-29 (7), July, 1044–50.

Pommier, D. and Wu, Y. (1986) 'Interleaving or spectrum spreading in digital radio intended for vehicles', *EBU Review*, 217, June, 128–42.

Priscoli, F.D. and Muratore, F. (1996) 'Radio and network comparisons for MSBN and GSM systems in a geostationary satellite environment', *International Journal of Satellite Communications*, **14**, 439–54.

Sayegh, S.I. (1986) 'A class of optimum block codes in signal space', *IEEE Transactions on Communications*, COM-34 (10), October, 1043–5.

Shelswell, P. (1995) 'The COFDM modulation system: the heart of digital audio broadcasting', *Electronics & Communication Engineering Journal*, June, 127–36.

Simon, M.K. (1986) 'Dual-pilot tone calibration technique', *IEEE Trans. on Vehicular Technology*, VT-35 (2), May, 63–70.

Sundberg, C.E. (1986) 'Continuous phase modulation: a class of jointly power and bandwidth efficient digital modulation schemes with constant amplitude', *IEEE Communication Magazine*, 24 (4), April, 25–38.

Taub, H. and Schilling, D.L. (1986) *Principles of Communication Systems*, McGraw-Hill, Singapore.

Trachtman, E. and Hart, T. (1999) 'Research elements leading to the development of Inmarsat's new mobile multimedia services', *IMSC '99, Sixth International Mobile Satellite Conference*, Ottawa, co-sponsored by Communications Research Centre/Industry Canada and Jet Propulsion Laboratory, 209–12.

Ungerboeck, G. (1987) 'Trellis-Coded modulation with redundant signal sets Part I: Introduction', *IEEE Communications Magazine*, 25 (2), 5–21.

Valadon, C.G.F.; Verelst, G.A.; Taaghol, P.; Tafazolli, R.; Evans, B.G. (1999) 'Code-division multiple access for provision of mobile multimedia services with a geostationary regenerative payload', *Int. Journal on Selected Areas in Communications*, 17 (2), February, 223–37.

Valadon, C.; Rosmansyah, Y.; Tafazolli, R.; Evans, B.G. (1999) 'The SWAID Project: Deriving powerful modulation and coding schemes for future satellite multimedia systems', *IMSC '99, Sixth International Mobile Satellite Conference*, Ottawa, co-sponsored by Communications Research Centre/Industry Canada and Jet Propulsion Laboratory, 262–7.

Viterbi, A.J. (1985) 'When not to spread spectrum – a sequel', *IEEE Communications Magazine*, 12–17.

Xiong, F. (1994) 'Modem techniques in satellite communications', *IEEE Communication Magazine*, 32 (8) August, 84–98.

Yasuda, Y.; Kashiki, K.; Hirata, Y. (1984) 'High-rate punctured convolution codes for soft-decision Viterbi decoding', *IEEE Transactions on Communications*, COM-32 (3), March.

Communicators

5

5.1 Introduction

In the context of this chapter, communicators are systems or devices which provide radio connectivity between users. Terrestrial users access an MSS network through *gateways*, which are large fixed Earth stations interfacing terrestrial system(s) with the satellite component; mobile users communicate through *mobile terminals*, which vary in size and appearance from pocket telephones to large terminals using 0.5–1 m antennas (see Figures 1.9(a)–(c) and 3.16). Depending on an individual's perception and background, mobile Earth stations are called mobile terminals, user terminals, satellite phones, communicators, ground terminals, etc. In this section, we will refer to them generically as mobile terminals (MT) unless a clear distinction is necessary. Gateways are essentially large fixed Earth stations similar to those used in FSS stations at the physical layer, but incorporate a number of higher layer functions specific to MSS. Only a brief description of the radio part of gateways is given here, as they are described in a number of textbooks (e.g. Richharia, 1999). Similarly, tracking, telemetry and command (TT&C) stations are also well described in the literature and are not included here.

The focus here is on mobile terminals as they are a relatively recent introduction; there has been a rapid evolution and phenomenal interest in this technology in the past decade. There is heightened public interest regarding the biological effects of RF radiation on people due to the proliferation of cellular phones and exaggerated publicity given to claims of adverse effects of radiation. A section is devoted to this topic summarizing the current state of knowledge.

5.2 Gateways

Gateways provide a radio connection between a fixed network and the mobile elements of an MSS system. They are the correct size to receive and process a large number of simultaneous calls, which necessitates a high degree of linearity in amplifiers and mixers to minimize adverse effects of intermodulation. Gateway receivers operate over a wide dynamic range, as signals transmitted from mobiles suffer from fading and are affected by variations in satellite antenna gain over the coverage area. Adaptive mobile power control and gain shaping of the antenna footprint alleviate demanding dynamic range requirements to a certain extent (Goldberg, 1996). Gateways support network-related functions such as call set-up, radio resource management, user database management, mobility management, switching, and support to network and business management centres (see section 8.5).

Figure 5.1 shows the main hardware support entities of a large Earth station. The antenna system is usually a large parabolic dish several metres across with low noise temperature.

The RF part of Earth stations comprises a coder, modulator, up-converter, high-power amplifier and a parabolic antenna typically 10 m in diameter for the C band feeder link. Signals are received by the same antenna/feed system, amplified in a low noise amplifier, down-converted, and demodulated/decoded to reproduce the message. In systems where a reference signal is transmitted for frequency synchronization, receivers incorporate an automatic frequency correction loop.

The tracking system is an essential feature of a gateway. Common tracking techniques include step-track, intelligent track and monopulse. Step-track systems are simple and low cost and therefore widely used in MSS gateways. They are susceptible to amplitude fluctuations which may be caused for a variety of reasons, such as scintillation, rain, etc. Step-track systems incorporating intelligence offer a good compromise between cost and accuracy. Monopulse systems are highly accurate and agile and therefore best suited for Earth stations supporting applications where rapid response with high accuracy is essential, such as spacecraft operations.

Gateways have to comply with various RF and network-related functional specifications to ensure that the performance of participating Earth stations is compatible with that of others. Typical RF parameters where compliance is mandatory include antenna side-lobe patterns, antenna gain/system noise temperature (G/T), EIRP, frequency and EIRP stability, phase noise, frequency response and group delay.

The demand assigned SCPC frequency division multiplex accessing scheme is a common MSS access method. When using such an accessing scheme, gateways install a bank of single channel units commensurate with traffic requirements. A channel unit typically comprises a voice codec/transcoder and data multiplexer, scrambler, FEC encoder, frame synchronizer and modulator. For a TDMA system, a timing generator is essential in accordance with the network's TDMA synchronization technique.

Gateways also support forward and return link signalling. Forward signalling to mobiles is usually done over a time division multiplexed (TDM) broadcast channel, which is suitable for demodulation at mobiles. Signalling supported in the return direction includes random access burst transmissions for call set-up and in-band signalling for call management. Gateways exchange signalling with the NCS for radio resource and mobility management. The method of communication differs; some operators communicate on radio channels, others on terrestrial links, and yet others combine the two.

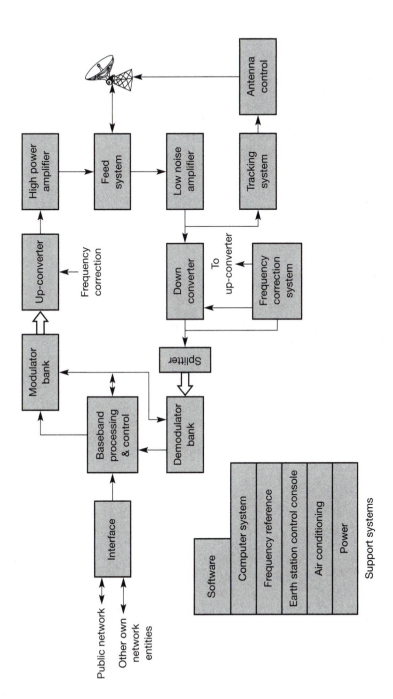

Figure 5.1
Main entities of a large
Earth station

Depending on the architecture, an MSS network may incorporate either a central or a distributed management. In a centrally managed system, an NCS manages a number of functions, such as transmission of bulletin board, spot beam identifiers and frequency reference pilots, radio resource and mobility management, performance monitoring and communication with systems' business management entities. Modern gateways rely heavily on software for control and system management.

A pilot signal is used by participating gateways in the forward direction for correcting their transmission frequency to avoid adjacent channel interference. The most significant sources of frequency error in the forward direction are long/short-term drift in Earth station and satellite oscillators, and Doppler frequency variations due to satellite motion. In the return direction, a pilot is used for cancelling frequency uncertainties caused by satellite oscillator drift and Doppler, but not frequency errors introduced by MTs due to large frequency deviations between mobiles.

5.3 Mobile terminals

MTs are low cost, low capacity and lightweight with a capability to operate in a multipath and interference environment. Terminals are generally aesthetically pleasing, easy to operate and incorporate power-saving features to prolong battery usage per charge. They are frequency agile, comply with electromagnetic interference standards, support signalling, network protocols and power control, and manage security. Terminals with speech capability use low bit rate voice coders in the range 2.4–4.8 kbps, compliant with acceptable speech quality. To support interworking with terrestrial systems, terminals have to operate in multimode. Terminal size must comply with physical limitations.

MTs may be categorized in a number of ways. They may be categorized according to their size as:

- personal communicators;
- portable;
- mobile mounted.

Personal communicators are MTs which are carried on the person; these may be pocket sized, similar to cellular telephones, or light desktop sized and easily carried by individuals (< 1 kg in weight). *Portable sets* are generally more suitable for fixed site application because of their directive antenna, which is typically the size of a briefcase/suitcase. *Mobile mounted* sets may be mounted on ships, land vehicles or aircraft with a capability to operate while the mobile is in motion.

Categorization may also be made by transmission rate as follows:

- low bit rate (0.02–4.8 kbps);
- medium bit rate (4.8–64 kb/s);

- high bit rate (64 kbps–2 Mb/s);
- broadband (> 2 Mb/s).

Terminals can also operate in an asymmetric mode whereby, depending on the application, one of the two service links operates at a higher bit rate.

MT antenna gain, G/T and radiation pattern are fundamental electrical properties which set limits on throughput. Antenna gain and transmitter power set limits on maximum EIRP transmitted from the mobile and hence through-put in the return direction. G/T sets limits on the minimum receiver noise power density and hence the throughput in the forward direction for a given satellite EIRP. Radiation pattern affects the magnitude of transmissions to and reception from the unwanted directions and hence interference received or caused by the mobile, as well as magnitude of multipath noise. Thus we note that the antenna system is a critical element.

Ship-borne antennas have the least constraint in this respect. The space constraint on terminals for small ships and yachts is more severe. Aircraft anten-nas must exhibit low air drag and comply with strict air safety regulations. Land mobile vehicle-mounted antennas have to comply with strict size, profile and cost constraints. Land portable antennas should be lightweight and low cost, and finally, hand-held terminal antennas must be a few centimetres in size, lightweight and low cost, with minimum leakage for health safety reasons.

5.3.1 Antennas

Before discussing various categories of terminal we will briefly review MSS termi-nal antennas and tracking systems.

Scattered and reflected radio signals, collectively known as multipath noise, from around a vehicle are picked up by mobile antennas, causing signal fluctuations. Due to the nature of multipath, their level increases with antenna beamwidth.

Tracking is required for medium gain antennas either when the satellite is non-stationary or in applications when calls are supported on moving vehicles. Tracking is desirable but not essential for fixed-site applications on geostation-ary MSS systems. In these applications, terminals can operate with a clear line of sight, taking full advantage of satellite power.

The ionosphere and multipath cause random polarization rotation to elec-tromagnetic waves. One method for minimizing polarization mismatch is to introduce polarization tracking; however, this poses demanding requirements on antenna design. A more effective solution is to use circular polarization, as random depolarization has an insignificant impact on it. In practice, antenna polarization exhibits some ellipticity, which is measured as axial ratios. Figure 5.2 shows maximum mismatch loss as a function of the axial ratios of the satel-lite and mobile antenna. Typical magnitudes of ellipticity for satellite antennas are of the order of 2 dB and so we note from the figure that for 0.5 dB mismatch loss, mobiles can exhibit an axial ratio of about 4 dB.

Figure 5.2

Maximum mismatch
loss for various axial
ratios of satellite and
mobile antennas
(Graphics: AR)

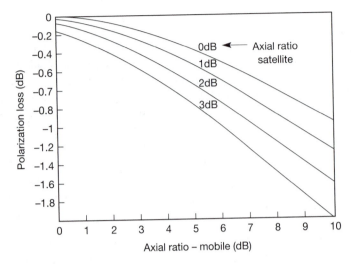

Figure 5.2

Maximum mismatch loss for various axial ratios of satellite and mobile antennas (Graphics: AR)

Typical antennas used in L-band mobile terminals, together with their characteristics, are summarized in Table 5.1. In an operational MSS, the antenna gain and side-lobe performance are specified generically to give manufacturers the freedom to optimize their terminal in the manner suited to them. Antenna gain is specified indirectly in terms of minimum EIRP and G/T and side-lobe performance is defined as a generic mask.

In the future, the use of higher frequencies such as the K_a band should enable smaller antennas in the 'high-gain' category. Note that the G/T of terminals oper-

Table 5.1

Main features of antennas and G/T of mobile terminals

Category	Tracking	Typical range of gain (dBi)	G/T (dB/K)	Common antenna type	Application examples
High gain	Step-track, program-track; intelligent track	15–21	–10 to –4	Parabolic dish on ships or microstrip array for land portables	High data rate service for ships and land portables
Medium gain	Fixed or tracking	4–15	–23 to –10	Short backfire, phased array, helical, microstrip	Voice and medium speed data for land portables, aircraft, etc.
Low gain	Fixed	0–4	–30 to –23	Dipole (drooping), quadrifilar, microstrip	Hand-held voice service, low data a rate applications on ships, aircraft, etc.

ating in the K_u and K_a bands are weather dependent because of the dependence of antenna noise temperature on water content along the signal path.

Common mobile antennas include crossed dipole antennas, helical antennas of various types, microstrip patch antennas, phased arrays and parabolic dishes. Parabolic dishes are generally used where space is not a major consideration, such as on large ships. Phased array antennas are used where aerodynamic drag should be minimized, and reliability with high tracking speed is essential. Therefore, they are used in aeronautical installations. Other types of antennas, such as crossed dipole, helical and patch antennas, are used for land mobile and maritime communications.

The crossed dipole is made of two half-wavelength antennas placed at right angles to each other and fed with equal amplitude signals, which are $\pi/2$ apart in phase. Therefore dipoles require a power divider and a 90° phase shifter for feeding the two dipoles. The antenna transmits in circular polarization with a near omni-directional pattern in the azimuth plane and an elevation angle pattern which has a maximum in the Z-axis or 90° elevation. By bending the dipoles and adjusting their distance from the ground plane, it is possible to tilt the boresight angle in any desired direction for optimizing the dipole to operate at any specific elevation. Figure 5.3 is a diagram of such a dipole. The bandwidth of such an antenna is relatively narrow.

The helical antenna consists of a number of turns of wire wrapped around a dielectric material or in air, mounted over a ground plane shown in Figure 5.4. The antenna operates with circularly polarized waves in the Z-axis direction and exhibits a wide bandwidth of around 200% with a medium gain. The gain and beamwidth of the antenna are proportional to the number of turns of the helix. A quadrifilar antenna comprises four helical antennas spaced equally around a cylinder. Typically, the ground plane has three times the diameter of the cylinder. The antenna exhibits a much wider bandwidth than a single helical antenna. The four helixes are fed with equal amplitude signal with phase shifts of 0, 90, 180 and 270°, which makes the arrangement relatively complex.

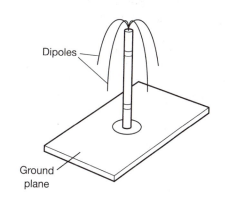

Figure 5.3
A crossed-dipole antenna

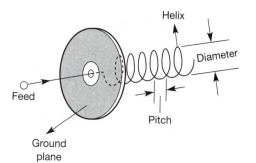

Figure 5.4
A helical antenna

Figure 5.5

A circular patch antenna

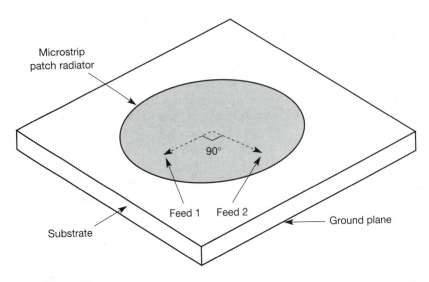

Microstrip
patch radiator

90°

Feed 1   Feed 2

Substrate

Ground plane

Antennas etched on microstrip, known as patch antennas, are useful when a low profile is essential, e.g. for vehicle mounting. Figure 5.5 depicts a single circular patch antenna for producing circular polarization.

The patch can be excited in basic or higher order mode from two feed points. The resonance frequency of the patch varies inversely with the radius of the circular patch and the relative dielectric constant of the substrate. For a substrate of dielectric constant 3, the patch diameter is 5 and 10 cms at 1 and 2 GHz, respectively. Higher order excitation can be used to gain maximization at any given elevation angle. Microstrip technology is suitable for mass production and offers a potentially low-cost solution for personal communications.

Phased arrays comprise an antenna array in which the amplitude and phases of exciting signals are varied electronically. The boresight of a far field pattern depends on the amplitude and phases of excitation, which may be changed in response to a tracking error signal to effect an enormously agile and reliable tracking system.

The design of land mobile antennas is the most challenging as they must be compact, low profile and low cost. Various types of antennas have been studied for land mobile applications (Milne 1995; Shafai *et al.*, 1995).

In general, antenna gain, multipath, blockage, polarization characteristics and antenna noise temperature of low profile antennas degrade as the elevation angle is reduced. Therefore, antennas are overdesigned at higher elevation angles, or alternatively, their gain is maximized in the elevation angle range where the operational satellite is likely to appear within the service area, a condition which suits a regional geostationary system well.

The gain of a low-profile antenna in the direction of the satellite depends on the effective projected area in the direction of the satellite and therefore varies as $\sin(\eta)$, where η is the elevation angle. In array antennas, commonly used in medium–gain terminals, electrical boundary conditions do not support transmission of circularly polarized waves close to the horizon. Furthermore, the antenna axial ratio and gain degrade at low elevation angles. For example, a low-profile antenna of two wavelengths in size has a maximum gain limit of 7 dBi and an axial ratio of 11 dB at an elevation of 15° (Milne, 1995). We have already mentioned the influence of antenna radiation on multipath.

Antenna performance is sensitive to vehicle structure and the height of the antenna above the ground plane. Reflection and diffraction, which depend on the curvature and effective area of the ground plane on the vehicle roof, cause ripples in the elevation antenna pattern. The magnitude of the ripple decreases as the elevation angle is increased and is more pronounced for large beamwidth antennas.

The antenna noise temperature of low-profile antennas depends on the elevation angle and environment around the antenna. Measured antenna temperature at elevation angles of 30–60° range between 30 and 50° K when there is a clear line of sight, increasing to 50–85° K in the presence of the wooded skyline (Milne, 1995). The system noise temperature is ≈ 200° K when diplexer and LNA units are mounted close to the antenna or made an integral part of the antenna assembly.

Portable antennas require maximum gain in the broadside direction without necessarily needing to have a low profile, whereas vehicle-mounted antennas must be low profile, compact and preferably tracking. Microstrip radiators offer a low-profile solution. It is possible to achieve near omni-directional patterns and gains of 2–6 dBi from a single microstrip radiator, while medium gain can be achieved by an array of radiators.

Hand-held terminals require an omni-directional pattern in the azimuth within the elevation angle range of interest, e.g. 5–90° for geostationary satellites and 30–90° for a non-geostationary satellite system and typical gains of 0–1 dB with an axial ratio < 5 dB. Quadrifilar helix and patch antennas can meet such requirements. An L-band quadrifilar helical antenna proposed for operation with a geostationary satellite is illustrated in Figure 5.6 (Caballero *et al.*, 1995).

The half-wavelength helix has an axial length of 0.4–0.5 wavelength in air, a total height of 15.5 cm and an external diameter of 2.4 cm without a radome. The antenna is fed from the bottom through a wideband balun to balance the antenna and a matching circuit, comprising a quarter-wavelength transformer and a circuit with two parallel stubs. The head–antenna interaction is an important consideration in antennas used in hand-held terminals (Chuang, 1994; Toftgard *et al.*, 1993). Studies show a 10 dB or more attenuation due to obstruction from the head and so the antennas were designed to operate above the head as shown in Figure 5.7.

In applications where a tracking antenna is necessary, either mechanical or electronic steering with phased array is used. Phased arrays, commonly used in aircraft, offer a low-profile solution giving high steering speed, adaptability to different environments, reliability and lower power consumption than their mechanical counterparts, but at present the technology is not amenable to low-

Figure 5.6
An L-band quadrifilar
helical antenna
(Caballero *et al.*, 1995)

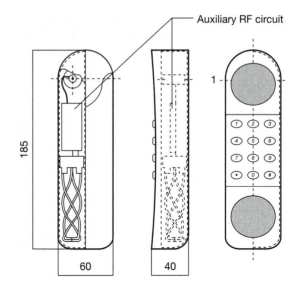

Figure 5.6
An L-band quadrifilar
helical antenna
(Caballero *et al.*, 1995)

Figure 5.7
Antenna position during
a call (Caballero *et al.*,
1995)

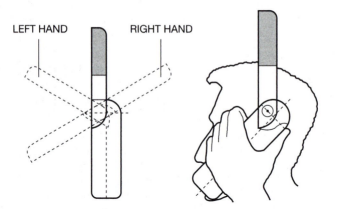

LEFT HAND   RIGHT HAND

cost applications. Phase shifters used in phased arrays are lossy and therefore increase the antenna noise temperature which has to be compensated by increasing their size. An increase in the number of phase shifters such as PIN diodes tends to increase the cost. However, the antenna design lends itself to high volume, repeatable mass production with potentially low cost. Costs of the order of $300 per unit (1995 base) may be achieved for a sustained requirement of around 10,000 units per year, but at present the use of such antennas is limited to applications where cost is not the primary consideration. Low-loss mechanically steered antennas hold an advantage in respect of cost. Typically such antennas comprise a planar microstrip array which is mechanically steered in azimuth by low-profile stepper motors. For regional geostationary satellite systems, the elevation pattern can be shaped to the desired elevation angle range.

Tracking and acquisition of satellites in a mobile environment are demanding due to the vehicle movement and satellite movement for non-geostationary satellite systems. Angular speeds of up to 60°/s and acceleration of more than 50°/s² may be required necessitating updates at rates in excess of 6 Hz. A simple tracking system is necessary to maintain low cost. Step-track is a simple method where the tracking receiver maximizes the received signal level by stepping towards it using small exploratory steps. The technique is unreliable in the presence of signal fluctuations. Medium-gain (10–12 dBi) L-band monopulse antennas for vehicle use have been developed by the ESA for a European mobile satellite system and MSAT programme (Garcia *et al.*, 1993). The ESA design used microstrip technology for its potential of low cost and mass production and the MSAT system is based on a modulated monopulse scheme developed by JPL for the MSAT programme (Jet Propulsion Laboratory, 1988). In a conventional monopulse technique, two balanced receiver chains and a dual-channel rotary joint are required – one for the sum signal (Σ) and the other for a difference signal (Δ), which makes the system bulky and expensive. In the *modulated monopulse scheme*, the sum and difference signals are multiplexed, thereby removing the need for two receiver chains. The monopulse beam former produces the sum and difference signals which are fed into a monopulse modulator comprising a phase shifter and a directional coupler. The difference signal is the amplitude modulated with a square wave and added to the sum signal in the directional coupler; the multiplexed signal comprises signals at $f_c \pm f_m$, where f_c is the centre frequency and f_m is the modulation frequency. The signals are down-converted, divided and detected separately to provide tracking correction in a microprocessor-based tracking processor. The error signals are used to drive stepper motors, which move the antennas to the desired direction in steps of 7.5° at speeds up to 40°/s. There is a provision in the tracking processor to obtain data from an external sensor to accommodate tracking when the signal is lost due to shadowing. Figure 5.8(a) shows a block diagram of the full system, including the receiver. The antenna comprises three patches providing a gain of 10–12 dBi etched on a low permitivity substrate to reduce antenna mass for low inertia, as well as to reduce the number of elements. The circularly polarized patches are fed into the RF comparator to provide sum and difference signals as shown in Figure 5.8(b). The tracking system was used to track signals from Inmarsat satellites providing accuracies of 0.72–3.90° during test runs.

Open-loop techniques are not susceptible to signal fluctuations as they derive error signals from external sensors such as a flux-gate compass and gyro-compass or a vehicle's inertial navigation system. The error signals for each update are derived by taking the difference between the current estimated position of the satellite and the previous location and knowledge of vehicle position.

A simple magnetic compass needle is the simplest sensor, but it is inaccurate, sluggish and more importantly does not include an electrical interface. A flux-gate sensor can provide electrical signals for estimating a north direction to an accuracy of a few degrees through a toroidal coil and a sensing coil. DC

Figure 5.8
(a) A block diagram of
the tracking system
(Garcia *et al.*, 1993). (b)
Sum and difference
signal (Garcia *et al.*,
1993)

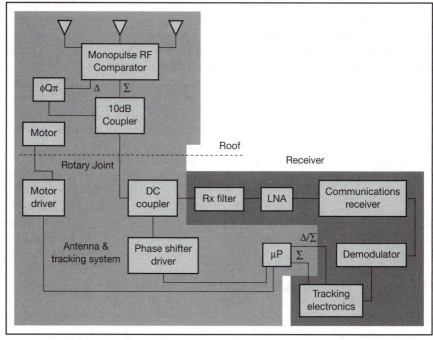

(a)

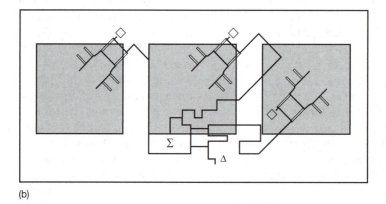

(b)

current is passed through a coil wound over a highly permeable core, causing
the toroid to be magnetized and produce magnetic flux. Two sensing coils are
wound orthogonally to each other on the toroid – one coil to sense the
north–south component and the other to sense the east–west component. Each
coil comprises two oppositely wound segments at opposite ends. A current flows
through the sensing coil whenever there is a difference in flux density between
the two segments. The sensor has to be calibrated for a difference between true
and magnetic north as well as for a vehicle's magnetic environment.

The fibre-optic gyro outputs angular velocity or a mobile's orientation using an optical phenomenon known as the Sagnac effects which states that the phase difference produced between lightwaves travelling in opposite directions of an optical loop is proportional to the angular velocity of the loop. The sensitivity of the sensor depends on the area of the loop and therefore multiwound coils are used in practice. The fibre-optic sensor is insensitive to acceleration, has a rapid response and a large dynamic range. A phase modulation-type optic sensor is a commonly used sensor. The sensor comprises a loop, a solid-state laser source, a phase modulator and a photodetector for phase detection (Ohmori *et al.*, 1998).

Open-loop systems tend to be more expensive and generally require some form of closed-loop tracking for initial acquisition. Step-track systems are often a cost-effective method of tracking for such a purpose. Combining both methods of tracking also provides a higher level of accuracy and reliability. The open-loop technique is well established in aeronautical systems, as inertial sensors onboard aircraft are readily available, except in some smaller planes and helicopters, where they are difficult to interface. In such cases, low-cost three-dimensional sensors are necessary. A low-cost technique for such applications developed by the Communications Research Centre proposes the use of a geomagnetic field together with satellite antenna pointing vectors (Sydor and Dufour, 1993). The magnetic field sensor provides the instantaneous geomagnetic field in a three-dimensional space by use of an orthogonal adaptive magnetic sensor. The raw output of the sensor is processed in a neural network which adapts the outputs to magnetic perturbations on the aircraft. The output provides changes to aircraft pitch, roll and yaw, which are mathematically transformed to satellite look angles. The correction is derived by estimating the difference between the wanted and actual look angles. Figure 5.9 illustrates the experimental set-up used for proof of this concept.

5.3.2 Hand-held communicators

A number of recent systems have targeted the hand-held telecommunication market in the expectation that satellite phones would be as appealing as cellular phones in certain geographical areas and to certain types of users. Typical services on offer include voice, circuit and packet-switched data, facsimile, paging, messaging and position location. Terminals may be required to support supplementary services, such as position location and others available in terrestrial networks, such as conditional call diversion, caller identification, voice messaging, charging and payment support, etc. Users are generally not interested in technology or systems and expect a similar quality of service on satellite systems to that offered by cellular phones. However, as satellite systems are unable to offer coverage in heavily shadowed environments, identical service quality is not feasible under all conditions. It is anticipated that some user cooperation and education will be essential, so that users can derive maximum benefits from satellite technology by being aware of its limitations. System designers on their part try to minimize the impact of shadowing by using features such as short

Figure 5.9

Experimental set-up of a
low-cost tracking system
for aeronautical
applications (Sydor and
Dufour, 1993)

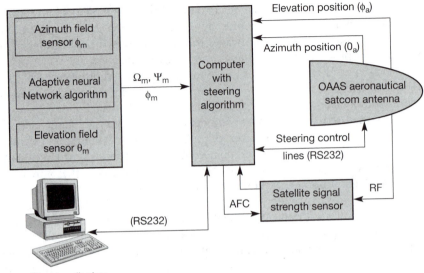

Tracking display
computer/data acquisition

messaging and alerting when the user terminal is heavily shadowed and increasing link margins as much as practical.

A number of unique requirements arise for the design and production of hand-held satellite telephones. Some of the critical drivers are the cost of terminals, service, multimode support, user interface, aesthetics, prevailing fashion, size, weight, battery lifetime and antenna pointing.

The EIRP and G/T requirements are fundamental RF parameters which determine the size of a user terminal. Low-cost terminals use small non-tracking antennas and transmitted power is limited due to the need to maintain transmitter power within safety limits (see section 5.4). The space segment is designed such that these limitations are accommodated. Table 5.2 lists a typical set of RF requirements for a TDMA system in which users operate at high elevation angles through a HEO satellite system serving Europe (Stojkovic and Alonso, 1995). Table 5.3 lists service and terminal requirements of a typical hand-held service.

Table 5.2

Parameter of a TDMA
terminal for operation
with an HEO system

	Terminal type	
	Hand-held	**Notebook**
Antenna gain (dBi) (Edge of coverage)	1–2	8.5
Average transmit power (mW)	<500	250
Peak EIRP (dBW)	7	8.5
G/T dB/K (Edge of coverage)	–22.5	–15

Service	PSTN voice;
	Satellite and terrestrial cellular support
	Optional service
	Duplex data at ≥2.4 kbps
	Facsimile at ≥2.4 kbps
	High penetration paging
	Navigation service
Band	L (1.5/1.6 GHz) and S (2.5 GHz)
Average transmit power	0.25 W (averaged over 6 minutes)
Antenna size and polarization	Diameter = 10 mm, height = 100 mm; circular polarization
Digital signal processing capability (MIPS)	100
Telephone size	Less than 300 cc
Talk time between charges	1 hour
Idle time between charges	24 hour
Cost	Less than $750 for dual-mode phones

Table 5.3

Typical service requirements of a hand-held terminal. Note that the target equipment price is sensitive to external influence such as business strategy and technology evolution; the trend is downwards

The talk time for each battery recharge is an important practical consideration, especially for users who may stay away from base for a long time. Current targets in cellular systems are for stand-by operation of about 100 hours. To improve talk and stand-by time, battery drainage should be reduced in addition to improving battery technology. A power amplifier consumes a significant amount of battery power and improving its efficiency is vital.

The rate of improvement of battery technology is much slower than that of electronic components. Nickel cadmium (NiCd) and nickel metal hydride (NiMH) batteries are common battery technologies for portable sets. The latter offers a higher capacity and is more environmentally friendly, but more expensive. A number of new technologies are under development. Lithium-ion batteries are an emerging technology, which has the potential to reduce battery sizes and weight considerably, offers a longer life and cycle time and is more environmentally friendly in terms of disposal; however, at present such batteries are expensive. Improvements in terms of energy density are possible when lithium polymer batteries have been developed. Leakage of harmful electrolytes is not possible with this technology, as it uses a polymer instead of liquid electrolyte and enables cells to be shaped. In the longer term, it is anticipated that use of NiCd will decline because of the toxity of cadmium and lower energy density. Table 5.4 compares AA-size batteries of three technologies (Scholey, 1995).

There is also interest in introducing replaceable alkaline batteries as the power requirements of personal communication products fall. This is already evident in pagers, many of which use a single replaceable battery. However, considerable power reduction will be necessary before it becomes a viable option in mobile phones.

In addition to improving battery life, manufacturers are improving battery power management by including features such as 'doze', 'sleep' and 'suspend', which reduce battery drainage, and introducing intelligent batteries, which include

	Nominal voltage (V)	Rated capacity mA h	Weight (g)	Gravimetric energy density Wh kg$^-$	Volumetric energy density Wh l^{-1}	Cycle life Number of cycles
Nickel cadmium	1.2	600	24	30	89	700
Nickel metal hydride	1.2	1,100	26	51	149	500
Lithium-ion	3.6	450	18	90	203	1,200

features such as visual communication of battery status to users or passing battery data on to host systems. Such features can extend battery life, e.g. the user can budget battery usage and reduce the probability of data loss in case of a shutdown.

The complexity of a terminal is increased by a need to compensate for the Doppler Effect and support active power control. The size, mass and cost of terminals are minimized by the use of VLSI and mass production techniques. One further consideration is the degree of user cooperation built into the system. If a user is willing to operate in the open, higher link margins are possible, but the user must be aware of the limitations and be willing to cooperate. For a situation where no other type of communication is available, the operator can expect a greater degree of cooperation.

Due to the large number of supplementary features available in a typical mobile phone, it is essential that the human-machine interface be as simple as possible. At present, most cellular systems incorporate a small keypad with a graphical display for menu-driven interaction; other features include voice-activated commands, memory, muting, auto-answer, hands-free operation, subscriber identity module (SIM) card support, integration with a personal digital assistant (PDA) or a personal computer for e-mail, web browsing, etc. Third generation telephones offer wider bandwidth and larger screens for image display, etc. Several satellite operators are likely to introduce satellite terminals capable of transmitting at speeds supported by terrestrial third generation systems.

Technology issue

Satellite phones are unlikely to see volumes similar to cellular phones in the foreseeable future and therefore manufacturers have tended to maximize use of cellular technology for satellite phones. The discussion in this section addresses technologies related to cellular phones where commonalities are possible. To support a multimode receiver, it is necessary to develop a generic receiver with a common front end, followed by a programmable modulator/demodulator and baseband units. Several new personal satellite communication systems have been developed to operate with partner GSM and/or CDMA systems in various parts of the world.

One possible architecture of a hand-held terminal is illustrated in Figure 5.10. Typical antenna sizes are expected to be about 100 mm in height and 10 mm in diameter, and typical EIRP and G/T requirements are about 500 mW and

–25 dB/K, respectively (see section 3.4). A diplexer is a critical unit in the front end of a receiver as it isolates the transmit and receive chain but it also introduces a loss at a point where loss must be minimized. The power amplifier and low noise amplifier use MMIC technology and it is possible to integrate the entire RF section in a single MMIC. HEMTS is a likely technology for a low noise amplifier.

Modem, voice processing including voice recognition, user and peripheral interface control and housekeeping functions are best performed by digital signal processing chips and microcontrollers. The digital signal processing chips require highly linear analogue to digital converters with a large dynamic range. The attempt is to convert analogue signals to digital as close as possible to the RF, which brings processing advantage at an early stage allowing more precise timing, etc.

Battery technology is important, not only for capacity but also because of weight considerations and other less obvious needs such as environment friendliness. The sleep mode is necessary to conserve battery power where only the most essential subsystems are active while in a standby mode. With the current state of technology, battery weight is expected to dominate the handset weight.

In terrestrial systems, the handsets have reached the lower limit of size. The chips are already the smallest part of the set compared to the size of the display or battery. However, further reduction in battery drainage can be achieved by reducing the power requirements of chips. Reducing the number of chips improves power consumption, as more power is lost in the PCB interconnections than within a chip itself. Furthermore, by integrating more functions in chips, chip packaging and fabrication costs reduce. Another target is to reduce supply voltage, which increases talk and standby times. Alternatively, retaining the same battery supply, it is possible to pack more circuits into the same volume, which can be used to improve performance by way of more flexible software.

Currently 2–3 chips are used in both RF and baseband. The intention is to introduce single-chip design using 0.35μ CMOS, which would allow easy customization as well as reduction in power. Making changes to the architecture of a customized design is easier if only a single chip is used. Increasing the number of chips would require more changes and hence increase the cost and time. Customization is being demanded by smaller phone manufacturers in order to

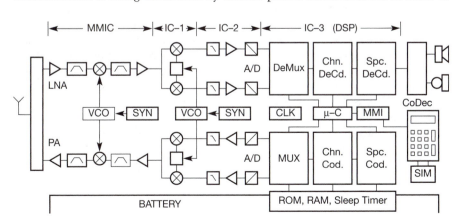

Figure 5.10
An architecture of a hand-held terminal (Stojkovic and Alonso, 1995)

distinguish their product for marketing, etc. On the RF side, the effort is to reduce the number of fixed components such as resistors, capacitors, etc., down to 100 from 300, making the RF part cheaper and easier to assemble. Various approaches are used for signal processing of analogue and digital signals. Some companies use separate chips for each; others mix functions. Separate digital and analogue signals avoid cross-talk and allow an independent process for each. CMOS technology, which has a leading edge, is easier to implement by separating analogue and digital technologies. It is believed that eventually RF and baseband may be combined into a single chip.

Technologies for integration of different RF bands such as GSM (900 MHz) and DCS 1800 (1,800 MHz) as well as a combination of different modes (GSM and DECT) are already in an advanced development stage. This is occurring in two stages. In the first stage, similar to the integration of PDA with a mobile phone, telephones using two different standards are assembled together in a single package. The real goal is to integrate them at a chip level. Integration of DCS with GSM is less problematic, as baseband technologies are similar – the main difference lies in the RF; whereas integration of GSM with DECT is more challenging, due to their entirely different technologies. Manufacturers are also integrating, on a single chip, baseband processing with PDA functions and GPS receivers to provide new functionality. Problems in the integration of PDA with GSM include the incompatibility of PDA operating systems with GSM, which operates in real-time, and the need of higher voltage for driving the PDA LCD, which negates the battery-saving features of a GSM set. Evolution of such terrestrial technologies has led to the introduction of dual-mode satellite/terrestrial terminals. Chip-level integration of satellite/terrestrial phones does not seem far away, provided demand rises to sufficient volumes.

Rapid advancements in cellular technology have led to very strong competition between manufacturers and rapidly changing handset designs because of changes in consumer tastes. A similar trend can occur in satellite phone markets, if volumes pick up. These changes place a heavy demand on manufacturers and in particular those in the VLSI area where typical turnaround times for chip manufacture are about 9–12 months. Each development poses stringent requirements on cost, size and power consumption and the need to deal with multiple standards. The typical obsolescence time of a chip reduced from 2–3 years in the mid-1990s to less than a year by 2000. One technique to manage such rapid changes is the use of development platforms which can provide a low-risk, rapid route for chip development.

A microprocessor card known as SIM is the personalized part of a mobile terminal. The SIM provides user authentication and radio transmission security, and stores user data. All the subscriber's information data, which the network authenticates for granting network access, is security protected. The SIM card can be ported to any terminal, giving the user the flexibility to use any terminal, as the network considers the SIM as the user's subscription account.

Figure 5.11 shows a breakdown of costs for a GSM-type cellular phone at several levels in the late 1990s (Mobile Europe, 1997). Material cost dominates the cost; electronics has the maximum material cost, out of which chips cost the maximum.

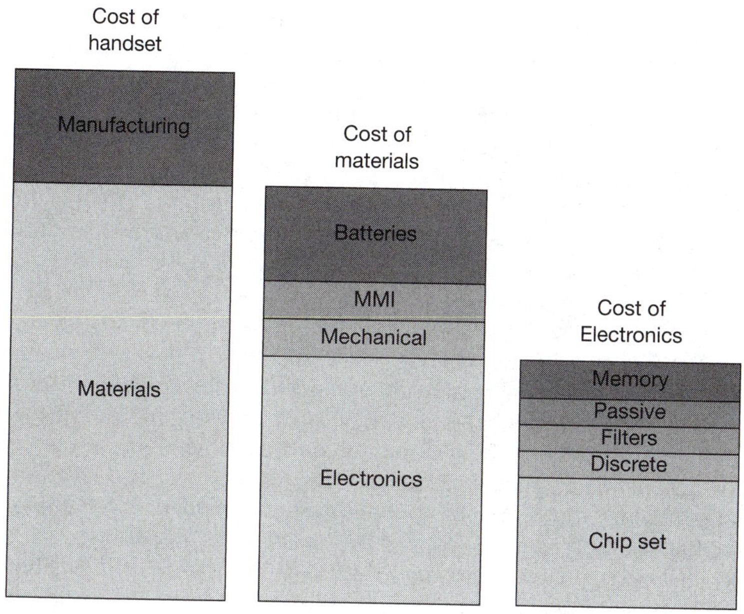

Cost of handset

Manufacturing

Materials

Cost of materials

Batteries

MMI

Mechanical

Electronics

Cost of Electronics

Memory

Passive

Filters

Discrete

Chip set

Figure 5.11
Breakdown of costs for a GSM cellular phone at several levels in the late 1990s (Mobile Europe, 1997)

Examples of message signals

Mobile terminals are used for a variety of communications, which include voice, facsimile, electronic mail, file transfer, compressed video, etc. Most modern MSS systems are digital transmissions with a provision to interface with existing analogue terrestrial systems.

In an analogue system, voice-band data is sent through a modulator-demodulator (modem), which modulates the incoming data stream to a compatible analogue signal; at the receiver, signals are demodulated to retrieve the digital stream. To harmonize their use throughout the world, the ITU has developed standards.

- V22bis is a standard for 2.4 kbps transmissions, suitable at best for hand-held terminals.
- V32bis is the standard for 9.6 kbps modems. The modems include error correction between modems as a standard.
- V32bis is the standard for 14.4 kbps modems. These modems have a lower handshake time and are more reliable.
- V34 is the 28.8 kbps standard.

Similarly, facsimile standards were introduced by the International Telegraph and Telephone Consultative Committee (CCITT). Analogue standards Group 1 and Group 2 fax standards were too slow for non-textual information, such as drawings and diagrams. In response, the CCITT adopted the Group 3 standard (in 1980) for the digital transmission of images through modems, which provided a threefold improvement in speed compared to Group 2 standard machines. Originally, the Group 3 fax was developed for transmission of 2.4–9.6 kbps data. With recent advances in voice-band modem technology, transmission rates of 14.4 kbps and above are common. In 1984, the CCITT adopted a Group 4 fax standard to supersede the Group 3 fax by permitting error-free transmission of documents over digital networks, such as ISDN, at speeds up to 64,000 bits/s. At such rates, transmission time for a single page is less than 10s.

Electronic-mail or e-mail comprises text, graphics and sometimes sounds and animated images exchanged between users through a computer network. Internet e-mail is used most extensively by remote mobile users who may also access their company intranet. The e-mail system is now an indispensable part of business and personal communications. A commonly used standard adapted for supporting e-mail is X.400.

There has been interest in two-way video communication over telephone lines for a considerable period of time. A videophone system comprises a video camera, a display, microphone and speaker at both ends. Analogue signals are converted to digital signals, compressed using a standard such as MPEG and transmitted over a digital line to the destination, where a reverse operation retrieves the transmitted signal. Such a system was demonstrated as early as 1927 in the USA by AT&T. A public service, called picture-phone, was introduced in the USA in 1971. Despite considerable development and marketing efforts, the market response to service was poor, resulting in withdrawal of the system in the late 1970s, proving it to be a concept searching for a non-existent market. Further developments during the 1980s culminated in the introduction by AT&T in 1992 of a videophone service over analogue lines. The system uses video compression to an extent that it is possible to transport full motion colour video transmissions over V.34 modems at rates of 16.8–19.2 kbps. The CCITT introduced a standard in 1990 for operating full motion colour videophones in the range 56 kbps–2 Mbps over the digital network. The standard supports full motion colour video with a resolution of 288 lines at 352 pixels per line or 144 lines at 176 pixels per line. Also, a standard for video-conferencing at rates of about 384 kbps was adopted. Video clips are sent regularly over standard MSS circuits, using high compression coupled with data storage. Live video transmissions from remote locations over 64 kbps MSS circuits are in regular use by journalists, etc. Video usage in MSS is expected to increase significantly as the public becomes more aware of such services through terrestrial third generation mobile systems. The ITU has set a target for standardizing videophones to operate in the range 9.6–28.8 kbps, well within MSS capabilities.

Television broadcasts targeted for MTs require a better quality and a larger transmission rate. Television transmissions in 400 kHz using the QPSK modulation scheme were made regularly to US ships via the Inmarsat network and television broadcasts for commercial airlines are about to be introduced. Video-conferencing also requires a higher quality of service and is available over fixed satellite networks.

Voice coding

Bandwidth efficiency is of paramount importance in an MSS system. Traditional voice coding techniques such as pulse code modulation (PCM) or adaptive differential PCM (ADPCM) are too inefficient for the purpose, offering bit rates in the range 16–64 kbps. Such coding techniques attempt to reproduce the transmitted waveform by sampling and coding and hence there is a lower limit on the sampling speed, in accordance with the Carson rule. By contrast, source coding or parametric coding extracts specific characteristics of speech for digital coding. Speech consists of a sound-carrier which is modulated by the intelligence component. The bandwidths of speech parameters are narrower, which allows the coding rate to reduce in the range 0.04–4 kbps. A hybrid method combines the best elements of waveform and source coding.

In general, the complexity of the coding technique increases as the coded bit rate is reduced. DSP chips reduce the cost to acceptable limits. Recent MSS speech systems operate at a coding rate in the range 2.4–8 kbps allowing RF bandwidth of 5 kHz. In linear predictive coding (LPC), commonly used in MSS, speech is modelled as a time-varying linear predictive filter. Examples of speech coding techniques used in mobile communications are adaptive predictive coding with maximum likelihood quantization (APC-MLQ) at 16 kbps used in the Inmarsat-B system; multi-pulse excited LPC (MPE-LPC) at a bit rate of 9.6 kbps in the Inmarsat aeronautical system; improved multi-band excitation (IMBE); code-excited linear prediction (CELP) and its family, such as VSELP, used in the IS-54 North American system at 7.95 kbps and 6.7 kbps, and PSI-CELP at 3.45 kbps in the Japanese PDC system; regular pulse excitation and long-term prediction (RPE-LTP) used in the GSM system at 13 kbps, etc.

A measure of distortion in the digitized version of speech is quantization distortion caused by speech digitization, measured as the quantization distortion unit (QDU). One unit is the equivalent of distortion resulting from converting speech to a 64 kbps PCM and then reconverting it back. QDU performance standards have been recommended for system designers, e.g. ITU specifies a limit of 14 QDU for any international connection. The distortion is distributed across the network, e.g. in a 5-4-5 distribution; the originating country is allowed 5 QDU, the international transit 4 and the terminating country 5. In an international mobile satellite service, multiple countries and operators are always involved. Furthermore, the interaction of low bit rate codecs in tandem and with other coders is not yet fully understood and hence existing standards do not necessarily address all aspects of mobile systems. New concepts, such as the introduction of an impairment factor in ITU-T G.113, are being introduced to address such concerns.

Effort is under way to develop a generic codec by programming a number of them on a single DSP, enabling access to multiple networks. It should also be possible to adjust the coding rate to suit channel conditions – thereby increasing the channel capacity through adjusting the code rate according to channel conditions.

As the performance of voice coders degrades due to increases in bit error, one consideration in the selection of a codec is its susceptibility to errors. Noise rejection of the handset is another important consideration. At the receiver, local noise can enter through the air gaps around the earpiece (and the other ear!). At the transmit end, noise entering along with speech can cause problems for a number of handset components – voice codec, voice activity sensors, speech interpolation systems, echo suppressors, etc. A voice activity detector can be falsely triggered, causing unnecessary RF transmission. Hands-free terminals have a compounded problem due to significantly higher noise pick-up. A noise reduction algorithm with echo-cancellation reduces the problem at the expense of processing delay. It is important that during type approval, the only time to capture such problems, mobiles are tested in environments close to reality.

5.3.3 Vehicle-mounted terminals

5.3.3.1 Ship-borne terminals

Ship-borne terminals were the first to be introduced for civilian use, as maritime safety was the dominant theme when the first civilian MSS was introduced in the 1980s. Depending on the size of ships, a variety of ship-borne terminals can be used. Large ships can deploy parabolic dishes typically 0.8–1.2 m, whereas yachts and small boats can accommodate only small low-gain antennas.

Two representative examples are presented here to demonstrate their salient features: the Inmarsat-B system introduced around 1993 to provide digital communications and safety services to ships via Inmarsat's global network of satellites and land Earth stations; and the Inmarsat-C system introduced in 1991 to support communication services to small ships and land mobiles.

Inmarsat-B terminals use large tracking antennas around 1 m in diameter, whereas Inmarsat-C terminals are lightweight with small non-tracking omni-directional antennas. Some of the main characteristics of the terminal are summarized in Table 5.5 (Inmarsat, 1991a, 1991b).

(a) Large terminals

There are various versions of the Inmarsat-B system – land transportable, vehicle mounted, ship-borne terminals, etc. Inmarsat-B ship-borne terminals comprise an outdoor deck unit consisting of a stabilized, tracking antenna system to which essential RF electronics are connected; and a below-deck indoor unit where the main signal processing is performed and to which peripherals such as a telephone, personal computer, etc. are attached either as an integral part or as a plug-in. A block diagram of the main units of a ship Earth station are shown in Figure 5.12.

The outdoor unit comprises a parabolic tracking antenna mounted on a stabilized antenna platform. The antenna is housed in a radome for protection

Parameter	Specifications	
	Inmarsat-B	**Inmarsat-C**
Transmit frequency range (MHz)	1,626.5–1,646.5	1,631.5–1,646.5
Receive frequency range (MHz)	1,525–1,545	1,530–1,545
Transmit EIRP (dBW)	25–33	14
Receive G/T (dB/K)	−4	−23
Typical antenna size (m)	0.9	0.1
Typical high power amplifier type	Class-C	Class-C
SCPC channel pairing	No	No
Synthesizer step (kHz)	10	2.5
Data service (kbps)	9.6/16	0.6
Voice-band data (kbps)	up to 2.4	Not Applicable (NA)
Voice coding	16 kbps; APC	NA
Voice channel modulation/coding	O-QPSK; $\frac{3}{4}$ FEC – constraint length =7	NA
Voice channel transmission rate	24	NA
Voice channel bandwidth	20	NA
Forward voice activation	Yes	NA
Forward power control	Yes	NA
Satellite forward link EIRP (SCPC) (dBW)	16	21 (messaging/signalling)
Telex channel (forward link)		
Access/modulation; coding; transmission rate (kbps)	TDM/BPSK; $\frac{1}{2}$ FEC; 6	NA
Telex channel (return link)		
Access/modulation; coding transmission rate (kbps)	TDMA/O-QPSK; $\frac{1}{2}$ FEC;24	NA
Forward signalling channel characteristics	As Telex	TDM/BPSK; $\frac{1}{2}$ FEC; 0.6 (messaging/signalling)
Return request channel		
Access/modulation; coding; transmission rate (kbps)	Aloha; O-QPSK; $\frac{1}{2}$ FEC; 24	Aloha BPSK; $\frac{1}{2}$ FEC; 0.6

Table 5.5

Main characteristics of Inmarsat-B and Inmarsat-C terminals (Inmarsat, 1991a, 1991b)

against the extremely harsh weather conditions experienced on ships. The electronics comprise subsystems which must be mounted close to the antenna to minimize insertion loss, i.e. diplexer, low-noise amplifier, down-converter, high-power amplifier and electric motors for tracking. Usually they are mounted on

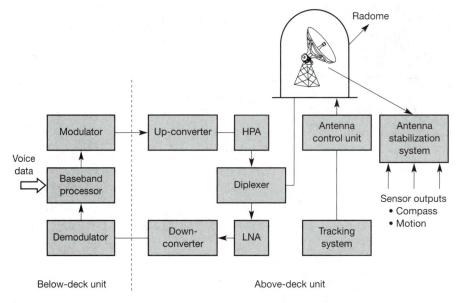

Figure 5.12
Main functional entities
of a ship Earth station

the antenna mount. Antenna tracking and stabilization are maintained through feedback from the ship's motion sensors and a tracking receiver. A step-track system is commonly used for satellite tracking. A four-axis stabilization comprising an azimuth/elevation axis mounted on the X–Y axis is generally used. Motion sensors comprising accelerometers, rate and level sensors provide roll and pitch information. In another type of stabilization scheme, flywheels are used to provide the necessary inertia for maintaining antenna stabilization, but flywheel systems are complex and heavy. A three-axis stabilization system is also used in some modern designs (Hoshikawa, 1988).

The main components of a channel unit housed in the below-deck equipment are shown in Figure 5.13. The scrambler removes discrete spectral lines from transmissions by using a pseudo-random signal. The initial state of the scrambler shift register is changed randomly during each call set-up to improve security and privacy.

The terminal can support frequency division demand assigned voice and data services. The voice service uses APC (adaptive predictive coder) at 16 kbps; transmissions are voice-activated, and accessed using demand assigned frequency division multiplexed 20 kHz spaced SCPC channels. The frequency synthesizer can tune in steps of 10 kHz for SCPC channel assignments within the ranges shown in Table 5.5. Telex communication is received from a TDM broadcast channel at 6 kbps and transmitted via a 24 kbps TDMA return channel. Signalling and bulletin board operate at 6 kbps and return signalling is transmitted at 24 kbps. Signalling and terminal control are managed by a microprocessor-controlled unit. Class-C non-linear amplifiers are used for high efficiency.

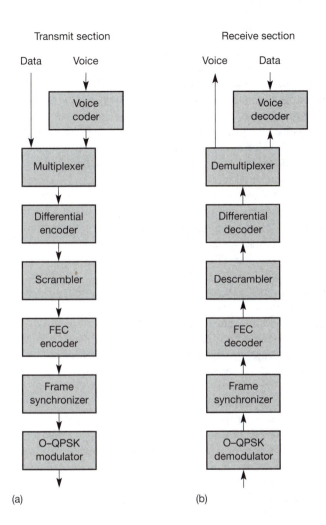

Transmit section   Receive section

Figure 5.13
Main radio
components of the
below-deck equipment

(a)   (b)

(b) Small terminals

The Inmarsat-C system was developed for low cost, low bit rate (600 bps) communication services to small ships, yachts and land vehicles. The system was later adapted to operate in aeronautical systems. The main characteristics of the terminal are summarized in Table 5.5.

The standard-C system, weighing 2–4 kg, uses a store and forward transmission scheme as well as circuit-mode-type end-to-end connections. In the store and forward mode, messages are formatted and transmitted in a simplex mode at a time when a channel becomes available. In the end-to-end mode, a permanent or a semi-permanent connection is established for a call. Other features of the system allow duplex circuits, polling, data reporting and group call reception. The transmission rate is 1,200 symbols/s with $\frac{1}{2}$ rate FEC convolution coding. As terminals deploy omni-directional antennas, multipath noise is severe. Interleaving with $\frac{1}{2}$ rate coding is used as countermeasures to multipath, while an ARQ scheme is used to combat deep fades which break continuity.

Due to their light weight, terminals can be easily mounted on vehicles and vessels; hand-carried versions are also available; there is usually a standard interface for connecting to a personal computer. Some models incorporate a built-in console for creating messages. Figure 5.14 (Inmarsat, 1988) represents a block diagram of an Inmarsat-C mobile terminal. The terminal comprises two main functional parts: data terminal equipment (DTE) and data communications equipment (DCE). The DTE is used to format messages and transfer them to the DCE for transmission to the network. The DCE stores the message until a channel is available for transmission. A microprocessor is used for various control functions; a scrambler is used for dispersing the signal; and a frequency synthesizer can tune the radio frequency in steps of 2.5 kHz. A reverse operation is performed in a receiver.

Various types of omni-directional antennas are used – quadrifilar helix, crossed drooping dipole, microstrip patch. Quadrifilar antennas, protected by a radome, are commonly used for ship terminals and microstrip patches are used where a low profile is desirable.

The terminal receives network information from a time division multiplex broadcast transmitted by an NCS. Land Earth stations of the network also broadcast signals to mobiles containing signalling and messages. Mobile terminals use the Slotted Aloha scheme to access the network.

5.3.3.2 Land-mobile portable sets

By far the greatest interest in recent years has been in land mobile communications due to the large demands envisaged for such a service. Land-mobile terminal and service costs are of paramount importance for the success of such a service. Vehicle-mounted, transportable and individual sets belong to this category. At present, such sets provide throughput ranging from a few hundred bits per second to 64 kbps, depending on their antenna size, the space segment capability of the MSS and the market segment. Portable and vehicle-mounted land-mobile sets were introduced in the early 1990s as a part of the MSS technological evolution and, while their size continues to shrink, throughput has increased. Desktop-size personal portable sets supporting voice and a few kbps data rate appeared around 1996, and within two years hand-held terminals were introduced, while throughput increased from 600 bps at the start of the decade to 64 kbps by 2000.

The design of such sets is technically challenging due to stringent size limitations and the severe propagation conditions experienced on land. Omni-directional or near omni-directional antennas with antenna gain <3–4 dBi are lightweight and low cost but provide limited throughput, whereas medium-gain antennas with gains of the order of 4–15 dBi are complex when tracking is incorporated, heavier and more expensive but can provide higher throughput (at present ~ 2.4–24 kbps). Demands on the complexity and func-

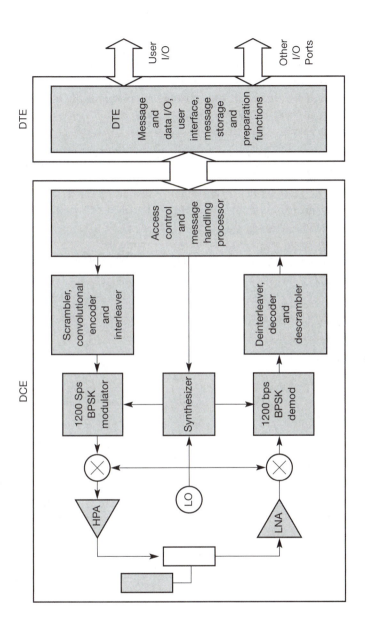

Figure 5.14
Inmarsat-C terminal block diagram (Inmarsat, 1988)

tionality of other electronic units are similar to other types of mobile terminals. In the remaining part of the section, a representative terminal of this category, capable of being used in a number of environments is described (Fuji *et al*, 1995), followed by a description of a K_a band mobile terminal used in the ACTS programme to illustrate an architecture being prepared for next generation mobile systems (Abbe *et al*., 1995).

The mobile terminal developed to operate in the AMSC/TMI network has the flexibility to operate on land or marine mobiles, as a transportable or fixed site installation, and deploys a medium-gain or a high-gain antenna system, as necessary. The specification of the mobile terminal is summarized in Table 5.6. The terminal can support circuit-switched voice and data services at 2.4 kbps and 4.8 kbps respectively, cellular roaming, net radio and Group 3 facsimile ser-

Table 5.6
Characteristics of mobile terminal (Fuji *et al*., 1995)

Parameter	Specifications
Transmit band (MHz)	1626.5–1660.5
Receive band (MHz)	1525–1559
Channel spacing (kHz)	6
Channel tuning increment (kHz)	0.5
Channel rate (kbps)	6.75
Modulation/filtering	QPSK/60% cosine roll-off
Scrambling	15 stage PN sequence (ITU-R report 384-5)
FEC coding; decoding	$\frac{1}{2}$ and $\frac{3}{4}$ punctured FEC, K = 7; Viterbi
Signalling	
Forward; Access/Modulation/Rate (kbps)	TDM/DQPSK/6.75
Return; Access/Modulation/Rate (kbps)	Random access/DBPSK/3.375
	TDMA/DBPSK/3.375
Data rate (kbps)	2.4 or 4.8
Interface	
Handset – voice	4-W with serial async keypad and LCD signals
Data interface	RS232 C (CCITT V24 or V28)
Facsimile interface	CCITT group 3, 2.4 kbps voice band signal
IB interface	CCITT X.25
Power supply (V)	
Volt/Amps	12 (Range 11–16)/ 6
Size/weight	
Trans-receiver	12"(D)×8"(W)×2"(H)
Unit	5 lbs

vices. A cellular portable phone can be plugged in to provide MSAT voice services. The terminal has a minimum EIRP of 12.5 dBW and a minimum G/T of –16 dB/K and –12 dB/K with medium- and high-gain antennas respectively.

The terminal comprises an antenna, a trans-receiver unit and a hand-held phone. Various types of antennas are used, each suited to a specific category, as listed in Table 5.7. Disc and dome antennas are used for roof-mounted applications on land; transportable antenna systems for trucking applications; flat antennas for briefcase; and fixed site antennas for remote fixed location applications.

Antenna type	Disk	Dome	Transportation	Fixed site	Transportable
Approximate size and weight	Diameter =13.7"; Height = 1.9" Weight = 5.3 lb	Diameter = 6.8"; Height = 6.6" Weight = 3 lb	Diameter = 11.5"; Height = 7.2" Weight = 5.5 lb	Width = 19.7" Depth = 19.7" Height = 2" Weight = 14 lb	Width = 14" Depth = 14" Height = 4.3" Weight = 20 lb
Maximum EIRP (dBW)	16.5	16.5	15.5	16.5	16.5
Receive G/T (dB/K)	–16	–16	–13	–10	–12
Gain category	Medium	Medium	High	High	High
Elevation angle range (degree)	25–60	15–60	25–60	5–90	15–90

Table 5.7

Characteristics of various types of antennas (Fuji *et al.*, 1995)

The main components of a dome-type antenna unit are shown in Figure 5.15. The antenna comprises an axial mode helical antenna mounted on an inclined plane to serve the desired elevation angle range. Other components of antenna units are a rotary joint, a LNA, HPA, diplexer (DIP) and motor for tracking.

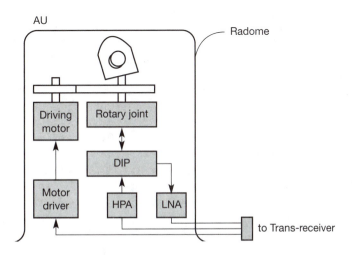

Figure 5.15

Main blocks of a dome antenna unit (Fuji *et al.*, 1995)

The main subsystem of a trans-receiver unit is shown in Figure 5.16.

The unit, common to all types of antenna, comprises up and down frequency converters, a baseband processor, a logic and signal processor, a power supply, and an echo canceller for hands-free operation. The frequency synthesizer is a dual tuning type PLL synthesizer driven by a direct digital synthesizer with low spurious and phase noise characteristics. The baseband processor, comprising a digital signal processor, performs modulation/demodulation plus framing/deframing. The demodulator uses differential detection with maximum likelihood sequence estimation with a Viterbi algorithm. The voice codec uses the IMBE technique which provides 6.4 kbps data. The logic and signalling processor controls the main functions of the mobile terminals, including network access and control.

A mobile terminal has been developed under the Advanced Communications Technology Satellite (ACTS) programme to demonstrate next generation K_a-band MSS technology. Known as the ACTS Mobile Terminal (AMT), the terminal is a proof of concept K_a band terminal developed by NASA at the Jet Propulsion Laboratory. The AMT is intended to demonstrate speech and data transmission and act as a precursor to the development of other types of terminals for aeronautical, maritime, land and personal communications. The AMT was also developed to support the characterization of K_a band propagation in a mobile environment. The terminal operates at 30 GHz in the uplink and 20 GHz in the downlink and at data rates of 2.4, 4.8, 9.6 and 64 kbps.

Some vital system-level considerations are as follows:

- geostationary system – location 100° W;
- high gain spot beams;
- regenerative transponders;
- terminal operational elevation angle – 30°;
- FDMA scheme;
- estimated 98% link margin for an average year – 1.2 dB at 30 GHz and 1 dB at 20 GHz; far higher attenuation can occur during heavy rain;
- high shadowing loss;
- high Doppler and Doppler change rate due to vehicle motion – of the order of 3 kHz and 370 Hz/s, respectively;
- high phase noise in the communication channel due to high noise closer to carrier centre frequency – requires robust modulation scheme against phase noise.

The key technical challenges were:

- development of high gain tracking antennas;
- compensation for rain attenuation;
- compensation for high Doppler, frequency uncertainties and phase noise.

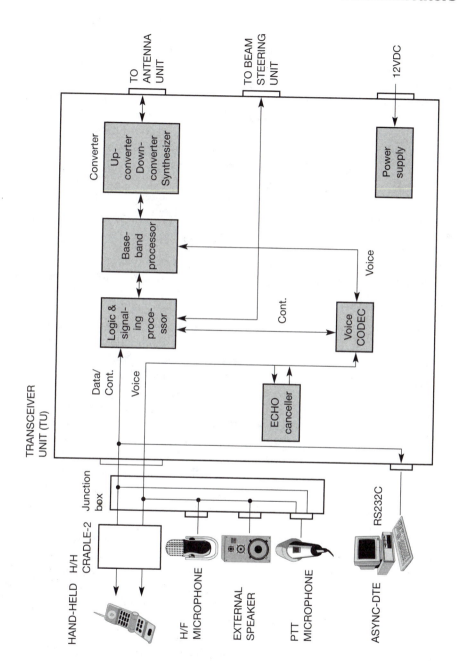

Figure 5.16
The main units of trans-receivers (Fuji *et al.*, 1995)

To counter the rain attenuation, a rain compensation algorithm (RCA) is used. The dynamic algorithm involves pilot power measurement and beacon measurement at the fixed station followed by an exchange of information between the terminal and a decision to lower the data rate from 9.6 to 4.8 or 2.4 kbps in the presence of rain, thereby increasing the link margin from 3 to 6 dB.

Figure 5.17 shows a block diagram of the AMT.

The terminal controller is responsible for all the coordination mentioned below:

- translation of communication protocols to executable instructions, e.g. timing and handshake between speech coder, modem, user interface, external devices during call set-up, end of call and data rate change;
- operation and execution of the RCA algorithm;
- control of intermediate frequency (IF), RF and antenna electronics;
- system monitoring and interface to data acquisition system;
- support of test functions during experimentation, e.g. correlation, bit error measurement.

The AMT modem has been developed with the aim of minimizing the effects of the phase noise of the satellite. A DPSK modulation scheme is used because of its simplicity and robustness to phase noise and deep short-term fading. A $\frac{1}{2}$ rate convolution coding with interleaving is used. An offset of ±10 kHz is tolerated by the modem. The required E_b/N_0 for a BER of 10^{-3} at 9.6 kbps is 6.6 dB in Gaussian noise, degrading by 1 dB in the presence of phase noise.

In addition to up and down conversion, the IF converters track the pilot and pre-compensate for Doppler. The pilot provides a frequency reference for the terminal and is useful for Doppler pre-compensation of the transmitted signal. The pilot signal is also used in the RCA unit, for antenna tracking, and by the terminal propagation data acquisition system. An IF frequency of 70 MHz is used in the receiver and 3.373 GHz in the transmitter in compliance with fixed Earth station hardware.

Two types of vehicle antenna were developed – a reflector-type elliptical antenna and an active array comprising integrated HPA and LNA units as MMIC. The reflector antenna is enclosed within an ellipsoidal radome of 9" base diameter and a height of 3.5". The active array minimizes RF loss. The terminal has a minimum EIRP of 22 dBW, a G/T of –8 dB/K and a bandwidth of 300 MHz. The reflector permits TWTA to operate at powers lower than 1.5 W and operates with a mechanical tracking system. The antenna is dithered around the boresight by about a degree at a rate of 2 Hz while the pilot signal strength is measured. The signal level, together with the vehicle's inertial information extracted from a turn rate sensor, is used to estimate the tracking error. The system is able to maintain tracking in the presence of fade lasting up to 10 s. The small antenna diameter and narrow beamwidth enable the use of such a tracking scheme.

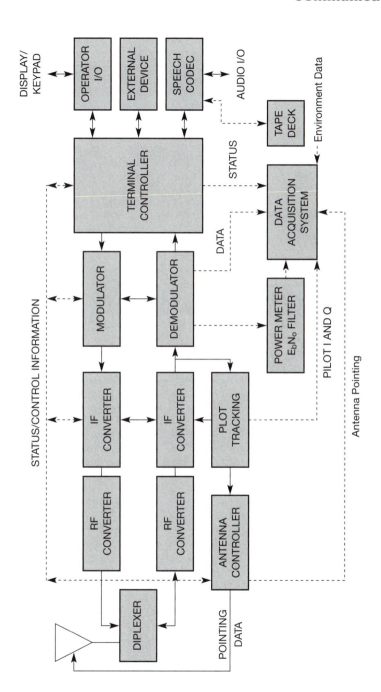

Figure 5.17
Block diagram of ACTS
mobile terminal (Abbe *et
al.*, 1995)

Finally, the data acquisition system acquires and displays measured data related to propagation, communication link, pilot and data signal state, noise, antenna pointing, vehicle heading, etc. for performance evaluation.

5.3.3.3 Aeronautical terminals

Aeronautical terminals have to comply with strict regulatory guidelines to ensure that the installation does not cause any detrimental effects to onboard electronics. The Aeronautical Radio Incorporation (ARINC) specifies terminal characteristics for civil aviation in the ARINC 741 standard (Aeronautical Radio Incorporation, 1990). Aeronautical terminals, often called Aeronautical Earth Stations (AES) for historic reasons, are categorized into four classes. Class 1 AES support low speed service with low gain antenna; class 2 AES support passenger voice service with high gain antennas; class 3 AES support voice and high data speed; and class 4 AES offer voice, high speed and low speed data services. The standard also specifies the characteristics of AES subsystems – satellite data unit, RF unit, diplexer, low noise amplifier, high power amplifier, antenna and beam steering electronics.

Aeronautical terminals used in the Inmarsat network are described here as they are representative of this category of mobiles. There are two basic types of services supported – within each there are various service options compliant with ARINC standards, i.e. low data rate, high data rate, voice or a combination of these. Low data rate terminals deploy omni-directional antennas and high data rate terminals deploy medium gain antennas of about 12 dB. Table 5.8 summarizes the main characteristics of these two types of terminals.

Table 5.8
The main characteristics of two classes of Inmarsat aeronautical service terminal

Parameter	Terminal characteristics	
	High gain	**Low gain**
Transmit frequency range (MHz)	1,626.5–1,660.5	1,626.5–1,660.5
Receive frequency range (MHz)	1,525–1,559	1,525–1,559
Transmit EIRP (dBW) (Nominal)	25.5	15.5
Receive G/T (dB/K)	–13	–26
Antenna characteristics		
Typical antenna type	Phased array	Helical
Antenna gain (dBi)	~ 12	~ 0
Tracking	Program track	None
Hemispherical coverage %	>75	>85
Typical high power amplifier class; power (W)	Class-A; 60	Class-C; 40
SCPC channel pairing	Yes	NA
Synthesizer step (kHz)	2.5	2.5

Table 5.8
Continued

Low rate data service (kbps)	0.3	0.3
Voice channel modulation; coding	A-OQPSK; $^1/_2$ FEC	NA
Voice channel transmission rate	4.8/9.6	NA
Voice channel bandwidth (kHz)	17.5	NA
Forward voice activation	Yes	NA
Satellite forward link EIRP at start of call (SCPC) (dBW)	21.5	NA
SCPC channel		
Access/modulation; coding; transmission rate (kbps)	DA-FDMA/A-QPSK; $^1/_2$ FEC; 21	NA
Data channel (return link)		
Access/modulation; coding; transmission rate (kbps)	R-TDMA/A-BPSK; $^1/_2$ FEC; 0.6	R-TDMA/A-BPSK; $^1/_2$ FEC; 0.6
Forward signalling channel characteristics		
Access/modulation; coding; transmission rate (kbps)	Packet mode/ A-BPSK & A-QPSK; $^1/_2$ FEC; 0.6	Packet mode/A–BPSK; $^1/_2$ FEC; 0.6
Forward data channel characteristics:		
Access/modulation; coding; transmission rate (kbps)	Packet mode/TDM; $^1/_2$ FEC; 0.6 & 10.5	Packet mode/A–BPSK; $^1/_2$ FEC; 0.6
Satellite forward link EIRP (Data) (dBW)	12.5	21.5
Return data channel characteristics:		
Access; modulation; coding; transmission rate (kbps)	Random access or R-TDMA; A-BPSK/ A-QPSK; $^1/_2$ FEC; 0.6/10.5	Random access or R-TDMA; A-BPSK; 0.6
Return request channel		
Access/modulation; coding; transmission rate (kbps)	Aloha/A-BPSK; $^1/_2$ FEC; 0.3	Aloha/A-BPSK; $^1/_2$ FEC; 0.3
Typical application	Aeronautical administration communication (AAC)	Administration administration communication (AAC)
	Aeronautical operation control (AOC)	Aeronautical operation control (AOC)
	Air traffic control (ATC; anticipated)	Air traffic control (ATC; anticipated)
	Aeronautical passenger communication (APC): Voice	

Figure 5.18 depicts the main units of a high gain AES; for a low gain antenna the functional units are similar but with different characteristics. A phased array antenna provides electronic steering through a beam steering control unit and corrections are provided by the program track subsystem. The program track system, embedded within the data processing unit, derives tracking corrections by calculating the difference between the estimated satellite position and the aircraft's position obtained from the onboard inertial navigation system. To minimize intermodulation noise in the case of multi-carrier transmissions, the high power amplifier (HPA) comprises a linear amplifier of about 60 W or an EIRP of 30 dBW which can support a maximum of 2–3 voice channels. The RF processing unit interfaces with baseband signals and includes up–down conversion, modulation–demodulation and coding–decoding. Data processing involves reformatting, protocol conversion, etc. to suit the service and application. The ADS system inputs the required aircraft data for remote ground surveillance.

An AES receives signalling messages on a broadcast channel transmitted from the ground Earth station (GES) of the network and, in turn, uses a return (or R) channel at rates of 600 bps, 1.2 kbps or 10.5 kbps – 600 bps is used at logon. A GES responds at the same rate, informing the AES of the appropriate bit rate for further communication, derived from the signal quality of the request burst and depending on satellite in use. TDMA channels, operating at the same rate as R channels, support data applications.

A voice channel unit, shown in Figure 5.19, comprises a voice or data unit and interface, a scrambler, an encoder, an interleaver to minimize the impact of error bursts, a preamble and a unique word generator for timing and identification. A reverse set of operations is performed at the receiving unit.

High gain antennas are usually phased arrays installed either at the top of the fuselage or on both of its sides. Top-mounted antennas increase air drag to a certain extent but can view the satellite without the so-called blind zone (or 'keyhole') associated with phased arrays, whereas side-mounted antennas have

Figure 5.18
Main units of a high gain
AES

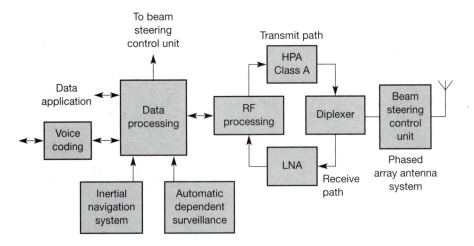

Figure 5.19
Voice channel unit
of AES (Inmarsat, 1990)

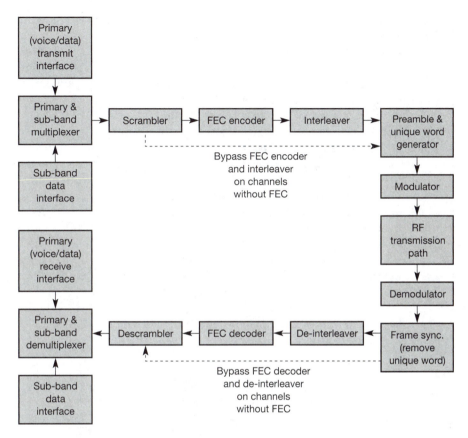

lower air drag but have keyholes. Keyhole areas can be reduced by positioning phased array units around an elevated mount.

5.4 Environmental issues

With heightened awareness of mobile communications, and in particular cellular systems, a number of environment-related issues and concerns have been raised:

● **Aesthetics**: Aesthetics relates to mobile terminals and network infrastructure such as base station antennas. In particular, base station antenna sites have undergone severe scrutiny due to their unsightly appearance affecting city skylines. A number of innovative ideas are under study and development. In one such experiment, an antenna system was camouflaged as a tree, blending unobtrusively with its surroundings. Fortunately satellite gateways are far fewer and generally sited well away from population centres and therefore do not pose concerns from this viewpoint. Mobile terminal designs are changing rapidly and they are now a fashion accessory – wristwatch telephones are under trial; other innovative ideas include pendant phones.

- **Electromagnetic interference (EMI) from mobile telephones to other equipment**: Concerns have been raised regarding the EMI effects on health and safety equipment such as hearing aids, cardiac pacemakers, equipment in hospitals, on aeroplanes, ships, etc., and in a more general sense, of everyday equipment such as fixed telephones, radio, television, etc.

  In situations where the EMI effects of mobile terminals on sensitive medical equipment are not known, the best approach has been to forbid their use in areas such as hospital theatres, etc. However, this approach is not feasible in public places, where a mobile user may answer a telephone next to an individual with an implanted device. The best course of action is to ensure that such medical devices have high resistance to EMI. The effects of EMI on equipment such as hearing aids is less critical, and controlled tests on more vital equipment such as implanted pacemakers do not show adverse effects of RF radiation; clearly, however, making such devices resistant to EMI is in the public interest and it is essential to introduce regulations for EMI immunity for such equipment.

- Concern regarding the effects of RF radiation on people has received significant public attention recently due to some highly publicized claims. For this reason, the next section reviews the current state of knowledge.

5.4.1 Biological effects

Tens of millions of people use mobile telephones world-wide and the user population is expected to increase dramatically in the next decade. Therefore a better understanding of mobile phones' biological effects on humans is imperative, and if adverse effects are observed, adequate safeguards in the design of mobile systems and phones are crucial.

At present, interaction mechanisms for low-level exposure to emissions from hand-held mobile phones and their biological effects are not well understood. In the absence of conclusive evidence, concerns regarding the adverse effects on humans of RF radiation from cellular phones continue to be raised in public by individual claims or through inconclusive results being reported by the press, etc. Claims pertain to the possibility of contracting cancer, DNA damage, heating of the brain, headaches and tiredness. However, none of these claims has been proven or reproduced under controlled test environments up to mid-2000. A number of research programmes are ongoing throughout the world at various institutions, such as the World Health Organization, to understand more quantitatively the effects on humans of RF emissions in the frequency bands ($\sim$ 800–2200 MHz) and power (10–1,000 mW) used by various cellular standards. Lin (1997) and Foster and Moulder (2000) give good summaries of the work done to date applicable to mobile communications. The results applicable to terrestrial mobile systems are equally applicable to mobile satellite communications, as the frequency ranges are similar and the same international safety standards apply.

In the western world, a commonly used health and safety RF exposure standard, ANSI/IEEE C95.1–1992 (updated in 1999 and currently under revision), promulgated by American National Standards Institute (ANSI), provides a guideline for exposure limits. Table 5.9 lists allowable specific absorption rate (SAR) in tissue, the incident power density causing it and the averaging time for estimating the power density for various frequencies for this standard within the frequency range of interest for mobile systems. SAR is a dosimetric measure defined by the National Council on Radiation Protection and Measurements (NCRP) of the USA as 'the time derivative of the incremental energy absorbed by (or dissipated in) an incremental mass contained in a volume of a given density'.

Frequency (MHz)	Controlled environment			Uncontrolled environment		
	Power density (mW/cm^2)	SAR (W/kg)	Averaging time (min)	Power density (mW/cm^2)	SAR (W/kg)	Averaging time (min)
800	2.7	8.0	6.0	0.53	1.6	30
900	3.0	8.0	6.0	0.6	1.6	30
1,800	6.0	8.0	6.0	1.2	1.6	30
2,200	7.3	8.0	6.0	1.47	1.6	30

Table 5.9 ANSI/IEEE C95.1–1992 guidelines in frequency range of interest for mobile satellite systems

The value for SAR is 0.4 W/kg averaged over the whole body or 8.0 W/kg for any one gramme of tissue for a controlled environment and 1/5 less for an uncontrolled environment (user has no control over RF source). The variation in power density is given as f/300 and f/1,500 mW/cm^2 (f = frequency in MHz) for the controlled and uncontrolled environments, respectively, within the frequency range 800–2,200 MHz. The permissible radiated power can be relaxed in the range 450–1,500 MHz for both controlled and uncontrolled environments when the body is at least 2.5 cm away from the radiating antenna and radiated power is ≤ 7(450/f) W for the controlled environment or ≤ 1.4(450/f) W for the uncontrolled environment, as long as the SAR limit is maintained for situations where exposure is either not under the control of or is without the user's knowledge.

SAR, along with other quantities including incident field and induced field, is used as a measure of RF exposure. Due to the significance of such measurements for research and radiation compliance of equipment and the difficulty in making accurate measurements, in particular, when the radiation source is close to the head, considerable effort is under way to improve the measurement technique (or dosimetry). The measurements can be done either experimentally, theoretically or a combination of the two. To date, a universally acceptable procedure for such measurements has not been agreed, due to which there tends to

Figure 5.20
SAR measurements
inside a phantom human
head from a cellular
telephone transmitting at
a frequency of 835 MHz
and a power of 0.6 W
(Balzano *et al*, 1995)
© 1995 IEEE

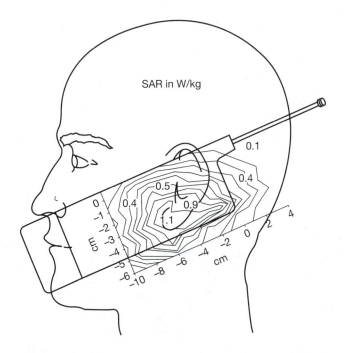

be variability in reported results. Figure 5.20 (Balzano *et al.*, 1995) illustrates SAR measurements made by isotropic electric probes inside a fibreglass human head filled with liquid dielectric brain-equivalent material, for radiation caused by a cellular telephone transmitting 0.6 W through an extended antenna, as shown. Note that the maximum SAR of 1.1 W/kg is at the ear lobe nearer to the antenna, close to the feed point of the antenna; the level is much lower elsewhere; and the maximum SAR is within acceptable limits.

A number of SAR measurements within the brain tissue of the phantom human head have been reported in the 800–900 MHz band (cellular band) using 0.6 W transmissions from communication devices placed 1–3 cm from the surface of a head model. There is considerable variability in SAR value reported (0.44–6 W/kg); however, most of the results comply with the ANSI/IEEE C-95.1 guidelines except for a few reports which show SAR in excess of the limits. The most pessimistic results reported are as high as 6 W/kg. Possible reasons for the wide variability are the difficulty in making accurate SAR measurements, the dependency of measurements on the position of the antenna with respect to the head, and the influence of a device's hardware on its radiation characteristics. It is noted that the SAR values are lower for longer antennas (in terms of wavelength) than shorter ones – this is due to higher current regions being higher up on the antenna and therefore further away from the head for longer antennas.

Interestingly, measurements on human models standing next to trunk-mounted mobile telephone antenna transmissions at 835 MHz transmitting at 1

W show exposure levels well within the ANSI standard at any point. Field strength measurements made by researchers on standard off-the-shelf cellular phones in the UK show compliance with radiation standards.

Existing measurement methods can be tedious for large-scale measurements such as when required for compliance and type approval of mobile phones. Hence computer-assisted numerical methods are under investigation. The difficulty here lies in modelling the human body and transceivers accurately. For example, to simplify computing requirements, mobile phones have been modelled as a metal box with an antenna, without considering the influence of internal circuitry, sub-assemblies, packaging or the external geometry of the cellular phone. Magnetic resonance imaging (MRI) based models have been used for simulating the human head; and finite-difference, time-domain numerical methods have been used for numerical computation. Simulated SAR results in the brain region for cellular telephones operating in 800–1,800 MHz using antennas of $\frac{1}{4}$ to $\frac{1}{2}$ wavelength vary considerably due to the differences in modelling assumptions, but reported results for 0.6 W transmissions (0.65 – 1.97 w/kg) generally show compliance with ANSI standards for radiation received within the brain region. However, power deposition in superficial tissues can exceed 1.6 W/kg. Results pertaining to the reduction in SAR with increased distance and increase in antenna size confirm experimental observations. The user's physical environment has been observed to influence SAR measurements. The presence of a reflective ceiling can redistribute SAR, reducing the peak SAR inside a human head. A reflective vertical wall, by contrast, can increase the peak SAR within the head by up to 100%. A number of problems need to be solved before a standard technique is developed; problems include the influence of tissue inhomogenity and size of volume element on SAR distribution in a model.

Adverse thermal effects on the functioning of the human body at high intensity is well known. At the levels at which mobile phones operate, it is very unlikely that radiation causes thermal effects. Heating of the brain by microwave radiation at the very worst is estimated to be no more than that caused by mild exercise. Hence research efforts have converged on the effects which may be caused by long-term use, including life-long exposure to low-level radio waves.

Tumour induction from mobile phone radiation has been a topic of particular attention. There were relatively few studies of this topic until recently. The results conducted on animals in the frequency range reported to date, with one exception, indicate that tumour growth is not caused by RF radiation; however, RF radiation in the range 915–2,450 MHz does accelerate tumour growth if the tumour is initiated first by a cancer promoter. In another study, life-long exposure of rats to pulsed microwaves showed no difference on their general health or life expectancy however, a statistically significant increase was observed in primary malignancies at the end of life, which does require further research. A study conducted specifically for mobile telephones operating at 915 MHz, using the same modulation scheme, reported no increase in tumour growth.

Several studies related to the nervous system have concentrated on the neurovascular construct known as the blood-brain barrier (BBB), which maintains the physiochemical environment of the brain within the narrow limits essential for life, following a 1990 report of radiation effects on the BBB. The conclusions of such studies remain controversial, in particular, for low-level radiation.

Microwave-induced auditory sensation, which pertains to auditory sensation in humans and animals from microwave pulses, is a well-known microwave effect. It has been observed that if a microwave pulse or pulse train is delivered above a threshold power density (e.g. 400 mJ/m2 at 2.45 GHz) to a human subject, it is heard as a click for a single pulse or as a series of clicks at the pulse repetition rate. At present, it is believed that the audio sensation is caused by acoustic pressure induced by thermo-elastic waves which are produced by soft tissues in the head on absorption of microwave radiation. The acoustic pressure travels to the inner ear by bone conduction where it is treated in the same manner as normal sound. It is believed that the threshold microwave auditory effect has insignificant impact on hearing apparatus. However, at present, there is little data regarding the long-term effects of the microwave auditory effect on either the central nervous system or the middle and inner ear hearing apparatus. Research is in progress to gain insight into the possible effects from mobile telephones.

The formation of cataracts through exposure to microwave radiation is another concern. It is generally accepted that prolonged high-level microwave radiation does cause cataracts in the eyes of laboratory animals, but below a threshold there is no such effect, no matter the duration of exposure. The conditions which trigger the cataract are debatable.

A valuable investigative tool for studies related to humans is epidemiology, in which the incidence of a disease is studied across communities. Studies conducted over specific groups such as military microwave workers and air force veterans, exposed to radiation in their occupation, are inconclusive – two studies showed statistically non-significant results and one indicated statistical significance. Other epidemiological studies, which have investigated the effects of microwaves on the growth and development of children from parents exposed to microwaves, the incidence of cancer, and mortality rates generally show no adverse effects. A study conducted in 1994 to investigate the near-term mortality rate of cellular phone users in 1994 showed no difference in relation to non-users. Studies conducted in the USA in 1999 on 300,000 mobile users show no effect on mortality. All other epidemiology studies to date have yielded negative results. One of the problems in all studies to date is the lack of measurements of microwave exposure levels, which lead to large uncertainties. In those studies in which a statistically significant rise of adverse effects was observed, the results were diluted by cross-impact from other sources. Furthermore such studies must be conducted over a decade or more, making them difficult to control and interpret. Thus it is difficult to conclude any positive or negative association from the studies so far.

Researchers in 1997 at the Royal Adelaide Hospital in Australia observed that mice exposed for 18 months to microwave emissions at the frequency and intensity used in digital phones were twice as susceptible to cancer of the lymph system than others which were not exposed. However, three other teams were unable to reproduce the results, including a team at Brooks Air Force Base in Texas which found no evidence of increase in cancer to exposed mice genetically made susceptible to breast tumours for 20 hours each day over 18 months.

In 1995, a research team at the University of Washington, Seattle, reported DNA breaks similar to the type caused by carcinogenic chemicals or X-rays in rats exposed to microwave radiation. Two independent studies conducted by research teams in St Louis, Missouri, and Belgium subsequently failed to reproduce the result. The World Health Organization, in collaboration with mobile telephone firms, is currently funding a study which will compare 3,000 brain tumour cases with 3,000 controls to isolate any difference due to mobile telephone use.

In 1998, a Swedish study demonstrated that people making more frequent telephone calls were more likely to complain of headaches and tiredness. However, it has also been observed that people's awareness biases such observations; for example, it has been reported that people in Norway were twice as likely to complain of such symptoms when compared to Sweden where the radiation effects of mobile phones attracted lower publicity.

While such investigations progress, mobile telephone manufacturers continue to improve techniques for minimizing exposure to radiation. Techniques include shielding radiation in the direction of the brain and mounting antennas away from the head. Trials conducted in the National Physical Laboratory (NPL), UK, for the science magazine *New Scientist* used a simulated human head – a human skull covered by simulated flesh – to study shielding provided by two commercially available shielding devices. Field strengths were measured inside the head using two makes of mobile telephones, one each from Motorola and Nokia. A reduction of 16–48% in field strength was observed when the antenna of the mobile phones was kept pointing downwards; the amount of reduction depended on the make of the mobile phone. The shielding was less effective in each case when the antenna was pointed upwards. In all cases, the quality of reception degraded when a shielding device was used. The NPL study concluded that the most effective method of reducing radiation to the brain was to use a communicator which has the radiator mounted on a belt which reduced the field strength within the head by 94%, though the radiation to other parts of the body increased. Cautious mobile phone users have simple solutions – use an external earpiece to keep phones away from the head, reduce continuous usage time, avoid use far away from base stations where mobile phones transmit higher power, use shields, etc.

While there is general agreement on certain adverse effects of high levels of microwave radiation, it is evident from data available to date that there is no cause for concern regarding safety aspects from low-level RF transmissions as encountered from existing mobile telephones. Several investigations are in progress to improve our understanding of the subject.

5.5 Next generation technology

The look, feel, cost and services offered by mobile terminals and service cost are paramount to the success of future mobile satellite systems. The next generation of mobile terminals are expected to consume enormous processing power coupled with large memories, huge integration of electronic components, and attractive packaging incorporating a user-friendly interface and high quality display. Palm-sized screens for internet browsing or displaying images, wrist-wearable communicators, plug-in RF units for laptop communicators, and units with a voice-operated interface are under development.

Moore's law is now reaching a limit due to the enormous investments required in enhancing computing power. Moore's law is the phenomenon witnessed in recent years wherein the power of computing doubles every 18 months to 2 years, with a proportional decrease in cost. There is a considerable scope for growth in displays and input devices such as the keyboards used in mobile terminals. Display technologies under research include plasma display panels and cholestric liquid-crystal screens and voice and handwriting recognition. Other futuristic technologies with the potential to improve the human-machine interface are holographic display technology and language processing systems. Such technologies are likely to be initially introduced in the fixed network due to the relatively higher power requirements of such displays.

The term 'software radio' is often used for radio systems which may be configured through software. Development of configurable radio systems is vital for multimode terminals capable of adapting the air-interface between terrestrial and satellite systems or between terrestrial systems with different air-interfaces. This approach also permits mobile operators to adapt to new standards without the need to invest large amounts in altering the infrastructure, and thereby facilitates the introduction of the next generation of mobile systems.

Agent and Java-type technology is expected to reduce the complexity and cost of terminals and systems through downloadable applications, permitting the benefits of multimedia services to be available to larger sections of the population. Increasing use of software technologies such as artificial intelligence, fuzzy logic and intelligent agents should provide improvement in applications such as information assimilation, entertainment, etc. Considerable enhancement in database technology is expected to handle the increasing amount of information digitization.

Data compression technologies of interest are MPEG-4 and its successor, CMTT and fractal coding. These technologies should increase efficiency in spectrum usage as well as data storage. Technical developments are also necessary in the field of service creation technology so as to permit the introduction of short lifecycle services with acceptable overheads, as it is possible to envisage such products being offered due to rapid technical evolution. Equally important will be technology for billing as more players join the market and the revenue is shared between more entities.

It is envisaged that next generation mobile technologies will be a subset of the fixed broadband services. Thus broadband technologies affecting the fixed network are likely to spur the demand for similar services in mobile systems. The technologies of interest here include Bluetooth and IEEE-1394 (FireWire), optical fibres, and copper loop technologies such as ADSL and VDSL.

Studies conducted in the European Union suggest that by 2015 there will be no distinction between the fixed and mobile services and users will expect services to be available anywhere at the lowest cost, to be able to match their communications needs to the service on offer and manage their contactability. Users will own one or more unique addresses, depending on their role, such as that of an executive or a parent. The mobile terminals will be small enough to be fixed unobtrusively in the user's clothing and, for some applications, to be surgically implanted. The visual displays could have several forms – some capable of image projection, others using displays on spectacles and yet others providing holographic displays. User interfaces will use voice recognition with the capability of understanding and interpreting commands in natural language. Keyboards as we know them today will be obsolete; manual interfaces may remain in a different form such as handwriting recognition. The trend to move from voice to data will continue as more applications are digitized. There will be a considerable shift from voice to data, as voice messaging over evolved forms of the internet will become the norm.

Virtual meetings through video telephony and video-conferencing will reduce the need for face-to-face meetings, reducing the need to travel. Teleworking will reduce commuting and may actually reduce demands for mobile communications. Similarly, with the advent of high quality television and other video entertainment devices, people may travel less for leisure activities. On the other hand, some types of travellers such as global business travellers will continue to travel due to the increasing globalization of business, making two-hour trans-Atlantic flights, and this would mean use of global satellite systems, increased need of roaming and service portability. Travelling for tourism may increase as people have more leisure time.

Satellites play a vital role in providing wide area coverage, to cover geographic areas which may never be economically served by terrestrial communications. Until satellite systems are fully integrated with terrestrial mobile systems, a large number of differing systems will serve the needs of international travellers, journalists and adventurers, and in general wherever there is a need for communications in remote areas. Most existing personal communications MSS systems offer dual-mode terminals capable of operating with local terrestrial mobile systems such as GSM, digital AMPS, CDMA, etc. Multimode terminals, which may operate with more than one type of terrestrial system or indeed with a combination of satellite and terrestrial systems, are already emerging.

Most satellite systems capable of operating with pocket telephones in the medium term (~2005) will be able to offer low bit rates of the order of ~2.4 to 9.6 kbps. High bit rates of the order of several hundred kbps will be offered via palm-sized communicators and bit rates of the order of 0.5 Mbps with relatively large-sized terminals at the start. Chapter 12 explores future evolution in further detail.

Further reading

Abbe, B.S.; Agan M.J.; Girardey, C.C.; Jedrey, T.C. (1995) 'Channel and terminal description of the ACTS mobile terminal', *IMSC 93, The Third International Mobile Satellite Conference*, Pasadena, California, 16–18 June: JPL Publication 93-009, pp. 387–92.

Aeronautical Radio Incorporation (1990) 'ARINC characteristics 741: Aviation Satellite Communication Systems 1990'.

Agrawal, A. (1998) 'Mobile chips: towards a common platform', *Telecommunications*, March, 86–8.

Balzano, Q.; Garay, O.; Manning, T.J. (1995) 'Electromagnetic energy exposure of simulated users of portable cellular telephones', *IEEE Trans. Vehicular Technology*, **44**, 390–403.

Bernardi, P. and Lin, J.C. (1997) 'Editorial', *Wireless Network*, **3**, 435–7.

Bernardi, P.; Cavagnaro, M.; Pis, S. (1996) 'Evaluation of the SAR distribution in the human head for cellular phones used in a partially closed environment', *IEEE Transactions on Electromagnetic Compatibility*, **38**, 357–66.

Caballero, J.E.; Badenes, J.; Fernandez, J.; Martin-Pascual, C.; Municio, F. (1995) 'Hand-held terminal antennas for personal satellite communications', *IMSC'95, The Fourth International Mobile Satellite Conference*, Ottawa, co-sponsored by Communications Research Centre/Industry Canada and Jet Propulsion Laboratory/NASA, pp. 351–6.

Chou, C.K.; Guy, A.W.; Kunz, L.L.; Johnson, R.B.; Crowley, J.J.; Krupp J.H. (1992) 'Long term, low level microwave irradiation of rats', *Bioelectromagnetics*, **13**, 469–96.

Chuang, H.R. (1994) 'Human operator coupling effects on radiation characteristics of a portable communication dipole antenna', *IEEE Trans on AP*, **42** (4), April, 556–60.

Ericsson Radio Systems A.B. (1999) 'Welcome to third generation', Stockholm, Sweden.

FCC, 'RF exposure compliance', **www.fcc.gov/oet/rfsafety**

Foster, K.R. and Moulder, E.J. (2000) 'Are mobiles safe?', *IEEE Spectrum*, August, 23–8.

Fuji, T.; Tsuchiya, M.; Isota, Y.; Aoki, K. (1995) 'Design and performance of mobile terminal for North American MSAT network', *IMSC'95, The Fourth International Mobile Satellite Conference*, Ottawa, co-sponsored by Communications Research Centre/Industry Canada and Jet Propulsion Laboratory/NASA, 365–9.

Fujimoto, K. and James, J.R. (1994) *Mobile Antenna System Handbook*, Norwood MA, Artech House.

Gandhi, O.M. (ed) (1990) *Biological effects and medical applications of electromagnetic fields*, Englewood Cliffs, NJ Prentice Hall.

Garcia, Q.; Martin, C.; Del Valle, J.C.; Jongejans, A.; Rinous, P.; Travers, M.N. (1993) 'Microstrip monopulse antenna for land mobile communications', *IMSC '93, The Third International Mobile Satellite Conference*, Pasadena, California, 16–18 June: JPL Publication 93-009, 575–80.

Goetz, I. (1996) 'Mobile network transmission quality', *BT Technology Journal*, 14 (3), July, 81–91.

Goldberg, B.G. (1996) 'Technical challenges for a crowded wireless environment', *Satellite Communications*, December, 32–5.

Guy, A.W. and Chou, C.K. (1986) 'Specific absorption rates of energy in man models exposed to cellular UHF mobile antenna fields', *IEEE Transactions on Microwave Theory Technology*, **34**, 671–80.

Haugli, H.; Hart, N.; Poskett, P. (1993) 'Inmarsat's future personal communicator system', *Space Communications*, **11**, 129–40.

Hawkes, N. (1999) 'Mobile phones quicken the brain', *The Times*, 8 April: 12.

Holmes, J.N. (1988) *Speech Synthesis and Recognition*, Van Nostrand Reinhold (UK), ISBN 0-278-00013-4.

Hombach, V.; Meier, K.; Burkhardt, M.; Kuhn, E.; Kuster, N. (1996) 'The dependence of EM energy absorption upon human head modeling at 900 MHz', *IEEE Trans. Microwave Theory Technology*, **44**, 1865–73.

Hoshikawa, T. (1988) 'Inmarsat ship earth station type RSS401A', *Anritsu Technical Journal* (56), Sept.

Inmarsat (1988) *Inmarsat-C System Definition Manual*.

Inmarsat (1990) *Inmarsat-Aeronautical System Definition Manual*.

Inmarsat (1991a) *Inmarsat-B System Definition Manual*.

Inmarsat (1991b) *Inmarsat-M System Definition Manual*.

Jet Propulsion Laboratory (1988) 'MSAT-X Quarterly No 13', January.

Lin, J.C. (1980) 'The microwave auditory phenomenon', *Proc IEEE*, **6**, 67–73.

Lin, J.C. (1997) 'Biological aspects of mobile communications fields', *Wireless Networks*, **3**, 439–53.

Michaelson, S.M. and Lin, J.C. (1987) *Biological Effects and Health Implications of Radio Frequency Radiation*, New York, Plenum.

Milne, R. (1995) 'Performance and operational considerations in the design of vehicle antennas for mobile satellite communications', *IMSC'95, The Fourth International Mobile Satellite Conference*, Ottawa, co-sponsored by Communications Research Centre/Industry Canada and Jet Propulsion Laboratory/NASA 329–33.

Mobile Europe (1997) 'When less is more', 7 (2), February, 27–30.

Moulder, E.; 'Internet site with cell phone safety information', **www.mcw.edu/gcrc/cop/cell-phone-health-FAQ/toc.html**

NCRP (1981) 'Radio frequency electromagnetic fields; properties, quantities and units, biological interaction and measurements', *Report 67*, National Council on Radiation Protection and Measurements, Bethesda MD.

Ohmori, S.; Wakana, H.; Kawase, S. (1998) 'Vehicle antennas', Chapter 4, *Mobile Satellite Communications*, Boston, Artech House, 87–123.

Phipps, J.H.; Lewis, B.V.; Roberts, T.; Prior, M.V.; Hand, J.W.; Elder, M.; Field, S. (1990) 'Treatment of functional menorrhagia by radio frequency induced thermal endometrial ablation', *Lancet*, 335, 374–6.

Polk, C. and Postow, E. (eds) (1996) *Handbook of Biological Effects of Electromagnetic Fields*, CRC Press, Fl.

Richharia, M. (1999) *Satellite Communication Systems*, Macmillan Press Ltd, Basingstoke and London (jointly published with McGraw-Hill, New York).

Richharia, M. (1986) 'Design considerations for an earth station step track system', *Space Communication and Broadcasting*, **4**, 215–28.

Rothman, K.J.; Loughlin, J.E.; Funch, D.P.; Dreyer, N.A. (1996) 'Overall mortality of cellular telephone customers', *Epidemiology*, **7**, 303–5.

Rudge, A.W.; Milne, K.; Oliver, A.D.; Knight, P. (1982) *The Handbook of Antenna Design*, Vols 1 and 2, Peter Peregrinus, Hitchin, Herts.

Salford, L.G., Bun, A., Eberhardt, J.L.; Persson, B.R.R. (1993) 'Permeability of the blood-brain barrier induced by 915 MHz electromagnetic radiation, continuous wave and modulated at 8, 16, 50 and 200 Hz', *Bioelectrochemistry & Bioenergetics*, **30**, 293–301.

Scholey, N. (1995) 'Rechargeable batteries for mobile communications', *Electronics & Communication Journal*, June, pp. 93–96.

Shafai, L.; Moheb, H.; Chamma, W. and Barakat, M. (1995) 'Low profile antennas MSAT applications', *IMSC'95, The Fourth International Mobile Satellite Conference*, Ottawa, co-sponsored by Communications Research Centre/Industry Canada and Jet Propulsion Laboratory/NASA, 334–9.

Stevens, C.L. (1996) 'Powered for future', *Wireless Business & Technology*, 2(4), May, 33–7.

Stojkovic, I. and Alonso, J.E. (1995) 'Key issues for user terminals operating with HEO satellites', *IMSC'95, The Fourth International Mobile Satellite Conference*, Ottawa, co-sponsored by Communications Research Centre/Industry Canada and Jet Propulsion Laboratory/NASA, 370–5.

Sydor, J. and Dufour, M. (1993) 'Aeronautical satellite antenna steering using magnetic field sensors', *IMSC 93, The Third International Mobile Satellite Conference*, Pasadena, California, 16–18 June, JPL Publication 93–009, 593–8.

Taira, S.; Tanaka, M.; Ohmori, S. (1991) 'High gain airborne antenna for satellite communications', *IEEE Transactions on Aerospace and Electronic Systems*, **27** (2), March, 354–60.

Toftgard, J.; Horusleth, S.N.; Andersen, J.B. (1993) 'Effect on portable antennas of the presence of a person', *IEEE Trans on AP*, **41** (6), June, 739–46.

World Health Organization, 'RF fields and health', **www.who.int/peh-emf**

Spacecraft 6

6.1 Introduction

High-power satellites with sensitive onboard receivers are central to the concept of a mobile satellite service, as these requirements are a prerequisite for supporting services to small mobile terminals. Therefore the advent of mobile satellite communications had to wait until satellite and Earth station technology had developed to an extent that satellite transmitters could provide the requisite power and Earth stations had shrunk to a size supportable on mobiles. Over the past two decades, both components of technology have progressed to an extent that it is possible to communicate with hand-held satellite telephones. In this chapter we will outline trends in satellite technology.

Figure 6.1 shows the main subsystems of a spacecraft.

The main communication unit of a satellite (or payload) comprises an antenna system and a repeater (or transponder). When an intersatellite link (ISL) is used, the payload also supports an ISL payload. A payload requires support from a number of subsystems collectively known as bus, which comprise a structural subsystem for housing components, a thermal subsystem for temperature maintenance, a propulsion system for orbit-raising and orbital maintenance and an electrical power subsystem. Spacecraft status monitoring, tracking and commanding are performed by the telemetry, tracking and command system – the spacecraft's link to the ground control centre.

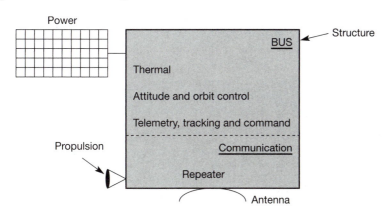

Figure 6.1

Main subsystems of a spacecraft

Keeping to the theme of the book, the majority of the chapter will address the payload, as it has a direct bearing on the capabilities and architecture of an MSS system. At present, transparent (or 'bent pipe') satellite transponders are the norm, but regenerative transponders with onboard computers are expected gradually to supersede them because of several advantages offered by the latter. The chapter includes: a brief overview of transparent transponders but with the emphasis on regenerative transponders, as the technology of the former is well documented in several textbooks; a look at the influence of orbital altitude on spacecraft design; system aspects of inter-satellite links; and a summary of the current state of technology. To begin with, some enabling spacecraft technologies are explored, followed by issues related specifically to MSS. Regenerative transponders have been addressed in greater detail because of their increasing importance to MSS needs.

6.2 Satellites for MSS

There are a number of system requirements which influence the architecture of an MSS payload:

Operational frequency: At present, L and/or S bands are prevalent but in future higher frequency bands are likely to be used;

Spacecraft EIRP and G/T in service link, must be adequately high to serve mobile terminals;

Spectrum must be used very efficiently;

Dynamic distribution of spacecraft EIRP across spot beams is a prerequisite to respond to changes in traffic pattern;

Full eclipse operation: this applies to LEO and MEO satellite systems and to international GEO; regional MSS operators using a GEO system may not operate at full power if they choose the orbital location so that the eclipse occurs late at night.

These top-level requirements cascade down to the following payload-related requirements:

1   Transponder
    - **Available spacecraft EIRP** must be high enough to be able to serve present and future traffic far enough into the future – ideally to the end of spacecraft life.
    - **Flexible real-time power and bandwidth allocation** between spot beams should be possible to permit allocation of space segment resources where traffic demands it.

- **Onboard processing technology** should be considered to benefit from the advantages of regenerative transponders such as improved link performance, packet switching, etc.
- **Intersatellite links** should be considered to enhance network connectivity.

2 Antenna system

- **Spot beam deployment** becomes necessary to enhance frequency reuse, reduce mobile EIRP and make effective use of spacecraft transmitter power.

3 General

- **Technology amenable to mass production** is essential for supporting large non-geostationary constellations.
- **Higher integration of payload components** such as output combiner with antennas, bandpass filters with LNA, solid state power amplifier stacks, etc.

Eclipse operation is a bus requirement but is included here as it is critical to LEO and MEO MSS systems in particular. The number of eclipses is large for LEO and MEO satellites due to their orbital geometry, unless the orbital design has minimized or eliminated the occurrence of eclipses (see Chapter 2). Spacecraft batteries undergo a considerable number of charge/discharge cycles, which reduces the lifetime of the batteries.

Figure 6.2(a) and (b) depicts the main functional blocks of an MSS payload in the forward direction; the return link performs an identical operation in the reverse direction.

MSS feeder links usually deploy only a few large fixed Earth stations and therefore a single or a few beams are enough to meet capacity and frequency coordination requirements, whereas service links serve a huge user population dispersed all over the service area, which necessitates a large number of spot beams – from a few in the 1990s to several hundreds in the first decade of the twenty-first century. The received signals are bandpass filtered to eliminate out-of-band signals, amplified and fed into the payload processor where the signals are converted to IF using standard techniques, and finally, mapped to channels, each routed to a specific spot beam. When regenerative transponders are used, IF signals are demodulated, decoded and processed at baseband where they may also be routed. The processed baseband signals are coded and modulated, then up-converted, amplified and fed into a beam-forming network and antenna system for transmission into an appropriate beam. On-board processing can be performed at baseband, IF or RF and hence satellites with onboard processing need not employ a regenerative transponder. The signal-routing and beam-forming network and the antenna system play a dominant role in MSS payloads due to the need for a large number of spot beams.

6.2.1 Transponders

Due to the simplicity, reliability, technological maturity and flexibility in changing signal format and accessing schemes, many first generation satellite systems developed for hand-held services use transparent transponders. Globalstar and the ICO system use a conventional transponder, whereas the Iridium system incorporates a regenerative payload.

There is a wealth of information on conventional transponders and hence the topic is treated here at a refresher level. However, as mentioned, regenerative transponders are discussed in greater detail because of their future potential as well as their relatively scant coverage in the literature.

Transparent transponder

Figure 6.2 (a) shows a conceptual block diagram of a conventional transponder.

A bandpass filter removes out-of-band signals received by the antenna; a low noise amplifier amplifies the signals; the signals are divided into a number of channels, each destined for a pre-selected spot beam, amplified, up-converted and fed into power amplifiers, a beam-forming network and transmitted to the desired spot beam. There are a number of beam-forming methods which are discussed later. Consider some recent developments applicable to MSS payloads.

Due to a need to serve a large number of spot beams flexibly, it becomes necessary to be able to segment the spectrum into narrow bands for flexible and spectrally efficient routing; a finer granularity is better suited in this respect. Narrowband SAW filter banks for de-multiplexing, followed by switches and an analogue beam-forming network, provide flexible allocation of power and bandwidth. But due to limited granularity of the SAW filters (~400 KHz), the channel bandwidths are coarse, which limits the beam traffic-matching capability.

De-multiplexing granularity can be improved by an analogue time domain Fourier transform technique called the Chirp Fourier Transform (CFT). The technique can be further augmented by digital methods to obtain channels narrower than 100 KHz, permitting flexible allocation of small numbers of FDMA channels.

In an advanced adaptation, the hybrid CFT and digital de-multiplexing technique use channel-by-channel satellite switching, thus achieving the ultimate by way of channel-to-beam routing. Consider a system where the outputs of such a de-multiplexer are fed into a digital narrowband beam-forming network capable of generating a very large number of repositionable beams peaking on individual users. An accurate user location is necessary, an algorithm for which can be included as a part of the digital processing required for digital beam forming. The technique can offer channel-level processing including power level control, active interference suppression, enormous adaptability to traffic variations and frequency reuse, high use of satellite and mobile RF as the user is at gain peak, and transparency for introduction of new services. This type of system is at a conceptual stage awaiting technological development.

Regenerative transponder

The advantages and theory of regenerative transponders have been well established for well over a decade, but their introduction in practice has been gradual. A regen-

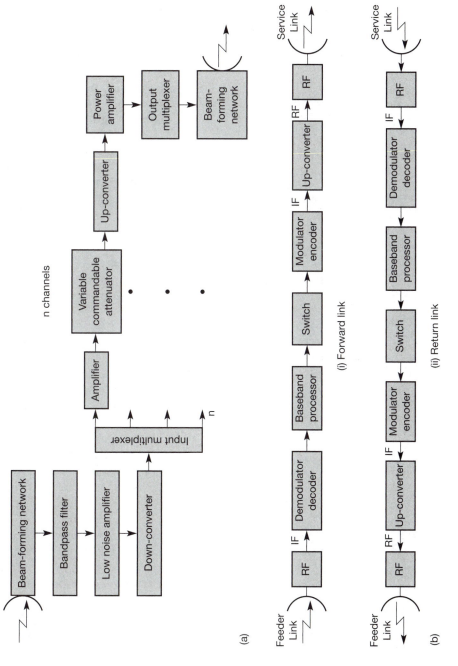

Figure 6.2
(a) A block diagram of a conventional transponder. (b) Block schematic of an MSS payload with onboard processor, showing (i) forward link (ii) return link; beam forming network is assumed to be a part of RF block

erative transponder performs essential signal processing such as demodulation, modulation, transponder or beam routing and switching, as well as advanced processing such as error correction/coding, reformatting of data, interference reduction, rain fade compensation, packet switching, etc. (see Figure 6.2(b)).

Figure 6.3 (a) Relationship between uplink and downlink carrier to noise ratio for conventional and regenerative transponders. (b) Relationship between BER and E_b/N_0 for conventional and regenerative transponders (Brugel 1994)

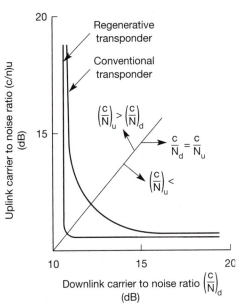

(a)

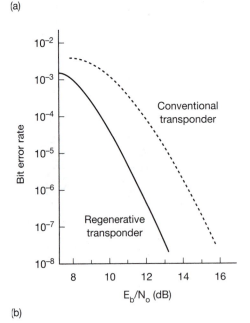

(b)

A regenerative transponder provides considerable performance advantage over its transparent counterpart under similar link conditions. In a conventional transponder, all types of link noise are amplified or attenuated by the same amount as the wanted signal, and therefore noise components add up. In a regenerative transponder, the baseband signal is regenerated in the uplink and thereby the noise components in the up and down links get de-coupled; the total BER in such a case is given as

$$e_t = e_u + e_d \qquad (6.1)$$

where

e_t = total bit error, e_u = uplink bit error rate and e_d = downlink bit error rate.

Figure 6.3(a) compares the behaviour of uplink and downlink carrier to noise ratios for these two classes of transponder for the same link quality, assuming a linear channel without coding.

When the uplink and downlink carrier to noise ratios are of the same order of magnitude, a regenerative transponder shows a clear advantage. This is generally the case for MSS links and, in particular, the service link when mobile terminals can transmit very low EIRP levels. Note that the advantage is reduced as uplink carrier to noise is increased, which would imply that for MSS the advantage is not very significant in the forward link where fixed stations are able to transmit high power. The advantage of the regenerative transponder is significantly increased when inherent non-linearity of the satellite channel is included; non-linearities are mainly introduced at the outputs of the Earth station and satellite transmitter. Between 2 and 5 dB gains are possible, even in the presence of high uplink carrier to noise ratio density (see Figure 6.3(b)). The advantage in the link margin can be traded off against interference in an interference-

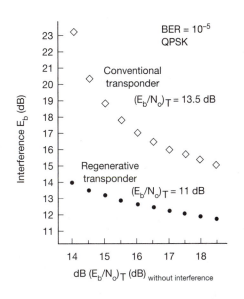

limited situation, and hence regenerative transponders can be used to increase system capacity in such an environment (see Figure 6.4). The decoupling property of regenerative transponders gives resistance to uplink or downlink signal fading. Further, up and down links can be matched separately, giving an advantage of several decibels in power resulting in an increase in overall system capacity.

Consider an MSS system using a K_a-band feeder link and L- or S-band service link. The feeder link is stable for the majority of the time (>95 %) when the predominant noise is Gaussian. For the remaining periods, the link fades to varying degrees due to rain or scintillation in periods of seconds, minutes; whereas the service link consists of a large number of short duration (2–3 minutes) randomly varying channels. As the characteristics of each segment are different, different types of coding schemes are desirable to maximise coding gain. A regenerative transponder allows such a scheme.

The main advantages and disadvantages of regenerative transponders are summarized in Table 6.1.

It has been estimated that, under average operating conditions, improvements of the order of 8 dB are possible by regenerative transponders over conventional transponders in the presence of severe uplink rain fading. The calculations assumed equal uplink and downlink carrier to noise ratios under nominal conditions; carrier to noise ratio of the downlink 10–15 dB above the detector threshold; and a fading of around 12 dB in the uplink. Under more realistic link conditions, improvements of the order of 10–12 dB have been demonstrated.

A regenerative transponder with baseband processing permits reformatting of data and therefore the uplink and downlink multiple access schemes can be different and optimized to suit each. For example, in a TDMA system operating

Table 6.1

Main advantages and
disadvantages of
regenerative
transponders

Advantages	Disadvantages	Comments
Improved channel quality	Increase payload complexity	Extent of disadvantages will reduce as technology matures
Interference resistance	Heavier payload	
Higher capacity	Larger power drainage	Recent breakthroughs include development of high volume MMIC capability and improvements in radiation hardening of digital components
Flexible and dynamic routing of messages	Technology comparatively new	
Better network interconnectivity	Difficult to change signal format, accessing schemes, modem/codec etc. with current technology	
Allows use of smaller terminals		
Permits optimization of up and downlink independently		Most K_a band systems plan use of onboard processors
Access security		

through a conventional transponder, the forward link burst rate is limited by the mobile Earth station's G/T and demodulator performance. A regenerative transponder eliminates such a limitation by allowing the downlink accessing scheme to be better suited to mobile Earth station capability, e.g. by grouping channels into a manageable set of TDM streams at a rate commensurate with the mobile Earth station's capability, while leaving the feeder station the flexibility to transmit in TDMA. Figure 6.5 represents a configuration of a data adaptation regenerative transponder which combines a number of high rate TDMA service links and distributes data at low rates to a large number of mobile users, thereby improving feeder link spectrum efficiency.

A regenerative transponder offers the possibility of traffic routing, switching and onboard processing and thus the name 'intelligent' satellite or a 'switch in the sky'. Switching is done at RF, IF or baseband. RF and IF switching have been in use for several years on Intelsat VI and VII satellites for supporting TDMA traffic. Incoming bursts of data from one or more spot beams are routed to the destined

Figure 6.5

Data rate adaptation
using a regenerative
transponder
(Brugel, 1994)

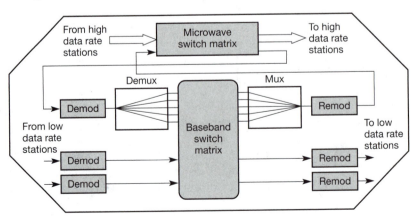

beam using a pre-programmed switch matrix; the switch control unit is programmable through ground commands in accordance with operational requirement, providing flexibility to respond to traffic demands and increasing the spacecraft capacity by ~ 30% compared to a TDMA system without satellite switching.

RF or IF switched TDMA systems provide coarse interconnectivity – when a TDMA slot is not filled, the capacity remains unutilized; if switching is performed at message and packet level, messages/packets from all beams destined for a given beam, can be combined and routed to the appropriate beam, thereby improving resource utilization and network connectivity. A conventional time-space-time switch consisting of input and output buffers interconnected with a switching matrix can be useful.

Call set-up and routing can be autonomous or managed by the ground network control centre (see Chapter 8). Onboard call set-up can minimize latency and reduce ground network complexity at the expense of a more complicated on-board architecture. In addition to switching, considerable processing is necessary for routing and call establishment. The onboard processor must be able to interpret signalling protocols, extract and decode destination addresses from packets, store messages on a temporary basis during call set-up, and establish a routing path, which may be transferred over an intersatellite link. In an MSS system, it is also necessary to participate in the network mobility management function, which may be achieved in a similar manner to that on the ground (e.g. use of HLR/VLR databases). An additional onboard task in such a scheme is to convey to the network business management centre call record data for billing, diagnostics, fraud monitoring and traffic data collection.

Radio resource management is another functionality which can be transferred, at least partially. Considerable capacity enhancement is possible in a multi-spot beam carrying non-uniformly distributed variable traffic across beams when resources are allocated dynamically. Consider a regional system where the coverage spreads across several time zones; the business traffic peak tends to ripple from eastern spot beams to the west, according to business hours. Traffic patterns may get severely distorted in the event of an unusual event such as a sports event, military action, etc. In such cases, allocation of fixed spectrum/power in spot beams is wasteful – a dynamic and flexible sharing of resources can improve the productivity (see Chapter 8).

The architecture of a baseband processing satellite to support a public network must be compatible with its transmission system, e.g. ISDN, DVB, IP, ATM, etc. The signalling and protocol within the satellite network should be transparent to underlying protocols. At present, there are no standards established for protocols within a satellite's processor; standardization of protocols should enable interconnection of satellite systems.

Consider a few representative (but not exhaustive) baseband switching techniques in a multiple carrier TDMA system. Three methods are proposed – carrier switching, simultaneous-slot switching and slot switching as illustrated in Figure 6.6 (Taira *et al.*, 1998). The TDMA scheme consists of three spot beams with one carrier per beam and three slots per carrier.

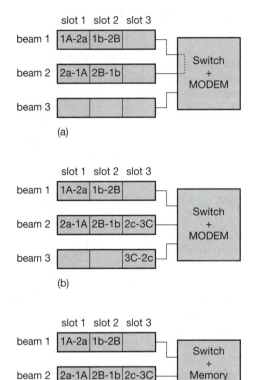

In a *carrier switched* system, the source and destination TDMA slots are interconnected via carriers switched to the source and destination beams, excluding connectivity with spot beam 3. A request originating in beam 3, destined for beam 2, is dropped because there are no carriers available in beam 2. The carrier switch can be controlled either by the ground or onboard network control centre.

In a *simultaneous-slot switched* system using the same call pattern, a call request from beam 3, destined for beam 2, is assigned to slot 3 even though slots 1 and 2 of beam 3 are empty. The assignment corresponds to the slot available in beam 2. In this method, the message is switched to the appropriate beam within the baseband processor.

In a *slot-switched* system, a beam 3 call request is assigned the first available slots in each beam, i.e. slot 1 in beam 3 and slot 3 in beam 2. Thus messages have to be stored onboard in order to direct them to the correct slot in each beam and hence onboard memory is necessary, as illustrated in the figure.

The authors compared the onboard traffic handling capacity of these schemes using a computer simulation assuming a single dimension 12-spot beam over Japan (13 m diameter), S-band mobile system and other characteristics summarized in Table 6.2, and compared the hardware size and power consumption of the onboard processor for each. The results revealed that the carrier switching scheme is the least efficient, while the performance of the other two schemes is very similar; the processor hardware size and power consumption are minimum for the simultaneous-slot method.

Regenerative transponder technology

The technology of the regenerative transponder is relatively new, which explains its limited use so far. However, many next generation MSS and FSS satellites plan to deploy such transponders. A general set of requirements for the equipment is summarized as follows:

Parameter	Characteristics
Air interface	
RF	2.6/2.5 GHz
Modem	π/4 shift QPSK/coherent demodulator
Multiple access	Multi-carrier TDMA; 5 slots/carrier
Transmission rate	70 kbps
Information rate	5.6 kbps (voice)
	32 kbps (data)
Error correction	Convolution coding (constraint length 7, coding rate $\frac{1}{2}$); Viterbi decoding (3bit, soft decision)
Capacity simulation parameters	
Frequency reuse	Reuse every third beam
Capacity/beam	120 voice, 20% mobile-mobile (i.e. 40 channels)
Call duration	Exponential with a mean of 120 s
Call arrival	Poisson distribution
Traffic density in each beam	Uniform
Congestion handling	Call blocked and cleared during congestion
Call blocking rate	3%

Table 6.2

Simulation conditions for comparing the capacity performance of various MC-TDMA baseband switching schemes

- lightweight to minimize weight penalty;
- radiation hardened, reliable and compact;
- compliant with performance specifications such as speed, accuracy, reliability, noise immunity, BER performance, level stability, etc.
- provision of autonomous checks to ensure correct functioning;
- fault tolerant hardware and software to minimize occurrence of faults/errors.

Baseband processor architecture can be implemented in various ways, but as illustrated in Figure 6.7, the main communications related functional blocks comprise:

- modulator/demodulator;
- coder/decoder;
- analogue to digital converters;
- call processing/control;
- switch.

Modulators can be implemented either directly at the downlink frequency or at IF; the techniques are relatively straightforward for the more conventional modulation schemes such as PSK, except that the hardware must be space

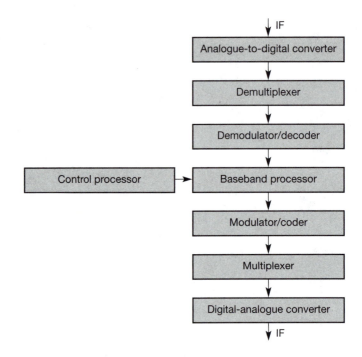

qualified. The most complex implementation is that of the demodulator and baseband signal processor.

Demodulators use conventional demodulation principles; coherent demodulation, being the most efficient, can therefore be expected as a preferred choice in most systems. In addition to managing the well-known problems of carrier and clock recovery, the implementation problem onboard a satellite is further complicated by the necessity to demodulate a number of incoming carriers. If a separate demodulator was deployed for each carrier, the size and power requirements of the demodulator bank would become prohibitive. Instead, multiplexed carriers are demodulated in a group, and the hardware size is thus reduced – this is achieved by *multi-carrier demodulation*. Frequency division multiplexed carriers are digitized and separated into individual carriers, which are then filtered and demodulated individually by standard digital signal processing techniques. Multi-carrier demodulators may be implemented in analogue form or digitally by the Fast Fourier Transform (FFT) technique in a device known as a trans-multiplexer.

Analogue domain FFT implementation can be achieved by a bank of contiguous tuneable surface acoustic wave (SAW) filters in conjunction with a chirp oscillator; thereby, transformed signals are channelized simultaneously. The signals are then digitized for further processing.

Due to differences in transmission sources, particular attention is given to synchronization. The differences can impose a limit on the data rate and frame duration of individual carriers.

Figure 6.8 (a) shows an example of an onboard processor proposed for an MSS payload to support a geostationary multiple carrier TDMA system for hand-held telephone and medium rate data service, as specified in Table 6.2 (Takeda *et al.*, 1995). Figure 6.8(b) shows the architecture of the forward and return processor for the proposed system.

The system incorporates baseband switching and mobile-to-mobile communication with signal regeneration and on-board call set-up. In calls to/from a fixed direction, onboard call setup (in cooperation with gateway) is used but without signal regeneration. Call set-up is managed by the control processor. De-multiplexers can demodulate up to 100 FDM carriers per beam, while the switch can handle more than 100 carriers per beam. All four processors are implemented in ASIC and utilize digital signal processing. The forward and return processors illustrated in Figure 6.8 (b) utilize polyphase FFT in preference to other techniques such as the bulk-FFT method and multistage method because it minimizes the number of multiplications per sample, thereby reducing the hardware size. The polyphase filters comprise FIR filters attenuating to 35 dB in the adjacent channel. The FFT used N = 128, which represents the number of carriers which may be de-multiplexed simultaneously and governs the size of the circuit. The pipeline junction method is used to minimize complexity. The polyphase inverse FFT method is used in the multiplexer to simplify the overall design. Modulator and demodulator units are implemented digitally. The control unit performs all demand assigned channel allocation. Its functions include analysis and processing of channel control packets, call set-up and mobility management, i.e. registration of the mobile terminal position, timing adjustment for each

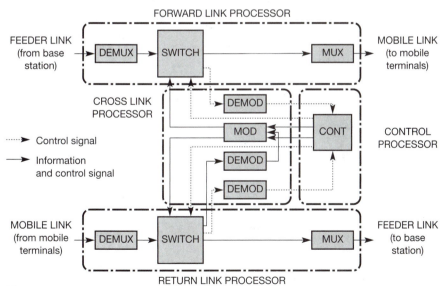

(a)

Figure 6.8(a) A block diagram of an onboard processor proposed for an MSS payload to support a geostationary multiple carrier TDMA system for hand-held telephone and medium rate data service, as specified in Table 6.2 (Takeda *et al.*, 1995). (b) Architecture of the forward and return processor for the proposed system (Takeda, *et al.*, 1995)

Figure 6.8
Continued

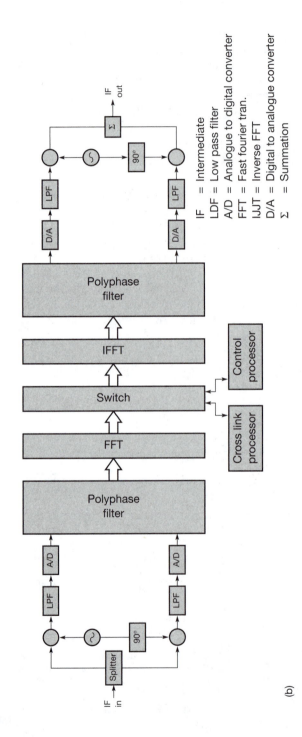

IF = Intermediate
LDF = Low pass filter
A/D = Analogue to digital converter
FFT = Fast fourier tran.
IJJT = Inverse FFT
D/A = Digital to analogue converter
Σ = Summation

(b)

mobile, call control including channel search, assignment, carrier monitoring, channel release and RF transmission management. The controller's software can be downloaded from the ground through the feeder link. The controller can also generate standard timing signals, provide statistics of channel connection, etc. A redundant processor is used due to the criticality of the subsystem.

Baseband processing is achieved by integrated circuit chips at various levels of integration, i.e. LSI, VLSI, etc. Most of the processing is implemented in the digital domain, but demodulation can use analogue processing. These devices can be fabricated by a number of solid state technologies. One of the main challenges in a space environment is radiation hardening of devices to minimize gradual performance degradation due to radiation dosage. In geostationary orbit, the main sources of radiation are electrons trapped in the Van Allen belt. Other sources of radiation are solar flares and cosmic rays. Even with good shielding (e.g. 4–5 mm aluminium shields), onboard devices must be able to withstand a total radiation dose of 3×10^5 rad (Si) over a 10-year period. Additionally, digital devices are susceptible to single event errors such as latch up and soft errors due to ionization and high energy particles contained in cosmic rays. An adjunct advantage of miniaturization is a reduction in radiation dosage per component and hence an improvement in radiation hardness. The radiation hardness for a regenerative transponder can be specified as follows (Kato *et al.*, 1987):

Latch up: free
Total dosage hardness: 1×10^5 (S_i)
Soft error rate: 1×10^{-5} error/bit day.

The main fabrication technologies expected to provide components for regenerative transponders are compared, vis-à-vis component density, power consumption, speed, susceptibility to single-event, soft errors and latch-up error due to radiation, and availability in the year 2000 time-frame, in Table 6.3 (Kato *et al.*, 1987). As regenerative transponder technology is evolving rapidly because of a number of regenerative satellites planned in the near future, particularly in the USA, the interested reader is encouraged to consult the most recent literature. Here we summarize the general trends as envisaged in the 1990s.

The most commonly used technologies are Silicon (Si) or Gallium Arsenide (GaAs) technologies. CMOS/SOS technology provides the best performance, except for immunity to total dosage. However, the component density of devices using this technology is lower and development cost is high; therefore they are better suited to special applications such as military. CMOS bulk devices have relatively poor latch-up performance, but several promising development approaches for hardening CMOS bulk devices are under way. GaAs and bipolar technologies are most susceptible to soft errors. It is expected that all the technologies will be able to provide the desired radiation hardness, but CMOS/SOS technology will continue to remain expensive.

Table 6.3
Si and GaAs
technologies

Material	Device	Total radiation dose [Rad (Si)]	Soft error performance	Latch-up performance
Si	Bipolar	10^5–10^7	Poor	Very good
Si	CMOS/Bulk	10^5–10^6	Good	Poor
Si	CMOS/SOS	~5×10^4	Very good	Very good
GaAs family	Metal Semiconductors (MES); Hetrojunction bipolar transistors (HBT); High Electron Mobility Transistors (HEMT)	>10^7	Poor	Very good

CMOS technology has a higher number of gates per chip, larger storage density in terms of bits per chip compared to bipolar and GaAs technologies and is expected to remain so in the near future.

CMOS devices have the lowest power consumption, but in terms of propagation delay–power product GaAs show better performance than other technologies. Gate delays are the least for GaAS, followed by bipolar and CMOS.

It can be expected that for low and medium speeds, CMOS devices will be preferred and for high speed, bipolar or GaAs may be preferred, leading eventually to GaAs as the preferred choice.

For analogue onboard modems, the choice of two fabrication techniques is available – monolithic microwave integrated circuit (MMIC) which are suited for lower frequency bands up to ~ 4 GHz, and microwave integrated circuits (MIC) suitable for higher frequency bands. The MMIC device choice is either an Si bipolar device processed by a technology called super self-aligned process technology (SST) or a GaAs device. Below ~2 GHz, SST processed Si bipolar devices were preferred in the 1990s due to their proven radiation hardness in terms of total dosage. Between ~1 and 6 GHz, GaAs MMIC can be used. For modems above ~4 GHz, MIC technology can reduce the hardware significantly.

A number of experimental missions have flown in recent years for technology validation and communication concepts. Examples of such missions are the USA's Advanced Communications Technology Satellite (ACTS), Japan's ETS-VI and COMET experiment, Italy's ITALSAT programme and ESA's ARTEMIS satellites. Onboard processing has been used in military satellites such as FLTSAT-7 and FLTSAT-8 for well over a decade. Amongst commercial systems, the Iridium system is a forerunner in extensive use of regenerative transponders including the ISL. A number of forthcoming K_a-band satellites intend to use regenerative transponders (see Chapter 11).

NASA's ACTS programme was intended to demonstrate advanced technologies for next generation satellites with four main technology thrusts – K_a-band system design and component development, onboard multibeam communication processing system, multiple dynamically-hopping high gain antenna system and advanced network control architecture combining Earth and space segments. Figure 6.9(a) and (b) (Campanella *et al.*, 1990) illustrates ACTS' communications payload and baseband processor unit respectively.

The ACTS multibeam communication package operates at 30 GHz in the uplink and 20 GHz in the downlink and includes a regenerative transponder comprising a baseband processor (BBP) connected to a multibeam antenna system. There are two modes of operation – in the baseband processor mode (BPM), communication is set up using the BBP, and in the microwave switch mode, the traditional satellite switched TDMA scheme is utilized. The BBP mode supports a lower bit rate and is called a low bit rate system (LBR) and the satellite switched mode, supporting a high bit rate, is known as a high bit rate (HBR) system.

The ACTS communication package uses a TDMA scheme which, for the LBR, operates at burst rates of 27.5 Mb/s (available in uplink only) or 110 Mb/s comprising respectively 1,728 and 432, 64 kb/s, channels assigned within 1 ms frames. There are three fixed spot beams directed towards Atlanta, Cleveland and Tampa; and two hopping beams covering two large north-eastern US regions which also provide spot coverage to 13 other states. The total capacity for each beam is 2,000, determined by the onboard memory. An adaptive coding scheme is used to make the link robust in the presence of rain fades or other degradation such as interference. Transmissions are normally uncoded; however, during adverse conditions the coded mode can be activated by ground commands. The LBR mode uses convolution coding with consequent data rate reduction, and can be effected by ground commands. When both links are coded, the half rate FEC code provides a margin of 10 dB at the expense of reduction in throughput by a quarter. In the HBR mode, link compensation is obtained by transmitter power control.

The baseband processor demodulates the serial minimum shift keyed (SMSK) signal and, if commanded, decodes the incoming TDMA bursts. Each channel is then routed through a 3X3 TST switch. The LBR system is controlled by a master control station which communicates with users via an orderwire (OW) for call set-up and with the ACTS multibeam communication package for controlling BBP and antenna scanning. To establish a call, the user sends a request via the OW contained in the reference burst to the master control station (MCS) giving its identity, the destination identity and the number of 64 kbps circuits required. The MCS determines TDMA bursts compliant with the request, routing, and beam dwell control message, and transmits the message to the BBP and hopping beam antenna control of ACTS in an uplink burst. The message is stored in an onboard memory, while the assignment information is sent by the MCS to the source and destination stations over the OW. The signalling messages are protected by FEC and repetition to minimize the

Figure 6.9
(a) ACTS multibeam
communication system
(Campenella *et al.*,
1990) (b) ACTS
baseband processor
(Campenella *et al.*,
1990) © 1990 by
American Institute of
Aeronautics and
Astronautics, Inc.
Reprinted with
permission

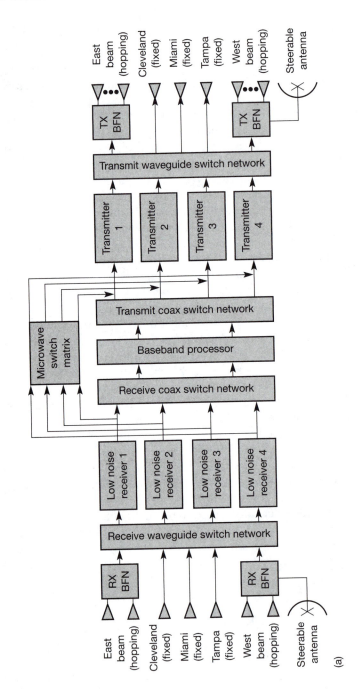

Figure 6.9
Continued

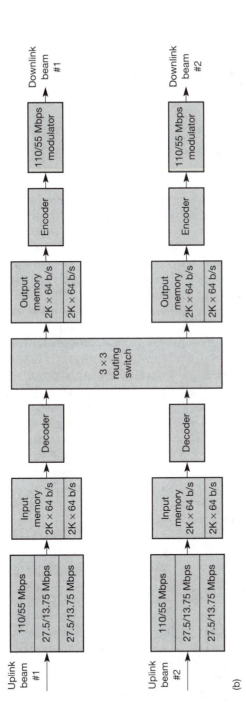

(b)

probability of error. A synchronized burst time plan change is then executed by the BBP, hopping beam system, and affected Earth stations. At the end of a call, the 'on-hook' message is sent by the calling or called party over the OW, which is used by the MCS to return the capacity to the space segment.

The satellite switched transmission mode, i.e. the HBR system, operates at 220 Mb/s to interconnect the three fixed spot beams mentioned earlier. The satellite switch can also be commanded to remain fixed to interconnected beams for conducting certain experiments.

ACTS baseband hardware makes use of custom LSI and SSI chips to minimize the size, weight and power of the BBP, combining a number of technologies depending on the complexity, size and speed needed for the application. Table 6.4 shows the type, number, processor function where each LSI is used, its dimension and total radiation dose (Gargione, 1990).

The hopping beam-forming network uses, as a basic element, fast ferrite switches capable of changing state in less than a microsecond. The 30 GHz LNA use high-electron-mobility transistors (HEMT) to achieve a 3.4 db noise figure. It is interesting to note that Government Electronics Group (a subdivision of Motorola, in Chandler, Arizona), the proponent of the Iridium system, built the BBP. The company conducted some ACTS experiments to evaluate the onboard demodulation technique at the K_a band to assist in development of high bit rate modems. The data assisted the company to proceed with development of the Iridium system.

NASDA (National Space Development Agency) and MPT (Ministry of Telecommunications) have a long-term programme for development of advanced satellite communications technology and systems through engineering test satellites (ETS) and communications and broadcasting engineering test satellites (COMETS) programmes. A number of onboard technologies, such as multiple narrow spot beams (0.3°) and their interconnectivity using onboard switching, an S-band mobile communications payload, S, K_a and optical wavelength intersatellite links, were flown on ETS-VI but unfortunately, due to a problem in the apogee-kick motor, the satellite did not reach geostationary orbit. However, ETS-VI did not include a regenerative transponder; call by call routing was designed to be routed at RF or IF through a satellite-switch controller which obtained routing information through the TT&C channel. Its follow-on satellites in the COMETS series are expected to extend the multibeam mobile technology to the K_a and millimetre band (50/40 GHz) with regeneration, and demonstrate S- and K_a-band LEO–GEO intersatellite links (Oshima, *et al.*, 1993).

ESA's Olympus satellite, launched in 1989, demonstrated a number of technologies – a 18/12 GHz high power direct broadcast, a 5-beam K_u-band communication payload using a 4×4 onboard switch matrix, a 12/20/30 GHz multiband propagation payload and the primary 30/20 GHz package for advanced communication experiments. The advanced communication experiment payload, using two narrowbeam spot beams of 1°, was used in a double-hop onboard simulator experiment, where the onboard simulator, using a SAW-based multi-carrier demodulator for demodulating a group of

Function	Application	Technology	Quantity	Radiation hardening level (total dose in RAD)	Comments
Maximum-likelihood convolution decoder	Input channel decoding	CMOS	4	10^4	3-micron gate length; ~ 30,000 – transistor array
Memory update controller (MUC)	Control memories	CMOS	24	10^6	5-micron gate length
Serial-parallel/ parallel-serial	Input channel, output channel	CMOS	28	10^6	5-micron gate length
Encoder	Output channel encoding	MOSAIC-ECL	2	10^6	3-micron emitter width; 10×10 cell array; ~2,000 – transistor array
Serial to parallel	Input channel	MOSAIC-CML	6	10^6	3-micron emitter width; 10×10 cell array; ~2,000 – transistor array
Parallel to serial	Input memory, routing switch, output channel	MOSAIC-CML	32	10^6	3-micron emitter width; 10×10 cell array; ~2,000 – transistor array
Correlator	Input channel synchronization	MOSAIC-CML	6	10^6	3-micron emitter width; 10×10 cell array; ~2,000 – transistor array
Timing and control: A	Input channel synchronization	MOSAIC-CML	4	10^6	3-micron emitter width; 10×10 cell array; ~2,000 – transistor array
Timing and control: B	Clock distribution	MOSAIC-CML	4	10^6	3-micron emitter width; 10×10 cell array; ~2,000 – transistor array

Table 6.4
Technology used in BBP custom LSI

24 QPSK/FDMA channels was placed on the ground. The demodulated signals were reformatted and transmitted back to the spacecraft as a TDM stream. The experience gained by this simulated OBP experiment was utilized for the ITALSAT project, a programme sponsored by Italy for advancing payload technology. ITALSAT has three main payloads. Two of the payloads use regenerative transponders; the third and most advanced payload consists of a K_a band regenerative transponder using an onboard baseband processor for routing calls to any of the six spot beams directed over Italy's main metropolitan areas. Signals from the six spot beams are multiplexed, grouped in threes, demodulated in coherent QPSK demodulators and fed into a 6×6 baseband switch which operates in channel increments of 32 kbps for being routed to the desired beam. The switch is controlled by an onboard demand assigned multiple access controller which obtains call set-up information from signals uplinked on the TT&C channel. The baseband switch, in addition, synchronizes the demodulated bursts and generates a reference burst for network synchronization. K_a band uplink operates at 147.456 Mbps. The satellite was launched in January 1991.

ARTEMIS (Advanced Relay and Technology Mission) system, sponsored by ESA, builds on some technologies developed for the ITALSAT programme. The main thrust is to test an optical intersatellite link and an L-band mobile communication system using regenerative transponders. The spacecraft uses a platform integrated control system to increase reliability, reduce weight and reduce power consumption through the use of autonomous onboard processing to provide functions such as processing of onboard telemetry, management of power distribution, attitude and orbit control management, thus simplifying ground control requirements.

Other notable developments have been in the defence sector but much of the technology remained classified. As an example, the US Department of Defense's Fleet satellite system flight-tested onboard processing technologies on FLTSAT-7 and FLTSAT-8 as long ago as 1986 and 1989 respectively. The onboard processing system consisted of a demodulator at the VHF operating frequency using a surface acoustic wave FFT technique. The onboard processor consisted of an intelligent controller capable of assigning and routing calls autonomously, i.e. without assistance from a ground station. Thus if we extrapolate the trend, it appears that satellite hardware technology for proposals such as Teledesic is already available or achievable within a few years.

Finally, Iridium satellites with extensive use of onboard processing and ISLs clearly demonstrate the maturity of regenerative transponder technology in the USA.

Table 6.5 summarizes the main regenerative technologies and antenna systems demonstrated for MSS during the 1990s (Brugel, 1994) (see also Chapter 11).

6.2.2 Antenna systems

Invariably, all systems being planned or in operation deploy multiple spot beam antennas to maximize spacecraft power usage and spectrum reuse. Technology for generating over 300 spot beams in the L band using large reflectors of 8–20 m diam-

Table 6.5
The main regenerative technologies and antenna systems demonstrated for MSS during the 1990s (Brugel, 1994)

	OLYMPUS	ITALSAT	ACTS	ETS VI	ARTEMIS	IRIDIUM	COMETS
Country/ Organization	European Space Agency	Italy	USA	Japan	ESA & Italy	USA/Motorola	Japan
Launch	1989	1991	1992	1994	1996	1998/99	1999 (planned)
Frequency band (GHz) communications	K_a band (30/20), K_u band (18/12 & 14/12)	K_a band	K_a band (30/20)	K_a band, K_u band (14/12); S band (2.6/2.5)	K_a band (30/20), L-band, K_u band	S band (K_a band ISLs)	K_a band (30/20), mm band (50/40)
Antenna system	DBS K_u band: two spots; specializes services payload: 5 spots; Advanced communications Payload: two spots (1°)	2 m, 6 spots, 5 & 3.5 m antenna	K_a band – 3 fixed spots, 2 pairs hopping spots, 1 fixed spot: various beam steering modes; beam size 0.27° and 1°	K_a band – 13 spots 1 (0.3°); C-band – 1 spot; S band 5-spots	3 m, multispot	3.2 m, 2.3 m, 0.8 m, multi-spot with scanning	S band 48-spots
Regeneration capability	4×4 IF switch; ground based onboard simulation tests	6×6 baseband switch	4×4 IF switch; 3×3 baseband switch	16×12 IF switch – with onboard switch controller	Baseband switch	Baseband processor	Baseband packet-mode switch with routing to ground or other satellites via ISL
Demodulator	SAW-CFT ground-based	Coherent QPSK at 147 Mbps	Digital demodulation (SMSK)	Not implemented	N/A	N/A	N/A
Traffic routing technique	SS/TDMA controlled through ground commands	Onboard DCP	BBP controlled from ground	Satellite switch controlled from ground	Onboard autonomous	Onboard	Onboard
Solid state technology	CMOS for SAW-CFT	CMOS for demodulator and DCP, GaAs for solar cells	CMOS, MOSASI-ECL and MOSASI-EML for BBP (see Table 6.4)	Satellite switch: CMOS;SSPA: GaAs, IF switch: GaAS	SAW-CFT	GaAs VLSI	N/A
ISL	GEO-LEO K_a band	None	None	S-Band (2.3/2.1 GHz); MM wave (43/38 GHz); optical	GEO-LEO data relay	S band and K_a band GEO-LEO	LEO-LEO K_a band

eter and multiple feed combinations is available. The peak gain of each spot beam is of the order of 30–40 dB with an out-of-band isolation of 20 dB within reusable spot beams which use a terrestrial cellular-like reuse grid. A number of techniques have been proposed to generate spot beams with such stringent requirements.

Direct radiating arrays use active radiating elements with considerable flexibility and failure resistance. The transmit and receive arrays can be mounted on the same aperture or separately. Separately mounted antennas are simpler to design but their deployment is complex, whereas the reverse is true for composite antennas. The latter may consist of dual-frequency patches or separate interleaved radiating elements.

A number of reflector antennas have been proposed. In a multi-feed reflector antenna, the transmit and receive feed arrangement may be optimized separately. For example, in the receive section, a focusing reflector antenna using beam synthesis requires the smallest feed arrangement. As amplitude control at feed does not have any efficiency impact, the technique uses optimal weighting of pre-amplified signals from a few feeds only (Le Normand *et al.*, 1988). On the transmit side, the antenna may be fed with clusters of feed each with one power amplifier. This arrangement requires feed switching to cope with changes in traffic load. Alternatively, a semi-active multi-matrix arrangement such as used in Inmarsat-3 satellites may be used. More recently, the technique has been used in the ACes (Asia Cellular Satellite) program for generating 140 spot beams using a planar feed assembly comprising 88 cup-dipole radiators. This type of arrangement has good power efficiency with small reflector and feed sizes. In such an arrangement (see Figure 6.10), feeds are shared between several beams and fed from the same power amplifiers via a Butler-like

Figure 6.10

Multi-matrix spot beam generation (Benedicto *et al.*, 1993)

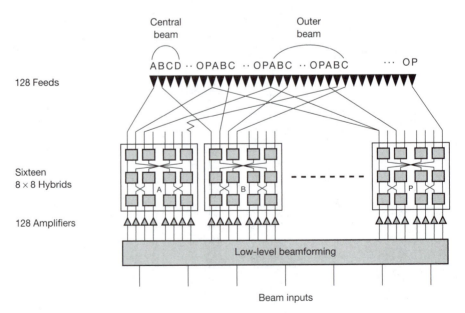

matrix. The matrix adjusts the amplitude and phase so as to create spot beams at the desired Earth locations. Another approach for shaping antenna beams has been the use of shaped reflectors.

Lightweight passive large deployable antennas have been introduced by Hughes in satellites such as AMSC. This lightweight single-piece flexible 7m × 5m antenna weighs around 20 kg and is passively deployed. The antenna is curled and fitted to the nose of a satellite during launch and when released on station, the 'spring-back' antenna springs into its original shape.

Astro Aerospace Corporation

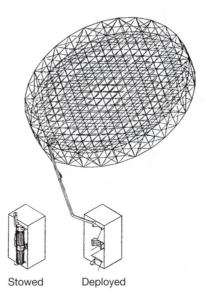

Stowed   Deployed

Figure 6.11
Stowed and deployed form of a reflector under development (Thompson, 1997)

have developed low-cost, lightweight, deployable reflector technology for all offset-fed antenna sizes used on modern MSS geostationary satellites, meeting the stringent passive intermodulation (PIM) and electrostatic discharge (ESD) specifications. The reflector can be stowed in a very compact form, giving an aperture size to stowed length ratio from 15 to 25 when the aperture varies from 6 to 25 m. Figure 6.11 shows the stowed and deployed form of the reflector.

Considerable effort is under way to introduce phased arrays on satellites. Static phased arrays have already been introduced in satellites such as used in the Globalstar network. Research is in progress towards the introduction of reconfig-urable phased arrays which should enable operators to change the shape of a beam according to demands or after spacecraft relocation. The Teledesic system plans to use phased array antennas to provide an Earth-fixed frequency assign-ment scheme which requires satellites to track Earth-fixed regions.

The development of a potentially useful class space structure known as an 'inflatable deployable structure', capable of deploying antennas of up to 30 m, is being sponsored by NASA. The structures could operate in the frequency range 0.3–88 GHz and may be ready for MSS applications by 2003. A successful space flight of an inflatable antenna has already been demonstrated, in May 1996. One of the main difficulties is to construct 10–20 m antennas at low weight and cost; the use of inflatable antennas could reduce by 'orders of magnitude the cost, mass and launch volume of large space structures'.

6.2.3 Effect of orbital altitude on spacecraft design (see also 8.3.5)

Up to the early 1990s, geostationary orbits were an obvious choice for MSS net-works. Since then, the introduction of non-geostationary orbits in MSS network architecture has caused a radical change in network design, leading to a huge demand for satellites capable of operating in low, medium, elliptical or hybrid orbits. Both orbital characteristic and network architecture influence non-geostationary spacecraft design, by imposing specific constraints. Orbital characteristics of interest are the number of eclipses, radiation environment and altitude. A network requirement of particular interest is the use or not of inter-satellite links and the extent of onboard participation in network features such as mobility management and routing.

Eclipse operation affects the design of a satellite's power system, as the power system must be designed to operate during an eclipse when solar cells cannot generate electricity. Rechargeable batteries are used to support the operation during such power blackouts; however, the problem specific to low/medium orbits arises due to the need for a much higher number of charge/discharge cycles and relatively shorter period for recharging the storage batteries. The number of eclipses in circular orbits increases as orbital altitude is reduced, with the exception of a Sun-synchronous orbit having an orbital plane normal to the direction of the Sun such that satellites are never in an eclipse (see section 2.2.3). A geostationary satellite undergoes around 90 eclipses a year, with a maximum eclipse duration of 72 minutes, whereas a LEO satellite at an orbit of 1,000 km which has an orbital period of around 100 minutes can undergo eclipses lasting about $\frac{1}{3}$ of the orbit, i.e. ~ 34 minutes when the orbital plane is parallel to the Earth–Sun vector. Up to 5,000 eclipses can occur in a year in a LEO system. In the absence of solar power, batteries have to supply power. The batteries discharge to a certain depth during an eclipse and are recharged when they come out of it. The recharging time available to LEO satellites is therefore considerably lower, whereas the number of charge–discharge cycles is much higher. The life-time of a battery is governed by the number of charge–discharge cycles and the depth of discharge. Thus a battery which can operate for 12 years at a discharge depth of 0.8 in a geostationary orbit may have a lifetime of only around 5 years at 50% discharge depth – the geostationary orbit would have a total of 1,080 eclipses in the period whereas the LEO would have up to 25,000 eclipses.

A number of battery technologies are available – Nickel-Cadmium (NiCd), Nickel-Hydrogen (NiH2), Silver-zinc (AgZn), Lithium ion (Li). Nickel-metal-hydride (NMH) has limited application and a Sodium-Sulphur (NaS) battery is under development. Ni-H is expected to be the technology of choice, with a capability to support 15–20 Kw (geostationary) satellites being considered.

The radiation environment depends on orbital altitude. In this respect, the HEO satellites are most susceptible as they could pass through a high radiation environment during parts of their orbit. LEO and MEO altitudes are selected so that high radiation environments are largely avoided. Satellites in low Earth orbits can be affected by drag, in particular if their array size is large, and hence there is an upper limit on DC power from LEO satellites. Note that the power requirements of a satellite should reduce with a reduction in orbital altitude for an identical terminal type and traffic, for two reasons: satellite EIRP per channel should reduce due to a lower path loss and a reduction in traffic captured within the satellite's footprint. A typical feature of electronic circuits is a sudden short failure in the presence of a radiation environment (see section 2.2.2). This can cause a sudden software failure and hence it becomes necessary to use fault-tolerant software, especially in critical aspects of spacecraft operation.

The number of gateways can be reduced considerably with intersatellite links (ISL). This feature is, in particular, useful for low

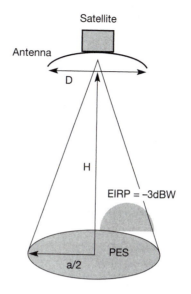

Figure 6.12
Geometry for estimating spacecraft antenna (adapted from Egami, 1995)

Earth satellite systems which may require hundreds of gateways for world-wide connectivity without intersatellite links. Associated with ISL is a need for signal regeneration and packet-switching for establishing routes, and this requires a considerable amount of signal processing and computing. Incorporating ISL in polar orbit is less difficult due to a relatively simple geometry. ISL have also been proposed for geostationary satellite systems. Mobility management is quite involved for non-geostationary orbits due to satellite movement with respect to users. If the network architecture requires the participation of satellites, this imposes considerable processing requirements on satellites. Similarly, when satellite participation is required for radio resource management, satellites must incorporate signal regeneration with adequate processing.

A significant effect of altitude on spacecraft requirement arises due to the strong influence of altitude on coverage area. Figure 6.12 depicts the geometry of the problem. It can be readily shown that the edge-of-coverage carrier to noise ratio (C/N) at the spacecraft from a user transmission is given as (Egami, 1995)

$$(C/N) = (\eta/32)(E_h/kT)(D/H)^2 \qquad \qquad \dots (6.2)$$

where
η = antenna aperture efficiency, E_h = EIRP of mobile terminal, k = Boltzman constant, and T = satellite's system noise temperature, D = antenna diameter and H = satellite altitude.

From equation 6.2, we note that for a specified EIRP and satellite system noise temperature, the uplink C/N depends on the $(D/H)^2$. As an example, Figure 6.13 (a) and (b) demonstrates the size of antenna for a hand-held system which typically has an EIRP of –3dBW, assuming satellite system noise of 300 K

Figure 6.13
Satellite altitude versus
spacecraft antenna
diameter to support
hand-held
communications at
altitude (a) up to 12,000
km (b) ranging between
10,000 and 40,000 km
(Graphics: AR)

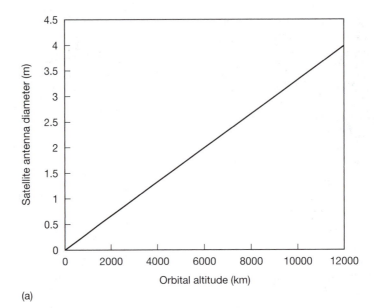

(a)

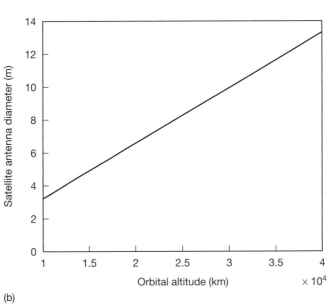

(b)

and $\eta = 60\%$. Note the large antenna size required for GSO. This was one reason why GSO systems were not favoured for hand-held service in the early 1990s when large antenna technology was not developed to acceptable standards.

The cell (or spot beam) diameter can be obtained by simple geometry as a = 2H tan θ/2 ... (6.3)

where θ = satellite antenna 3 dB beamwidth of receive antenna ~ 70λ/D.

LEO and MEO satellites have an advantage when it comes to the launching of satellites. It is essential to launch spacecraft in clusters for economy and to install a constellation as soon as possible (see section 6.5). Satellites have to be developed in accordance with the volume and shroud specifications of the launch vehicle.

Due to the need to deploy large constellations, traditional methods of satellite manufacturing required a major reorganization. Only a few geostationary satellites are required to attain a given coverage – typically 1–2 satellites for regional systems and 4–5 for international systems, and moreover, satellites can be deployed quite independently. Hence, traditionally, satellites would be assembled one at a time over months and years, with rigorous testing at each step; replacement and diagnosis of defective parts could be tedious when spacecraft required disassembly. On the contrary, LEO and MEO constellations require significant numbers if not all satellites are deployed for reliable service coverage, requiring a radically different spacecraft assembly approach. We will consider the satellite production approach of a number of recent systems as illustrations.

Iridium satellites were assembled with techniques used in automobile assembly lines, resulting in a satellite being produced every five days at the peak of activity in 1998, with up to 10 satellites being assembled in a day (Iridium website). The production time of each satellite was around 35 days. Motorola engineers took production into consideration from the initial design phase with the view that 80–90% of cost, quality and production time is influenced by satellite design. The production was viewed in its entirety from component construction through to spacecraft delivery to the launch site. Components were produced to allow just the quantities needed for the current assembly – this 'lean production' is used by Japanese car manufacturers. The technique allows continuous refinement of the assembly process. Other techniques involved use of off-the-shelf components, use of easily pluggable and interchangeable circuit boards, and rigorous low-level tests to minimize system-level problems. Before the start of production, extensive electrical, software, acoustic and vibration tests were performed on two full-scale satellite models to minimize the tests on production runs. Experience gained through assembly of the first batch of satellites was used for refining further production runs. The production line operated with about 50 engineers working in three shifts continuously at 15 different stations. Each satellite required about 160 printed boards which were assembled in one factory, sent to the next one where they were assembled and tested as subsystems, and then sent to the final assembly line where they were integrated as satellite units.

The ORBCOM satellite manufactured by Orbital Sciences Corporation and Teleglobe follows a different strategy. ORBCOM satellites are small 100 lb satellites, 34 in all, launched 8 at a time. The process begins by using a rigorous programme for accepting components so as to minimize failures. All satellites are identical, even though it was possible to make some satellites less sturdy than others due to lower mechanical stress on satellites kept on top during the stacked launch – this approach is believed to have saved considerable manufacturing effort and time. Like the Iridium approach, only the first two satellites of the series were tested rigorously. Some of the tests were eliminated from later models, as the spacecraft were already characterized. Removing such tests reduced the manufacturing time by an estimated 50%. The goal was to produce a satellite every 10 days.

TRW had geared their manufacturing techniques in support of their 12-satellite MEO system called Odyssey, which later merged with ICO (now New ICO). Their plans included mass production of MMIC (monolithic microwave integrated circuits) using automation, with the aim of speeding up manufacturing and improving the quality. The satellites were planned to be manufactured in TRW's standardized facility, which, in its expanded state, would handle up to 16 satellites at a time. Satellites would flow through an assembly process starting from the structure, propulsion system, while solar array and payload were assembled separately. All subsystems would then be integrated and tested. Innovative features were added to speed up testing – these included a thermal vacuum chamber with two doors, one for incoming units and the other for the tested satellite, thus speeding up the testing.

Similarly, new features were introduced for manufacturing the 48 Globalstar satellites manufactured by Loral Space and Communication. The company uses design engineers who remain assigned to a product through its full production run. Mechanical and electrical interfaces are standardized to allow quick changes of failed subsystems. The design is modular; each module is tested thoroughly before being assembled as a spacecraft. In the production line, satellites are moved from one assembly point to another and tested at each 'functional island'. The satellites are always moved forward. In the case of a defective satellite, it is moved out of the production line and therefore does not affect the production flow.

6.3 Intersatellite links

Many recent proposals have planned to incorporate intersatellite links (ISL). The Iridium system is a prime example in the use of intersatellite links. The topic of ISL has been revisited a number of times since the 1980s but it was only in the late 1990s that the technology was really deployed in commercial satellite systems. Several future K_a and V-band geostationary satellite systems also intend to use such links.

The idea of intersatellite links has been around for a considerable time, having been studied for the Intelsat-IVA application. At that time, the interest was in links between geostationary satellites or between LEO and geostationary satellites, such as in the TDRS (tracking and data relay satellite) programme. There were a number of disadvantages which precluded the use of ISL for commercial systems at that time. The ISL package would cause an unacceptable increase in the weight of the satellite and require onboard processing, which was not developed sufficiently in the 1980s. Moreover, ISL would add additional noise to the severely power-limited downlinks, and additional propagation delay introduced in the ISL was not desirable for many applications. Nevertheless, where necessary, ISLs have been used since 1976. Table 6.6 gives a summary of the systems using ISL (Morgan, 1998) (see also Table 6.5).

System	ISL frequency (GHz)	Year of launch	Application
LES-8 LES-9	36–38	1976	10-100 kbps military communication
TDRSS	S band, 13.8 and 15	1983–93	Data collection and relay
Iridium	23 GHz	1998–99	Mobile communication ISLs at 7 Mbps
COMETS	2, 23 and 26	1998	Limited due to launch failure
ADEOS	23 and 25 GHz	1998	Limited due to launch failure

Table 6.6

Examples of usage of intersatellite links

Recent interest in the use of ISL stems from the fact that the technology has advanced to a sufficient extent, coupled with the awareness that ISL can provide distinct advantages in terms of network connectivity. Some of the advantages of deploying ISL are as follows:

- Intersatellite links can provide the shortest route between source and destination.
- The cost of terrestrial routing can be minimized.
- It provides a means of bypassing other carriers.
- Intersatellite links allow re-routing of packets if a node is congested; this may be either gateways or satellites.
- Spare capacity can be redistributed between satellites which may be co-located or distanced. This feature also improves redundancy.
- The use of ISL on geostationary satellites can eliminate the need for multiple-hop satellite transmissions, or alternatively terrestrial connection, when connecting widely separated sites. This approach also reduces propagation delay, and saves valuable spectrum by avoiding satellite–satellite hops.
- ISL allows an increase in the elevation angle of the Earth station by allowing it to use a higher elevation satellite and completing the link to the low elevation satellite. Such an arrangement could, for example, connect London and Tokyo with 15° elevation visibility using an ISL between two satellites separated by 30°. Alternatively, such an arrangement can provide high elevation coverage for MSS within a region by creating a single virtual

satellite by connecting two well-positioned satellites. Such a scheme could also allow two widely separated MSS regions to be connected as a single network and thus open interesting opportunities for new MSS architecture.

- ISL can be used to create a satellite cluster where, for example, an MSS satellite could connect directly to an FSS satellite, thereby bypassing the normal terrestrial tail.
- Intersatellite links can increase the instantaneous coverage of low orbit satellite systems by connecting the LEO satellite to a satellite at a higher orbital altitude. An example of such a scheme is NASA's TDRS system. In the system, LEO satellites gather weather and Earth resources data and transmit them to the ground via geostationary satellites for instant analysis. The ISL are also used for LEO satellite command and control.
- ISL can also be used for interconnecting different satellite systems for lease or for mergers; if an ISL facility is built in, it reduces the risk to investors if the full capacity of a satellite is not utilized, as it is easier for the satellite to become a part of another consortium. One could envisage the use of a low-cost buffer satellite to interconnect different operators' satellites, if the ISLs are not compatible, as there is no agreed standard on ISL architecture (Morgan, 1998).

Frequency bands

Most of the frequency bands used for ISL have been allocated around the oxygen and water absorption bands, which are not useful for Earth–satellite paths because of the high path attenuation. Figure 6.14 shows the attenuation of signals as a function of frequency. The first absorption band is due to water vapours centred around 22.2 GHz, where theoretical vertical one-way attenuation is of the order of 0.5 dB for moderate humidity and the second absorption peak of ~ 150 dB at moderate humidity at around 60 GHz is caused by oxygen. The attenuation is proportional to the cosec of the elevation angle. Even if this band is used in terrestrial systems, the signals suffer considerable attenuation, thus minimizing the risk of receiving or causing interference to intersatellite links. Table 6.7 gives a summary of the main ISL allocation. The reader should refer to the latest issue of Article 8 of the Radio Regulations for the most recent allocation.

Optical frequencies are also used for intersatellite links. Optical links provide a much greater bandwidth but require a more precise tracking between satellites. At present, the ITU does not regulate optical frequency for intersatellite links. Optical wavelengths in use depend on the laser source. Typical examples of lasers are Diode pumped Nd:YAG (wavelength 1.064–0.503μm), Indium Gallium Arsenide Phosphide laser (wavelength 1.3–1.5μm) and Aluminium Gallium Arsenide laser (wavelength 0.8 μm).

Implementation issues

ISLs may use tracking antennas/telescopes or fixed antennas/telescopes, and as mentioned above, microwaves or lasers. The TDRS system uses moving antennas with tracking and so do the Lincoln satellites, even though they operate in an

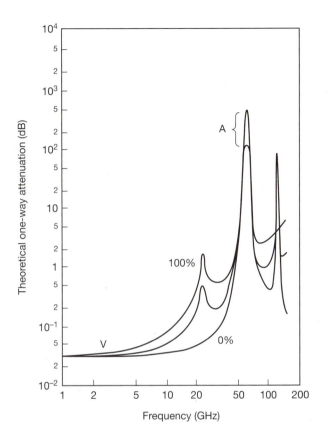

Figure 6.14
Theoretical one-way attenuation for vertical paths through the atmosphere (assumptions: US standard atmosphere, for July, 45° latitude). Solid curves represent moderate atmosphere, dashed curves the limits for 0% and 100% relative humidity. V – vertical polarization; A – limit of uncertainty (CCIR Rep 390-4) © ITU

Frequency (GHz)	Bandwidth (MHz)	Allocation type	Note
22.55–23.55	1,000	Co-primary	Radio astronomy use
24.45–24.75	300	Co-primary	
25.25–27.5	2,250	Co-primary	Geostationary uses limited
32–33	1,000	Co-primary	Used for radio navigation and deep space communication
54.25–58.2	3,950	Co-primary	
59–64	5,000	Co-primary	
65–71	6,000	Co-primary	
116–134	18,000	Co-primary	
170–182	12,000	Co-primary	
185–190	5,000	Co-primary	

Table 6.7
Frequencies used for ISL (Note 1: All bands are shared; Note 2: Consult latest RR tables, as allocations are revised regularly)

inclined orbit at about 14° separation. Fixed antennas simplify the ISL system. Such links are in use in the US government's geostationary Milstar system and Iridium LEO system. The tracking system reduces the station-keeping requirement of satellites. In contrast, fixed antennas require the satellites to remain within the beamwidth of the antenna if continuous communication is needed, imposing tight orbital maintenance; however, it may be that station-keeping requirements for other system needs are even more stringent.

A large number of the forthcoming K_a band systems have opted for intersatellite links. Optical lasers are planned for the Silex experiment (semiconductor laser ISL experiment) which will use AlGaAs lasers for connecting SPOT-4 remote sensing satellites with the Artemis satellite. The Japanese Optical Inter-orbit Communications Engineering Satellite may also link with Artemis in a cooperative arrangement (planned for 2000). ETS-6 had planned S- and K-band intersatellite experiments but the satellite failed to reach a geostationary orbit for which the mission was intended. Experiments include acquisition, tracking and control of laser links. Links will be established with ESA's Artemis satellite. The proposed Wideband European Satellite (WEST) system was intended to use optical ISLs (planned for 2001). Laser links can provide high EIRP due to their cohesive nature but the receiver must track the source. This puts stringent station-keeping and stability requirements on spacecraft. One method proposed for this type of link is to deploy a two-stage tracking system wherein a wider beam is used for initial acquisition, followed by finer tracking. Another consideration in optical links is that of Sun interference, as sun rays could destroy the receivers. Thus the receiver must be isolated when such events are expected – for example, during satellites' sun acquisition. Teledesic intends to use 60 GHz intersatellite links connecting up to eight adjacent satellites.

6.4 Emerging technologies

A number of spacecraft technologies have emerged from laboratories for use in commercial applications in the past few years (Foley, 1995). These technologies offer a number of improvements applicable to all types of communication satellites. However, satellite systems have now been commercialized to an extent that a satellite is seen as an asset, and buyers may be reluctant to accept a new technology to minimize delay and risk a project. Recent declassification of military technology has enabled developments conducted in military programmes such as Milstar to reach commercial markets. These emerging satellite technologies are expected to make satellites smaller for a given performance requirement, smarter, more complex and yet cost effective.

A number of electrical propulsion technologies have been introduced recently, enabling an increase in satellite life by reducing dependence on chemical fuel for orbital maintenance. As an example, the Hughes Galaxy 3R satellite required 800 lbs less station-keeping fuel by using a Xenon engine – an electrical technique which has 10 times the specific impulse of existing chemical fuels. A Xenon ion propulsion engine ionizes xenon gas by passing it through an electrified metal grid resulting in an efficient generation of thrust. Other recent techniques are the arc-jet system and stationary plasma thrusters. A plasma thruster allows a reduction of in-orbit mass by 15–20%.

Weight reduction is being introduced by deploying lightweight antennas and lighter material on the spacecraft body and payload. For example, aluminium honeycomb and graphite for the spacecraft body have been introduced by the Hughes aircraft company. In the electronics section, alloys that combine aluminium with lighter metals such as Lithium and Beryllium are under consideration. Use of newly developed stronger graphite fibre in composites is expected to reduce the overall weight by providing desired structural strength with less material.

Space Systems/Loral have introduced GPS receivers on LEO satellites such as Globalstar for autonomous attitude control and position determination. Thereby, the need for onboard sensors for attitude determination is eliminated and ground operations cost is reduced as the satellites are able to estimate their ephemeris. Research is under way to extend the concept to geostationary satellites which lie above the GPS constellation. Other innovations introduced by the company include better packaging with improved onboard processing which permits faster processing in a smaller space, use of static plasma thrusters for propulsion and Gallium Arsenide solar cells for higher power.

The time for introducing space-hardened desktop computer technology has been reduced from 15 years to about two years, which gives the satellite industry a considerable advantage in terms of making the most of research efforts in the computing field. This rapid turnaround is expected to be of paramount importance for future projects, such as Teledesic, which envisage the use of powerful onboard computers requiring up to 300 million operations per second for functions such as fast packet-switching for supporting ATM technology.

Considerable progress has been made in monolithic microwave integrated circuits (MMIC) and application specific integrated circuits (ASIC) technologies, enabling a reduction in size, mass and costs of these circuits by a factor of two to five and, additionally, making satellites more amenable to mass production techniques, an essential ingredient for rapid constellation deployment (discussed later). MMIC are integrated circuits on low-loss dielectric and used for functions such as amplification, mixing, etc., and ASIC are integrated circuits developed to perform specific tasks such as computation or electronic functions of attitude control subsystems. Weight reduction of up to 100 kg has been reported. ASIC and MMIC also enable the design of subsystems in modules that may be reused for different missions.

Other areas where significant development has occurred are the Gallium Arsenide solar cell, Travelling Wave Tube Amplifier (TWTA) and battery technologies, which together enable introduction of very high power satellites (~10 kW). Gallium Arsenide cells are more efficient than Silicon cells, enabling generation of higher power in the same surface area. Due to a large dissipation in TWTA, of the order of 50%, the devices become very hot, therefore cooling becomes essential. At higher power the traditional conductive cooling is not effective, therefore radiation cooling is used instead. In this method, heat is dissipated directly to space by using long cylindrical TWT and fins which protrude outside the spacecraft into cold space. Satellites of 15–20 kW, capable of 4.5–6 kW EIRP, are already under development and industrialists are forecasting power levels of 30 kW capable of transmitting EIRP of >9 kW by 2005.

We have already discussed a number of developments progressing in payload technologies. Some developments in antennas and related technologies of interest are large lightweight antennas, phased arrays, multi-port tracking amplifiers, advanced beam-forming techniques, etc. Developments related to transponder technology include onboard processing hardware and software and, as already mentioned, in the areas of MMIC and ASIC, packaging and material technologies.

Spacecraft K_a band technology is expected to be used for next generation fixed and mobile personal communications. Currently, there is a huge academic research and industrial activity in this area. We have already discussed a number of enabling K_a band MSS-related technologies demonstrated in missions such as NASA's ACTS programme, which provided a useful testbed.

6.5 Launching satellite constellations

While the launching techniques for geostationary satellites are well known and documented, techniques used in launching non-geostationary satellite constellations are less so. Here we briefly review the launch sequence of geostationary satellites for completeness, followed by a more detailed deployment description of two specific LEO satellite constellations – Iridium and Globalstar.

A satellite is launched into an orbit by injecting it into space at the required altitude in a direction and velocity appropriate for the altitude. Figure 2.5 (Chapter 2) shows the theoretical relationship between altitude and velocity for circular orbits, for altitudes up to the geostationary orbit.

As mentioned before, orbits used for mobile satellite services lie within specific altitude ranges above ~700. Thus a launch vehicle must impart a velocity of 7.4 km/s for an altitude of ~900 km (LEO), 4.6 km/s at an altitude of 12,460 km (MEO) and 3.0747 km/s at a geostationary altitude. The velocity increment of a launch vehicle can be optimized by maximizing the ratio m_f/m_0 where m_f is the mass of expanded rocket fuel and m_0 is the total mass of the launch vehicle. Therefore, most conventional launchers use multiple stages; each stage is jettisoned after spending all its fuel reducing m_0 to maximize thrust from the subsequent stage. This type of technique is most common for geostationary launches due to the altitude to which satellites have to be lifted. Launches to low Earth orbits can additionally be achieved by a number of alternative techniques such as through space shuttles and by launch from aeroplanes. Yet another consideration is the inclination of the orbit. The inclination of an orbit is governed by the latitude of the launch station, and is given as cos (i) = sin (ε) cos (θ)   ... (6.4)

where,
i = inclination, ε = azimuth of launch and θ = latitude of launch site.

The minimum inclination equal to the latitude of the launch site is obtained by launching a satellite in an easterly direction (ε = 90°). Therefore, if the required inclination of the final orbit is less than the latitude of the launch station, orbit corrections have to be applied, expending valuable fuel. The problem is specifically applicable to equatorial orbit, and in particular, to the geostationary orbit. Therefore, launch sites are chosen to lie as close to the equator as possible. A novel solution to this problem is to launch a satellite from a transportable vessel at sea, thereby allowing a launch from any chosen latitude, including a location on the equator. This approach has been adopted by a company named Sea Launch.

Geostationary satellites may be launched directly into geostationary orbit or via a number of distinct phases – the latter being more common. A satellite is initially launched in a parking orbit; transferred into a highly elliptical orbit called the *transfer orbit*; injected into geo-synchronous orbit, at which stage the satellite is still inclined and therefore a readjustment is made so it is close to zero; finally, the satellite is allowed to drift to the desired orbital location. Velocity increment is least when a satellite is launched from a parking orbit to geo-synchronous orbit via an elliptical orbit. This principle is called *Hohmann Transfer*, named after Hohmann who recognized it in 1925. Once in orbit, the solar panels are deployed for 3-axis stabilized satellites, Sun and Earth are acquired, communication antennas are deployed, and the satellite is checked out and finally commissioned for operation provided that it is found to be fully functional.

Figure 6.15 shows the launch sequence of a geostationary satellite and the typical time required to achieve various mission milestones.

A satellite is launched in a vertical direction to minimize air friction; its guidance system tilts the rocket to 90° to achieve an easterly direction for obtaining maximum advantage of the Earth's motion. The rotational velocity of the Earth at the equator is 450 m/s. To launch a satellite in a retrograde orbit would require an additional 600 m/s velocity. The precise launch flight plan depends on the launcher. To ensure that the position of a satellite is favourable in terms of its position relative to the sun and that the satellite is visible from satellite ground stations, satellites must be launched within a specific time window called the *launch window*.

Deployment of non-geostationary constellations quickly and economically is vital for commercial success. Therefore multiple satellite launches are common – current vehicles can launch 3–12 satellites per launch in low Earth orbit. Satellites may be placed in low or medium Earth orbit directly or via a parking orbit. Typically, a launcher deposits clusters of satellites at regular intervals in a parking orbit. When a precise position has been determined, each satellite is commanded

Figure 6.15

Launch sequence of a
geostationary satellite
(Richharia, 1999)

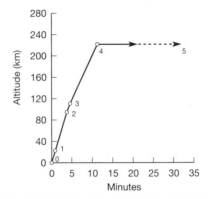

Event number	Event
0	Vertical lift-off
1	Guidance system begins tilting rocket towards east
2	First-stage drop-off
3	Second-stage ignition
4	Horizontal insertion into parking orbit 185 to 250 km
5	Second and third stages fired at equator to acquire transfer orbit

(a)

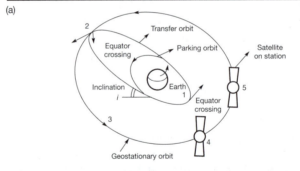

Event number	Event
1	Velocity increment to acquire transfer orbit; satellite spun for stabilization: attitude manoeuvres done before apogee kick-motor firing
2	Apogee kick-motor fired to give necessary velocity increment; orbit circularized and inclination reduced to near zero
3	Satellite despun
4	Three-axis stabilization acquired; antennas deployed/unfurled
5	Minor orbit corrections performed to correct residual orbital errors in orbit tests and position satellite on station

(b)

to move to the desired altitude and inclination, by firing onboard thrusters for a specific duration at the appropriate time, through ground commands.

Support of a network of tracking stations dispersed throughout the world is essential for the purpose.

The Iridium constellation comprises 66 satellites in six 780 km altitude circular polar orbital planes (see section 9.2.2.1) and the Globalstar constellation comprises 56 (includes eight in-orbit spares) satellites in eight 1,414 km altitude inclined orbital planes (see section 9.2.2.2). Both consortia targeted the deployment of the full constellation in about 15–18 months.

Consider the Delta II launch of Iridium satellites from Vandenberg Airforce Base in California (Kiernan 1997–1). Satellites are launched in groups of three. The first satellite was jettisoned at an altitude of 638 km in 3,130 seconds (about 52 minutes) after lift-off, followed by the other two at intervals of 200 seconds. The satellites gradually drifted apart over several days. Around an hour and forty minutes after the launch, the first radio contact with each satellite of the cluster is made by controllers stationed at Motorola's satellite control centre in Chandler, Arizona, when satellites are in view of Iridium's tracking station in Oahu, Hawaii. The contact is meant to check whether satellites are in the designated orbit and functioning normally. There are, in all, four tracking stations, located in Hawaii, Yellowknife and Iqualit in the Northwest territory of Canada, and Snjoholt in Iceland. To avoid the possibility of exhausting onboard batteries, solar arrays are deployed at an early stage, followed by deployment of communication antennas for ground and intersatellite communications. About three hours into the mission, satellites test their secondary antennas used in the initial checkout. The satellites are switched from the secondary antennas to the main antennas in the fifth orbit, giving an increased communications throughput. In the first two days, the satellites' primary antennas and modems are checked out, batteries recharged, software upgraded if required, and feeder link and intersatellite link performance checked out. About 48 hours into the mission, satellites begin ascending to the final location at an altitude of 780 km, firing low-thrust electro-thermal hydrazine thrusters. The final orbit is achieved gradually in about two weeks. Thrusters are fired first over poles to raise the orbit and then over the equator to circularize the orbit. Once on station, L-band communication links are activated and tested. Functions such as subscriber call handling, call handover between beams and satellites are checked out. Other functions such as paging, call forwarding and billing are tested subsequently. This is followed by beta testing, where average users tested the system. Software upgrades were anticipated every six months or so, as practised by terrestrial cellular system operators. A system simulation facility set up at the Chandler plant was expected to be utilized for solving technical problems. The facility uses an Iridium satellite and radio links allowing tests of satellite functionality under different conditions including different commands.

Globalstar planned deployment through a number of different launchers (Kiernan, 1997–2). The first two launches in batches of four, were to be made from Cape Canaveral in Florida, each on a single Delta 2 booster. The next three launches were planned in batches of 12 on Zenit boosters. The final three launches were to be made on Soyuz boosters, supplied by a Russian-French consortium, ferrying four satellites per launch. Some readjustments had to be done due to Globalstar launch failures. As the intention here is to present the launch process rather than the sequence of deployment or the company's strategy in this respect, we will not attempt to follow the chronology. Note that the launch of a large number of satellites at once can have a significant impact on the schedule and cost in the case of a launch failure. One Zenit launch failure resulted in a loss of 25% of the Globalstar constellation, but the company did well to recover from the setback.

For Delta 2 launch, four satellites held in a canister in the launch vehicle are jettisoned almost simultaneously at an altitude of 1,250 km. The first four Globalstar satellites were placed in the same orbital plane to simplify the launch. The initial few manoeuvres are initiated by each satellite autonomously using onboard computers. The manoeuvres include extension of the magnetometer boom, acquisition of Sun and the Earth, stabilization to avoid tumbling, and employment of solar arrays to avoid battery depletion. When in orbit, the satellites are controlled by Globalstar's Operations Control Centre in San Jose, California, using tracking stations in Texas, France, South Korea and Australia. A preliminary health checkout of each spacecraft is made to ensure that the vital satellite functions, such as attitude control and propulsion, behave normally. Within a few hours, each satellite is commanded to fire thrusters to jettison itself to its final altitude of 1,400 km. The satellite injected last is boosted first to minimize the risk of collision.

Zenit's vehicle launches 12 satellites simultaneously, delivering them to an altitude of 920 km. The satellites are held in canisters as with the Delta launch, and ejected within 4 s in rapid succession. Speed is essential to minimize the risk of placing satellites in the incorrect orbit due to movement of the canister. The initial manoeuvres are identical to the Delta launch. But as satellites are at a lower orbit, they travel faster, giving only 10–12 minutes of visibility from ground stations. Moreover, as only six satellites can be placed in each plane, the satellites are grouped and injected in three separate planes. A further consideration is that the satellites experience different radiation and thermal conditions at 920 km to those experienced at 1,400 km, for which they are optimized. Hence the satellites must be moved to the higher altitude as soon as possible. Satellite altitude is altered in groups of two or three to minimize the workload on ground controllers. Satellites are allowed to orbit until they reach their respective orbital plane, when they are jettisoned to their final altitude of 1,400 km.

The Soyuz launch also sends satellites to the same altitude as Zenit but only four satellites are launched at a time. The eight satellites launched from Soyuz were intended to remain partially activated as spares in a 920 km orbit.

After a satellite has acquired its final orbit, the communications payload is checked out, which may last several weeks. As there are no intersatellite links, satellites can be checked out independently one at a time.

Further reading

Bellaccini, S.; Rozera, G. (1991) 'The ITALSAT system', *Journal of the British Interplanetary Society*, **44 (9)**, 447–52.

Benedicto, J.; Rinous, P.; Roberts, I.; Roederer; Stojkovic, (1993) 'Geostationary payload concepts for personal satellite communications', *IMSC*, 16–18 June, Pasadena, California, JPL Publication 93-009, 169–74.

Brugel, E.W. (1994) 'On-board processing and future satellite communications services', *Space Communication*, **12**, 121–74.

Campanella, J.S.; Pontano, B.A.; Chitre, D.M. (1990) 'A user's perspective of the ACTS hopping beam TDMA system', *AIAA Conference Record Paper*, AIAA-90-0833-CP, 484–9.

Chmielewski, A.B.; Freeland, R. (1997) 'Low cost large space antennas', *International Mobile Satellite Conference*, 375–80.

Craig, A. *et al.*, (1993) 'Applicability of different on-board routing and processing techniques applied to a mobile satellite system', *Third International Mobile Satellite Conference*, June.

Egami, S. (1995) 'Applications of multi port amplifier to personal satellite communications', *IMSC '95, Fourth International Mobile Satellite Conference*, Ottawa, co-sponsored by Communications Research Centre/Industry Canada and Jet Propulsion Laboratory/NASA, 67–72.

Foley, T. (1995) 'Satellite technology – decades of development pay off', *Via Satellite*, (7), July, 32–9.

Gargione, F. (1990) 'ACTS hardware', *AIAA Conference Record Paper*, AIAA-90-0834-CP, 490–6.

Gupta, R.K.; Narayanan, J.N.; Nakamura, A.M.; Assai, F.T.; Gibson, B. (1991) 'INTELSAT VI on-board SS/TDMA subsystem design and performance', *COMSAT Technical Review*, **21 (1)**, 149–89.

International Journal of Satellite Communication (IJSC) Special issue on intersatellite links 1988.

Iridium website ***http://www.Iridium.com***.

Kato, S.; Arita, T.; Morita, K. (1987) 'Onboard digital signal processing technologies for present and future TDMA and SCPC systems', *IEEE Journal on selected areas in communication* (SAC), 5 (4), May, 685–700.

Kawai, M.; Tanaka, M.; Ohtomo, I. (1991) 'ETS-VI multi-beam satellite communication systems', *Acta Astronautica*, **25 (7)**, 407–14.

Kiernan, V. (1997–1) 'Planting telephone poles in the sky', *Satellite Communications*, April, 38–42.

Kiernan, V. (1997–2) 'Globalstar at the starting gate', *Satellite Communications*, July, 36–39.

Kiernan, V. (1996) 'Cheaper by the dozen', *Satellite Communications*, November, 29–32.

Loo, C.; Hayes, E.J. (1993) 'Double hop on-board signal processing using Olympus', *IEEE Globecom Telecommunications Conference*, 599–603.

Maral, G.; Bousquet, M. (1984) 'Performance of regenerative/conventional satellite systems', *International Journal of Satellite Communications*, **2**, 199–207.

Melnick, M.; Hadinger, P. (2000) 'On-board processing: enabling broadband satellites', *Satellite Communications*, 28–34.

Morgan, W.L. (1998) 'Pass it along', *Satellite Communications*, May, 50–4.

Muratani, T. (1981) 'Satellite switched time domain multiple access', *Fifth International Conference on Digital Satellite Communications*.

Naderi, M.; Campanella, J. (1988) 'NASA's Advanced Communications Technology Satellite (ACTS): An overview of the satellite, the network and the underlying technologies,' *12th International Communications Satellite Systems Conference*, AIAA, March, Paper no. 88–0797, 204–23.

Nelson, R.A. (1999) 'Spacecraft battery technology', *Via Satellite*, February, 104–18.

Normand, Le *et al.* (1988) 'A versatile array fed reflector antenna', *IEEE AP-S International Symposium Digest*.

Oshima, T.; Morikawa, H.; Oshida, K.; Ohuchi, C.; Shimada, M.; Kameda, K. (1993) 'Planned Experiments using the Communications and Broadcasting Satellites', *3rd European Conference on Satellite Communications*, Manchester, UK, 236–40.

Peach, R.C.; Miller, N.; Lee, M. (1993) 'SAW-based system for mobile communications satellites', *Proc Third International Mobile Satellite Conference*, Pasadena, California, 53–8.

Peter, R.A. (1988) 'Intersatellite links', *International Journal of Satellite Communications*, **6**, 79–80.

Riccharia, M. (1999) 'Satellite communication systems: design principles', Macmillan Press, Basingstoke and London.

Samejima, S.; Tanaka, M.; Ohtomo, I. (1992) 'Overview of satellite on-board multi-beam communication systems for ETS-VI', *Acta Astronautica*, **28**, 293–300.

Taira, S.; Takeda, O.; Otsu, Y.; Katagiri, H. (1998) 'Comparative study of switching methods for on-board processor in a mobile satellite communications system', *International Journal of Satellite Communications*, 16, 131–6.

Takeda, O.; Taira, S.; Kawakami, Y.; Yamasaki, N.; Otsu, Y.; Katagiri, H.; Senba, S. (1997) 'A study on on-board processor for advanced mobile satellite communications', *Proceedings of the Fifth International Mobile Satellite Conference*, Pasadena, California, June 16–18, co-sponsored by NASA/JPL and DOC/CRC; JPL Publication 97–11, Jet Propulsion Laboratory, Pasadena, June 16, 1997, 399–404.

Thompson, M.W. (1997) 'The AstroMesh deployable reflector', *IMSC 97*, *Proceedings of the Fifth International Mobile Satellite Conference*, Pasadena, California, 16–18 June, co-sponsored by NASA/JPL and DOC/CRC, JPL Publication 97-11, Jet Propulsion Laboratory, Pasadena, California, 16 June, 393–8.

Commercial issues

7

7.1 Introduction

A commercial satellite communications venture is no different from any other high-technology business and therefore its business model is also based broadly on the same principles. Figure 7.1 attempts to capture various network and business entities of a commercial MSS venture, along with their interaction.

The network entities of an MSS comprise a space segment and a ground segment; the latter can be further partitioned into fixed and mobile components. The business components comprise a number of specialist vendors for provision of space and ground hardware interacting with the system operator. The gateways may be owned by the operator or, depending on the business strategy, the service provider. The service provider offers service to the end user directly or through local retailers. Mobile terminal manufacturers, after obtaining approval from the operator, provide terminals to the user through their distribution chain. Embedded within such a framework are several ancillary businesses which may be hired by vendors and system operators for specialist services such as application development, hardware and software maintenance, etc.

The space segment and supporting ground infrastructure, usually owned by a single entity, constitutes the most significant up-front cost, as a glance at Table 7.1 should testify – investments range from $0.4bn–$9bn. Therefore the significance of a realistic business plan cannot be understated.

Estimates in Table 7.1 were revised as the project developed, but the table is indicative of the order of magnitude of the initial investment. When an MSS

Figure 7.1
Entities of a commercial MSS venture

Main assets	*Business entities*	*Product*	*Market (example)*
• Satellites	• Space segment provider	a) End user	• Individuals
• System infrastructure	• Gateway operator	• Telecommunication	• Fleet owners
• Gateways	• Manufacturers	services	• Ship owners
	• Vendors	b) Business to business	• Airlines
	• Service provider	• Satellites	• Government
		• Earth stations	• Journalists
		• Software	• Broadcasters
		• Hardware	
		• Operation support	
		• Satellite time	

System	Orbit/coverage	Initial estimate of cost by operator ($bn)
Aries	LEO/World	0.4–0.5
MSAT	GEO/Regional	0.55
Agrani	GEO/Regional	0.7
Thuraya	GEO/Regional	1.1
Globalstar	LEO/Global	1.3
Iridium	LEO/Global	3.7
ICO Global Communications Ltd	MEO/Global	4.5
Teledesic	LEO/Global	9

has achieved a successful operating service, introduction of new products requires much lower investment, unless, of course, the product can only be supported by a new space/ground segment architecture.

Suffice to state that market research conducted by recent operators at the outset concluded that the potential personal mobile satellite service market would support and sustain their individual system profitably. Figure 7.2 shows the anticipated number of big-LEO subscribers perceived by market surveys of the late 1990s (Corbley, 1996). Market analysts' views varied regarding the number of systems which could be sustained, ranging between 2 and 4. Iridium, ICO and Globalstar generally figured in the final reckoning.

Similarly, another study estimates that the subscriber base for little-LEO satellite systems, which provide messaging services and other services such as asset tracking, meter reading, weather monitoring and SCADA through inexpensive pager-like receivers, are of the order of 168 million units, comprising 42 million tracking units, 36 million monitoring units and 90 million messaging units (Crossman, 1999).

The current state of affairs in the big-LEO and little-LEO segments of MSS illustrates the significance and uncertainties associated with a business model. As is now well known, both Iridium and ICO have faltered in their stride, filing for bankruptcy. While ICO has managed to retrieve its finance and is now called New ICO, Iridium's system in its original form is deemed a spectacular failure. Further, New ICO has revised its business plans.

The intention here is not to analyse any specific system but to extract factors influencing recent developments in this particular MSS segment. Predominant influences appear to be changes in market conditions and rapid evolution of competing technology during the long gestation period of projects, ranging from 8 to 12 years. Ten years ago, when many of the big-LEO systems were conceived, cellular telephone sizes were much larger, large areas of the world were not covered by cellular systems, and international roaming arrange-

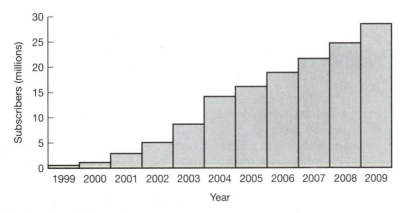

Figure 7.2
Anticipated number of big-LEO subscribers perceived by market surveys of the late 1990s

ments were in their infancy. Today, cellular phones are smaller (wristwatch telephones!), smarter, lighter and fashionable; the areas not served by cellular systems have shrunk and continue to shrink rapidly while data rates continue to increase, as third generation systems roll out. A user can travel to several parts of the world using the same phone, thanks to interoperator global roaming arrangements and increasing standardization in the cellular industry. Why would a business user not expect from satellite systems a similar quality and type of services as from a cellular system? This brings in the need to address the correct market segment. Someone without any means of communication is far more receptive to the idea of a satellite service – does the person care about cellular system offerings? Such a user is likely to be more cooperative while using the phone, e.g. ensuring that the satellite view is unobstructed. The product must be available to such users, which brings in the need for a realistic marketing strategy. Lastly, investors are anxious to receive returns within a stipulated time, beyond which funds begin to dry up. So how can operations be sustained and ageing satellites replaced? This stresses the importance of marketing research and financial analysis.

A business model must consider:

- revenue and returns;
- funding (capital, operating costs, source: own, stock market, bank loans);
- risk (delays, increased cost, regulatory hurdles, political difficulties, funding difficulties, launch failures, slow market penetration);
- operating cost;
- maintenance cost.

Consider some basic issues applicable to a mobile satellite personal communications service. These world-wide or regional services are likely to offer voice, data, facsimile, e-mail, messaging etc. on a personal telephone or a pluggable SIM card.

Satellite personal mobile systems are generally best targeted for regions where existing services are unavailable or unreliable. Examples are large areas of under-developed parts of Africa or Asia, expanses of developing countries such as India or China, and remote areas of developed countries such as Canada or the USA.

Consider the growth model of cellular systems. The world-wide cellular system penetration pattern clearly illustrates the growth spearheaded by developed countries, followed by uptake of the service by the developing countries when costs and technology mature and become affordable. Can this model be applied to mobile satellite services? At the time of introduction, terminal costs and call charges (plus subscription charges, if applicable) are relatively high, which may put the service beyond the reach of individuals who most need it. Individuals from richer countries and specialists such as journalists, explorers, etc. are likely to be the key consumers – but this begs the question of whether there are enough of them. While at the outset cellular systems address densely populated regions with much lower investment and a rich client base, the satellite market addresses the thin-route regions where the number of users is difficult to quantify and the population may not be as affluent and sophisticated. This is one reason why several new operators offer interoperability with terrestrial systems, with dual-mode handsets and commercial agreements with terrestrial operators. In the next generation, integration is expected to extend to the network level, which would provide economies of scale and open a wider user base.

7.2 System planning

To be viable, satellite system technology targeting the mass market must be market led rather than technology driven. In such an environment, commercial satellite systems can benefit greatly in terms of schedule and cost, leading to better returns, by using a systematic planning approach. Figure 7.3 shows the main entities and their interaction in an exercise of this nature (adapted from Lazear *et al.*, 1997).

A new satellite system or a product within an existing venture is initiated through marketing studies and based on telecommunication needs, preferred size and type of hardware, quality of service expectations, geographical areas, consumers' purchasing power, preferred cost of service, market size and anticipated growth, etc. User requirements have to be converted to revenue and returns through a business model which benefits by incorporating the entire life cycle of the product (more later). A number of iterations may be necessary between system designers, marketing and business strategists before a realistic set of requirements, technology and schedule materializes. During the design process, there is also a need to incorporate regulatory issues, such as spectrum availability, licensing policy in the target, time lag, possibility of obtaining the preferred satellite location and frequency bands, etc. The initial cost model –

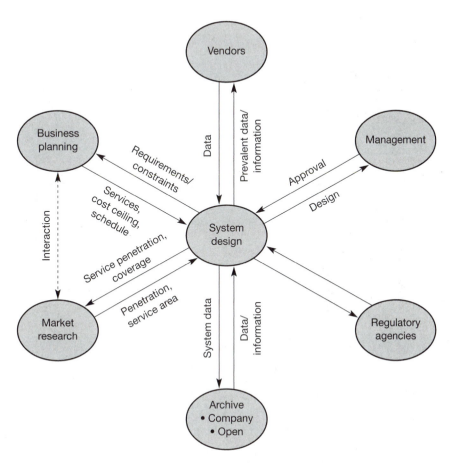

Figure 7.3
Main entities and their
interaction in system
planning

which may be based on the operator's own historic data, published reports, government business initiatives, documents in the public domain, empirical models extracted from the literature, etc. – has to be refined by interaction with manufacturers and vendors. Furthermore, system design concepts require industrial validation by way of technology, risk and schedule. At the end of such an interactive design process, a business plan, encapsulating system specifications, operational philosophy, a detailed programme schedule, capital/revenue etc., is ready.

Commercial satellite systems are planned with the goal of achieving the maximum return. Using a system approach, which includes the whole life of the venture, can identify areas of potential savings. This type of analysis also provides a clearer view of the commercial venture for possible investors and shareholders (especially when billions of dollars are at stake!). Such an approach has found favour with organizations such as ESA and NASA, as is evident from their studies and initiatives in this area (Sultan and Groepper, 1999). There are several generic cost estimations for space segment cost estimation, which divide the cost into recurring and non-recurring and apply regression fit to the existing

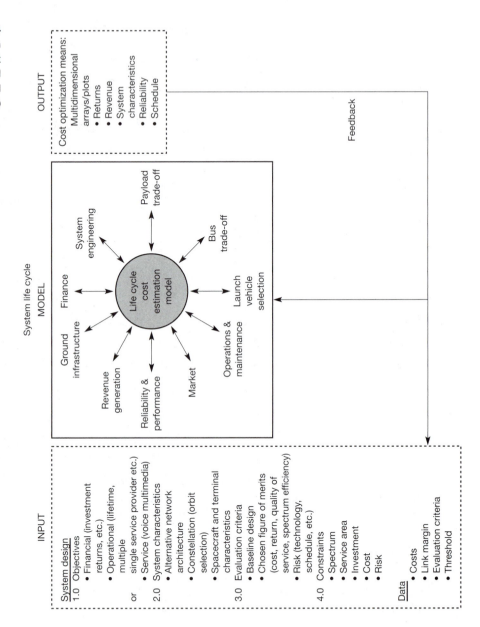

Figure 7.4

A top-level model of such a lifecycle cost–benefit analysis model (adapted from Sultan and Groepper, 1999)

database for estimating the cost of a new mission. A lifecycle programme cost estimation, however, requires, in addition, ground segment costs, operation and
maintenance cost for replenishing the space/ground segment, etc. When such a
model is developed, returns can be estimated in terms of profitability. Sensitivity
analysis for various parameters may be done in terms of profit returned. An example would be the profit/unit spacecraft EIRP. ($/W) or profit/unit bandwidth
($/kHz) or profit in terms of weighted combination of spectrum and satellite EIRP.
Figure 7.4 shows a top-level model of such a lifecycle cost–benefit analysis model.

The input to the model may comprise:

1   mission objectives: main financial objective, lifetime of the mission based
    on operational considerations, product lifetime, etc., together with system
    constraints;
2   system characteristics: alternative concepts, architecture, orbital characteris-
    tics, system drivers (e.g. user expectations in terms of terminal size,
    throughput);
3   evaluation criteria;
4   system requirements;
5   knowledge: cost models, spacecraft power/mass estimation model, risk strat-
    egy, growth trend, inflation trend, traffic distribution;
6   data: link margin, frequency of operation, available finance, expected
    return, historic data such as unit cost, etc.

The output expected from the model should be: performance evaluation in the
chosen figure of merit; profit; revenue as a function of time or bandwidth or
EIRP or schedule reduction; reliability; quality of service; etc.

The feedback loops offer the planner the opportunity to vary inputs and
evaluate the performance.

The ESA study identified and grouped the main cost drivers of a space mission as illustrated in Figure 7.5 (Sultan and Groepper, 1999).

Engineering trade-off analysis is used by planners at the outset to cut costs
with minimal risk and expenditure. The ESA study estimates a cost reduction by
a factor of two for 15 spacecraft when using economies of scale. A cost reduction of up to 35% is possible in medium and large space missions with good
management and implementation methods, although quantifying measures for
efficiency, motivation, management, etc. is difficult. A higher cost reduction can
be achieved where better control over teams, resources and facilities is possible.
Research and development requires initial investment but can offer up to 35%
cost reduction over periods of about 10 years when research costs are excluded.
Obviously, the magnitudes of savings mentioned here would depend on the various assumptions used in the study, but should be indicative of the sensitivity of
costs to these cost-saving measures.

Initial engineering studies benefit greatly through iterations as a better
appreciation of requirements, conceptual designs and cost sensitivities is gained.

Figure 7.5

Main cost drivers of a
space mission (adapted
from Sultan and
Groepper, 1999)

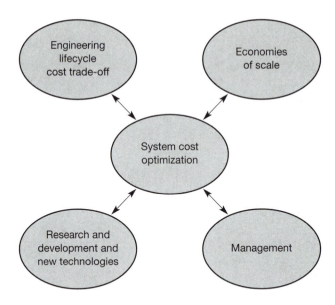

For example, the perception that the lowest spacecraft cost is the best option may not necessarily hold when the venture is viewed in its entirety. This phase involves close interaction between the business and engineering teams and is key to the eventual success of a venture.

The study applied the methodology for Inmarsat-3 lifetime optimization using data available in the literature and from previous studies. The results are presented here as illustrative of the approach, without claiming any resemblance to implementation. Geostationary systems using 6, 19, 18 and 52 spot beam systems were investigated to serve a range of services to various types of terminals for maritime, land and aeronautical services. A wide range of parameters were used in the trade-off analysis. Figure 7.6(a) compares four spacecraft designs for global coverage by three or four identical geostationary satellites, Figure 7.6 (b) shows spacecraft designs versus payload and mass for these designs and Figure 7.6 (c) demonstrates the cost and revenue of the system over a 10-year period. Design three shows the best configuration from a total revenue point of view and also offers the lowest mass, even though it is not the simplest or lowest cost design. This result reinforces the need for a parametric system modelling trade-off analysis.

Table 7.2 shows an estimate of the system breakdown cost of an FSS service as an illustration (Lazear *et al.*, 1997).

Satellite communication systems tend to be risky in several ways, and in particular, risks increase when introducing new technology or addressing a new market. Technical risks include: systems engineering risk in terms of architecture, complexity, etc.; space segment risks in terms of software and hardware development, speed of production, launch, operations, etc; risks in the ground segment in terms of hardware and software development, external interfaces with PSTN, etc. and operations; regulatory risks in terms of obtaining spectrum,

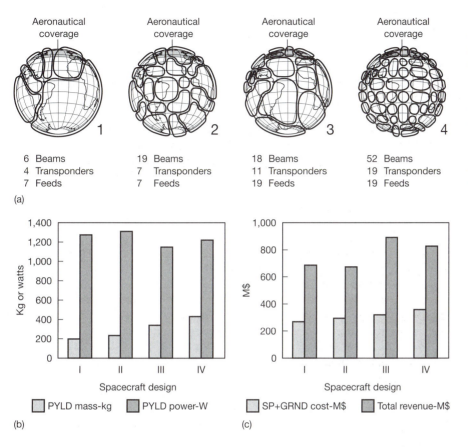

(a)

(b)

(c)

Figure 7.6
(a) Four configurations for global coverage by three or four identical geostationary satellites (b) spacecraft design versus payload power (watts) and mass (kg) (c) cost and revenue over a 10-year period for each design (Sultan and Groepper, 1999)

Segment	Percent of total cost
Satellite	18
Launch	13
Spacecraft control	5
Network control	5
Ground terminals	35
Operation and maintenance	24

Table 7.2
System breakdown cost of an FSS service (Lazear *et al.*, 1997)

operating licences in service regions, etc. In terms of programme, the risks involve cost, schedule, financing, programme management, risk management, marketing in terms of terminal costs, competition, call cost, etc. Gaffney *et al.*, (1995) performed an independent risk taxonomy of a number of non-geostationary MSS satellite systems. Risk parameters in the analysis included:

1 **system engineering:** interference control, forward error correction, and speech coding technology;
2 **space segment:** satellite manufacturing, spacecraft antennas, onboard processing, intersatellite links, effects of radiation belts;
3 **ground segment:** Earth station technology and hand-held terminal antenna;
4 **non-technical issues:** cost, schedule, regulatory, financing and market.

The study included the evolution of these systems – Odyssey, Iridium, Globalstar, ELLIPSO and Constellation – with regard to satellite launch, system cost and schedule. It was noted that each system had undergone significant revisions since its initial filing. Figures 7.7(a) and (b) illustrate as examples respectively the escalation in satellite launch mass and system cost using data in open literature. Note the escalation in launch mass in all systems, which was due to revision in system performance in terms of availability, which required an increase in spot beam gains. This caused a corresponding increase in system cost and a knock-on effect on the schedule as well. System costs cover all major cost drivers, i.e. design, development, production, launch, some ground segment costs and operations costs, but a comparison across systems is not possible with any certainty because of differences in the way each proposer estimated costs. The study concludes that the highest risks were in the areas of cost, schedule, financing, interference control, complexity escalation, and cost. Quite evidently, risks were well founded, as we know that the Odyssey system and Iridium system, in their original form, are now defunct (see above).

A business plan has to be approved by financiers, investors and creditors to ensure the viability of the project. It also lays out firm guidelines for development and informs the employees of the stated goals and plans. The next phase of the project involves design, development, launch and deployment of satellites, product development, introduction of product advertisement and operations. Careful monitoring of product growth, pricing, etc. is essential to keep the product competitive and viable. The entire programme requires careful management at all levels. The design flow of an MSS product using such a methodology is illustrated in Figure 7.8 (adapted from Lazear *et al.*, 1997).

Figure 7.9 portrays some of the main elements and interaction of the product marketing plan (Lazear *et al.*, 1997).

The success of an MSS business is strongly influenced by user equipment cost and service cost. Equipment cost and complexity can be minimized by using more powerful satellites; in addition, economies of scale are achievable after the introduction of a service. The service cost usually comprises two components – subscription charge and call charges. The subscription charge is included as a measure for the operator to cover some of its maintenance costs in keeping the network operational for each user. The main element of call cost is satellite resource usage. The trade-off is between terminal size, bandwidth and EIRP. In systems supporting small terminals, such as hand-held units, EIRP requirement is high but low/moderate bandwidth is used as spectrum is very limited. In larger terminals, EIRP requirement is low for an identical service. Initial trade-off analy-

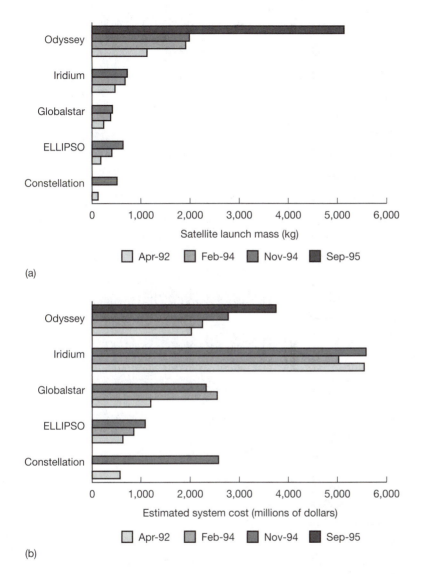

Figure 7.7
(a) Launch mass
evolution of non-
geostationary MSS
satellite systems
(Gaffney *et al.*, 1995).
(b) System cost
escalation of non-
geostationary MSS
satellite systems
(Gaffney *et al.*, 1995)

sis, discussed above, takes such factors into consideration to arrive at an optimum service cost. Figure 7.10 shows the factors which are likely to affect usage and hence revenue to the operator. MSS service areas are quite large, covering vast geographical expanses: countries, continents and for international service, the whole world. The requirement of the service is dependent on the infrastructure of the region, affordability of the population, government policy/licence, competition, marketing support such as equipment availability and after-sales service, quality of offered service, etc. The revenue is monitored by strategists who take corrective action to increase sales. The service charge is one of the most important factors

Figure 7.8
Design flow of an MSS
product (adapted from
Lazear *et al.*, 1997)

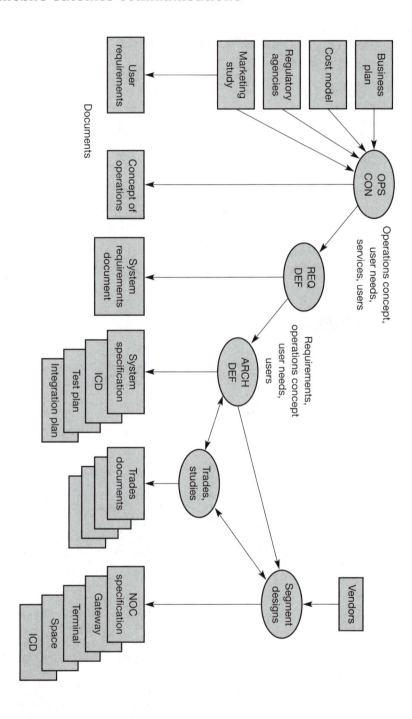

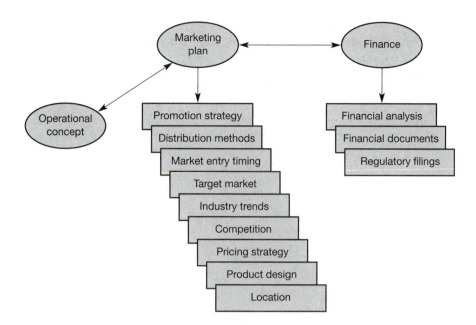

Figure 7.9
Some of the elements
and interactions of the
product marketing plan
(Lazear *et al.*, 1997)

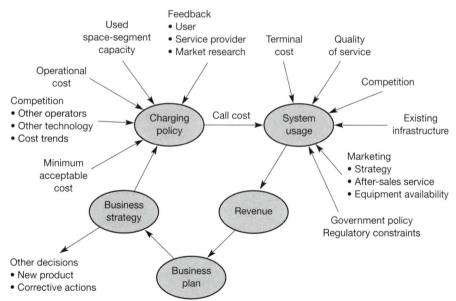

Figure 7.10
Factors which are likely
to affect usage and the
revenue to an operator

governing the amount of usage. If usage is not up to the desired level, strategists take corrective action, such as ensuring availability of user equipment, altering charges to a more acceptable level, etc. Charging policy, however, is governed by a number of independent considerations, which may be a minimum threshold for business viability, geographical dependence (i.e. may be based on user affordability), telecommunication trend (reducing call charges), competition from other operators or services, consumer feedback, investment return, operational and fixed costs, etc.

Figure 7.11 demonstrates the sensitivity of space segment call cost on the revenue of the space segment operator for a regional geostationary satellite system for a spacecraft with an EIRP of 70 dBW, total available bandwidth of 5 MHz and bandwidth per channel of 5 kHz, average EIRP/channel of 25 dBW for various fill factors and frequency reuse.

Figure 7.11
Sensitivity of space segment call cost on revenue of the space segment operator for a regional geostationary satellite system for various fill and reuse factors. Spacecraft EIRP = 70 dBW, total available bandwidth = 5 MHz, bandwidth per channel = 5 kHz, average EIRP/channel = 25 dBW. (a) Theoretical maximum (b) Weekday and weekend use of 50% and 25% of capacity respectively (c) Weekday and weekend use of 30% and 15% of capacity respectively (d) Weekday and weekend use of 15% and 5% of capacity respectively (Graphics: AR)

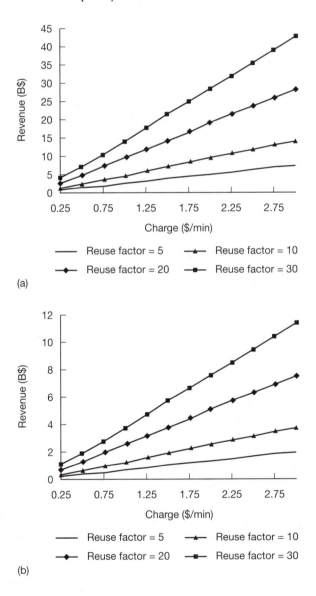

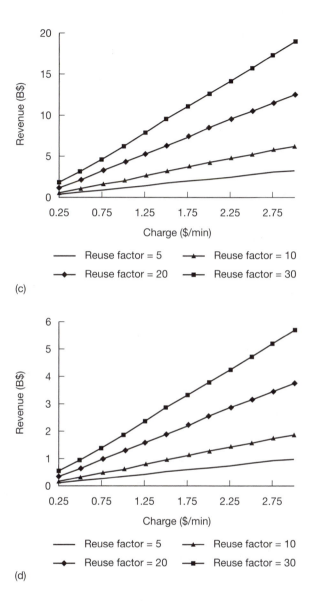

Figure 7.11
Continued

7.3 Service distribution model

Charges paid by a user consist of a number of components derived from the service distribution system which in turn is conditioned by the system architecture and the operational philosophy of the company. The components may consist of:

- space segment charge;
- home gateway charge;

- gateway charge of another service provider, when applicable;
- terrestrial charge (PSTN/Mobile/PDN, etc.);
- service provider's charge;
- distributor's charge if the service provider re-sells.

To illustrate various types of distribution system, let us consider the schemes used in practice. We will consider representative GEO, LEO, MEO systems and the little-LEO system – Inmarsat, Iridium, Globalstar, ICO and ORBCOM (Ingley, 1999).

- **Inmarsat's** service is distributed through gateway operators as shown in Figure 7.12. Inmarsat is the owner of the network, providing the space segment and the infrastructure to run its services. The services are offered by gateway operators whose call charge comprises fees paid to Inmarsat, the terrestrial operator, their own operational cost and a profit. There is a degree of competition between land Earth station operators and hence a variability. Users select a gateway on the basis of cost effectiveness and affiliation. Call charges vary, depending on the service. For some services, Inmarsat sells its satellite time wholesale at a reduced price, which is attractive for operators who are able to sell the purchased satellite time to users at a discounted rate. The distribution method of the next generation system (Inmarsat-4) may be revised following the privatization of the organization.
- Although the Iridium system is not operational in its original state, it is worth reviewing its distribution system, illustrated in Figure 7.13. Iridium LLC billed the operating company for the end-to-end call charges, consisting of space segment charges (operating cost and profit), non-resident gateway charges and terrestrial charges. The operating company then added the regional gateway charges and profit to set the call charge; and finally, the service provider, which could be the operating company itself, added its own costs before billing the user. Service providers comprised PTTs, cellular operators and, in some instances, the operating company was also a service provider. The Iridium system was able to reduce the cost of the terrestrial

Figure 7.12
A distribution system where services are offered by gateway operators

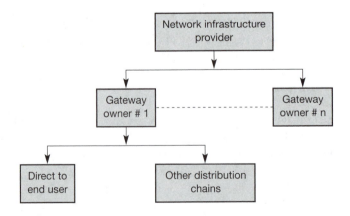

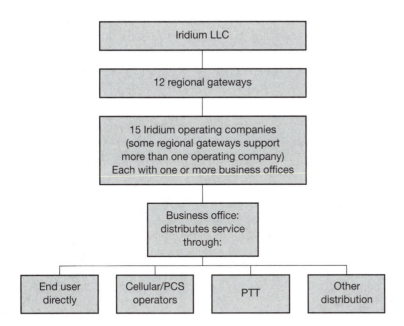

Figure 7.13
Iridium's distribution
system (Ingley, 1999)

network by being able to route calls via intersatellite links. Unfortunately, the system never took off, for various reasons.

Dual-mode handsets were able to communicate with various cellular standards by snapping in a cassette which supported a local cellular standard. This service, called global cellular, was expected to provide revenues to Iridium as the customer bill consisted of the cellular components and a component paid to Iridium.

- Globalstar's distribution system is illustrated in Figure 7.14. The system has been designed to support a decentralized distribution system allowing a country or a group of countries to exercise full control of local distribution; initially, up to 200 gateways were planned but this has been scaled down to 50, which are adequate for global coverage. Service providers own and manage gateways and arrange for terrestrial interconnections. They purchase satellite time in bulk from Globalstar; either the service provider or the local distributor, which may be a cellular or PCS operator or a PTT, bills the customer. Thus the bill consists of a satellite component, an overseas gateway component when applicable, service provider's operating cost and profit, and a terrestrial component. Globalstar bills the service provider for the satellite time and overseas gateway usage.

- To minimize dependencies on terrestrial systems, ICO Global Communication architecture (now New ICO: note: we will not consider the latest commercial changes here, as the purpose is to illustrate various possible approaches rather than specific commercial arrangements) consists of a backbone terrestrial optical fibre network interconnecting all satellite access

Figure 7.14
Globalstar distribution
system (Ingley, 1999)

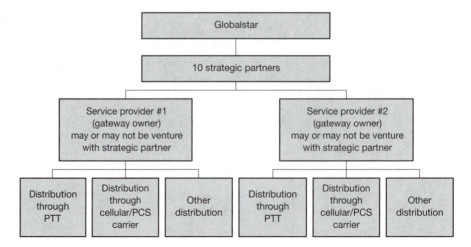

Earth stations. The satellite access nodes and the terrestrial network (ICONET) are owned by the company. ICO originally had about 60 service partners, each of whom could have one or more service provision arrangements. The service partners and service providers manage customer relations, set call charges, distribute terminals through various channels and bill users in some cases. The user charge comprises satellite and terrestrial usage and service partner/service provider operation cost plus profit. ICO charges service partners/providers for satellite and terrestrial usage. Figure 7.15 shows the service distribution model of the ICO system.

- The ORBCOM system provides low bit data-only service and hence has a different user base to the big-LEO systems. ORBCOM distributes its services directly to users, through value added sellers and through international licensees. The cost components are similar to the ones in big-LEO systems but the cost is based on used data bytes.
- Universal mobile telecommunication system (UMTS) business model: UMTS is the next generation mobile system for Europe (see Chapter 11), which can

Figure 7.15
ICO's service distribution
hierarchy (Ingley, 1999)

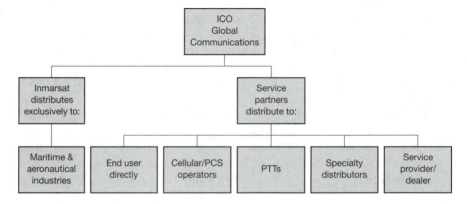

include a satellite component. The UMTS business model is envisaged to comprise the space segment owners, network operators, service providers, the subscriber and user. It is envisaged that the satellite component may be an extension to the public land mobile network comprising a system which serves both cellular and satellite cells or a system serving satellite cells only. In the case of the former, two possible business models are illustrated in Figures 7.16 and 7.17 (adapted from SMG5 TD 224/93rev1). Figure 7.16 illustrates a scenario where the satellite component is an integral part of a multiple public land mobile network (PLMN). Figure 7.17 illustrates business roles in a system consisting of multiple PLMNs sharing a satellite component of UMTS.

Yet another model applies to an architecture where the satellite component resides outside the terrestrial PSTN business arrangement. In such a case, the model reduces to that of the present types of arrangement, such as used by Inmarsat.

7.4 Billing issues

Customer care and billing are important and complex aspects of commercial MSS. The problem becomes compounded for international systems which must have gateways in a number of countries and whose service providers serve customers in a variety of currencies and bill them in local languages. The billing method depends on the business architecture of the system (see section 7.3).

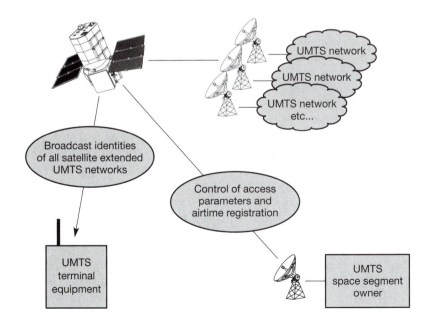

Figure 7.16
Satellite component integrated with multiple UMTS PLMN (Donald, 1995; IEEE Pers Com, Oct. 1995) © 1995 IEEE

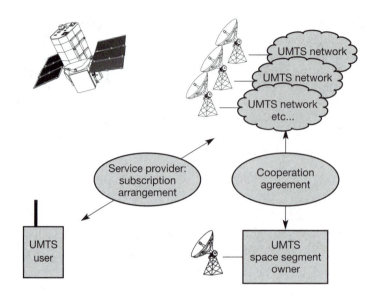

- In a **segmented billing scheme**, the system operator bills service providers, leaving the task of managing the customer billing to the service provider. The satellite usage time for each service provider is monitored by the system provider, who bills the service provider for space segment usage. The service operator has its own billing system for managing the terrestrial component and individual billing. In a variant of this scheme, the system operator may utilize the call monitoring system of service providers, to reduce the complexity and cost of billing. Each service provider sends bills to customers in a format and language understood locally. Thus the system operator's billing system is simplified as the operator needs to bill a relatively lower number of customers who can all work in an agreed language/currency.
- In a **semi-centralized billing system**, end-to-end call monitoring is done by the system operator. The operator then sends call records with space segment charges and non-resident gateway charges to the caller's home gateway operator, who cascades them down to the service provider after adding its cost; and finally, the service provider bills the customer with a mark-up consisting of its cost/profit.
- In a **centralized billing** system, the end-to-end call monitoring is performed by the system operator. The operator owns the entire system, selling the service through local service providers. The call records, together with space segment, gateway and terrestrial charges, are passed on to the service provider's business centre, which marks up the cost to account for its operating cost and profit before passing it over to the customer.

7.5 Investment routes

Commercial satellite projects arrange funding from a variety of sources, loans from investment banks, public equity markets, private investors, high yield debt notes, bonds, vendor money and strategic equity. In 1996, half of the capital or $11.4 billion for satellite communications raised in the USA went to only eight companies, and most of the funding went to MSS operators – Iridium, ICO Global Communications, Globalstar, Orbital Science and AMSC. Thirty-six per cent was raised in public debt and equity financing; other sources were bank financing (13%), vendor financing (9%) and strategic equity (42%). Generally a loan requires some security/guarantee due to the high risks perceived by financiers, and its availability depends on the state of the creditor's market. A short-term loan is arranged through incremental banking arrangements. Capital also flows through mergers, acquisitions and spin-offs. Sometimes funding takes on complex dimensions, involving financiers who may themselves take loans from several banks, such as used by Arianespace Finance for financing the launch customers of Arianespace (Foley, 1997). Far fewer analysts cover the satellite industry for purchase recommendations, etc. compared with other telecommunications technologies. Financial market resources are limited; the finance is divided between various sectors, such as FSS, MSS, DAR, satellite imaging, etc., according to the confidence the investors have in the proposals and the trend. Obviously, such forecasts are revised regularly, affecting stock prices and investments; when financing dries up, satellites have been among the first to lose funding. Also, satellite stocks have tended to be volatile to events and interest rates.

Bankers favour companies which reduce regulatory, development and business risks by having strong sponsors and a good management team, obtain as many licences as possible, and show evidence of competitive advantage. An industrial estimate gives around $42.9 billion for satellite-related financing between 1999 and 2007, of which $10.9 billion is estimated for MSS and $4.3 billion for digital audio radio services (Philips Satellite Group Editors, 1999). Incidentally, the majority of funding, i.e. $21.5 billion, is anticipated for multimedia systems. Financiers find the funding of satellite projects rather challenging, because of perceived risks such as launch failures. The space insurance industry was at a breakeven point at the end of 1998, that year being the worst on record when losses totalled $1.9 billion as opposed to earnings of $950 million. Recent setbacks, such as Iridium's difficulties and the 12-satellite launch failure of Globalstar, tend to dent the confidence of the financial community. However, investors tend to calm down if companies display confidence in the product – take, for example, the instance when the CEO of a satellite company recently volunteered to take the burden of all debts if the financial community did not invest; yet another company spent considerable effort in proving the technology by launching satellites in orbit to win investor confidence. In this respect, well-established companies with deployed networks and earnings are preferable from a financier's viewpoint. Aligning a

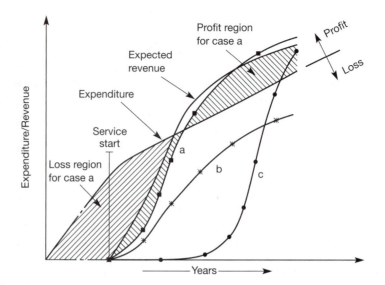

Figure 7.18
Sensitivity of business to rate of penetration. Case (a) closely matches expectations; case (b) never makes a profit; and case (c) represents late profit

company with major players, finding partners and corporate backing from large companies are other ways companies gain confidence.

As already mentioned, one of the biggest problems in raising the required funding for MSSs is their long gestation period. It was also noted that the cost of each system runs into billions and tends to escalate due to unavoidable delays in projects of such complexity. Figure 7.18 shows expenditure and return on capital for various rates of penetration.

7.6 Regulatory issues

Regulatory issues address compliance with national and international Radio Regulations which include issues such as intersystem coordination as well as licensing issues, which may vary between countries. These procedures are very lengthy, circuitous and therefore not very conducive to a commercial environment. A necessity to streamline regulatory procedures is generally felt by commercial operators.

The regulatory process for the licensing of MSSs is rather complex and therefore takes several years to complete. Furthermore, even within the ITU's global framework, regulations tend to differ by country. Typically, the process involves spectrum allocation, national licensing and coordination of the allocated frequency. Spectrum and frequency allocations are likely to incorporate international involvement. Spectrum allocation refers to the ITU Radio Regulations, which list the spectrum available to each service. Frequency coordination is required to guarantee that the operators do not cause interference to each other. The guidelines are laid out in ITU articles 11 and 13 of the Radio Regulations.

An operating licence is necessary for an operator to market services in a country. Some countries who may not license certain operators are concerned that terminals belonging to such unlicensed operators may be used within their country. MSS operators are addressing such concerns by building technical features capable of preventing this type of fraudulent use. For instance, the system may prohibit communication if a call is requested from a communicator located within a 'barred' country. Strict discipline, cooperation and agreement between parties are necessary. Infringement clauses may be necessary.

Other matters, in addition to spectrum allocation and frequency assignments, are related to type approval, global roaming and non-discriminatory access to market. Local partners are recognized as a necessity for best marketing access through local support and marketing skills, and at the same time, make it easier for domestic regulators to grant operating licences. Partners could benefit from subsidies and participation in the operator's system design/operations. For these reasons, some companies, such as Globalstar, have built systems which can support gateways in individual countries or in a region. However, other operators prefer to own all gateways to facilitate interoperability and maintain better overall control.

While the service markets may be segmented geographically, there are obvious advantages when the space segment and operations are maintained by a single entity – therefore MSS operators tend to keep operations within their jurisdiction. Benefits include network-wide performance evaluation, anomaly investigation, and data collection for operational, market and strategic planning, etc.

Operators often find it difficult to obtain an operating licence in some countries. The concerns of the telecommunication authorities of such countries may relate to their share of revenue, security concerns, political considerations, and government policy, etc. The revenue concerns stem from the fact that users may bypass the country's network using a foreign gateway, thus depriving the country of the revenue. The problem could be exacerbated if the service is offered from a regional gateway located in another (unfriendly) country. Roaming to other countries may also result in loss of revenue to the home service provider. From the operator's viewpoint, the difficulty lies in the fact that the telecommunication costs in certain countries are so large as to make the service unattractive, and then they are faced with the problem of billing a visiting terminal. Such problems have been managed in the cellular systems such as GSM by signing mutual agreements.

Many of the issues addressed above are under discussion at the ITU. Such debates may play an important role in harmonizing the growth of this technology and in the protection of sovereign rights of nations to regulate telecommunications by laying down international standards for intersystem operations, promoting international cooperation, etc.

7.7 Traffic forecast

A crucial requirement of an MSS business is communication traffic forecasting to enable planners to make timely and informed operational and strategic decisions. In Chapter 3, we discussed methods of converting forecasts to spectrum requirements. Chapter 8 addresses methods for short-term forecasts using measured traffic data flow through a network and is therefore well suited for the short-term business and operational needs of prevalent operators. Long-term market forecasting, on the other hand, is necessary by regulatory authorities for long-term spectrum planning; standardization bodies require forecasts for dimensioning satellite components of long-term initiatives such as UMTS; an MSS operator uses forecasts for long-term strategic planning; mobile terminal manufacturers may use them to plan their manufacturing activities, etc.

Here we will consider a methodology for forecasting the demands of the next generation of mobile satellite communication systems. Because of the uncertainties associated with each underlying assumption, a number of variants to the forecast model are possible.

The main inputs to a forecast model applicable to mobile communication systems are:

- general technology growth trends likely to influence the MSS market in the next 5–10 years;
- state of technology in the period to support the service needs;
- influence of competing technology such as terrestrial mobile systems, other MSS operators and personal communications technology such as FSS personal communicators;
- evolution path from existing MSS to the next generation;
- market penetration of similar service and rate of penetration;
- quality of service;
- number of operators and traffic segmentation;
- tariffs and terminal cost;
- geographical areas of interest and their gross domestic product (the whole world; a region; a country) and demographic data for the regions of interest;
- age, occupation and GDP of the target group, expenditure per head in similar technologies and acceptability of such services;
- sociological trends in region(s) of interest;
- influence of regulatory constraints such as spectrum segmentation;
- risks in obtaining licensing of services, especially for international systems.

Assumptions include market sizing, acceptable levels of tariffs, impact of competition, acceptable terminal costs and feasibility in terms of manufacturing costs, component costs, production volume, terminal complexity, service take-up, usage per terminal and market penetration rate derived from existing trends.

A notable technological development in the past decade has been the growing convergence of telecommunications, information exchange, computing, entertainment systems and business needs. The so-called information society of this decade is therefore expected to require technologies which offer these services to individuals transparently. This premise forms the basis of forecasts applicable to mobile communication systems.

Consider, as an example, the methodology used by the UMTS forum in forecasting the next generation mobile communication market in Europe and the world (UMTS forum, 1999). The purpose of the study was:

1   to determine the level and nature for third generation mobile services in the world, with emphasis on the European Union;
2   to determine key issues influencing the evolution and their impact on the market analysis;
3   to develop a robust analysis to forecast the world-wide mobile services market up to 2015;
4   to forecast emerging requirements and intermediate steps towards the introduction of UMTS/IMT-2000.

The methodology is summarized in Figure 7.19. Social trends and technology evolution trends were identified, using which a number of scenarios for market development were constructed and a forecast of each produced. The scenarios and forecasts were validated and industrial opinions regarding future market requirements and development were gathered through a series of workshops for industrialists. Finally, all research findings were consolidated, identifying the key drivers, enablers, barriers and uncertainties which would influence the future market. Clearly, the scope of the study was wide, with far-reaching implications for the future of mobile telecommunications. The study undertook forecasts for both terrestrial and satellite markets.

Based on technology trends, a market was defined. It was anticipated that the third generation mobile systems would offer broadband services, such as voice, graphics and video, to mobile users transparently, taking advantage of the convergence of fixed and mobile networks. It was assumed that broadband ser-

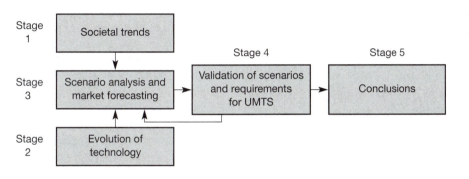

Figure 7.19
Forecasting methodology used by UMTS forum (adapted from UMTS report #8)

vices would be introduced as enhancements to existing technologies, such as DECT, before the commercial launch of third generation systems.

Behaviour and attitudes in society which are likely to affect acceptance of third generation systems were identified and their rate of development was assessed. Three key trends were identified. The market for fixed networked multimedia, such as pay-TV, video/audio-on-demand, interactive entertainment, educational and information services, and communication services such as video-telephone and large file transfer, has been growing rapidly at rates of over 60 per cent generally; computer information retrieval and communications technology is being assimilated rapidly, evidenced by phenomenal internet uptake, anticipated to reach 500 million world-wide users by 2005; there is a trend for accessing information and entertainment services by mobile users. An underlying assumption was that the demand for multimedia services in the mobile environment will be influenced significantly by the growth in the fixed sector and hence will follow it. The services considered were: passive audio-visual, e.g. direct to home satellite, video-on-demand, etc; passive audio service, e.g. audio-on- demand as an alternative to CD.

Over 30 applicable technologies were studied; of these, critical technologies which could influence the functionality, attractiveness and cost of mobile multimedia terminals were identified as semiconductor, display and interface technologies. The relevant technologies for the mobile multimedia services were identified as delivery and management technologies such as the internet, Java, database, spectrum-enhancing and service creation technologies.

It was assumed that factors influencing service take-up will be service affordability, governed by the price of the service and terminal, and service attractiveness. It was further assumed that the price of the service will be determined by competitiveness in service delivery and the cost of network access; the price of terminals will be influenced by component cost and production volumes; and service attractiveness will be affected by variety offered, usability and utility.

Four scenarios were modelled, comprising the permutation of two key factors – the take-up of fixed multimedia service in mass markets and the primary location of intelligence whether in the network or in the terminal – which were likely to affect multimedia service evolution up to 2005 (see Table 7.3).

Each scenario depicted the interaction between elements of a conceptual market dynamics model under different assumptions in social trends, technology evolution, regulatory/political environment and industry structure/competition. The scenarios were then characterized in terms of market size, its drivers, enablers and barriers. Some of the main characteristics are summarized in Table 7.4. The scenarios were useful in making a sensitivity analysis and thereby arriving at the specific issues and industrial development likely to influence the market forecast which could then be used for forward planning by the industry, market etc. The scenarios could also reveal the effects of

Scenario	Main features
Slow evolution	Mobile multimedia slow to evolve; unsuccessful liberalization discouraging service providers from entering the market; failure of standardization; attractiveness of service limited due to network security fears.
Business centric	Rapid uptake of multimedia in business sector; consumer market uptake slow; limited liberalization discouraging service providers from entering the market; limited success in development of global standards; high service and terminal prices; limited attractiveness of service due to low IT literacy.
Evolved mass market	Mobile multimedia accepted by masses after a slow initial development; successful liberalization; global standards developed resulting in reduced service/ terminal prices; good service attractiveness due to high levels of IT literacy, intelligent user interfaces, agent technology and continuation of distributed intelligence model (see Figure 7.20).
Commoditized mass market	Rapid growth in market; liberalization encourages service providers; global standards reduce service and terminal prices; centralized intelligence network model increases attractiveness through simple interfaces; stable regulatory environment encourages competition.

Table 7.3
Main features of the scenario developed by UMTS forum (UMTS, 1999)

uncertainties, such as slower than expected growth, on the forecast. The possibility of marked deviation from trend was emphasized. Such departures could occur through unforeseen developments or breakthroughs such as internet growth, technical breakthroughs in providing ISDN and video over copper pairs, etc. Table 7.4 lists the factors which may influence the evolution of these scenarios.

Figure 7.20 portrays the influencing forces in the evolved mass market model leading to encouragement in service take-up and entry of service providers. Successful liberalization, development of global standards, high levels of IT literacy, use of artificial intelligence and intelligent agents all have a positive influence, encouraging entry of service providers and increased service take up.

Table 7.5 lists the forecast produced by the UMTS forum (1999) for worldwide UMTS/IMT-2000 compatible users. The forecast for non-multimedia subscribers does not include the non-compliant MSS market, typically the handheld MSS sectors being introduced now. The forecast does not include installation of the fixed satellite service which could provide a personal multimedia service, as such systems are outside the UMTS/IMT-2000 systems.

Drivers	Growth of related services in the fixed network due to increase in internet use and reduction in usage costs. Demand for rapid remote access to information for business and personal use.
Enablers	Encouragement to competition and cheap access to services. Few world-wide radio or service standards. Development of IMT-2000 specification by ITU. Possibility to exploit GPRS in the multimedia delivery evolution path. Improvements in relevant technologies such as interface design, display, spectrum efficiency enhancements, semiconductor device.
Barriers	High cost and limited spectrum. User concerns on security and failure to resolve them. Slow IT literacy penetration in mass market.
Uncertainties	Growth rate in mobile multimedia market, network capacity and spectrum. Network model – network intelligence or device intelligence.

Figure 7.20

Forces which influence uptake of service in the evolved mass market (adapted from UMTS 1999)

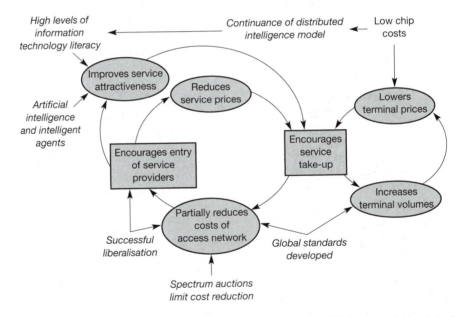

One problem experienced by planners is lack of published market-related data. In such cases, indirect methods have to be used. Regional population and market potential are quite indicative of the likely demand. In the absence of any data, market potential can be judged by analyzing the telephone density in an area. Table 1.8 (Chapter 1) shows the global population by region and telephones per 100 inhabitants. In reality, several additional factors must be considered;

Year	2005	2010
MSS subscribers (000s)		
Non-multimedia	4,875	7,500
Multimedia	6,585	10,975
Total	11,460	18,475
Average usage per subscriber (kB/month)		
Non-multimedia		
Voice	8,709	8,491
Low speed data	6,208	5,587
Multimedia		
Voice	1,194	1,561
Low-speed data	2,584	3,380
Asymmetric	26,154	34,247
Interactive	1,781	2,334
Total annual traffic (million MB)		
Non-multimedia		
Voice	509	764
Low-speed data	491	736
Multimedia		
Voice	94	206
Low-speed data	204	445
Asymmetric	2,067	4,510
Interactive	141	307
Total	3,506	6,968

Table 7.5
World-wide satellite market (adapted from UMTS report #8)

these include individual affordability, the commitment of governments to development in telecommunications, and the economic status of the region.

A traffic dimensioning methodology, developed within the Satellite Integration in the Future Mobile Network (SAINT) project of the European RACE II programme and suited for such a scenario, is summarized here as an illustration (Hu and Sheriff, 1995, 1997). The algorithm, developed for dimensioning the satellite component of UMTS (S-UMTS), uses logical assumptions and demographic data to predict satellite voice traffic in Erlangs for over 210 countries, presenting them on 36×72 traffic grids on world maps. An earlier study conducted for the European Commission was used as the basis for developing the model (KPMG, 1994).

The model presented here is as reported in the more recent of the two publications, which is believed to be an extension of the earlier work. Based on S-UMTS terminal types (five in all) and the role of S-UMTS, two categories of users were identified – mobile users and fixed users. Then traffic volumes for each user group were predicted as follows:

- assess S-UMTS feasibility in a region based on existence of terrestrial coverage;
- determine gross potential market (GPM);
- estimate S-UMTS service penetration P factoring in the take-up rate T_p.

The penetration P is defined as the percentage of people subscribing for S-UMTS. Parameter P reflects the ability of the population to subscribe for the service, which depends on the ratio of per capita income to the tariff of the service. Relative ability to pay versus penetration percent graphs used were based on terrestrial cellular service data. Penetration rate was assumed to follow the well-known product cycle comprising a slow introductory phase, followed by rapid growth, saturation and decline. The penetration rate at time t was modelled by curve-fitting historic data as follows:

$$T_p(t) = M/(1+ae^{bt}) \qquad\qquad ..\ (7.1)$$

where $T_p(t)$ = penetration rate at time t, parameters a and b are obtained by regression analysis and parameter M describes the saturation level.

The parameters M, a, b differ between countries and are therefore calculated for each country on the basis of cellular growth rate. Penetration in year t, $P_r(t)$, is obtained by factoring P and $T_p(t)$.

The number of subscribers N_t in the year of interest for a gross potential market, GPM, is then given as

$$N_t = GPM\ P_\gamma(t) \qquad\qquad (7.2)$$

A S-UMTS service becomes feasible only when terrestrial services become uneconomic, i.e. where the population density is below a threshold D_t. The threshold depends on the per capita income of the region. Hence the first step is to determine those regions where S-UMTS could be introduced. D_t was determined on the basis of per capita income for the years of interest (1998–2010). For 2010, D_t for high (>20,000 ECU per capita), medium (6,000–20,000 ECU per capita) and low GDP (<6,000 ECU per capita) were estimated as 3 people/km^2, 30 people/km^2 and cities of 1 million respectively.

Mobile users were categorized as cellular fill-in users and international travellers, i.e. people who travel to areas where there is no terrestrial coverage.

The GPM of the cellular fill-in, G_f, is defined as the number of people not served by the terrestrial UMTS. The number of cellular fill-in customers N_c in year t is then given as

$$N_c = N_{pop}\ (1-p/100)P_r(t) \qquad\qquad ...\ (7.3)$$

where N_{pop} = population of the region under consideration and p = percentage of population covered by the terrestrial UMTS.

The gross potential market for international travellers was estimated as the number of air travellers leaving a country, N_a. The call charge per minute was set at a higher level; the service take-up rate was set the same as cellular fill-in but penetration was lower due to the higher tariff. The number of subscribers N_i in year t, for this category, is then given as

$$N_i = N_a\, P_a(t) \tag{7.4}$$

where $P_a(t)$ = penetration rate in year t.

The gross potential market for fixed S-UMTS users, N_f is given as

$$N_{fg} = N_h - N_T \tag{7.5}$$

where N_h is the number of houses and N_T is the number of telephone mainlines in residential areas. Factoring the penetration of telephone lines per household in a manner similar to mobile users and factoring it by the take-up rate, $P_f(t)$, the number of subscribers N_f is given as

$$N_f = N_{g.}\, P_f(t) \tag{7.6}$$

Data of up to 225 countries, obtained from UN and ITU publications, were used. The number of S-UMTS subscribers depended strongly on handset subscription rate and call cost. Figures 7.21(a) and (b) show the subscriber forecast for low and high tariffs. Table 7.6 shows the cost and annual price fall assumed for these two scenarios. The number of subscribers for the lower cost entry is about 90 million against only 35 million for the high cost entry.

Cost component	Low cost entry (ECU)	High cost entry (ECU)	Annual price fall (%)
Handset	500	1,000	0.7
Subscription/year	60	240	0.7
Call rate/minute	0.1	0.5	0.5

Table 7.6
Cost and annual price fall assumed for the two scenarios mentioned.

Figure 7.22 (a) and (b) shows the split in S-UMTS users by region for 2010 when S-UMTS is targeted to have the maximum penetration. Figure 7.22 (a) illustrates that mobile users will be dominated by North American, former USSR, European and Asian users; and Figure 7.22 (b) shows that fixed users will be dominated by Asia, South America and Europe. It was also observed that 52% of the market will be for fixed users and the remainder for mobile users.

In the past few years, internet usage has increased very rapidly. Circuit-mode access for web browsing and other internet applications is inefficient due

Figure 7.21 (a) Subscriber forecast for low rate tariff (Hu and Sheriff, 1997) (b) Subscriber forecast for high tariff (Hu and Sheriff, 1997)

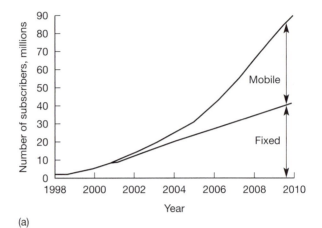

(a)

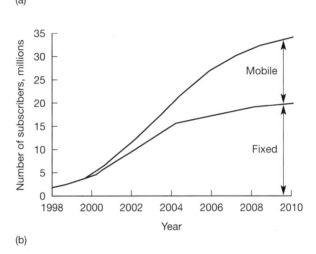

(b)

Figure 7.22 (a) Regional distribution of S-UMTS mobile users (Hu and Sheriff, 1997). (b) Regional distribution of S-UMTS fixed users (Hu and Sheriff, 1997)

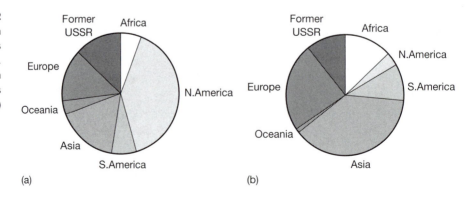

(a)   (b)

to the sporadic nature of transactions. Packet-mode transmissions with the call charge based on packets transiting the network are efficient and fair to users. Forecasts for this type of traffic are done in terms of traffic volumes measured in units of bytes. The number of physical channels is therefore derived as D_t/D_c, where D_t is the total forecast data volume and D_c is the capacity/channel. Considerable effort is under way to characterize internet traffic, which should enable more efficient channel resource utilization (e.g. Matarasso, 1999).

Further reading

Baker, M.J. (1975) *Marketing New Industrial Products*, Macmillan Press.

Corbley, K.P. (1996) 'Accessing satellite and cellular systems – Dual mode handsets provide the option', *Via Satellite*, February, 76–90.

Crossman, M. (1999) 'The little LEO scorecard', *Satellite Communications*, January, 70.

Donald, P. (1995) 'Standardisation of the satellite component of the UMTS', *IEEE Personal Comm.*, 68–74.

Foley, T. (1997) 'Satellite Financing – which direction will the market take', *Via Satellite*, 22–30.

Gaffney, L.M.; Hulkower, N.D.; Klein, L. (1995) 'Non-GEO mobile satellite systems: A risk assessment', IMSC '95, *The Fourth International Mobile Satellite Conference*, Ottawa, co-sponsored by Communications Research Centre/Industry Canada and Jet Propulsion Laboratory/NASA, A23–A27.

Hu, Y.F.; Sheriff, R.E. (1995) 'Traffic model for the satellite component of UMTS', IMSC 1995, *The Fourth International Mobile Satellite Conference*, Ottawa, co-sponsored by Communications Research Centre/Industry Canada and Jet Propulsion Laboratory/NASA, 508–11.

Hu, Y.F.; Sheriff, R.E. (1997) 'The potential demand for the satellite component of the universal mobile telecommunication system', *Electronic & Communication Engineering Journal*, April, 59–67.

Ingley, C. (1999) 'Global vision: Making the right connections', *Satellite Communications*, February, 38–45.

Ingley, C. (1996) 'Racing into the new frontier', *Satellite Communications*, December, 28–31.

KPMG (1994) 'Satellite communications and their consequences for European telecommunications trade and industry', *Report to European Commission*, March.

Lazear, Y. M.; Capots, L.H.; Langlais, R.D.; Price, K.M. (1997) 'System planning methodology for satellite communication services', *International Mobile Satellite Conference*.

Matarasso, C. (1999) 'Observation characterisation and modelling of World Wide Web traffic', IMSC '99, *Sixth International Mobile Satellite Conference*, Ottawa, co-sponsored by Communications Research Centre and the Jet Propulsion Laboratory, 310–15.

Philips Satellite Group Editors (1999) 'Satellite Finance – A year in review', *Via Satellite*, April, 18–24.

Sultan, N. and Groepper, P.H. (1999) 'Mobile satellite life cycle cost reduction: A new quantifiable system approach', *IMSC '99, Sixth International Mobile Satellite Conference*, Ottawa, co-sponsored by Communications Research Centre and the Jet Propulsion Laboratory, 246–51.

UMTS forum (1999) 'The future mobile market', Report (8), March.

System architecture 8

8.1 Introduction

In earlier chapters, various parameters essential for characterizing the MSS air interface were introduced. In Chapter 2, relevant orbital characteristics were outlined, and Chapters 3 and 4 addressed essential characteristics of the radio link. In this chapter, the MSS system is viewed in its totality. The topics include space segment architecture, network connectivity and network management.

The development of an MSS system is an iterative process involving a number of complex parameters. Service requirements, space segment architecture, operational concepts and business plans are evolved iteratively until an acceptable solution has been found. Interrelationships between these entities are complex and non-linear, requiring human judgement, and therefore a diversity of solutions is possible. The most critical and expensive part of an MSS is the space segment. Having successfully completed this first hurdle, the development can progress to other challenges – project financing, detailed system design, infrastructure development, deployment of constellation and finally the vital operational phase.

The most critical component of an MSS network is the service link – the cordless tether which gives mobility. The role of the network is to ensure that a real or virtual connection is maintained during a call, and calls can be directed to fixed or mobile users under normal conditions. Connectivity during a call may require handover between spot beams, satellites or fixed stations. There is also a practical need to minimize the terrestrial tail in order to reduce the terrestrial component of the user call charge. The interconnectivities between system entities are discussed and explained, again focusing on the space segment issues.

When an MSS system has been deployed, the operator must orchestrate network operations and management efficiently to ensure that the performance objectives are met, business goals kept under scrutiny and revised as necessary. The main issues related to operation and management of an MSS network are addressed; these include, amongst others, management of space segment resources, analysis of traffic trends, spectrum monitoring and quality of service assessment.

The chapter begins with considerations related to system optimization issues, the air interface, and network connectivity aspects. This is followed by explanation of the functional relationships between the network elements. Finally, the important topic of operational management is addressed.

8.2 Air interface

An MSS system comprises a space segment interfaced to a terrestrial segment, each of which comprises a variety of functional entities interconnected with each other by radio or terrestrial links. Feeder and service links are the most crucial as they provide mobility (see Figure 8.1), whereas other links support network functions which are independent of media, and hence use a cost-effective and reliable transport facility such as a private backbone transport network, PSTN lines, leased circuits, system's own spare space segment capacity, etc.

Figure 8.1 depicts various types of links used in support of an MSS system. The service link (SL) provides a radio connection to the terrestrial network; mobile-to-mobile communication is routed through a direct mobile–mobile service link (MMSL); intersatellite links (ISL) are used for routing calls in space in preference to terrestrial; feeder links (FL) connect a gateway to the satellite constellation; intra-gateway links (GLN), inter-gateway links (GGL) and network management links (NML) are essential for exchanging vital information in support of management of the network; and terrestrial links (TL) connect an MSS system to a terrestrial fixed or mobile network.

An air interface comprises the gateway and service links. As mentioned, the service link is the most crucial in MSS, as it imposes fundamental limitations on the capability of the mobile satellite system because of a number of factors, summarized as follows:

- a need to serve small mobile terminals;
- limited effective isotropic radiated power (EIRP) of satellites and mobiles;
- limited spectrum; and
- hostile propagation environment.

Figure 8.2 identifies the most critical components of the service link and the limitations attributed to each.

Satellite EIRP and G/T are limited by the state of technology. There has been a steady improvement in technology in the past two decades, which has enhanced MSS service capabilities proportionately. In the late 1970s, geostationary satellites could at best transmit EIRP of ~30 dBW; by contrast, satellites being launched at the beginning of this decade can transmit a 43 dB higher EIRP level (see Figure 1.2). Similarly, spacecraft G/T has been improving as spot beam size reduces. In the early 1990s, single beam geostationary satellites were the norm; a decade later, 140 spot beam geostationary satellites have been launched; the corresponding increase in G/T ranges between –10 dB/K and +15 dB/K.

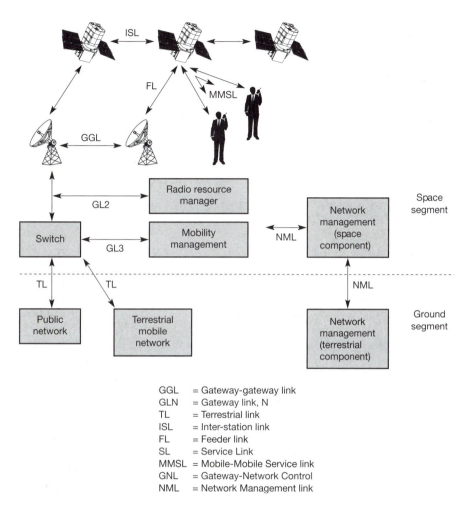

Figure 8.1
Main components and
associated links of an
MSS network

GGL	= Gateway-gateway link
GLN	= Gateway link, N
TL	= Terrestrial link
ISL	= Inter-station link
FL	= Feeder link
SL	= Service Link
MMSL	= Mobile-Mobile Service link
GNL	= Gateway-Network Control
NML	= Network Management link

Considerable technical effort is under way to alleviate the crucial problem of spectrum shortage caused by intense commercial competition for mobile communications service. An effective method is to narrow spot beam's size, thereby improving frequency reusability – for a single geostationary satellite, frequency reusability has increased by a factor of 20 within a decade. A combination of a low altitude satellite and narrow spot beams can give an even greater global frequency reuse capability due to the narrower field of view of low altitude satellites. The Iridium system is stated to reuse frequencies 180 times globally with 2,150 active spot beams. Modulation and multiple access methods also influence the system capacity to varying extents (see Chapter 4).

Shadowing and multipath effects restrict the throughput of mobile satellite channels, in particular the land mobile channel. From Chapter 3, we know that antenna directivity, vehicle speed and environmental characteristics influence

Figure 8.2

Inhibitors of MSS
service links

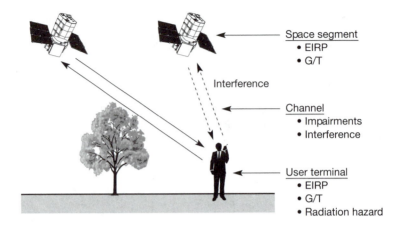

channel characteristics. Throughput can be increased by deploying directive mobile antennas and operating in clear environments but this is a trade-off against terminal cost/complexity and link reliability. Effects caused by shadowing/multipath can be partially mitigated by selection of robust modulation, coding and multiple access schemes as discussed in Chapter 3. A more radical solution is to use path diversity, but it requires visibility to more than one satellite in conjunction with network support. A cost-effective solution is to promote user cooperation. Because maritime and aeronautical systems suffer less severe shadowing and impose lower constraints on antenna size, a higher throughput can be supported. Similarly, high throughputs are achievable for fixed land mobile applications, while for non-real-time applications, store and forward techniques provide a reliable countermeasure.

Radiation health risks impose a bound on system throughput for hand-held applications, as power transmitted from units must be severely restricted (see Chapter 5), resulting in extremely low received signal levels on satellites. The problem is mitigated by improving satellite G/T and/or path loss reduction by deploying low altitude satellites. Hence, invariably, all the systems providing a hand-held service use a large number of spot beams and many of them use low or medium Earth orbits to reduce path loss.

In Chapter 2 we discussed constellation design solely in terms of orbital considerations. It must be amply clear by now that such an exercise cannot be considered in isolation as the constellation has a profound influence on system architecture. The influence of the orbital characteristics may be perceived differently by operators, leading to a variety of constellations and architectures. For example, an operator familiar with intersatellite communication technology may prefer the use of ISL; for the same reason, an operator may select a more complex geostationary satellite in preference to a LEO or MEO constellation. In the next few sections, factors influencing the system architecture are reviewed and a structured approach to system optimization is developed. The reader may refer to Chapter 7 for a business-oriented approach to such an optimization; it will become apparent that in a commercial MSS, technical and financial issues are intimately entwined.

8.3 System development

A commercial venture can either be technology driven, wherein innovative products are marketed with the assumption that the product will be appealing to consumers, or products may be developed for real or perceived market needs established through market research. Most of the recent MSS proposals rest on extensive market analysis by individual operators (see Chapter 7), which would appear to be a more balanced approach in view of the huge investments required for MSS.

The main considerations influencing a commercial MSS system architecture can be summarized as follows:

- communication service;
- user terminal characteristics;
- service area;
- traffic distribution;
- system capacity;
- quality of service;
- network connectivity;
- spacecraft technology;
- frequency band;
- orbital characteristics;
- launch considerations;
- schedule and financial risk;
- cost and revenue goals.

Due to the large number of intrasystem dependencies, MSS system development is well visualized as a structured top-down multi-layered iterative optimization, wherein system synthesis cascades down progressively to lower layers iteratively, as design implications are better understood and real-world constraints applied. The concept of such a market-driven system evolution is illustrated in Figure 8.3 (also see Chapter 7).

Market research provides a basic set of telecommunication requirements, terminal characteristics and service area which are used to synthesize a top-level system architecture and estimate approximate costs. Some iterations may be necessary until a realistic solution is achieved. This preliminary design is evolved into a detailed design taking into consideration the state of technology, costs, schedule, etc. If the detailed design gives an unrealistic specification, the requirements are modified until a more acceptable solution is found; the detailed design specifications are used for developing all system elements, again

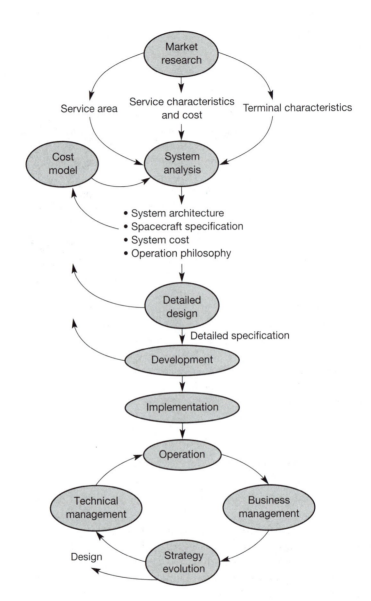

Figure 8.3
Concept of a market-
driven design
methodology

with a possibility of iteration. The system elements are integrated and tested, leading to the operational phase which incorporates both technical and business elements; finally, operational experience, coupled with market trends, is used for developing future strategy.

Due to large variability in constraints and assumptions, MSS development is not bound by rigid rules, nor does a set of requirements have a unique solution. Examples below should illustrate vast variations in rationale and assumptions used by system developers.

8.3.1 Influences
Communication service, market, service area and the cost model form the basis of a preliminary design. Consider each in some detail.

1. Type of communication service
    The main categories of services are:

    - low bit rate (a few bit/s – 1 kbps), real-time (e.g. aircraft automatic surveillance) or non-real-time (e-mail, SCADA);
    - medium bit rate (1–20 kbps) – voice, facsimile, data, etc.;
    - high bit rate (20 kbps – 64 kbps);
    - broadband (> 64 kbps).

    The size and orbital characteristics of a constellation depend on the type of service on offer. For non-real-time applications, intermittent coverage of the service area is adequate; however, for interactive services where time delay is critical, it is necessary to provide seamless coverage and to reduce the orbital altitude sufficiently. Lower orbital altitude also reduces path loss but, on the other hand, constellation size becomes larger, the network more complex and constellation maintenance more involved. Raising the altitude to force the delay to its upper bound reduces the constellation size at the expense of satellite transmitter power and reduction in spot beam size.

    Thus an operator such as ORBCOM, marketing delay-insensitive products, was selling with a partially deployed constellation. Yet others such as Teledesic have chosen low orbital altitude to achieve transmission delay comparable to the terrestrial system at the expense of a large constellation and complexity, while several regional operators have preferred a geostationary orbit because of reduced costs, risk, more stable link quality, etc. Iridium system designers preferred a low Earth orbit, incorporating intersatellite links for network connectivity to reduce ground segment complexity; ICO Global systems achieved similar objectives using a medium Earth orbit using terrestrial interconnections. Service throughput requirement additionally has a strong influence on terminal size and cost, together with spacecraft EIRP and complexity.

2. Market
    Market size and scope govern space segment capacity, size and cost of terminals, service cost and the service area. Some of the current marketing opportunities are:

    - personal communication for developing countries;
    - remote area communications such as rural telephony or public booths in remote localities or to serve government agencies;
    - communication for the transport industry;
    - regional or global paging service;
    - global/regional messaging such as e-mail, SCADA-type applications;
    - broadband services such as internet download, television, ISDN access.

    The ability of users to pay and the type of use – business, pleasure, distress or remote area communications – govern the portfolio of services, system cost/complexity and service cost. A system targeted for personal communications must support a low service cost hand-held service at the expense of a complex space segment, whereas a system targeting the transport industry, where size and cost of terminals or services are less critical, can deploy a simple space segment.

Profitability depends on the revenue versus total investment and operating costs of the system. Space segment capacity, cost and anticipated revenue must be matched to break even as quickly as possible. Over-sizing the system capacity of constellations can be expensive and risky; capacity can be enhanced as the constellation is upgraded should demand outstrip capacity. Unforeseen technical problems or complexity may necessitate a revision to constellation size/complexity later, after a detailed engineering study, as for the Iridium and Teledesic constellations which were each reduced from their original proposals of 77 and 840 to 66 and 288 satellites, respectively.

3. Service area

    Vital inputs to architecture design are service area, traffic distribution and specific needs such as a region-dependent call-barring feature arising from regional licensing/political problems. The information is used to determine the orbital inclination and eccentricity of non-geostationary constellations, spot beam distribution in the case of geostationary spacecraft and call-barring system features. Examples include spot beam coverage of Inmarsat's third generation satellites, which have been positioned over land masses to cater for land mobile traffic; coverage of the Globalstar system is optimized for mid-latitude regions and Ellipsat's for mid/high latitude regions; while the Iridium system includes features for region-specific call barring.

4. Cost model

    For developing a preliminary business plan, an approximate cost model is quite adequate, as precise costs can only be obtained after a system has been specified in detail. Planners can investigate the sensitivities of the business case to system attributes such as spacecraft cost, schedule, traffic growth, service cost, etc. using such a model. Empirical cost models are available in the literature for estimating spacecraft and launch costs from system-level requirements.

8.3.2 Constraints and considerations

System entities and their relationships must be synthesized applying real-world constraints and considerations. Such constraints may result in solutions which are, at best, sub-optimal in an academic sense but acceptable in the real world.

1. **Business**

    Financial viability is fundamental to a commercial venture. A number of proposed MSS have failed to see the light of day due to financial shortcomings; others such as Odyssey have merged with competitors; some, such as ICO, have experienced financial problems midway through the project, ending up being taken over by new investors; and yet others, such as Iridium, have failed. Consider some important business issues:

- Financial risk

  In the initial reckoning, a number of assumptions impacting on profitability would have been made. This would include traffic forecast, geographical distribution of traffic, service penetration rate, user requirements, user's ability to pay, impact of competition, etc. Their applicability may become questionable for a variety of reasons, such as delay in the introduction of the service, unexpected competition, poor service quality due to technical problems, delay in availability of user terminals, high user costs, etc. If perceived by potential investors, such risks are detrimental to raising initial capital or cash flow at the start of operation.

- Capital for an MSS venture is raised from a number of sources; success in this respect depends on the state of economy, the performance of the satellite industry in general and MSS in particular, the credibility of the proposers and the soundness of the business plan (see section 8.3.2 and chapter 7).

- Commercial issues, such as cost per call, segmentation of revenue between entities – infrastructure provider, service provider, etc. – must be well understood to avoid perception of exaggerated revenue.

2. **Network**

   Network connectivity design requires, amongst other things, consideration with regard to the location and management of gateways. System designers may opt for intersatellite links for increased flexibility in siting gateways. Systems such as Iridium, which deployed intersatellite links, are quite robust in this respect; by contrast, the Globalstar system is developed to operate with a large number of gateways to encourage regional participation; and the ICO system achieves interconnectivity through interlinked gateways.

   The extent of synergy with the terrestrial system affects a number of system parameters and the network architecture. A highly synergistic approach would require as many commonalities as feasible – this may include proximity of operational frequency, identical multiple access schemes, common network protocol, etc.

   The technique used in satellite–satellite and beam–beam handover can influence spacecraft design, constellation architecture and air-interface design.

   Other network considerations are mobility management techniques, and interfaces to network and business management systems and public networks.

3. **Hardware realization**

   First generation LEO and MEO systems are based on a number of novel technologies, such as integrated handset design, multiple launch, advanced spot beam technology, onboard processing, intersatellite links, etc. Depending on factors such as familiarity with technology, risk attitude, etc., a number of constellation architectures have emerged – operators have usually combined mature technologies with novel ones, though the extent of combination differs widely.

4. **Regulatory considerations**

   Regulatory issues pertain to the selection of frequency and the procurement of operating licences. The regulatory authorities of some countries impose technical and/or financial conditions for granting a licence. In such cases, an operator may have to prove the venture's financial health, modify the design for acceptability, alternatively accept a tentative licence to prove an innovative concept, etc. Designers should have included in their plans realistic estimates of their share of spectrum, in view of competition, etc.

5. **Orbital characteristics**
   Orbital characteristics are determined by coverage and service requirements. The constraints and detailed optimization techniques are discussed in Chapter 2. Suffice to state here that space environment, transmission delay and constellation size play a crucial role in determining the altitude, whereas coverage area is influenced by orbital inclination and eccentricity.

8.3.3 System synthesis

System requirements, considerations and constraints are synthesized into a baseline design and preliminary cost estimates, resulting in the following system definitions:

1. space segment architecture;
2. top-level spacecraft specifications;
3. space segment capacity;
4. user terminal specifications;
5. approximate system cost;
6. an initial business plan;

Due to the large number of variables, interdependence and constraints, optimization benefits from a combination of approaches – heuristic decisions may be preferred in some instances, whereas other aspects may benefit from mathematical analysis and simulation such as that developed for planning cost-effective VSAT systems (Dutta and Rama, 1992). This particular model allows a planner to choose key physical and operational parameters and perform important sensitivity analyses such as profit/cost versus traffic volume, technical parameters versus traffic, profit/cost versus price, operating parameters versus price, profit versus demand, etc. The model captures trade-offs between technical variables; estimates capacity and operational parameters through the geometric programming optimization method; and includes physical and regulatory constraints as well as the influence of transmission costs on planning variables. Figures 8.4 and 8.5 respectively demonstrate sensitivity analysis of operating parameters versus traffic volume and percent profit versus change in demand for the system considered by the authors.

8.3.4 Technical trade-off analysis

The reader may note that the groundwork for system trade-offs has already been established in preceding chapters and in parts of this chapter, as summarized in Table 8.1.

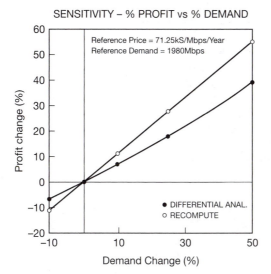

Figure 8.4
Sensitivity of operating
parameters to traffic
volume (Dutta and
Rama, 1992)
© 1992 IEEE

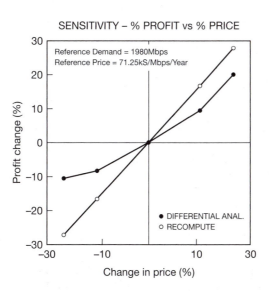

Figure 8.5
Profit versus price for the
system investigated
(Dutta and Rama, 1992)
© 1992 IEEE

Closely linked to the constellation design is service link optimization, which sets a bound on service throughput, as discussed above. Satellite altitude plays a vital role in specifying the G/T of the satellite. Assuming an EIRP of –3 dBW, typical of a hand-held system, a geostationary satellite system would require around 24 dB better G/T than a LEO satellite system under identical link conditions due to the difference in the path loss for these two cases – compare the G/T of the Iridium satellite (–3 to –10 dB/K) with that of the ACes geostationary satellite (18 dB/K).

System element	Chapter
Constellation size and capacity	2
Multiple access, modulation and coding	3
Link design	4
Satellite EIRP and G/T	4
Mobile EIRP and G/T	4
Network aspects	8
Business aspects	7

Provided that the EIRP and G/T of the satellite are adequate to support the service in terms of quality and capacity, then the most critical parameter in a link design is the fade margin. System capacity can be traded off against fade margin – a 3 dB drop in link margin can increase the capacity by a factor of two. This is a choice made by the system designers at the outset. For example, Iridium allowed a propagation loss of 15.7 dB whereas the ACes system uses a link margin of 6–10 dB. In practice, the link margin translates to user perception of service quality. Note that cooperative users are willing to learn to extract the best signal quality (e.g. by moving to a more favourable location while making a call), as evidenced in the cellular environment.

Constellations may be designed for diversity advantage (see Chapters 2 and 3). Systems such as ICO and Globalstar utilize diversity, while the Iridium system relied on a considerably high fade margin built into the link design without diversity. Diversity reduces the EIRP demands on satellite EIRP, at the expense of an increase in complexity of the network and a limited spectrum loss – except, arguably, when CDMA is used (see Chapter 4).

Modulation and coding and multiple access schemes have been discussed at length in Chapter 3. The system designers compromise between the spectral efficiency, noise resistance and cost-effectiveness of the various modulation and coding schemes. Similarly, multiple access schemes are chosen for the highest capacity and noise rejection and other features such as capability to support soft handover.

Assuming that the constellation has been optimized and the radio link developed, it is necessary to size the capacity required per satellite. For geostationary satellites, the problem is relatively straightforward. A simple but effective method is to estimate the user distribution and traffic originating per terminal; then the total traffic served by the satellite is the product of the number of users and usage per terminal. The fundamental problem for a new service is to estimate accurately the traffic distribution and usage per terminal!

It is essential that the radio resource of each satellite be utilized as flexibly as possible. Hence spacecraft power must be distributed dynamically between spot beams, and spectrum allocated as efficiently as possible (see Chapter 6).

Capacity estimation and radio resource management are more demanding for non-geostationary satellite systems because of the continually changing service area below each satellite. One approach is to size the capacity of each satellite for the densest service area, which then enables the use of only a single satellite design – resulting in economies of scale, flexibility in constellation deployment and in-orbit spacecraft redundancy needs.

Consider a global non-geostationary satellite system comprising p satellites and a uniform traffic distribution. If the total global traffic is E, then the capacity per satellite can be approximated as: T/p. In a real situation, traffic tends to concentrate along highways, ship routes, air corridors, cities, etc., with a diurnal variation, and hence a more refined traffic model is essential. In Chapter 7, a methodology used for estimating traffic distribution is presented, which may offer a more accurate estimate of geographic traffic dependence. Diurnal variation of traffic is given some attention in the later part of this chapter. The capacity required per satellite can be used to synthesize a baseline spacecraft design, given the required EIRP per channel and G/T derived from the link analysis.

We have observed that performance analysis and optimization require modelling of a number of system elements, their interdependence and estimation of performance statistics as a function of a variety of variables. The entities and scenarios modelled depend on the scope and intention of the system analysis, e.g. capacity evaluation, network performance such as time delay or handover statistics estimation, resource optimization, interference analysis, etc. System parameters to be modelled include diurnal traffic variation, traffic distribution, constellation dynamics, gateway distribution, intersatellite link connectivity, frequency reuse matrix, etc. Due to the inherent non-linearity and complexity of such a model and the need for repetitive analysis and flexibility, computer simulation is commonly used in performance evaluation. Figure 8.6 illustrates a flow chart of a simulation program used for estimating capacity, utilization, propagation delay statistics and length of PSTN lines of LEO or MEO constellations (Böttcher *et al*, 1994). The simulation was set up with the regional subscriber distribution shown in Table 8.2, but can be set up with any set of inputs, and calculations performed for successive instants of time until the necessary statistics have been estimated for different instants in time.

8.3.5 Impact of satellite altitude

In this section we will review the impact of altitude on system design, taking a LEO satellite system as a basis. The maximum revenue achievable from a LEO system depends on the system capacity, while maintaining the specified quality in terms of measures such as delay and bit error rate. System capacity is defined as the number of channels per satellite and the constellation in its entirety, and depends on available satellite EIRP, spectrum and frequency reuses. While available spectrum depends on regulatory considerations, satellite EIRP and frequency reuses are influenced by satellite altitude – required satellite power per channel increases with altitude and spectrum usage reduces given the same

Figure 8.6
A flow chart of a
simulation program for
estimating capacity,
utilization, propagation
delay statistics and
length of PSTN lines of
LEO or MEO
constellations (Böttcher
et al., 1994).
Reproduced by
permission of John Wiley
& Sons Ltd

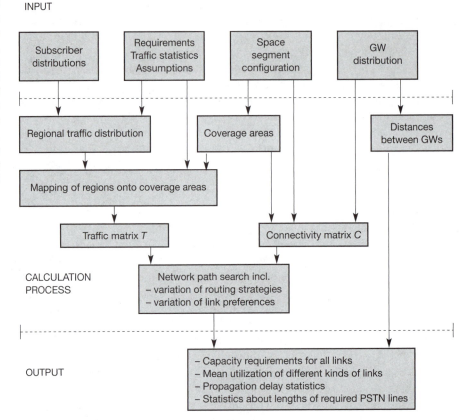

Table 8.2
Regional subscriber
distribution used in
simulation

Region	Percentage	Number of subscribers – land only (thousands)
North America	25	250
Europe	25	250
Asia	20	200
South America	10	100
Africa	10	100
Australia/New Zealand	10	100

number of spot beams, therefore it appears that a reduction in altitude can potentially provide increased capacity in addition to a reduction in delay.

Some of the system parameters of interest in determining the orbital altitude are:

- transmission delay;
- spectrum utilization;
- spacecraft power;
- user terminal EIRP.

Gavish and Kalvenes (1998) studied the influence of satellite altitude lying in range 500–7,500 km on these parameters, producing a number of interesting conclusions. As expected, the study demonstrated that low altitude increases frequency reusability, capacity and power. However, results pertaining to estimated transmission delays were not as obvious. Ground routing exhibited a shorter delay with an increase with altitude, as expected. When using intersatellite links, however, medium altitude within the range studied proved to offer the shortest delay because the number of switching nodes in low Earth orbit increased, which increased the switching delays, but it was noted that the delay depended on the type of traffic, i.e. local or long distance, and it was noted that, with evolution in technology, switching times should reduce. Furthermore, it was noted that frequency reuse increases as altitude is reduced; however, the limit in this case is the power capacity of the spacecraft. As expected, handset power increased with altitude, impacting the size and weight of the handset and the available throughput.

Figure 8.7(a) shows the number of frequency reuses possible over the Earth as a function of altitude for a number of spacecraft antenna beam sizes. Reuses at an altitude range from about 125 for spot beam size of 10° to about 20 for spot beams of 30°. Figure 8.7(b) shows the number of spot beams possible with various spot beam sizes at altitudes up to 5,000 km.

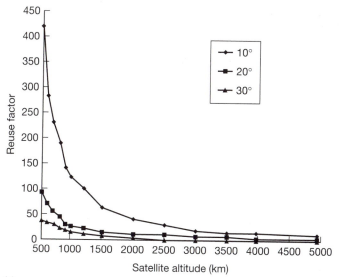

(a)

Figure 8.7

(a) Altitude versus possible frequency reuses for spot beam sizes of 10, 20 and 30° (Graphics: AR)

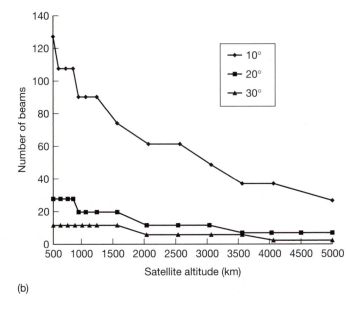

(b)

Figure 8.8(a) demonstrates the sensitivity of the solar panel size and battery mass with altitude, illustrating that the mass of the power system increases with altitude, even though capacity and eclipse period reduce; this occurs because capacity increases linearly but power increases as the square of the altitude, as portrayed in Figure 8.8 (b) which shows the increase in satellite power per channel with altitude.

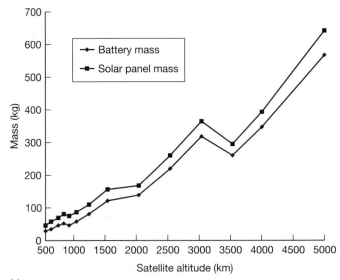

(a)

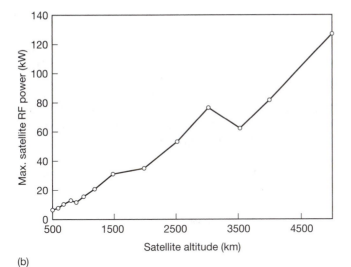

(b)

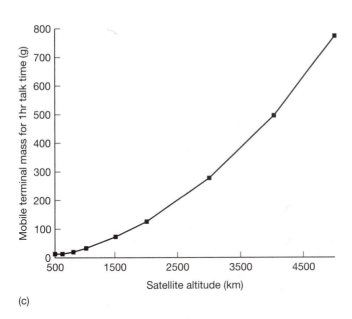

(c)

Figure 8.8
(b) Altitude versus satellite power requirement using the same assumptions as in (a) (Graphics: AR). (c) Altitude versus mobile terminal mass for 1 hour talk time for the system (Graphics: AR)

8.4 Network considerations

Today's mobile personal satellite communication systems have been developed with the objective of extending the terrestrial fixed and mobile services and providing supplementary services unique to satellite systems. They complement terrestrial mobile systems in areas where these do not exist or are unreliable.

Within such a network, there may be a need to create closed user groups or private networks by suitable arrangements such as leasing of space segment capacity. MSS networks are developed around these requirements, and because of differences in services and considerations discussed earlier, there are variations in network architecture.

Conventional network topology consists of mesh and star configurations. A *mesh network* provides full interconnectivity between users, whereas in a *star network* a large station communicates with a number of users who are not interconnected to each other through the central node. Earlier MSS systems tended to have a star topology because mobile-to-mobile links could not be sustained in power-limited satellites, but recent systems provide a direct mobile-to-mobile connectivity. MSS may consist of a *closed user group network* where the services are confined to operate within a group, such as a fleet of vehicles, or the system may be a part of the *public network*. Within a network, services may be provided as *circuit switched connections* such as used for voice, or *packet switched connections* for data networking. In a circuit switched network, an end-to-end call connection is set up before communication begins and the channel is released at the end of the call. A packet-mode network is connectionless and hence a user may begin transmission without a permanent connection to the destination; the network delivers the information on a packet basis to the recipient through a virtual path based on its routing strategy.

Traditionally, voice communication has been dominant in the public network and therefore all MSS networks have a circuit switched facility; due to the recent rapid growth in data traffic and the necessity to carry data economically, use of packet switched networks is increasing. Network configuration depends on the extent of integration between the satellite mobile and terrestrial cellular networks. There is little integration at network level for first and second generation MSS, but future systems will have commonalities. Dual-mode mobile terminals of existing MSS consequently operate with satellite and terrestrial networks separately, and are therefore integrated only in packaging and common hardware.

A satellite network comprises radio links for user mobility, a user mobility management system, a mobile switching centre, and an interface to the public network. Within the MSS itself there are two parts – a component for communication between the fixed parts of the MSS, such as between gateways, and the mobile component. The signalling for accessing the terrestrial network depends on the nature of the terrestrial service. Examples of terrestrial services, standards and protocols interfaced to mobile satellite services include telephony and facsimile X25, ISDN, IP, ATM – in each case, the satellite layer signalling is wrapped around those of the fixed network.

The signalling associated with the mobile component establishes a stable connection – real or virtual, and manages user mobility as well as radio resource and a reliable link, whereas signalling between the fixed parts of the MSS network establishes communication between the fixed elements for functions such as mobility management, billing, etc. Terrestrial systems such as GSM have chosen a well-established signalling method, signalling system number 7 (SS7), as the basis for specifying signalling between fixed elements of their mobile network, e.g. between the mobile switching centre and visitor location registers. SS7 is a CCITT specified signalling standard originally meant for signalling in the fixed terrestrial network between telephone exchanges, and is the basis for ISDN signalling. In the particular example of the GSM system, SS7 was modified to account for user mobility by adding features such as the mobile application part (MAP).

Note

The Open System Interconnect (OSI) model is a useful reference for understanding the functionality and architecture of a network. The model was proposed by the Internal Standards Organization (ISO) as a conceptual architecture for developing a telecommunications network. Figure 8.9 illustrates the model.

The model divides a network node into seven logical layers. Each layer performs a distinct function and communicates vertically. A layer offers service to its next higher layer, which can request specific parameters during the negotia-

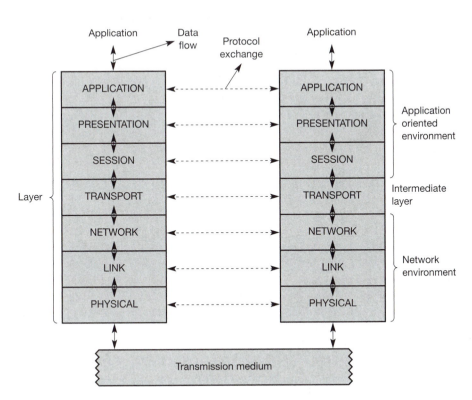

Figure 8.9
OSI seven-layer model

tion phase. During transmission, a message from a higher layer is encapsulated by the lower layer, with its descriptive header containing information for its peer layer on another node, as shown in Figure 8.9. The process is continued until the message reaches the physical layer, which transmits the signal through a communication medium, such as a satellite link, to the next node. When in communication with another node, each layer communicates with its peer using a peer-to-peer protocol, as illustrated in Figure 8.9. Table 8.3 summarizes the main features of the OSI hierarchy.

Table 8.3

Layers and functions of the OSI-7 network model

Layer number	Layer name	Function
7	Application	Provides application protocol; defines interaction between user and communication system; example of services provided: e-mail, facsimile, file transfer.
6	Presentation	Manages data description and data structure syntax, thereby converting abstract data syntax for applications to transport data syntax for transporting over the network and vice versa; example: conversion between different data sets, encryption, data compression.
5	Session	Connects application process which, for example, includes synchronization to enable restart in case of failure, remote log-in, etc.
4	Transport	Interfaces physical and logical parts of the network; provides transparent transfer of information between two entities in five service classes depending on application and network quality.
3	Network	Network supervision and flow control; e.g. monitoring of users and their addresses for maintaining routing table; ordering of received packets in the correct sequence.
2	Link	Detection of transmission errors in frames sent by physical layer; guarantees required error threshold.
1	Physical	Physical transmission; characteristics include transmission channel, type of channel usage, such as duplex.

Other standards used in distributed systems and computer communications are TCP (Transmission Control Protocol)/IP (Internet Protocol), now widely used for the internet; and the X and I series of standards for the telephone network laid down by the ITU. TCP/IP was developed before the OSI model and differs from it in a number of ways. It does not define the data link and physical layer and only partly the network layer; the upper three layers of OSI are compressed into a single application layer. The transport layer of the protocol supports two alternative protocols – TCP and User Datagram Protocol (UDP) – and the layer below it, known as the internet layer, is implemented by the Internet Protocol (IP).

A network architecture based on the OSI model is well suited for defining and standardizing a mobile system. An agreed standard allows manufacturers, network operators and service providers to develop products independently, yet allows them to be integrated into any operating system. Consider the example of the terrestrial GSM system. Its network standards/interfaces are used by several operators in various parts of the world, allowing subscribers to use their phone in partner networks. Due to wide acceptability of standards, many MSS operators, such as Iridium, Globalstar and ICO, adopted a similar network model, simplify-

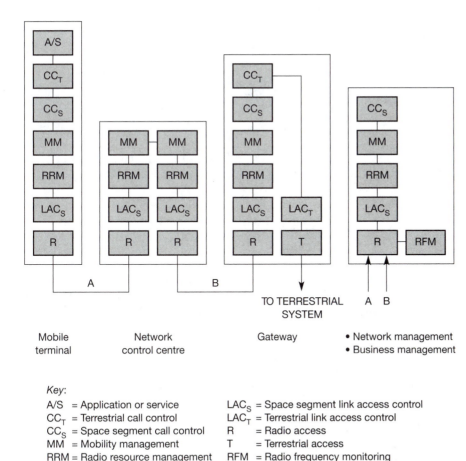

Figure 8.10
Network model of a generic MSS system in an OSI hierarchy

Mobile terminal   Network control centre   Gateway   • Network management
                                                                            • Business management

Key:
A/S = Application or service LAC_S = Space segment link access control
CC_T = Terrestrial call control LAC_T = Terrestrial link access control
CC_S = Space segment call control   R = Radio access
MM = Mobility management   T = Terrestrial access
RRM = Radio resource management   RFM = Radio frequency monitoring

ing integration of their system with terrestrial GSM systems. Figure 8.10 shows a generic network model of an MSS system in an OSI hierarchy.

At present, there is no equivalent MSS standard, as most operators have built their systems independently; but international standardization effort continues even though it has proved rather elusive (see Chapter 11).

The satellite networks must provide functions such as:

- call handling;
- switching;
- mobility management;
- radio resource management;
- network management;
- privacy, security management.

Figure 8.11 depicts the main functional entities of a commercial MSS, segmented broadly by their physical association.

Figure 8.11
Main functional entities of a commercial MSS, segmented broadly by their physical association

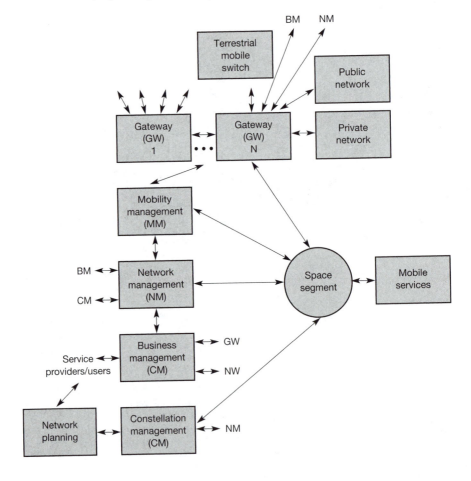

The *space segment*, comprising one or more satellites, provides the desired connectivity between the fixed and mobile segments. Some of the functions shown belonging to the terrestrial part can be either wholly or partially performed by the space segment. When regenerative transponders are used, some of the network management functions, such as call routing, may be transferred to the space segment. *Constellation management* involves standard telemetry and telecommand (TT&C) functions for satellite health monitoring, ephemeris generation, spacecraft orbit raising, orbital adjustment in the case of a spacecraft failure, launch support during initial deployment or replacement of failed spacecraft, etc. The *network management* (NM) function deals with real-time radio resource management, monitoring the radio spectrum, signal quality, network traffic flow, collection and dispatch of call data records to the business management system, interaction with the mobility management system for call set-up, etc., network traffic trend analysis to assist radio resource management, fault finding/diagnosis and fraud detection. NM functions provide call data records, user profiles and other user-related information to the business management system. The mobility management (MM) system maintains the user location in its database and interacts with the NM and gateways for call connection and user authentication, user profile, etc. In a GSM-based network, the MM involves the visitor location register (VLR) and home location register (HLR), with associated protocols for data exchange. All gateways of a network may either belong to the network provider, such as in the ICO system, or be owned by individual operators, as in the Globalstar system. The *business management system* constitutes a company's business centre and is responsible for customer billing for space segment usage, interfacing with the gateways or the network management system to obtain call records, updating user profiles of existing subscribers and introducing new subscribers. *Mobile services* represent various services on offer to mobile users, including support of associated protocols. The current portfolio of services include voice, data, facsimile, paging, message delivery and emergency calls; supplementary services such as call transfer, call forwarding, call waiting, call hold, conference calls, etc. The *network planning centre* oversees network management, capacity and business trends and specific events of interest, and develops strategies for changes to the network, such as expansion of capacity, introduction of new services, redeployment of satellites in the network, etc.

Prevalent MSS systems differ vastly in services, space and ground segment architecture, business arrangement, etc. Figure 8.12 portrays one possible network architecture to illustrate the interaction between the network, entities mentioned above. Differences may lie in the techniques used for mobility and resource management. For example, radio resource management may be distributed, contrary to the centralized scheme illustrated in the figure.

The application layer of the mobile terminal provides the desired service to the user; this may be a service through a public or a private network – consider PSTN as an example. The service is set up using a standard call control protocol with the given PSTN service at the application level. This, for example, involves sending a

Figure 8.12

A typical MSS network architecture

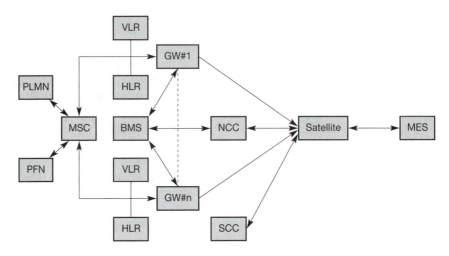

key:
VLR = Visitor location register   HLR = Home location register
GW#n = nth gateway   PLMN = Public land mobile network
PFN = Public fixed network   MSC = Mobile switching centre
BMS = Business management system   NCC = Network control centre
SCC = Satellite control centre   MES = Mobile Earth station

ringing tone to the user phone. Prior to this event, a radio frequency path and routing should be established via the space segment's call control procedures.

The mobility management functionality involves locating the position of the called mobile and establishing a route to maintain the call. In a packet switched network, a route is established for each packet – example protocols for establishing the route include centralized, distributed and flooding schemes. In a GSM-type architecture, this involves interrogation of the home and visitor location registers to determine the location of the mobile, authentication and user service authorization.

In a centralized routing strategy, a database comprising user and spacecraft locations is maintained centrally, which is interrogated on a call-by-call basis by entities setting up the call for establishing a suitable route. The scheme is susceptible to single point failure and incorporates considerable signalling overheads. In a distributed routing strategy, the database is distributed, for example, to gateways or satellites, and thus the scheme is more resistant to failure. Each node can establish a route on its own. In both schemes, mobiles transmit their location on a regular basis. In the flooding scheme, packets are sent to all visible nodes; the message is accepted if a node recognizes the address, otherwise it retransmits the message; the process continues until the packet is received at the destination. The scheme is simple and robust as there is no need to maintain a database, but it is wasteful of network resources.

Radio resource management (RRM) involves maximizing the use of available spectrum and spacecraft power, ensuring that all practical constraints are met; inter and intrasystem interference are kept within acceptable limits, space-craft/mobile hardware limitations are respected, etc. This layer also accounts for dynamic power control and diversity management of the radio link to ensure that the link offers acceptable fidelity.

8.4.1 Mobility management

A major goal of MSS systems is universal, unrestricted personal mobility. This network feature demands continuity of service in the short term during calls, and in the long term involves large geographical mobility, which may necessitate migration across networks on a single number or address. Encapsulated within these lie concepts of wireless access, terminal mobility, personal mobility and management of user service profiles.

A variety of wireless systems exist, each best matched to a specific environment; therefore, a universal system cannot be based on satellite system(s) alone. This leads us to the concept of a universal personal communication network, a topic which has received considerable attention from the international community in the past decade. This section focuses on existing and evolving satellite communications networks; the concept of a unified system is addressed in Chapter 11.

Mobility management of mobile systems requires a number of unique features, summarized as follows:

- **Network connectivity**, whereby calls are delivered to a mobile anywhere within the service area. The functionality is generally achieved through location registration, paging and call establishment. Location registration involves logging-on and regular location reporting by a terminal to facilitate call establishment. Paging involves transmitting a message to verify the current location of a mobile prior to the establishment of a call. In general, a call connection comprises terrestrial and satellite components. We will see in a following section that there are wide variations in routing strategies.
- **Roaming** is a network function which allows users to migrate to other networks or to a mobile switching centre (MSC) other than its home MSC.
- **Handover** involves handing live calls from one spot beam (or cell) to another or from one satellite to another. Handover may be necessary at a mobile or a gateway.

The architecture of a mobility management system for MSS depends on a number of factors:

- **Number of spot beams**: The complexity of a mobility management system increases as the number of spot beams increases.
- **Satellite motion**: Mobility management for LEO/MEO systems, where satellites themselves are moving, requires special features.
- **Multiple operators**: When gateways belong to different operators, the mobility management system must provide suitable provision for roaming between them.

- **Extent of integration with terrestrial system**: Depending on the extent of integration, architectural features to manage internetwork mobility have to be introduced.

Network connectivity and roaming

An MSS system connects with other networks through one or more gateways; the number of gateways depends on the space segment architecture, service area and terrestrial routing arrangements; and their location depends on proximity to necessary terrestrial traffic, economics, logistics, political considerations, operator's operational and business plan, etc.

Space segment connectivity is dictated by the characteristics of the service, orbital characteristics of the space segment and service area, i.e. regional, world-wide, etc. For interactive low delay services, each user requires continuous visibility of satellites, whereas delay-tolerant services can operate with intermittent connections. In a non-geostationary orbit, connectivity through the space segment changes with time and therefore appropriate network features are necessary to accommodate satellite movement; geostationary satellites have significant advantage in this respect. The extent of terrestrial routing is yet another consideration during the development of network topology.

Figures 1.8(a) and (b) (Chapter 1) portray two types of store and forward systems – satellite and Earth station based systems, respectively; while Figures 1.7 (a) and (b) represent architecture for carrying interactive traffic through terrestrial and intersatellite routing, respectively.

In a satellite store and forward system, a user transmits messages whenever a satellite appears within visibility. The message is stored in a satellite buffer and transmitted when the satellite arrives within visibility of the destination, which may be a gateway interfaced to a terrestrial network or another mobile terminal. System capacity in this case is CN bytes, where C is the storage capacity per satellite and N is the total number of satellites in the constellation. Throughput is therefore a trade-off between the storage capacity of a satellite, bound by space-qualified storage technology, and constellation size, which governs the message transfer delay and drives the cost of the system.

System throughput can be increased if store and forward functionality is transferred to a ground Earth station, where management of message storage and routing is easier and cost-effective. In this type of system, messages are transferred by satellites to ground Earth stations as soon as a gateway is visible; the gateway routes the message to its destination by the most efficient method, either through the space segment or terrestrially.

Interactive services exhibit low tolerance to delay and therefore near-continuous contact between end users becomes imperative, which essentially implies that both the gateway and the mobile terminal remain in continuous contact. The key difference between architectures lies in the way network routing is established.

In the simplest approach, the service area is confined to an area covered by the footprint of satellites visible from a single gateway. Each gateway is connected only to the local/regional fixed network; as the users are confined within a specific area, this arrangement is best suited for a regional system. When all gateways are interlinked and a mobility management system introduced, the coverage can be extended to the entire service area. A call from a user in the fixed network can be routed terrestrially (or through a satellite hop) to the gateway covering the mobile user; a route is derived from location information obtained through the home location register/visitor location register (HLR/VLR). Similarly, a mobile user can connect to a fixed party through the gateway to which it is connected. A suitable point for connection to the fixed party can be derived by the mobility management system through the called party's number. Dedicated inter-gateway links can offer lower delay and cost; routing through public networks, on the contrary, removes the burden of installing and maintaining a link but can increase delay and call costs disproportionately.

Figure 8.13 shows connectivity when the service area is confined to a single country. One or more gateways can be deployed, depending on the size of the country. The figure illustrates four gateways covering four areas of a country.

Figure 8.14 represents the interconnectivity applicable to a regional MSS deploying a single gateway which serves several countries. It is necessary to interface the gateway to an international switching centre. The terrestrial link can be quite lengthy for some participating countries. Commercial concerns may arise, as each section of terrestrial links may add costs as each participating entity draws its share of revenue; furthermore, there may be security concerns when calls are routed through a hostile country.

In passing, we note that the end user charges consist of a number of components (see Chapter 7 for details):

Figure 8.13
Routing in a single-country MSS network

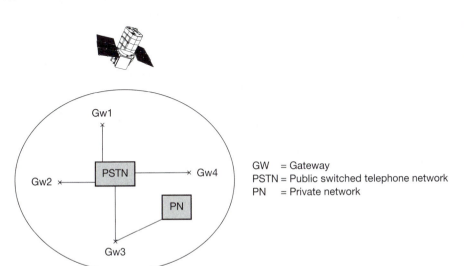

GW   = Gateway
PSTN = Public switched telephone network
PN   = Private network

Figure 8.14

Inter-connectivity in a
regional/global system

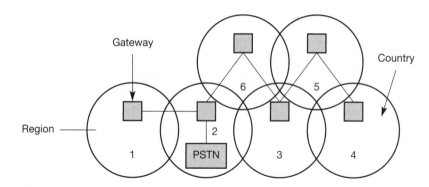

- space segment charge for utilizing spacecraft power and bandwidth, including profits;
- gateway charges, which may have a non-resident component;
- PSTN/PDN and, when applicable, cellular operator's charges;
- service provider's operating cost and profit;
- lease line charges, if used.

The extent of terrestrial routing can be reduced by deploying intersatellite links (ISL) as shown in Figure 1.7(b). In this type of architecture, a major part of the route can be supported on intersatellite links and therefore the system can operate with a relatively lower number of gateways and less dependence on terrestrial routing.

As the number of gateways is increased, a network becomes more resistant to single node failures due to path redundancy, and additionally, it is possible for traffic to flow more uniformly; however, an increase in the number of nodes adversely affects cost and complexity. In practice, the number and location of gateways is dependent on practical considerations, such as:

1. The position and number of gateways depends on whether the service area covers a country, a few countries (region) or the world.
2. The location of a gateway depends on its proximity to terrestrial traffic as well as logistics. The closer a gateway is to traffic centres, the lower the tail-end cost.
3. Political factors have considerable influence in deciding the location of gateways – for example, a local operator may be preferred.
4. Space segment capacity overheads.

Consider the routing arrangements of a few recent systems – Iridium, Globalstar and ICO. The Iridium system comprised 12 regional gateways, distributed throughout the world, which were shared by 15 operating companies who owned, operated and managed the gateways. The Iridium system had considerable flexibility in choosing the number and site of gateways, as it deployed intersatellite links. Theoretically, a single gateway can support this type of network, but this would lead to an increase in size of the terrestrial component, thus increasing terrestrial line costs.

The Globalstar system has been designed to support single country as well as regional gateways, each of which may or may not be interconnected. Initially about 50–60 gateways are planned, with gateways to be introduced as the system evolves. Globalstar uses a decentralized gateway architecture, giving the operators independence in operating their part of the network, which also allows service areas to expand as necessary.

The ICO system comprises 12 self-owned regional gateways, called satellite access nodes (SAN), interconnected to each other and interfaced with PSTN/PDN, and two network management centres. The SAN and the network management centres are connected through a dedicated high capacity optical-fibre link to minimize dependence on public networks. The Appendix gives a list of gateway locations for ICO and Iridium systems.

Call handling

Call establishment techniques, standardized in the GSM (global system for mobile communications) terrestrial system, have been emulated in many recent mobile satellite systems such as ICO, Iridium and ELLIPSO, thereby simplifying integration with partner GSM terrestrial systems and maximizing benefits through improvements to the standard.

In the GSM system, each mobile registered to the network is assigned a home area. User service profile, location, billing and other data are maintained in the home location register (HLR) in the home area of each mobile. When a mobile migrates outside its home mobile switching centre (MSC), it registers itself with the visited MSC. The mobile identification is entered in the visitor location register (VLR) of the visited MSC and the information is communicated to the HLR, which updates the mobile's location details. Whenever the mobile is called, the gateway where the call is placed interrogates the mobile's HLR for location and other information for authentication and establishing appropriate routing.

Typical call handling sequences for terrestrial and mobile-originated calls at mobile and network level are shown in Figures 8.15 (a) to 8.15 (d). Note that there may be variations in the specific details, but the process should essentially be similar. A GSM-type network architecture is assumed, as many recent satellite systems have modelled their architecture around it.

A call originated by a terrestrial party is received by the MSC; the MSC interrogates the HLR for service validation, authentication, location of the mobile and mobile status – busy or not (see Figure 8.15 (a)). Assuming that the mobile is authorized and available, the interrogation provides the location of the mobile, and if the mobile has roamed to another area, the address of the VLR which may be the address of a gateway. The radio resource manager sends a call announcement on the broadcast channel at the appropriate location of the network. If there is no response by the mobile within a specific time, the terrestrial party gets an 'unobtainable' message by return signalling message. When the mobile responds to the call announcement, the radio resource manager assigns resources to the gateway and the mobile. Following allocation, the mobile receives a notification such as a ring; when the device is off-hook, a handshake protocol ensures radio connectivity. The call then proceeds, after exchange of application-specific protocols. Any further exchange of signalling for functions such as power control, handover, end of call, etc. are conducted 'in-band' i.e. within the assigned channel. When the device is kept on-hook at the end of a call by either party, depending on which party has terminated the call, the gateway or the mobile informs the resource manager, and the resource manager then returns the radio resources to the common pool.

When a call originates at a mobile, a request is sent on a shared pre-assigned channel; the mobile's identity and service request are authenticated by the network by interrogation of the mobile's HLR, which maintains its service profile. After authentication, the radio resource manager proceeds to allocate the appropriate resources to a gateway which may, depending on the network arrangement, be identified by the mobile or the network on the basis of some criterion, e.g. closest gateway to the called party. Establishment of radio connectivity between the gateway and the mobile is confirmed by a simple technique such as a 'loop back'. Simultaneously, the gateway sends a call request to the public network, which initiates a ring at the called party's number using the network's standard procedures. When the called party has answered, applicable end–end protocols are exchanged, followed by communications.

Handover

Handover is the process of changing beams, satellite or gateway during a call. It may be necessary for a number of reasons – the communicating satellite may move below the specified elevation; a better satellite path may be available when diversity is used; the user may move out of a spot beam coverage; the network resource manager may decide to divert capacity to a less congested beam; satellites may move out of visibility of feeder stations; when intersatellite links are used, handover may occur between satellites. Due to a static link geometry, handover management is simpler for a geostationary satellite system than for a non- geostationary satellite system. To minimize or eliminate handover between spot beams, non-geostationary satellites may deploy a quasi-stationary spot beam system, wherein the spot beams of each satellite remain fixed to predefined points on the surface of the Earth until the satellite moves below visibility, at which time a new satellite replaces the area vacated by the old beam. This type of Earth-fixed arrangement simplifies frequency planning but requires rapid, accurate and synchronized steering of spot beams on satellites. Figure 8.16 shows various types of handover.

Beam–beam or satellite–satellite handover need not be applied to geostationary satellite systems deploying only a few spot beams, as coverage boundaries are fuzzy and it may take hours for slow-moving mobiles – ships, land vehicles, pedestrians, etc. – to move across satellites or beams. Often, aero-

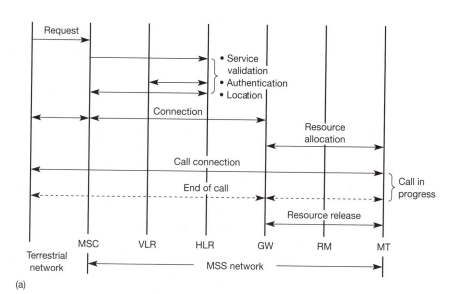

(a)

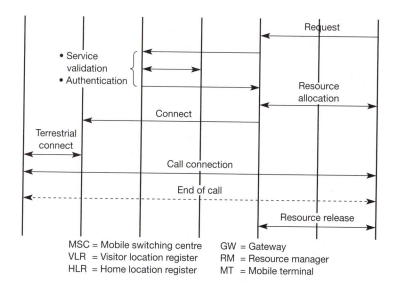

(b)

MSC = Mobile switching centre GW = Gateway
VLR = Visitor location register RM = Resource manager
HLR = Home location register MT = Mobile terminal

Figure 8.15

(a) and (b) Main network functions for (a) terrestrial originated call (b) mobile originated call (c) terrestrial originated call and (d) mobile originated call

Figure 8.15
Continued

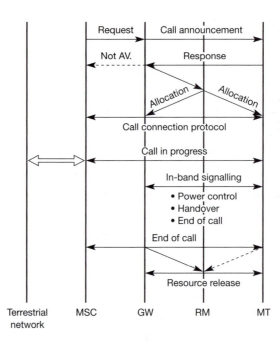

(c)

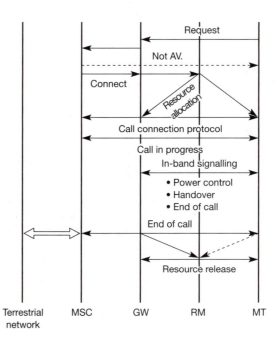

(d)

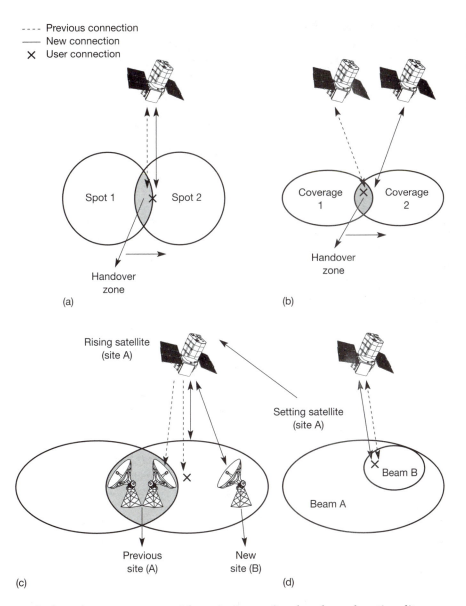

- - - - Previous connection
——— New connection
✕ User connection

Figure 8.16
Various types of
handover: (a) spot–spot
(b) satellite–satellite
(c) feeder link–feeder link
(d) congested beam-
non–congested beam

Spot 1   Spot 2

Handover
zone

(a)

Coverage
1   Coverage
2

Handover
zone

(b)

Rising satellite
(site A)

Setting satellite
(site A)

Beam B

Beam A

Previous
site (A)   New
site (B)

(c)   (d)

nautical services can operate without incorporating handover functionality – as,
for example, in Inmarsat third generation systems. Also, regional systems operating in a single beam do not require handover.

Mobility management examples
In this section, we will highlight some of the issues discussed in the preceding
sections with specific examples.

Geostationary satellite systems

Satellites of the MSAT system consist of six spot beams. Four beams cover contiguous territories – east, east/central, west central and western regions of the USA and Canada; the remaining two cover the states of Alaska and Hawaii, as well as southern reaches of the USA and Caribbean. A mobile can make and receive calls from anywhere within these regions.

Each mobile is assigned a home region and a beam at the time of initial registration. At the same time, information regarding other beams is introduced into the mobile's memory. When a mobile is switched on, it searches for the home broadcast channel, locks to it, extracts relevant network information, and is then ready to make or receive calls. If the home channel is absent, the mobile determines the identity of the broadcast channel being received; this information is transmitted to the network control station and thereafter this beam becomes the new home beam of the mobile until it moves to another beam.

A mobility management system has recently been introduced in the Inmarsat system to enable use of a single number in all regions. The system requires each mobile to indicate its location. The information is received and maintained in databases located within specific land Earth stations for call routing purposes.

The third generation Inmarsat system uses 4–5 spot beams per satellite. Therefore it becomes essential to locate a mobile for placing a call. Land and maritime systems use similar techniques, whereas the aeronautical system, which has a distributed architecture (i.e. each ground Earth station operates individually with minimal central control), uses a different scheme. In the land and maritime systems (Inmarsat B and M systems), a system bulletin board is transmitted in the global beam, from which mobiles read frequencies of spot beam identifiers. The frequencies of each broadcast channel are hard-coded into mobiles and hence are fixed, whereas changes to the spot beam identifying frequencies are possible through updates to the bulletin board. For fixed to mobile calls, the mobile is paged by the network control station (NCS) on the global broadcast channel to which all idle mobiles are tuned. On receiving a call announcement, the mobile responds with its spot beam identity, which it derives from the spot beam identifiers. The NCS then assigns a channel in the mobile's spot beam. The size of a spot beam is governed by the level of spot beam identifier. When a mobile makes a call, it requests a channel in its present spot beam. The NCS then proceeds to establish the call in the selected beam. If all the channels for the spot in the channel are busy, a channel is borrowed from the global beam, provided the service can be supported in the global beam. There are no beam–beam or satellite–satellite during-call handovers, as the spot beam roll-off is gradual.

The aeronautical system is based on an Earth-fixed spot beam system in which spot beam contours are broadcast on a global beam broadcast channel. An aeronautical mobile establishes its current spot beam by autonomously checking the beam within which it lies. The preferred spot beam is signalled to the communicating Earth station for call establishment. There is no beam–beam or satellite–satellite handover. A timer is used to drop a call if the spot beam crosses a spot beam boundary during a call.

Let us consider some techniques applied to more recent proposals. Many of these systems use hundreds of beams and others, in addition, also operate in non-stationary orbits.

Super-geostationary systems

An approach similar to the one discussed above for MTSAT has been proposed, with the difference that spot beam broadcast frequencies are reusable rather than unique to avoid the need for a large number of frequencies and consequent spectrum wastage (Johanson, 1995). This necessitates the mobile extracting spot beam/region information from the message rather than from the frequency. When a mobile migrates outside a beam, it sends an update of the new location to the control centre. In a system which serves a region partially, i.e. use of terminals is disallowed in certain countries, it becomes necessary for the mobile to send more precise location information, e.g. by using a GPS receiver. It would then be possible to disallow calls when the mobile migrates to forbidden territories. When this type of system comprises a number of gateways and/or operators, the GSM concept of VLR and HLR can be used for call management, provided that the operators incorporate a cooperative arrangement.

In a super-geostationary satellite, beam hand-offs are a necessity, in particular for high-speed mobiles. Similarly, handover is mandatory in non-geostationary satellite systems due to satellite motion. Beam–beam hand-offs should ensure no break in connection, which is more demanding in satellite systems than for their terrestrial counterpart due to the time delay involved. Soft handover is a technique which eliminates any potential break by using a combination of diversity and CDMA. The technique was pioneered in the Globalstar system (see next section).

Non-geostationary satellite systems

In non-geostationary systems, the mobility management system requires management of satellite movement in addition to that of the mobile. Satellite movement is predictable and therefore routing to any mobile can be established through the appropriate satellite with relatively straightforward calculations, provided satellite orbital parameters are available at each gateway. In a GSM-type architecture, the mobile periodically transmits its location over a passing satellite which relays it to the HLR through the gateway connected to the satellite at the time, or via an intersatellite link (ISL) if used. Beam and/or the satellite handover is similar to the cell–cell and base station–base station handover experienced in terrestrial cellular systems.

Beam–beam handover can be initiated either by the fixed Earth station or by a mobile. In many systems, the service links operate in different up and down link bands, e.g. L and S bands, which have different transmission characteristics.

As a consequence, signal monitoring at a ground station for the purpose of beam handover is unreliable due to differences in propagation behaviour. In these cases, mobile-initiated handover is preferred. One proposed solution is to use a dual receiver at the mobile (Johanson, 1995). While a call progresses through one receiver, the second receiver monitors other spot beams. When a better quality of signal is received from another beam, the mobile requests a beam handover from the gateway, which assigns radio resources to the new beam and transfers the call to the new beam in synchronism with the mobile transceiver. Satellite–satellite handover requires the mobile transceiver to compensate for differences in path loss and Doppler between the old and new radio connections, which is not necessary for beam–beam handover of the same satellite.

Let us consider examples of some non-geostationary satellites with regard to the handover methods. Although the Odyssey system has now been shelved after its merger with ICO, it pioneered the Earth-fixed frequency reuse technique. The service area is divided into predefined regions; satellite antennas utilize a steering antenna system which locks each of its spot beams to these predefined Earth-fixed contours until the elevation angle moves below a predefined threshold, when the Earth-fixed contours are handed over to a new satellite. This system had no provision for spot beam handover because satellites in medium Earth orbit have a long visibility of about two hours, during which spots remain fixed. The signal quality near the beams undergoes a graceful degradation, much like geo-stationary satellite systems, and hence inter-beam handover is not necessary – the size of each spot is of the order of 800 km diameter for this MEO system. The probability of a high number of calls remaining active at the instant of satellite handover is low because, during the coverage overlap of about 10 minutes between satellites, most existing calls, which typically last 2–3 minutes, terminate normally; new calls during this interval are established via the new spot beam. Furthermore, the network was capable of handing over calls between satellites, thereby offering the capability of re-establishing a call on a new satellite.

The Globalstar system incorporates a soft handover technique enabled by the system's CDMA scheme. Feeder-link handovers are not used. When a gateway receives signals transmitted by a mobile from two beams of either the same or an adjacent satellite, the gateway begins transmission of a time-shifted version of the code though the new beam. The mobile's rake receiver is able to track both signals, combining them to provide diversity advantage, and finally drops the signal from the old beam when the signal is unusable; following this handover, the receiver begins to search for signals from the next beam. In this technique, handover is not instantaneous or 'hard', as the receivers are able to exercise an autonomous decision to initiate the handover at the most appropriate instant. By contrast, in a 'hard' handover where the network commands a mobile to switch to a new gateway, there are occurrences of dropped calls and hence loss of information.

By eliminating the need for feeder-link station handover, Globalstar system coverage is limited to the area around the Earth station visibility arc – a cell of about 650 km. Therefore a honeycomb of Earth stations is necessary for global coverage.

An earlier proposal was to reduce the number of spot beam handovers by elongating the beam shape in the direction of satellite motion, though this would have been achieved through loss in antenna gain. The arrangement allows calls to be carried for longer, thereby minimizing the need for spot beam handover.

Finally, the Iridium system used all types of handover – spot beam, feeder link, and, by virtue of ISL with onboard processors, intersatellite handover.

Administrative and commercial network management

This administrative and commercial management system performs the following functions:

- subscriber management – commissioning of new subscribers, modification of existing user service profiles and billing;
- user equipment identity database management;
- user terminal status list management for checking fraudulent use;
- traffic statistics generation for network provisioning, planning, etc.;
- report generation, including incident, fault and network change reports.

Security management

Security measures are defined by each network operator as per their own operational philosophy. Example of security features are:

- maintaining subscriber confidentiality;
- subscriber or/and mobile authentication;
- data and signalling security.

Subscriber and gateway commissioning

Gateways

To avoid corrupting radio frequency transmissions and other network functions it is essential that, prior to its introduction into a network, the station is tested in all respects. This includes radio frequency compatibility tests and functional checks for the service to be supported by the Earth station. Each operator produces a detailed list of tests and methods used for testing. Many of these mandatory tests are witnessed by representatives of the operator and assessed for compliance. The Earth station is certified for introduction into the network when all tests have passed successfully.

Mobile Earth stations

To ensure that a mobile station meets the performance standards specified by the network operator, it is essential that exhaustive tests are performed. The network operator certifies a model for production after the terminal has undergone thorough testing in accordance with the operator's recommendations. Tests are designed to check that the terminal responds correctly to network control signals and protocols, its ability to log on, quality of voice facsimile and data transmission, distress/safety-related functions, satisfactory reception of system management messages, end-to-end connection with public/private networks, transmitted power compliance, regulatory requirements such as compliance with distress messages, etc.

When a new subscriber starts service, the mobile terminal is commissioned to ensure that proper routing, billing and accounting have been set up and that the terminal identification has been entered for the purpose of validation and billing. The commissioning procedures depend on the operator and the service. Aeronautical or maritime terminals required to comply with international safety standards would require a more rigorous commissioning procedure, whereas a satellite telephone for personal use has relatively simple requirements.

Note: End-to-end voice quality

To maintain end-to-end voice signal quality, the following considerations apply in the transmission link design of mobile systems:

- mobile Earth station speech codec performance;
- impact of cascaded echo control devices in a transmission path, as they can interact with each other;
- performance of the fixed link, which includes the fixed link of the mobile system as well as the connected fixed network.

Transmission parameters affecting end-to-end quality include propagation delay, speech level at transmitter and receiver, quantization distortion introduced by the speech codec, thermal noise, multipath noise, co-channel interference and group delay. It is necessary to apportion degradation amongst various components of the link, including other network and international connections. It is therefore necessary to follow an agreed planning standard, and if necessary, contractual agreement for guaranteed performance. In many countries, such as the UK, guidelines to facilitate link design and interconnect agreement between operators have been laid down by the operators' interest group.

Long propagation delays incurred in mobile networks require good echo control, because perceived degradation increases as propagation delay increases. Delay is added in voice codec processing, in channel coding and in transmissions. Mobile terminals usually operate under conditions of high ambient noise and therefore must exhibit good acoustic and echo characteristics.

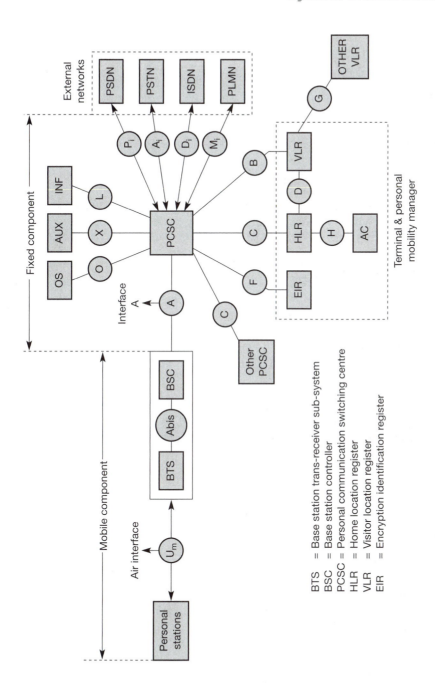

Figure 8.17
Reference model TR46
simplified (adapted
from Dean and
Estabrook, 1995)

BTS   = Base station trans-receiver sub-system
BSC   = Base station controller
PCSC = Personal communication switching centre
HLR   = Home location register
VLR   = Visitor location register
EIR   = Encryption identification register

Send loudness rating (SLR) and receive loudness rating (RLR), determined by acoustic transducers and the analogue–digital/digital–analogue converters, are specified to meet user loudness requirements under operational conditions. To meet a given LR requirement, acoustic loss between the mouth and microphone must be low, and good contact must be possible between the ear and the earpiece of the handset. The shape and size of the handset also affect echo level. Significant echo is caused by coupling of the received signal to the microphone. The permitted echo coupling within a network is specified by 'terminal coupling loss'; ITU-T G.131 (ITU 1987) specifies echo protection versus transmission delay of up to 300 ms. The smaller handsets preferred by users therefore cause difficulty in meeting the echo coupling loss specifications.

Echo cancellers connected in tandem tend to reduce the net advantage and therefore intelligent cancellers are preferred. In their interoperating arrangements, UK network operators specify a particular set of rules for disabling/enabling echo cancellers. The scheme has been implemented using CCITT No. 7 signalling system.

Fixed networks usually incorporate speech codecs at various points of their link, such as for rate adaptation, voice messaging and call forwarding, causing degradation to end-to-end quality. Quality can also be degraded by digital speech interpolation (DSI) systems along the transmission route. Tandeming an MSS service with a terrestrial mobile system may stretch the propagation delay to its limit. Introducing a GSM circuit to a geostationary MSS circuit would result in a delay of the order of 355 ms, which when added to non-linear speech coding, echo and link bit errors can degrade signal quality quite severely (ITU, 1989). Therefore some terrestrial systems prefer cable routes for GSM–GSM links rather than satellite routes. A careful assessment of degradation along the route is essential.

Some practical measures can be adopted to achieve high end-to-end voice quality – a good terminal type approval test ensures good voice quality/echo performance; apportioning of link quality with other operators, using measures such as a standard contract, can guarantee the specified quality; following sound end-to-end link design practices based on ITU-T recommendation, etc., ensures a reliable basis. Figure 8.17 depicts a possible network model.

8.5 Network management

An MSS network comprises a number of physical items and associated software which together perform various functions. The framework comprises the following logical elements:

- a physical layer, consisting of various pieces of equipment and their software; e.g. HPA in an Earth station;
- a transmission layer, comprising a group of physical equipment performing a network function; e.g. a gateway; and,
- a service layer which provides the desired circuit- or packet-based transport.

Management of a network, or telecommunications management network (TMN), has been formally defined in the CCITT recommendation M.30. This is a useful concept in defining the management function of an MSS network. The concept has been applied in the management of terrestrial mobile systems such as the GSM and in emerging MSS systems. The framework provides a structured approach for overall management of a network, which makes it easier for network operators, managers of specific elements and equipment manufacturers to communicate. The functions include administrative management, configuration management, network operations, maintenance, security management and evolution management. Figure 8.18 represents application of the concept to an MSS network.

The transmission network used for a management system is an independent telecommunications network and interfaces with various elements of the managed network, depending on specific monitoring need.

Network management is key to the success of an operational MSS system. Management is done at a global level by viewing a system in its entirety. Information flow is managed such that relevant information is available to the manager at the appropriate time and corrective actions are managed to maintain the desired quality. The main aims of the operational management for an MSS are to:

- guarantee a specified quality of service by direct or inferred end-to-end quality measurement;
- guarantee the specified reliability by taking adequate counter-measures in case of failure of network components;
- supervise various equipment, involving the collection of anomaly reports, their analysis and action for fault repairs;
- maximize network productivity, e.g. by allocating resource where and when necessary.

Network operations supervision is an extremely complex matter due to the huge number of parameters to be monitored. Most operators attempt to present the data in a comprehensible manner on a friendly interface with a powerful graphics capability. Some of the desirable features of a monitoring system are:

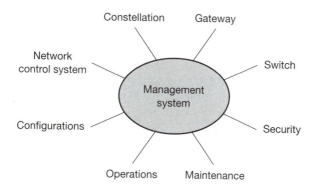

Figure 8.18

A telecommunications management network structure of an MSS

- centralized monitoring, in preference to a number of scattered facilities;
- display and monitoring of the network topology so that faults may be isolated easily; capability to display lower-level entities provides a quicker fault diagnosis;
- capability to provide statistical processing to indicate trends to facilitate fault analysis; statistical data also provides useful evaluation of network fault trends, variations of parameters, information for refinement in setting alarm limits, etc.

By maintaining the performance standards of specific parameters of the network, the network operator is able to guarantee the end-to-end quality. The parameters may be satellite EIRP variations with traffic load, frequency of each radio carrier, received carrier to noise ratio, network congestion, numbers of repeat requests in an ARQ scheme, message delivery time observed at various gateways, number of failed calls, holding time, etc. The MSS network provider often has no control over the quality of service in another network from where the call is originating or for which it is destined. Thus it is necessary to use periodic end-to-end call monitoring and keep a watch on user feedback which can provide an invaluable insight into network behaviour in a large network.

Configuration management involves sending network information to mobiles and gateways; typically these may include the transmission frequency of signalling channels, participating gateway station identification, etc. The management function accounts for subscriber creation, deletion, modification and billing; database management – mobile identity, including its status (authorized, fraudulent, etc.); traffic measurements for network provision, etc.; and event report management.

Measuring signal quality is vital, particularly during regions' busy hours when the network is fully loaded. During these periods, satellites are heavily loaded, resulting in increased inter-modulation noise, possible degradation in carrier EIRP, etc. Other quality measures are grade of service and end-to-end connection time. Grade of service is estimated at the busiest period. The congestion observed by a user can be caused either in the terrestrial network, in the gateway or in the space segment resource management system. The system has to be provisioned to ensure that congestion is managed at all sources.

8.5.1 Radio resource management

The revenue earning capability of an operator is bound by the operator's satellite resources, i.e. satellites' total EIRP and the spectrum available to an operator. Therefore it is in an operator's interest that spectrum utilization is maximized through efficient radio resource management. The key to spectrum efficiency lies in the way RF channels are arranged within a satellite transponder and in the network as a whole, respecting all practical constraints (Richharia, 1992). We refer to this process here as frequency planning. Another practical consideration is the management of transmitted power from each satellite, which is essentially limited by the technology available at the time of satellite manufacture. MSS systems require a large amount of power per channel for supporting personal services and hence the satellite radio frequency power must be managed efficiently for maximizing revenue from the limited resource.

8.5.1.1 Spectrum management

Spectrum management involves assigning RF carriers on each satellite and in the network as efficiently as possible, frequency list maintenance and spectrum assessment for future needs, as shown in Figure 8.19 (Richharia, 1992).

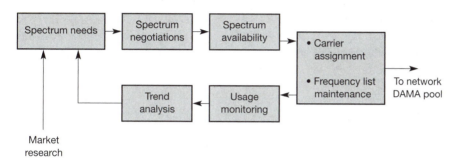

Figure 8.19
Main entities of an MSS frequency planning process

MSS operators assess their spectrum requirements through market research, and when historic data are available, through trend analysis, or a combination of these two processes. The spectrum requirement is presented to operators' coordination meetings or to the regulatory authority within the operators' jurisdiction. Depending on a number of factors, as outlined in Chapter 3, an operator may be granted access to the spectrum, though not necessarily to the entire demand.

Spectrum management then becomes more of an operational issue, requiring detailed planning wherein frequencies are assigned to signalling and traffic as efficiently as possible in response to real-time traffic needs, applying constraints, rules and optimization, and evolving into a frequency plan which subsequently must be maintained accurately. Frequency lists are disseminated to various network entities, such as network control stations, the network control centre, traffic monitoring systems, system analysts, etc. Thus suitable export interfaces are built to transfer frequency plans in a timely manner. Most of the operational tasks related to spectrum management are dynamic, due to rapidly changing requirements. A related task is monitoring spectrum usage and its trend analysis for planning.

The remainder of the section discusses the process of generating frequency plans – one of the more technically challenging parts of spectrum management. Figure 8.20 (a) to (c) (Richharia, 1992) represents the main processes used for managing spectrum in an operational system. Figure 8.20 (a) captures the process at the top level. Channel requirements are generated by trend analysis of recent traffic usage and events which may distort the trend. Traffic usage is monitored as call holding time, data volumes, and call failures due to congestion and other reasons. This data is analyzed for trends, resulting in a set of detailed requirements in terms of the number of channels for traffic carrying and signal carriers. The frequency assignment process generates a frequency list respecting the constraints and complying with the desired grade of service. Figure 8.20 (b) presents the frequency assignment process, which maps the requirements to a frequency list, taking available spectrum, reuse matrix and other rules as the basis. The mapping function, shown in Figure 8.20 (b), creates a frequency plan applicable at time t by taking into consideration the relative positions of satellites and spot beams, and applying constraints to ensure that all practical considerations, such as receiver tuning range, signalling carriers and location-specific interference, are satisfied. Carriers are finally laid based on applicable optimization rules. The process can repeat for time t_1, if necessary. Figure 8.20 (c) shows the real-time frequency assignment process per call or session.

It is also necessary to monitor the quality of service in terms of user satisfaction and the performance integrity of each channel. The outputs of such performance measures can be introduced into the frequency planning as specific rules. For example, the report could mention signal degradation to one or more specific channels, problems reported in specific parts of the service area, etc.

Depending on the chosen optimization method and criteria, different carrier arrangements are possible. In addition to inter-modulation noise, there are a number of other sources of degradation; these may arise from spurious or out-of-band emissions from other satellite operators, terrestrial interference within a specific region, intrasystem interference, etc. A system could become interference limited when a large number of spot beams are involved, much like their terrestrial cellular counterpart. The most complex frequency planning scenario is that of a non-geostationary satellite constellation comprising satellites with a large number of spot beams due to spatial variations of interference.

Several authors have used inter-modulation noise as an optimization criterion; channels are assigned in ascending order, from the best to worst channel in terms of inter-modulation noise. If the radio resource manager assigns carriers in an ascending order then degradation gets worse as loading increases, as early choice assignments have the better quality. As the system is not fully loaded at all times of day, the effect of inter-modulation noise perceived by users is low for most of the day. Voice-activated carriers provide a further 2–3 dB advantage in terms of inter-modulation noise. Such schemes do not consider interference as an optimization criterion, and furthermore, inter-modulation noise is usually a small component of the noise budget.

Several techniques have been investigated for frequency assignments in cellular radio, such as generalized graph colouring (Cozzens and Wang, 1984; Hale, 1980). Recently there has been some interest in applying heuristic optimization methods to frequency assignments (Holland, 1992, 1975; Goldberg, 1989, Kirkpatrick *et al.*, 1983; Castelino *et al.*, 1996). In particular, genetic algorithms based on Darwin's theory of evolution, simulated annealing derived from the concept of minimal entropy, and tabu search have been investigated for their

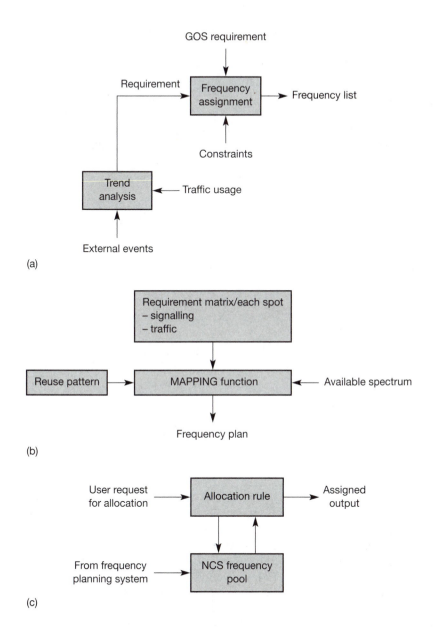

Figure 8.20
Processes used for
managing spectrum
resource: (a) overall
process (b) process for
generating frequency
list (c) real-time
channel allocation

(a)

(b)

(c)

applicability to mobile satellite communications (Jahn, 1999; Sammut, 1999; Pujante and Haro, 1997). Their work demonstrates the potential application of such schemes to operational satellite systems.

For geostationary satellite systems, antenna patterns remain static on the surface of the Earth and hence interference geometry is stationary, whereas in non-geostationary satellite systems interference geometry changes continuously.

From this perspective, frequency management is simpler for geostationary systems compared to the management of non-geostationary satellite systems. While, for geostationary satellite systems, frequency plan adjustments are mainly required in response to traffic variations, for non-geostationary satellite systems adjustments are necessary simultaneously to accommodate changes to traffic load and changing interference profile; the rate of adjustments to frequency plan is matched to traffic requirements. Figure 8.21 represents coverage at two instants t_0 and t_1 of a LEO satellite system, assuming satellite-fixed reusability as shown. Note that the reusability applicable at t_0 does not apply at instant t_1, necessitating changes to the frequency plan.

Figure 8.21
Satellite reference frequency plan demonstrating the need for frequency plan change. The coverage patterns of satellites in two orbital planes at two instants t_0 and t_1 are shown. Spot beams reusing frequencies are marked with indentical numbers. At instant t_1, co-channel frequencies in patterns 1, 3 and 4 will interfere, requiring changes to frequency plan

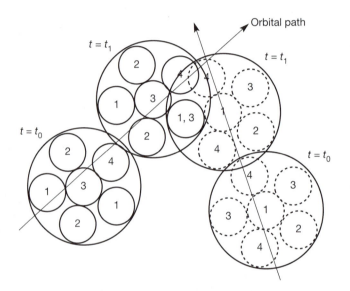

Traditionally, geostationary satellite systems use fixed assignment, i.e. frequency lists remain unchanged and radio resource managers manage real-time frequency assignment applying predefined rules. A fixed assigned approach is also used in first and second generation terrestrial cellular systems. The approach is suitable when traffic demands are well characterized, traffic is spread uniformly across spot beams, traffic variations across spot beams are low and the capacity requirements are easily manageable. Each spot beam is allocated channels to satisfy its peak demands, and frequency lists are updated to meet changes to traffic demands. These schemes may also be applied to non-geostationary satellite systems. Fixed assignment schemes become increasingly inefficient as traffic distributions become non-uniform and time dependent.

In a dynamic frequency assignment scheme, frequency lists are changed in near real time in response to demands, ensuring that interference is maintained within specification at any time. In a centrally managed system, frequency lists are prepared centrally and transferred to each radio resource manager. The rate of transfer depends on various system variables – traffic dynamics, real-time optimization capability, acceptable number of call handovers for reliable operation, etc. In a distributed architecture, frequency plans are prepared by gateways autonomously, with some form of feedback from other gateways or real-time measurements. The resource manager may be resident on the ground or in space.

Based on these possibilities, resource management may assume the following architecture:

- Frequency plans are prepared so each gateway can implement its plan without communicating with others. The architecture is well suited for geo-stationary satellite systems, can be used in MEO systems, but may become unmanageable for LEO satellite systems due to the rapidly changing interference and traffic scenario.
- Frequency plans are prepared so gateways may borrow channels from each other in real time, ensuring that they do not infringe interference limits.
- Frequency plans are prepared in each gateway autonomously through real-time interference measurements or feedback from other gateways or network management systems, etc.
- Spacecraft-based resource management: Frequency plans are prepared on the ground and downloaded regularly to the resource manager of each spacecraft.
- Spacecraft-based resource management: Satellites compute frequency plans autonomously based on rules/algorithms downloaded from the ground.
- Hybrid schemes: Fixed and dynamic allocations are combined to obtain optimum performance. The fixed pool is sized for a specific traffic threshold, with the dynamic pool serving traffic beyond the threshold and shared by contention or rules, such as mentioned below.

Many of these schemes have been studied extensively for terrestrial cellular systems and reviewed for their applicability to non-geostationary satellite systems. For example, consider the following schemes investigated by Sammut (1999).

- *Satellite-based decision* (Bjelajac, 1996): In this distributed scheme, assignments are made autonomously on satellites, based on carrier to interference ratio assessment on candidate channels and supplementing this information with measurements on other satellites. The scheme can be simplified by removing intersatellite links, as in the ORBCOM system. The possibility of user-initiated allocation, wherein the user sends its preferred channel in its assignment request through its own measurements, is also considered. An Iridium-type system is evaluated through a computer simulation.
- *Ground station based autonomous decision* (Finean, 1996–2): In this scheme, channels are managed at ground stations instead of on satellites. Channels are assigned on the basis of interference measurements on candidate

channels at the ground station as well as the called mobile. The scheme was evaluated for an Iridium-type LEO system. In this type of scheme, the number of attempts to find an unoccupied channel can become large in high load conditions, causing delay in call set-up and a reduction in handover reliability.

● *Cost function minimization* (Del Re *et al.*, 1994): In this scheme, satellites assign channels autonomously without any coordination with other satellites or gateways, through minimization of a grade of service cost function. The optimum channel at any instant is chosen to be the one that minimizes the number of spot beams where the channel will be unusable because of interference caused by it. As proposed, the scheme does not consider intersatellite interference and therefore requires refinement for practical purposes where the constellation comprises more than one satellite – though it should be recalled that regional geostationary satellite systems can operate with a single satellite.

One approach to simplifying frequency plan development for non-geostationary satellite systems is to use an Earth-fixed reuse pattern as illustrated in Figure 8.22.

In this scheme, spacecraft antenna beams are steered to fixed regions on Earth in synchronism, thereby emulating a geostationary satellite system. In addition to simplifying frequency plan development, the number of satellite

Figure 8.22
Earth-referenced reuse scheme. Blocks of spectrum are assigned to Earth-referenced cells within clusters. Satellites dwell on each cluster for as long as possible and switch to the next cluster in synchronism

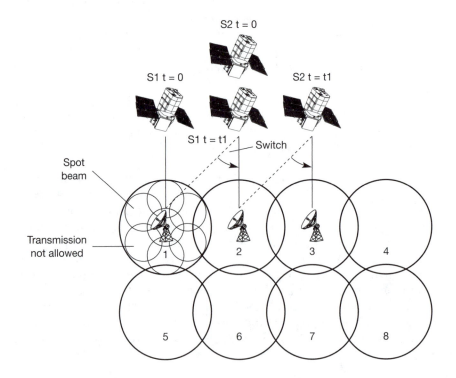

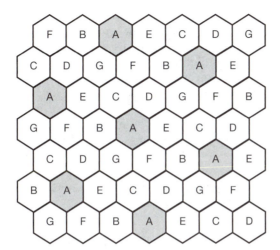

Figure 8.23
Seven-cell reuse pattern

and user terminal handovers are reduced. Figure 8.23 portrays an Earth-fixed cluster with seven-cell frequency reuse.

We will briefly review the channel management strategies of the Iridium, Globalstar and ICO systems – each system has an independent approach due to differences in constellation geometry and signal characteristics.

Iridium satellites consist of 48 spot beams per satellite (Figure A-3, Appendix A). This results in 2,150 active beams on the surface of the Earth. The system uses frequency division multiplexed TDMA with seven-cell reuse, enabling an effective frequency reuse 180 times around the world. The frequency allocation process is not available in open literature as it is the subject of a US patent filing. The constellation is polar and therefore satellites converge as they approach the pole, causing an increase in coverage overlap. The reuse pattern is maintained for the entire orbit by progressively switching off overlapping spot beams; this is possible due to the predictability of satellite paths. As an aside, note that this system intentionally avoids diversity in favour of spectrum efficiency.

The Globalstar system uses frequency division access with wideband code division multiple access (CDMA) and transparent transponders (see also Chapter 9). This scheme minimizes restrictions in frequency reuse because of the interference rejection property of spread spectrum modulation. A direct sequence frequency division spread spectrum CDMA scheme with a 1.23 MHz spread and QPSK modulation is used and the transmission band is spread over thirteen 1.23 MHz bands. Each gateway uses a unique code, allowing signal isolation between different ground stations, beams and satellites. To avoid their code overlapping in different beams, gateways apply a different time off-set to each beam within its view. In the return direction, mobiles use the same binary sequence as the forward channel but with user-dependent time-offsets to allow user identification.

The ICO system is a 10-satellite MEO system placed in two inclined planes (see Chapter 9). Details of the frequency management scheme are scarce in the literature. As far as is known, frequency plans are generated centrally, taking into consideration the orbital and spot beam characteristics, daily traffic variability on each satellite and other factors such as interference. Channels are assigned such that inter and intra-satellite reused channels have sufficient isolation. Spectrum is divided into non-overlapping blocks which are then divided between two orbital planes, thus avoiding spectral overlap between orbital planes; within a plane, each block is further subdivided into sets; within each satellite, sets are assigned so that they do not cause interference to each other; intersatellite interference is managed by ensuring that spectrum assigned to leading spot beams does not overlap with the set assigned to the trailing spot beams of the satellite in front, and similarly, the spectrum assigned to the trailing edge of the satellite does not overlap with the set assigned to the leading edge of the satellite behind.

8.5.1.2 Traffic analysis

Frequency lists used by the resource manager should match the traffic needs of the network to comply with the specified grade of service. In an MSS network, traffic varies in the short term over minutes or hours, with an underlying trend which manifests itself over weeks and months. These variations are modelled in different ways due to differences in their underlying mechanism.

In a circuit-mode network, traffic is measured in Erlangs, which can be converted to the number of circuits using a Poisson or Erlang model (see technical note in the next section). Estimates of the number of RF circuits for packet-mode transmission requires assessment of packet delivery time, throughput perceived by users, average load on the channel, etc.

Short-term traffic trend

It is well known that traffic carried over satellite networks varies diurnally, with a peak during business hours (Morgan and Gordon, 1989). When a service is offered in an area covered by several time zones, the diurnal profile of each time zone varies in accordance with the zone's business hours, hence the traffic peak over a satellite migrates from the most easterly time zone towards the west. Figure 8.24 (a) portrays a hypothetical scenario where a geostationary satellite is located at 60° E, and provides coverage over regions lying within a 5° elevation contour. Dominant traffic-generating regions, their peak traffic in Erlangs and time zones are also marked on the figure.

Figure 8.24 (b) represents a 24-hour traffic profile within a single time zone for typical weekday business traffic, assuming a normal working day between 9 a.m. and 5 p.m. with about an hour's lunch break. Traffic rises rapidly at the start of business; there is a dip in traffic load as people disperse for lunch over an hour, followed by another surge in traffic. The traffic subsides rapidly after the end of the business day. Weekday traffic for social needs is expected to rise gradually over the day, with a rapid increase between 6 and 10 p.m. and then tapering off. A weekend traffic profile would show a small amount of background business traffic whereas the social traffic would be higher.

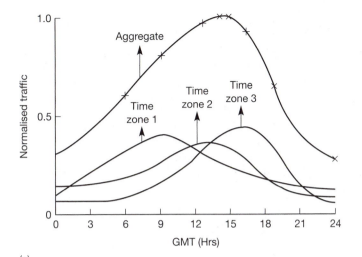

(a)

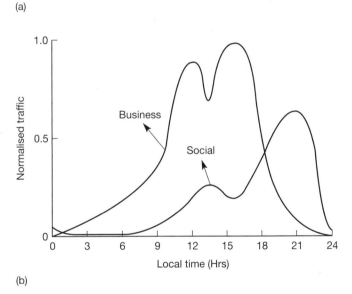

(b)

Figure 8.24
(a) A model used for obtaining diurnal variation in traffic load in a system servicing multiple time zones (b) Model of business and social traffic for generating diurnal traffic profile on satellite. The dip and rise in early afternoon can be caused by a lunch break

A refinement to the model would be to segment traffic by market sector, as shown in Table 8.4.

Business traffic for land, maritime and aeronautical services would continue to peak in business hours. Traffic for the aeronautical environment would, in addition, be influenced by aircraft route, flight time and duration; land traffic distribution has a strong relationship to cities, motorway routes and telecommunications infrastructure; and maritime systems' spatial distribution will have strong correlation to sea routes. This type of segmentation can potentially allow effective time-of-day-dependent radio resource management.

Table 8.4
Market segmentation
by sector

Sector	Variants
Land	Terminal type: Vehicle mounted, hand-held, paging, broadband, rural telephony, pay booth
	Traffic type: Social, business, tourism, sport
Maritime	Terminal type: Installed on cargo ships, yachts, cruise ships.
	Traffic type: Business, social, pleasure, sport
Aeronautical	Traffic Source: Air traffic control, cockpit communication, passenger communication
Distress and safety	Aeronautical, maritime or land

Carried traffic drops to low levels at night, during weekends and public holidays, when satellite capacity remains under-utilized. To encourage the use of unutilized transponder capacity, tariffs are lowered. Traffic spikes occur on special occasions such as Christmas or New Year, or when a notable event of public interest occurs for which extra space segment capacity may have to be installed.

In summary, traffic carried by a satellite is influenced by:

1. spatial distribution of traffic in each time zone;
2. magnitude of traffic in each time zone;
3. day of the week; i.e. a working day or a weekend;
4. time of day;
5. type of user base i.e. business, social, voice, data;
6. user environment – aeronautical, land, maritime.

Medium-term traffic trend

Knowledge of short-term traffic variation is essential for real-time radio resource and network management. For long-term planning covering months and years, the underlying trend is of greater interest.

Trend analysis through curve fitting is a simple and effective technique and is used quite commonly in telecommunication networks. Its accuracy can be improved by superimposing seasonal trends such as the effects of a holiday season, a traffic surge on New Year's day, knowledge of new commissioning, etc. Figure 8.25 illustrates the approach using a number of curves fitted to the same set of data. The error resulting from various estimates is also shown in each instance.

The main limitations of this approach are:

1. **Need for historic database**: Historic data are not always available. Data are non-existent when a new operator introduces a service or an existing operator launches a new product. In such cases, theoretical assumptions have to be made or trends estimated (see section 7.7 for a methodology).
2. **Limited accuracy in long-term prediction**: External influences are not included in this method; this assumption, while applicable for short-term fore-

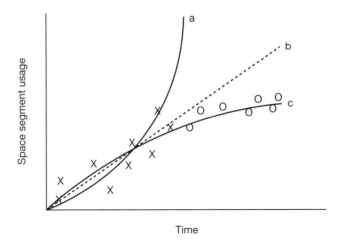

Figure 8.25
Trend analysis using
curve fitting. Various
types of curves have
been fitted to a
hypothetical set of data
and extrapolated for
forecast. Note the
sensitivity of selecting
various types of curve for
forecasting (x = historic
data; O = actual growth;
dashed line =
extrapolation)

casts, is unlikely to apply in the longer term to the mobile communications market due to considerable activity in the mobile communications arena as new and/or improved services and products are introduced regularly and competition increases. As mentioned earlier, refinements can be made by including external influences, such as loss of traffic to a competing product, etc.

Product life cycles follow the well-known S curve whose time-scales are influenced by a number of factors – evidently, the accuracy of long-term projections depends on how well a product life cycle has been modelled, as illustrated in Figure 8.26. The trend in the decline of analogue systems was, for example, much slower than anticipated – operators of analogue systems were able to retain customers by call cost reduction, aided by the availability of low-cost handsets enabled by mature technology.

Generally, the accuracy of simple trend analysis should be adequate for periods of a few months to a year, depending on the stage of the product cycle.

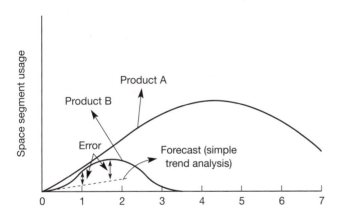

Figure 8.26
Life cycles of two
products – A and B. Note
the forecast error for
product B when a simple
linear extrapolation is
used. The model could
have been improved by
incorporating additional
knowledge. Note also the
differences in the life
cycle of the product

Forecast models should have adequate granularity for geostationary satellite systems as resources have to be planned for each beam or beam cluster when beam sizes are smaller, as represented in Figure 8.24 (a). Note that spot beams in regions within the same time zone would have identical time profiles but spatial distribution may vary; the largest time-offset in the profile will be experienced between beams which are longitudinally furthest apart.

In an operational environment, the grade of service of the space segment can degrade under unusual network conditions. Some possibilities include:

1. **Unusual events**: Events such as a natural disaster, war, an election, a sporting or social event, etc. cause unusual flow of traffic through an MSS space segment. This type of traffic surge is also experienced in terrestrial mobile systems in conditions such as a road traffic jam, etc.
2. **System malfunction**: System architectures using signal strength detection techniques for spot beam (or cell) identification are sensitive to variations in the signal strength of the cell identifier.

Figures 8.27 and 8.28 portray the scenario when the signal strength of the cell identifier in spot beam f either increases (case 1) or reduces (case 2) by, say, 1 dB due to a system malfunction. In case 1, traffic from adjacent beam is captured by beam f, as a number of mobile stations originally locked to other beams move to beam f as its signal level is now higher, causing an increase in traffic carried by the beam f (Figure 8.28). A traffic hot spot, such as marked in spot beam g (see Figure 8.27), can further increase the traffic captured by beam f relative to the scenario of uniformly distributed traffic, as depicted by the dotted line in Figure 8.28, while for case 2, the spot beam f shrinks in size, causing migration of its traffic to other beams and a corresponding increase in traffic carried by other beams – as demonstrated for beam f in Figure 8.28.

Figure 8.27

Spot beam size of beam f is changed by +1 dB (case 1) and –1 dB (case 2) due to change in level of spot beam identifier

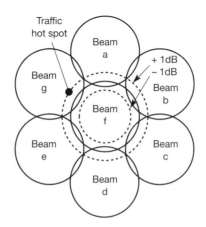

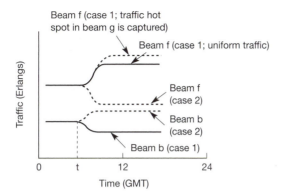

Figure 8.28
Change in traffic captured
by beams 'f' and 'b' for
cases shown in Figure
8.27 including a situation
where beam f captures a
traffic hot spot

Figure 8.29 depicts a scenario when the spot beam identifier of the central spot beam has failed. Traffic of the beam in common areas migrates to the other beam whereas the central dark area is left unserved as its traffic migrates to an umbrella 'global' beam.

The scenarios depicted above become time and spatially variant for non-geostationary satellite systems. Figure 8.30 represents a hypothetical traffic flow as a function of longitude on a low Earth equatorial orbit satellite for one pass applicable at the time shown.

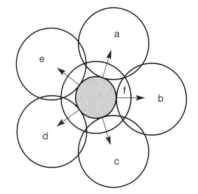

Figure 8.29 Migration of
traffic from beam f in the
case of failure of its cell
identifier to other beams.
The mobiles in the
central area will either
have no service or
migrate to the umbrella
cell if supported in the
system

When we consider systems based on Earth-fixed spot beams, spot beam identification is made by comparing the spot beam boundary with the position determined by each mobile. Here the impact and type of system malfunction differ – an anomaly occurs due to a malfunction in the vehicle's position determination system or a mismatch between real and assumed spot beams. The consequence in either case is signal quality degradation in the affected mobile, but the problem of traffic migration is absent. The operator can change the size of the spot beam by announcing the change to all mobiles.

Applicability of traffic forecast to non-geostationary satellite systems

Because of satellite motion relative to the Earth and the lower orbital altitude, the coverage pattern of satellites in a non-geostationary satellite system is time variant and illuminates a smaller area, implying lower traffic per satellite and a composite diurnal and spatial variation traffic profile through each satellite. A simple model is developed here to illustrate the concept (see Figure 8.30). The surface of the Earth is divided into 15° segments, which gives a difference of an hour between local time at the centre of lateral neighbouring cells. Time zones lying in business hours are shown for 12:00 GMT. Traffic is assumed to peak

Figure 8.30

Hypothetical traffic flow
as a function of longitude
on a low Earth equatorial
orbit satellite for a pass
for mid-afternoon in the
European region.

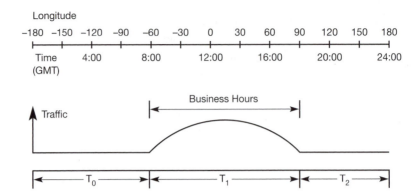

around midday, tapering on either side. Using this model, it is possible to estimate the traffic on each satellite of an equatorial constellation such as ECCO as it orbits the Earth. Traffic will be maximum for $T_1/(T_0+T_1+T_2)$ of the orbital period on each satellite. In reality, traffic distribution is likely to be non-uniform and constellations could use an inclined, polar or hybrid orbit.

Signalling channel requirements

The discussion above dealt with the derivation of requirements for traffic channels. An MSS network also requires a number of signalling channels to transfer network information to mobiles and allow mobiles to initiate and maintain communication. The number of signalling channels depends on the network topology. The numbers are higher for distributed architecture than for a centralized architecture. Typically, it is necessary to have a broadcast channel for each spot beam and one or more signalling circuits for signalling from mobiles. As signalling channels do not earn revenue and consume satellite resources, an attempt is made to minimize their number.

Note: Congestion theory

Congestion theory is a branch of traffic engineering dealing with a telecommunication system's ability to manage traffic offered to a network. Traffic in this context is defined as an aggregate of messages or calls which may pass through the network's shared facilities, such as trunks. Traffic intensity is measured as the product of the number of calls C and their average duration or holding time during a specified time. Thus if C calls pass through the circuit switched network in T hours, the traffic carried is CT Erlangs, i.e. if a call lasts for 1 hour, the traffic carried by the network is 1 Erlang. In the USA and Canada, traffic intensity is measured as hundreds of call-seconds or CCS per hour; thus CCS/hour = C.H, where C is the average call rate per hour and H the average call duration in hundreds of seconds. The measure of a call's duration is holding time, i.e. the time when a circuit is occupied by a call. Other definitions of holding time may be used by operators for calculating revenue, taking into consideration signalling overheads during call set-up.

A network is usually sized to offer an acceptable quality of service at the busiest hour of the network. There are a number of definitions of busiest-hour traffic; for example, CCITT recommendations Q80 and Q87 define it as the average of busy hour traffic for the five busiest days of the year. Grade of service is a measure of congestion in a network specified for the busiest period. Grade of service is defined as the probability that a call offered to the network fails to obtain a circuit at its first attempt in the busiest hour. A 2% grade of service implies that out of 100 call attempts made, two of them were blocked due to congestion. A 0% GOS is ideal but would require a large number of circuits and is therefore wasteful of network resources. Operators therefore allow some calls to be lost due to congestion. Mathematical statistics are used to estimate the number of circuits required to achieve a specified grade of service, and therefore their applicability to networks depends on the accuracy of the model compared to real traffic.

The Poisson model assumes that there are infinite sources, lost calls are held, calls arrive randomly and holding-time distribution is negative exponential. In practice, the condition of infinite sources is satisfied approximately when the ratio of number of sources to number of circuits is more than ~ 20. Lost calls held implies that when a call is not satisfied at the first attempt, it is held for a duration equal to the holding time, and if it is satisfied subsequently, the call is cleared at the end of the holding time; however, the call is considered lost in any case. In practice, it has been observed that the call holding time of telephone conversations follows a negative exponential distribution. The probability of exactly N circuits being busy for a mean offered load of P is given as

$$P = \frac{A^N}{N!} \, e^{-A} \tag{8.1}$$

where N = number of trunks and A = offered traffic (Erlangs).

The Erlang B model assumptions are identical to the Poisson model assumptions except that here lost calls are cleared. If a call is not satisfied at the first attempt, it is cleared and does not appear again during the busy hour. The probability of exactly N circuits being busy for a mean offered load of P is given as:

$$P = \frac{A^N / N!}{1 + A + A^2 / 2! + A^3 / 3! + \ldots + A^N / N!} \tag{8.2}$$

The Erlang C model assumptions are also identical to the Poisson model assumptions except that here the lost calls are delayed. If a call is not satisfied at the first attempt, the source continues to demand a circuit until a circuit becomes available, at which time the idle circuit is seized for the duration of the holding time. The probability of exactly N circuits being busy for a mean offered in this model is given as:

$$P = \frac{(A^N / N!)(N /(N - A))}{1 + A + A^2 / 2! + A^3 / 3! \ldots + (A^N / N!)(N /(N - A))} \tag{8.3}$$

Figure 8.31 compares the number of channels versus grade of service for various levels of traffic for these three models.

8.5.2 EIRP management

We have observed in an earlier section that EIRP demands on MSS service links are high to compensate for the low sensitivity of mobile receivers. This imposes a fundamental technology-dependent limit on space segment capacity – particularly for geostationary satellites – due to limitations of spacecraft antenna, power generation and high power amplifier technologies. A typical average forward link power requirement of an L-band geostationary satellite to support voice personal communications is of the order of 25 dBW, thus a 60 dBW spacecraft, assuming 10% EIRP overhead for signalling, etc., would carry around 2,850 carriers or 285,000 users, assuming a traffic intensity of 10 mill-Erlangs per user. Now consider the scenario when EIRP is not maintained within limits, resulting in an over-drive of 1 dB; for this example, this would cause a reduction in capacity by around 20%! In practice, spacecraft power is not hard limited and therefore the same capacity could be supported with a reduced signal quality caused by increased inter-modulation noise and carrier suppression – the extent of each depending on whether automatic level control is incorporated in the transponder.

Furthermore, a higher signal level has a direct influence on the amount of interference caused to other users, and a low signal level makes a carrier susceptible to interference.

Figure 8.31
A comparison of Poisson, Erlang B and Erlang C traffic models
(Graphics: AR)

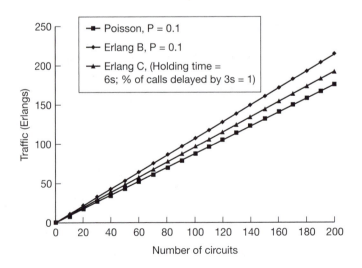

The return link satellite EIRP requirements are significantly lower – for the example considered above, depending on the ground Earth station size, the requirement would be of the order 20–40 dB lower. Transponder power stealing is an inherent characteristic of the MSS return link of non-regenerative transponders. Transmissions from users operating through adjacent satellites are received at relatively high levels, due to the low directivity of user terminals, and retransmitted by all satellites visible from the mobiles, particularly if there is some coverage overlap. This type of power loss can be minimized by switching off satellite channels lying within other operator bands, by deploying regenerative transponders which inherently reject unwanted signals, or, where possible, providing high antenna discrimination towards the footprint of other satellite(s).

From preceding discussions, we note that it is necessary to utilize, monitor and maintain EIRP efficiently to maximize spacecraft resource, maintain signal quality and minimize RF interference.

A number of techniques are incorporated in radio link design to maximize EIRP utilization. Dynamic power control ensures optimum satellite EIRP utilization to achieve the desired signal quality, though the latency caused by feedback can limit the advantage in applications intolerant to delay. The advantage gained through power control depends on signal fade distribution amongst users. If the majority of users operate under fading conditions, then a larger EIRP is drawn from the satellites; but the reverse is true when the majority of users operate in clear sky conditions. Voice (or data) activated carriers reduce average power, together with a reduction in average inter-modulation noise. Signals are transmitted only when information is present; the carrier remains switched off during the remaining time, thus reducing the average EIRP per carrier. Other widely used EIRP reduction techniques include coding, and store and forward communication, each of which essentially reduces EIRP by introducing redundancy to the signal to make it robust to noise.

EIRP variation over 24 hours, which can be derived from Figure 8.24, provides a measure of total spacecraft load as well as spare EIRP capacity versus time of day. On a longer-term view, the trend provides information regarding the satellite's EIRP saturation. EIRP management can be introduced in ground Earth stations by budgeting specific amounts of EIRP to each Earth station, according to its requirement. A level-monitoring feedback loop at the ground station can ensure that a ground station has enough budgeted power.

8.5.3 Radio frequency monitoring

To keep track of transmission levels and frequency, intra and intersystem interference, and unauthorized transmissions, and to identify faulty mobiles, satellite transmissions must be monitored continuously. RF emissions are readily monitored on spectrum analyzers at suitably sited monitoring stations; the most flexible method is through a computer-controlled measurement system which offers operators the capability to measure a variety of parameters through a user-friendly graphical user interface. Some of the most common measurements are:

- EIRP of a single carrier, a group of carriers or a full satellite transponder;
- carrier frequency;
- unauthorized transmissions.

A number of additional operational tasks are essential for maintaining proper RF operation in the network; these include calibration of reference carriers, assistance to Earth stations at the time of commissioning for measurement of their antenna radiation pattern, checking transmission stability and other RF-related assistance.

One of the problems in managing the RF integrity of a modern MSS network is the difficulty in monitoring forward link emissions of all spot beams, because of a need to access each spot beam. The scope of the problem depends on the number of spot beams and whether a geostationary or a non-geostationary system is being considered. Emissions in the return link are received at a few large Earth stations in one or a few spot beams and are thus relatively straightforward to monitor and synthesize on the ground. Unfortunately, EIRP monitoring and control are more critical in the forward direction.

Consider first a geostationary satellite system; when the number of spot beams is small, say < 10, it is possible to deploy measurement stations in each spot beam and retrieve measured data via a land or satellite link at the central site where signals are reconstructed and synthesized. However, this technique is not practically suited for super-geostationary systems deploying hundreds of beams. In such cases, the number of monitoring sites can be reduced by locating each site so as to maximize the number of spot beams visible from each location, as illustrated in Figure 8.32. In practice, such ideal locations may not always be available; for example, it would be extremely difficult to install monitoring stations within spot beams which illuminate oceans or inhospitable territories. It is however possible to conceive a scheme where monitoring information is transmitted from low cost ruggedized transreceiver to a central site.

Alternative monitoring schemes mentioned below can either be used to augment the approach outlined above or deployed on their own.

- Transmissions can be monitored at each fixed station through a directional coupler at the output of the transmitter and measurements transferred to the central site through a land or satellite link, where signals are processed and synthesized to obtain the RF profile of participating stations and the network. A simple mathematical transformation can provide a reasonably accurate estimate of satellite EIRP.
  This technique can provide pseudo-monitoring of satellite transmissions in the forward link and is adequate for policing transmissions from the network's participating ground stations but does not monitor real satellite emissions, thus missing satellite-generated or extraneous spurious transmissions and interference.
- Statistical monitoring of spot beams can be achieved in systems which use power control and position reporting by extracting from each active mobile's signalling message a measure of signal quality. By compiling this information statistically, it is then possible to build a statistical model of

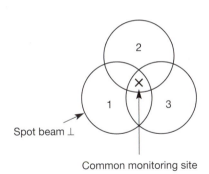

Figure 8.32
The concept of minimizing the number of monitoring sites by maximizing the number of simultaneously visible spot beams. In this example, three spot beams can be monitored from the same site

EIRP distribution over each spot beam as well as its size. Any subsequent deviation from the model would indicate a potential problem. Failure of a spot beam identifier or signalling channels would be noted by absence of signalling; an increase in spot beam signalling or identifier channel EIRP would show mobiles signalling from outside of spot beam limits, etc. Such a monitoring scheme is non-real time.

- In regenerative spacecraft architecture, it should be possible to measure the signal strength of carriers or groups of carriers and send the measured data to the ground on special carriers. Simple and coarse monitoring schemes could be based on detection of groups of carriers, monitoring of current through RF amplifiers and transmitting them over telemetry, etc.
- Optimal combination of monitoring information from various sources using schemes such as Kalman filter.

Measurement of total satellite EIRP over 24 hours requires periodic sampling of the entire spacecraft eirp in each beam and appropriate summation. when antenna gains differ between beams, summation of EIRP is not meaningful. In such cases, eirp is referenced to a suitable point, such as the input of the spacecraft antenna.

$$P_n = EIRP_n - G_n \tag{8.4}$$

$$P_t = \Sigma P_n$$

where P_n = EIRP of beam n at the reference point, $EIRP_n$ = total edge of coverage EIRP of carriers in spot beam n and G_n = edge of coverage gain of spot beam n. P_t = total satellite EIRP at the reference point.

Due to the presence of voice-activated demand assigned carriers, EIRP measurements must be averaged adequately.

Continuous monitoring of the emissions of non-geostationary satellite systems would require a number of monitoring stations dispersed throughout the world. This would seem rather expensive and an impractical solution.

Intermittent monitoring in such systems will therefore be more practical, wherein one or more monitoring stations are placed in each orbital plane. Monitoring at gateways, and statistical monitoring through information gathered from signalling, as discussed above, should offer a more acceptable solution. Off-line monitoring schemes using signalling data should also be applicable.

Radio frequency interference

MSS systems are susceptible to radio frequency interference (RFI) for the following reasons:

- Mobiles can move freely within the coverage area and hence are susceptible to local radio interference. MSS allocations are used by other satellite and terrestrial services such as FSS, radio relays, etc. in some regions as secondary allocations. In practice, it may be extremely difficult to enforce the ceasing of transmissions from such offending sources.
- Harmonics from local transmissions can cause interference or saturation of the mobile receiver front end. Saturation occurs because the front end of a mobile is sensitive and often wideband, allowing strong unwanted radio signals to pass through.
- Due to the large number of mobiles and possibly gateways, as well as the number of carriers transmitted from each gateway, the probability of interference from equipment malfunction is high.
- The probability of intrasystem interference is more likely because of the large number of intrasystem frequency reuses typical of modern MSS networks. If, for example, spot beam size increases due to a higher level of spot beam identifier, or a specific carrier EIRP exceeds specifications, then co-channel interference levels are certain to increase.
- The probability of intersystem interference is high due to spectrum sharing.

Radio frequency interference management

There are two elements to radio frequency interference (RFI) management – detection and elimination. RFI generally causes an increase in noise in affected channels, causing poor BER or signal to noise ratio. If interference emanates from similar channels of the network, it may cause an annoying intelligible cross-talk.

RFI can affect a single carrier or a group of carriers, it can be local or global. It is quite difficult to detect RFI to demand assigned carriers as RFI occurrence is confined to single calls, which typically last two minutes for circuit-mode transmissions and may not be noticeable in packet-mode transmission using ARQ or bandwidth on demand. Chances are that on obtaining a noisy channel, the user terminates the call and attempts another call, which is likely to be on a clean channel! Thus the impact of RFI on one or a few channels can remain undetected. RFI in such cases manifests itself in call statistics, as a reduction in holding time, an increase in call failures, etc. on the interfered RF channel. It is relatively easy to detect interference to signalling channels as it affects all or a significant number of users. Furthermore, the performance of signalling carriers in a system is closely monitored and any degradation in their performance can be detected quite rapidly. Local interference affects only a few users, is difficult to detect and may leave a user confused. An operator can only obtain such information by user feedback and careful assessment by maintaining a database of previous reports. In most cases, such reports can be traced to faulty user equipment and thus careful cataloguing of such information is vital. Interference occurring globally, for example due to spurious transmissions through a satellite, can be detected by regular monitoring of carriers. One method is to check each detected carrier against the frequency list, and if the detected carrier is absent from the frequency list, then the transmission is declared as interference. In general, the majority of interference is intrasystem. It can be expected that as the number of operators increases, instances of intersystem interference will increase.

Having discovered the presence of interference, it becomes necessary to eliminate the source. Intrasystem interference can be managed by each operator through internal procedures to eliminate known types and sources of interference.

Interference location techniques have been studied quite extensively. These may be based on triangulation, Doppler or other spectral signature detection techniques, such as those used in cellular systems for fraud detection. In an MSS network, locating the interference does not necessarily give an operator the means to eliminate the source, as the source may be in a remote or foreign hostile territory; a more tractable solution is to introduce features in signalling to force interfering sources to switch off on command.

8.5.4 Quality of service

In a general sense, quality of service (QOS) is a user perception of signal quality. Network managers require some quantifiable measures, some of which are as follows:

1. Grade of service (see technical note for definition): The end user perceives the grade of service which is the net effect of the terrestrial and space segments. Due to recent advances in terrestrial land lines, degradation to the grade of service of the terrestrial component is insignificant. However, the gateways do not always have enough capacity, especially if the traffic has built up very rapidly in a short period of a few months, or when unforeseen traffic surges occur. Economics do not warrant Earth station operators installing extra capacity which remains idle for most of the time. We have already discussed issues related to space segment capacity, and mentioned that it is governed by spectrum availability and the efficiency with which the spectrum is utilized by an operator.
2. Regular bit error rate, signal to noise ratio measurement, ARQ repeat assessment on test transmissions and randomly chosen traffic channels can provide a useful indicator of signal quality.
3. Carrier to noise power spectral density is another indicator of signal quality and can be readily incorporated into the RF monitoring equipment.

4. Customer feedback is a vital source of information, as it is extremely diffi-
    cult to monitor each location covered by MSS systems.
5. Statistical monitoring of call data records provides another measure of off-
    line quality assessment.
6. Packet data transmission defines quality of service in a number of ways,
    such as average throughput and delay.

8.6 Licensing issues

A vital requirement of a prospective operator is to obtain a licence from the regulatory authorities of the countries to which service is targeted. In most cases,
the MSS service areas extend to a number of countries and therefore operating
licence has to be obtained from various authorities, which makes this process
complex and time-consuming. Operators such as Inmarsat, ICO, Globalstar and
Iridium spent several years on this issue. The problem is fraught with political
and regulatory hurdles. Countries are often concerned about security and
potential loss of revenue to the local telecommunications operator.

Licences may often be awarded to more than one operator to encourage
competition; operators may belong to the country of jurisdiction or to another
country. Examples of resident companies for the USA would be Globalstar and
ORBCOM and of a foreign company, ICO. In such cases, the country may
favour local operators over foreign operators, or license both to encourage a
more competitive market environment.

Traditionally, the approach adopted by licence-granting authorities has
been to award licences on a first-come, first-served basis, following established
ITU procedures (see Chapter 3). However, due to increasing competition, some
regulatory authorities, in particular in the USA, are using other means to select
candidates and encourage efficient use of spectrum. In addition to obtaining a
licence to operate the space segment, MSS operators also have to obtain licences
to operate their gateways and mobile terminals in individual countries.

There may be a charge for granting licences to recover administrative costs.
Sometimes the costs are increased intentionally to encourage more efficient use
of spectrum, the rationale being that the operator will try to maximize revenue
by using spectrum more efficiently. When spectrum shortage is acute, the licensing costs can be increased to reduce the number of competitors. Authorities may
invite potential operators to participate in competitive bidding and select the
licensee on the basis of selection criteria such as technical excellence, funding
arrangement, level of risk in the proposal, etc. A licence may only be granted on
a conditional basis, such as proof of technical concept or funding. More
recently, spectrum auctions have been used for licensing terrestrial mobile systems. A major consideration when developing a charging policy for MSS
licences is that if each country charges an operator a licensing fee, the amounts
may become significant and prohibitive, as satellite operators usually provide
services to a large number of countries. Some of the possible approaches and
their merits and demerits are summarized in Table 8.5.

Type of licensing	Advantage	Disadvantage
First come, first served	Well proven	Favours early entrants
	Simple to implement	Does not encourage competition
		Accommodating late entrants becomes increasingly difficult
		Not well suited when there is acute spectrum shortage
		Spectrum usage not well controlled
Bidding	Possible to select the best applicants	Favours established operators
	Encourages efficient use of spectrum	Favours those with committed financial resources
	Encourages innovative technology	May increase the cost of the space segment, thus pricing some operators out of business
Auction	Offers best price when spectrum is scarce	High initial cost may lead to investment uncertainty
	Spectrum awarded to the operator willing to maximize spectrum usage	Negative impact on long-term growth as the operator may concentrate on recovering costs on profitable segments
		Favours existing operators, as newcomers need to establish an infrastructure in addition to funding the licence
		Increases cost to users
		Service may be offered only to profitable service areas, thus users requiring full coverage may be at a disadvantage
		Care necessary to keep spectrum fragmentation within acceptable bounds or operators may not be able to profit
Conditional [licence subject to conditions such as proof of concept, financial resources, etc.]	Safe approach, as spectrum is given after techniques/finance have been demonstrated	Operator may require up-front investments
	Spectrum not tied up	Risky if there is delay in the project, e.g. launch failure, financiers backing out

Table 8.5

Possible approaches for licensing, with their merits and demerits

A study conducted by the UMTS forum, using an economic business model for terrestrial systems, demonstrated that increasing the price beyond the administration cost is likely to have a negative impact on the development of future UMTS services (UMTS, 1998). The analysis for terrestrial UMTS indicated that profitability and pay-back periods will increase and become marked beyond a threshold.

Further reading

Ananasso, F. and Carosi, M. (1994) 'Architecture and networking issues in satellite systems for personal communications', *International Journal of Satellite Communications*, **12**, 33–44.

Bains, N. (1999) 'The ICO system for personal communications by satellite', *IMSC '99 Sixth International Mobile Satellite Conference*, Ottawa, co-sponsored by Communications Research Centre and the Jet Propulsion Laboratory, 88–93.

Bjelajac, B. (1996) 'CIR based dynamic channel allocation schemes and handover prioritisation for mobile satellite systems', *Vehicular Technology Conference*, Atlanta, Georgia, 28 April – 1 May, 313–18.

Bjelajac, B.; Cohen, B.; Fantacci, R.; Huggins, G.; Sammut, A. (1995) 'Dynamic channel allocation schemes for satellite-UMTS', *RACE Mobile Telecommunication Summit*, Cascais, Portugal, Vol II, November, 313–18.

Böttcher, A.; Jahn, A.; Lutz, E.; Werner, M. (1994) 'Analysis of basic system parameters of communication networks based on low earth orbit satellites', *International Journal of Satellite Communications*, 12, 85–93.

Castelino, D.J.; Hurley, S.; Stephens, N.M. (1996) 'A tabu search algorithm for frequency assignment', Ann. Oper. Res; 63, 223–250.

Castiel and Draim (1995) 'The Ellipso™ mobile satellite system', *International Mobile Satellite Conference*, Ottawa, IMSC 1995, The Fourth International Mobile Satellite Conference, Ottawa, co-sponsored by Communications Research Centre/Industry Canada and Jet Propulsion Laboratory/NASA, 409–18.

Cozzens, M.B.; Wang, D.I. (1984) 'The general channel assignment problem', Congr. Numer. 41, 115–129.

Dean, R.A. and Estabrook, P. (1995) 'Issues in PCS interoperability and Interworking', *IMSC '95, The Fourth International Mobile Satellite Conference*, Ottawa, co-sponsored by Communications Research Centre and the Jet Propulsion Laboratory/NASA, 497–502.

Del Re, E.; Fantacci, R.; Giambene, G. (1994) 'Performance analysis of a dynamic channel allocation technique for satellite mobile cellular networks', *International Journal of Satellite Communications (IJSC)*, 12, 25–32.

Dran, J.E. and Davidson, C.S. (1999) 'The ELLIPSO™ system – An optimal solution for the Canadian mobile satellite communications market', *Sixth*

International Mobile Satellite Conference, Ottawa, co-sponsored by Communications Research Centre and the Jet Propulsion Laboratory, 76–80.

Dutta, A. and Rama, D.V. (1992) 'An optimisation model of communications satellite planning', *IEEE Transactions on Communications*, **40** (9), September, 1463–73.

Finean, R. (1996) 'The role of satellites in UMTS', Ph-D Thesis, School of Electronic Engineering, Information Technology and Mathematics, University of Surrey, UK.

Freeman, R.L. (1996) *Reference Manual for Telecommunications Engineering*, John Wiley & Sons, New York.

Gavish, B. and Kalvenes, J. (1998) 'The impact of satellite altitude on the performance of LEOS based communication systems', *Wireless Networks*, No. 4, 199–213.

Goetz, I. (1996) 'Mobile network transmission quality', BT *Technol. J.*, **14** (3), July, 81–91.

Goldberg, D.E. (1989) *Genetic Algorithms in Search, Optimisation, and Machine Learning*, Addison-Wesley Publishing Company Inc.

Grandhi, S.A.; Yates R.D.; Goodman, D.J. (1997) 'Resource allocation for cellular radio systems', *IEEE Transactions on Vehicular Technology*, 46, 581–7.

Hale, W.K. (1980) 'Frequency assignment: theory and applications', Proceedings of IEEE, 68, 1497–1514.

Holland, J.H. (1975) *Adaption in Natural and Artificial Systems*, University of Michigan Press, Ann Arbor.

Holland, J.H. (1992) 'Genetic algorithms', *Scientific American*, 267, 66–72.

Hutcheson, J. and Laurin, M. (1995) 'Network flexibility of the Iridium[R] global mobile satellite system', *International Mobile Satellite Conference*, Ottawa – IMSC 1995, The Fourth International Mobile Satellite Conference, Ottawa, co-sponsored by Communications Research Centre/Industry Canada and Jet Propulsion Laboratory/NASA 503–7.

Inmarsat (1992) 'Project 21: The development of personal mobile satellite communication', June.

ITU (1987) *Handbook on Telephonometry*, Geneva.

ITU (1989) 'ITU-T G114', Transmission impairments, delay'.

Jahn, A. (1999) 'Stochastic optimisation of satellite frequency assignment', *IMSC '99, Sixth International Mobile Satellite Conference*, Ottawa, co-sponsored by Communications Research Centre and the Jet Propulsion Laboratory, 172–9.

Johanson, G.A. (1995) 'Mobility management in satellite networks', *IMSC '95, The Fourth International Mobile Satellite Conference*, Ottawa, co-sponsored by Communications Research Centre/Industry Canada and Jet Propulsion Laboratory/NASA, 29–34.

Kirkpatrick, S.; Gelatt, C.D.; Vecchi, M.P. (1983) 'Optimisation by simulated annealing', *Science*, 220, 671–80.

Morgan, W.L. and Gordon, G.D. (1989) *Communications Satellite Handbook*, John Wiley & Sons, New York.

Priscoli, F.D. and Muratore, F. (1996) 'Radio and network comparisons for MSBN and GSM systems in a geostationary satellite environment', *International Journal of Satellite Communications*, **14**, 439–54.

Pujante, A. and Haro, L. de (1997) 'Optimisation of satellite frequency plans with balanced carriers by simulated annealing', *Electronic Letters*, 33, 934–5.

Ramasastry, J. and Wiedeman, B. (1995) 'Use of access technology in mobile satellite systems', *IMSC '95, Fourth International Mobile Satellite Conference*, Ottawa, co-sponsored by Communications Research Centre and the Jet Propulsion Laboratory, 488–93.

Richharia, M. (1992) 'Satellite operations: management of satellite resources for mobile satellite services', University of Surrey, MSc seminar.

Sammut, A.J.R. (1999) 'Radio resource management for satellite personal communication networks', PhD Thesis, School of Electronic Engineering, Information Technology and Mathematics, University of Surrey, UK.

Schindall, J. (1995) 'Concept and implementation of the GLOBALSTAR mobile satellite system', *International Mobile Satellite Conference*, Ottawa, IMSC 1995, The Fourth International Mobile Satellite Conference, Ottawa, co-sponsored by Communications Research Centre/Industry Canada and Jet Propulsion Laboratory/NASA, A-11–A-16.

Schoen, D.C. and Locke, P.A. (1990) 'The ORBCOM data communication systems', *IMSC 1990*, The Second International Mobile Satellite Conference, Ottawa, co-sponsored by NASA/JPL and Communications Canada, 267–72.

Schoen, D.C. and Locke, P.A. (1995) 'ORBCOM – initial operations', *IMSC 1995*, *The Fourth International Mobile Satellite Conference*, Ottawa, co-sponsored by Communications Research Centre/Industry Canada and Jet Propulsion Laboratory/NASA, 397–400.

Singh, R.; Elnoubi, M.; Gupta, C. (1982) 'A new frequency channel assignment algorithm in high capacity mobile communication systems', *IEEE Trans. on Vehicular Technology*, VT-31.

Tisal, J. (1997) *GSM Cellular Radio Telephony*, John Wiley & Sons, New York, Section 7.3.

UMTS Forum (1998) *Cost impacts*, Report number 3.

Wiedeman, R.A. and Viterbi, A.J. (1993) 'The GLOBALSTAR mobile satellite system for world-wide personal communications', *IMSC 1993*, 16–18 June, Pasadena, California, JPL Publication 93-009, 291–6.

Representative MSS systems

9

9.1 Introduction

In this chapter, we illustrate practical applications of concepts discussed in previous chapters through examples of a few technically interesting MSS systems of recent years. The review is done only on the basis of the technical content of these proposals and their technical diversity, and has no bearing on commercial aspects. We will consider representative examples of GEO, MEO and LEO satellite systems. We have chosen Inmarsat and EUTELTRACS as geostationary satellite systems, ACes as a super-geostationary satellite system, the ICO system (not New ICO) as a representative MEO system, and the Iridium and Globalstar systems as examples of LEO satellite systems. In each example, the baseline design has been described rather than the evolved system because of a continuum of adjustments to business strategies and partners by companies in this dynamic market sector, leading in some cases even to changes to company names. The Iridium system has been selected as, despite its commercial failure, it remains one of the most technically advanced systems. Table 9.1 lists distinct features of each of these systems.

9.2 Representative examples

9.2.1 Geostationary satellite systems

Inmarsat system

The International Mobile Satellite Organization (Inmarsat) is an international mobile satellite service (MSS) provider, offering global communications and safety-related services to maritime, land and aeronautical mobiles throughout the world via a constellation of geostationary satellites. High latitude regions of the Earth out of view of geostationary satellites are not covered. The organization has recently been privatized, and is now known as Inmarsat Ventures Ltd. To ensure that the organization's public obligations related to safety and distress are satisfied, an affiliated organization has been created. Here we will not make any distinction and refer to them as Inmarsat.

Prior to the advent of satellite technology, communication with ships on the high seas was unreliable. Recognizing the advantages offered by satellite communications, the International Maritime Organization (IMO) initiated the formation of an organization called the International Maritime Satellite

System	Specific feature
Inmarsat (third generation)	First spot beam system with broadband (64 kbps) and desktop-size personal communicator; global service.
American Mobile Satellite Corporation (AMSC)	First generation regional geostationary satellite system offering medium bit rate communication.
EUTELTRACS	Regional, low bit rate geostationary satellite system.
Asia Cellular Satellite System (ACes)	A representative regional super-geostationary satellite system with 140 spot beams.
ICO Global Communication Systems (now new ICO)	A medium Earth satellite system for global hand-held service; uses diversity and transparent transponder; its evolved version supports broadband communication.
Globalstar	LEO satellite system for global hand-held communication; uses transparent transponder, inclined orbits, CDMA, diversity and soft handover.
Iridium	LEO satellite system for global hand-held service; uses regenerative transponder, intersatellite link and polar orbit.
Ellipsat	A hybrid orbit satellite system with an interesting constellation design for regional/global hand-held service; uses transparent transponder, allows gradual evolution of coverage and favours coverage of populated regions.
ORBCOM	Little-LEO low bit rate satellite communication system using simple satellites.

Organization for provision of safety and commercial public correspondence services to ships for peaceful purposes. Inmarsat was founded in 1979 and the maritime communications services were available in 1982, using satellites leased from a number of satellite operators. The success of the maritime services led to the introduction of services to other mobile environments, i.e. air and land.

The company continued to expand its service portfolio, user base and revenue throughout the past decade. It recently contracted the procurement of fourth generation satellites to support broadband communications to small portable sets and has diversified into VSAT and content sectors through acquisitions.

Technical features

Inmarsat's ground segment comprises a large number of fixed Earth stations, a wide variety of mobile Earth stations, and a space segment consisting of a constellation of transparent geostationary satellites deploying up to five spot beams per satellite and supported by older second generation satellites.

The satellites offer a seamless coverage of the Earth barring the sparsely populated upper latitudes, which are out of view of geostationary satellites. Each operational satellite has an in-orbit spare satellite ready to take over operation in the case of failure of the operational satellite. Capacity on spare satellites is leased, with the proviso that most of them, if not all, may be pre-empted when a spare satellite is pressed into service.

Gateways – known as coast Earth stations (CES) or land Earth stations (LES) by the maritime community, and ground Earth stations (GES) by the aeronautical community – serve as interfaces to the terrestrial public switched networks.

To maintain the integrity of the network, each LES has to comply with a specific set of RF requirements. Table 9.2 gives an example of some of the parameters; these specifications are revised by Inmarsat as necessary and the table should not be used for any serious design. Feeder links operate in the C band whereas service links operate in the L band. Each LES requires a number of verification tests before being authorized to operate within the network. Currently, there are around 40 LESs distributed around the world, owned by a number of operators.

Table 9. 2

Technical parameters of fixed Earth station

Parameter	C band	L band
Transmit pass band **Receive pass band**	6,417.5–6,454.0 MHz 3,599.0–3,629.0 MHz and 4,192.5–4,200 MHz (it was meant for first generation satellites)	1,626.5–1,660.5 MHz 1,525.0–1,559.0 MHz
Polarization: **Transmit** **Receive**	Right-hand circular (RHC) and left-hand circular (LHC) LHC and RHC	RHC RHC
Typical antenna gain: **Transmit** **Receive**	54.0 dBi 50.5 dBi	29.5 dBi 29.0 dBi
Ground station **EIRP/per carrier** **[Various types of carriers are transmitted]**	45–70 dBW	30–36 dBW
Receive G/T	30.7 dB/K (for operation with second and third generation satellites)	2.0 dB/K
EIRP stability **Frequency tolerance**	± 0. 7 dB ± 100 Hz	+ 1 to –2 dB ±200 Hz

Service	Approximate year of introduction	Communication Capability	Receiver pass band (MHZ) (Note 1)	MES G/T (dB/K)	MES EIRP nominal (dBW)	Antenna type	Applications
A	1982	Analogue telephony & voice band data, group call	1,535.0 –1,543.5	–4	36	Parabolic steerable, 0.8–1 m	Innumerable commercial, social, safety-related applications – both maritime & land
B	1993	Digital telephony & data transmission (16 kbps), fax (9.6 kbps), group call	1,525 –1,545	–4	33	Parabolic steerable, 0.8–1m	As above, with added advantages of digital technology
C	1990	Low bit rate store and forward communication (600 bps) with interfaces to X.400 and X.25 PSDN networks, group call	1,530 –1,545	–23		Omni-directional	International access to electronic mail, database access, global telex, X.400 & X.25 fax delivery
D and D+	1996–97	Low bit rate two-way data communications; tone (4 alerts), up to 32 characters numeric, 128 characters alphanumeric, longer messages sent as multiple messages; 2,000-bit transparent data per message; group calls, acknowledgement burst; short/long user data bursts up to 8 bytes	1,525 –1,559	–25 dB/k (nominal)	0–3 dBW	Omni-directional	Broadcast (e.g. financial data), vehicle tracking, personal messaging, supervisory control and data acquisition (SCADA)
M	1993	Digital telephony (4.8 kbps), Group 3 fax (2.4 kbps), group call	1,525 –1,559	–10 and –12	25–27	Steerable for vehicle mounting, parabolic or patch for fixed or portable use	Remote and rural fixed communications, communications from trunks, business travellers, police & emergency services

Table 9.3 Continued

Portable mini-M (Inmarsat Phone)	1996	Digital voice and data	1,525 –1,559	–17	15	Various types, patch, dish (rural tele-phony app-lications), gyro-stabilized mounted version, roof-mounted	Smallest terminals, ~ 2 kg weight, aeronautical version available ~ 4.5 kg; SIM facility; used by journalists, aid workers, business people, emergency services, etc.; rural telephony
M-Global area network and mobile packet radio network	1999	High-speed data (64 kbps)	1,525 –1,559	–7	28	Patch	Wide band applications
Aero-C	1992	As Inmarsat-C	1,530 –1,545	–23	11	Omni-directional	Messages up to 32,000 characters, non-safety-related communications on corporate, GA aircraft or helicopters
Aero-L	1990	Real-time duplex communications, 600 bps	1,530 –1,559	–26	13.5	Omni-directional	Real-time flight & passenger-related communication, e.g. engine monitoring
Aero-H	1990	Circuit & packet mode communication, (9.6 kbps)	1,530 –1,559	–13	10.5 –25.5	Steerable mechanical or phased arrays (12 dB gain)	Real-time and medium bit rate communication, e.g. voice, fax
Aero-I	1998	Cockpit/ passenger voice telephony and data in spot beams; packet-mode ISO 8208 data 600 bps– 4.8 kbps in global beam	1,530 –1,559	–19	19.5 (max)	Steerable mechanical or phased arrays (6 dB gain)	Passenger voice telephone, facsimile, PC data within Inmarsat-3 spot beam footprint, packet-mode data, cockpit voice and data, air traffic control, secure voice access to major air traffic control centres.
Inmarsat–E	1982	Distress and safety	1645.5 –1646.5	–23	11	Omni-directional antenna	Distress and safety – GMDSS compliant

Mobile Earth stations (MES) are categorized according to the environment in which they operate – aeronautical, land or maritime. Within each category, the mobiles are further subdivided into 'standards', according to the specific range of service provided.

Each MES manufacturer must obtain a one-off type approval from Inmarsat before marketing a specific type of terminal. Before introducing a terminal into the network, it is commissioned into the network by Inmarsat.

Salient features of some of the main services in current use are summarized in Table 9.3. However, as the products are evolving, the interested reader should visit their website for the current portfolio.

Inmarsat's primary role is to provide space segment capacity, charging each LES operator for space segment use. As of now, LESs are the service providers. They include their costs/profit on space segment charges when charging the end user.

Services

Services currently offered include A, Aero, B, C, M and paging – called Inmarsat-A, Inmarsat-Aero, etc, respectively.

Inmarsat also assists navigation services by way of distribution of differential correction for GPS through its point-to-multi-point data distribution services. Inmarsat's third generation satellites include a navigation transponder which can be used to enhance the accuracy, availability and integrity of GPS and GLONASS navigation signals. It is anticipated that the navigation signals transmitted at GPS frequency will be able to provide:

- global navigation satellite system integrity channel which will carry, amongst other useful navigation signals, the status of GPS and GLONASS satellites to ensure that data from faulty satellites are not used at the receiver;
- additional ranging signals to improve GPS availability;
- wide area differential correction signals which can be used by receivers to improve accuracy.

An integrated synergistic package of navigation and communication offers a vista of interesting and useful solutions, amongst which the application to aviation is notable. Accurate location information transmitted by each aircraft to the aircraft control centre allows a more effective use of trans-oceanic and other long-haul air corridors by permitting lower distance between aircraft, leading to large commercial benefits.

The next generation services will target broadband services up to 432 kbps to individuals, through affordable portable personal communicators, within the next few years, using leased capacity from Thuraya and subsequently via fourth generation satellites capable of transmitting several hundred spot beams. The network being developed to support broadband services incorporates features to integrate with terrestrial mobile networks.

Earlier, the Inmarsat-P programme conceived a space segment comprising an intermediate circular orbit (ICO) constellation to support hand-held global communication services. The implementation of the system was transferred to a spin-off company called ICO Global Communications (see section 9.2.4).

The following sections summarize the main features of each service (see the Inmarsat website for details and up-to-date information).

Maritime and land services

The Inmarsat-A service is its first generation analogue system supporting direct-dial telephone, telex, facsimile and data services. Terminals can also be equipped to support high-speed data (HSD) at 56–64 kbps. Although the system was originally developed for maritime applications, transportable versions of the terminals for use on land were in wide use. The system is being phased out in favour of more efficient digital services.

Inmarsat-B is a digital standard offering a wide range of services at a facsimile rate of 9.6 kbps and a data rate of up to 64 kbps, with terminal size similar to Inmarsat-A terminals, and is available in single and multiple channel versions. A digital modulation/coding technique enables more efficient use of the space segment. Inmarsat A, B and C services are a part of the Global Maritime Distress and Safety Service (GMDSS), an international safety and distress service under the jurisdiction of the IMO (see Chapter 10).

The Inmarsat-M service provides digital voice and 2.4 kbps data to briefcase-size terminals, heralding the era of personal communication. The service has now largely been replaced by a smaller version, known as Mini-M or Inmarsat-phone service, available through Inmarsat-3 spot beams.

Inmarsat-phone is a popular digital standard which provides direct-dial voice communication, 2.4 kbps data and facsimile for maritime and land users from anywhere within spot beam coverage of Inmarsat satellites. Currently, Inmarsat-phone terminals are about the size of a desktop telephone, weighing 2–3 kg. Terminal and user charges are considerably lower than both the A and B systems. Subscriber identity module (SIM) card capability gives subscribers the flexibility of using different terminals. Various versions of terminals are in use – maritime, land-portable, fixed, aeronautical, etc.

Global Area Network (GAN) is a broadband (currently, 64 kbps) circuit- and packet-mode service operating to small portable personal communicators about the size of a briefcase, and is the precursor of Inmarsat's next generation wide-band product, which is planned to offer throughput of over 400 kbps.

The Inmarsat-C system supports two-way, store and forward communication at a data rate of 600 bps for land and maritime users, with terminals weighing only a few kilograms. Inmarsat-C terminals use small non-tracking antennas, simplifying installation and maintenance. The service is suited for non-real-time short data messages such as position reporting or about a half-page of text (~ 250 characters). Around 90% of messages are delivered in less than 10 minutes. It is also possible to send telex, electronic mail and facsimile. The system can provide specialized communications – data reporting and polling. Data reporting allows short reports to be sent with lower delay and at a lower cost. Polling allows group message delivery either within a specific geo-graphical area or to a specific set. The system was introduced to complement Inmarsat-A, especially to support lower size ships and yachts. It is not necessary to be logged on to receive a message, as the message is saved in an electronic mailbox, from which the message is downloaded at log-on.

Inmarsat-E is a service for the provision of distress and safety communications at sea. Its terminal can be activated manually or automatically in the case of a distress emergency, following which the terminal transmits a distress message which continues to be relayed periodically for several hours, via fixed Earth stations, to a regional search and rescue centre.

Aeronautical services

A number of services are available for the aeronautical environment – at present, they include Aero-L, Aero-I, Aero-H, Aero-H+ and Aero-C.

Aero-L provides 600 bps *real-time* data communication using terminals with omni-directional antennas. The service complies with proposed International Civil Aviation Organization (ICAO) requirements to support future air navigation systems. The service is used mainly for air traffic control, operational and administrative purposes.

The Aero-H and H+ services provide real-time, 10.5 kbps data communications for flight-deck and airline operations using medium gain (10–12 dB) antenna terminals. The Aero-H service also offers flight deck voice and passenger services. The H+ service is an evolution of Aero-H which additionally provides a lower cost service within the spot beam coverage of Inmarsat-3.

The Aero-I service offers medium bit rate voice and data service through small, low-cost terminals within spot beam coverage of Inmarsat-3 satellites. Packet data services are available world-wide via the global beam.

Aero-C provides 600 bps store and forward data communications to aircraft and is an extension of the Inmarsat-C system. The service, excluding flight safety communications, is suitable for regional aircraft operators, in particular those operating in remote regions and aircraft operators in need of reliable service, but who do not envisage operating in the ICAO's proposed Air Traffic Management System.

Although Inmarsat's main role is to provide space segment capacity to commercial and safety traffic, spare space segment capacity is leased for specific applications, such as closed user group communication.

On a limited scale, and on a case-by-case basis, Inmarsat permits free use of space segment capacity to experimenters sponsored by appropriate authorities to encourage the development of mutually beneficial new and innovative applications in the area of mobile satellite services.

Space segment

In early 2001 the space segment comprises a constellation of nine geostationary satellites around the world. There are four main operational satellites, each backed by a spare satellite ready to take over operation in the case of failure of an operational satellite. Several spare satellites are kept separate from the main satellite in order to reuse the spectrum and offer the capacity generated this way for leasing. A network operations centre, located at Inmarsat headquarters in London, maintains the integrity of the network through continuous monitoring of satellite transmissions and traffic flow through the network. TT&C centres, placed around the world, can be managed centrally from its satellite control centre.

Satellites Inmarsat began operations by leasing capacity on various satellites – Communications Satellite Organization's MARISAT satellites, European Space Segment's MARECS satellite and International Telecommunications Satellite Organization's (INTELSAT) Maritime Communications System (MCS) packages on several INTELSAT series 5 satellites. The first generation satellites were phased out during 1991–2, and replaced by second generation satellites to meet the increased traffic demand, at least up to around the mid-1990s. Beginning in 1995, the second generation satellites have been phased out and replaced by third generation (Inmarsat-3) satellites which can deploy up to seven spot beams. Tables 9.4–9.5 summarize salient characteristics of Inmarsat satellites.

Generation	Identification (generic)	Frequency band	Frequency (MHz) and polarization (RHCP or LHCP) transmit	Frequency (MHz) and polarization (RHCP or LHCP) receive	Specified G/T (dB/K)	Specified EIRP (dBW)
II	INMARSAT-2	C	3,600.0– 3,623.0 (LHCP)	6,425.0 –6,443.0 (RHCP)	–14	24.0 (each polarization)
		L	1,530.0 –1,548.0 (RHCP)	1,626.5 –1,649.5 (RHCP)	–12.5	38.75
III	INMARSAT-3 (Note 1)	C	3,600.0 –3,629.0 (LHCP & RHCP)	6,425.0 –6,454.0 (RHCP & LHCP)	–13	27 (each polarization)
		L	1,525.0 –1,559.0 (RHCP)	1,626.5 –1,660.5 (RHCP)	–11.5 (global) –5.5 (spot)	39 (global) 47.4 (spot total) [Exchangeable]

Table 9.4

Inmarsat satellites: the main parameters of the mobile communications transponder as specified (Note 1: C × C and L × L transponder specifications are not included)

Parameter	II generation	III generation
Owners	Inmarsat/North Sea Marine Leasing Company	Inmarsat
First launch	Nov. 1990	Dec 1995
Numbers deployed (mid-2000)	4	5
Design lifetime (years)	10	13
Launch weight (kg)	1,230–1,320	1,900
Solar power (kW)	1.4	~2
Type of stabilization	3-axis	3-axis
Type of coverage	Global	Global and spot
Number of spot beams	None	Up to 7

Inmarsat-2 satellites provided around four times the EIRP and bandwidth of the first generation satellites. The monitoring and control of all spacecraft are performed at the Satellite Control Centre (SCC) located at Inmarsat headquarters in London. The telemetry and tele-command stations located at Fucino (Italy), Beijing (China) and Southbury/Santa Paula in the USA are linked with the SCC to provide real-time data for monitoring and control.

Third generation satellites were built by Martin Marietta Astro Space (the prime contractor responsible for the overall spacecraft design, manufacture, integration and testing) and Matra Marconi Space (responsible for the mobile communications payload) and launched during 1996–7. Third generation satellites are capable of delivering about eight times the power of Inmarsat-2 satellites and have 16 MHz wider bandwidth in forward direction and 11 MHz in return direction. Moreover, they provide spot beam coverage, which permits a more effective use of satellite transmitter power and offers better spectrum utilization through frequency reuse in spot beams. The enhanced G/T of spot beams supports smaller/cheaper mobile terminals. Another useful capability of the satellite is the flexibility in allocating both the EIRP and spectrum to spot beams, depending on traffic requirements, thus allowing effective use of satellite resources throughout their lifetime. The satellites also include a navigation payload to provide an international complement to the global position system and a C-to-C transponder to carry point-to-point traffic.

Figures 9.1 (a) and (b) (Richharia, 1994) show respectively the block diagram of the forward and return communication transponders. Figure 9.2 (Richharia, 1994) shows the EIRP and bandwidth range of each Inmarsat satellite generation.

Inmarsat third generation satellites include two sets of identical C-band frequencies in opposite circular polarization to achieve the desired isolation [X= right-hand circular polarization (RHCP) up, left-hand circular polarization (LHCP) down; Y=LHCP up, RHCP down]. However, L-band frequencies for each C-band polarization are RHCP and can therefore only be reused between spot beams which exhibit adequate isolation. The spectrum is divided broadly into L (land/maritime), M (maritime) and A (aeronautical) bands, for historic reasons (the bands are now generic). The G band corresponds to the global beam frequency band. Each band is further segmented into sub-bands, the smallest band being 450 kHz. Each sub-band can independently be switched to any beam, except the G1 band, which is fixed. The gain of each sub-band can be controlled independently.

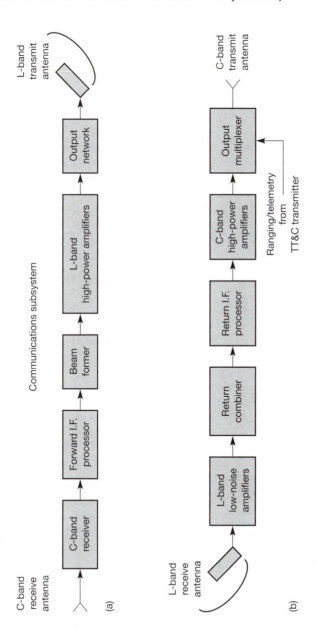

Figure 9.1
(a) Block diagram of forward transponder (Richharia, 1994).
(b) Block diagram of return transponder (Richharia, 1994)

Figure 9.2
Bandwidth and EIRP of
each satellite generation
(Richharia, 1994)

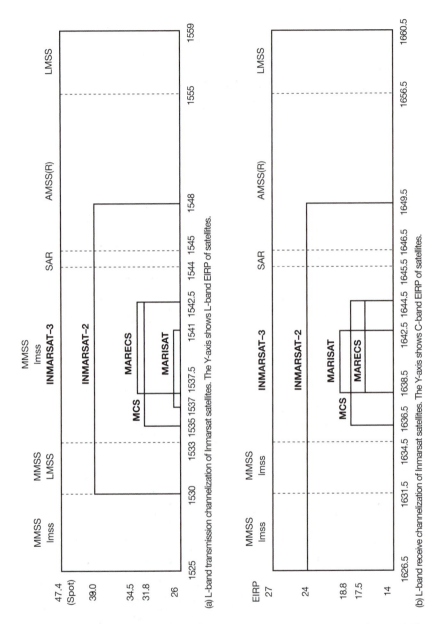

Figure 9.2
Bandwidth and EIRP of each satellite generation (Richharia, 1994)

There are three 'secondary' transponders. The C×C transponder provides a direct C-band to C-band link using Y polarization. It is used for communication between LESs. The L×L transponder provides direct L-band to L-band links and was meant to link mobiles directly, but has never been used because of operational considerations. The navigation transponder provides navigation signals in both L and C bands, as described earlier.

Coverage

The Earth has been segmented into four 'ocean' regions designated as Atlantic Ocean Regions – East and West (AOR-E and AOR-W), Indian Ocean Region (IOR) and Pacific Ocean Region (POR). Each region is served by a single satellite or a satellite cluster. Initially, the Atlantic Ocean region consisted of only one part – today's Atlantic East region, which left a small coverage gap in the Pacific Ocean. The Atlantic West region was added to fill the coverage gap. Associated advantages included addition of space segment capacity to share the heavy traffic in this part of the world, and coverage redundancy. Table 9.6 lists space segment deployment of operational satellites as in early 2001.

Ocean region	Operational satellite identification	Nominal operational satellite location
AOR (W)	3F4	54°W
AOR (E)	3F2	15.5°W
IOR	3F1	64°E
POR	3F3	178°E

Table 9.6
Space segment deployment: operational satellites (2001)

The service area in each ocean region is generally limited to areas within which a satellite appears above about 5° elevation from mobiles, corresponding to about ±76° latitude. Interestingly, aeronautical services via omni-directional antennas are known to provide coverage up to almost 86° latitude because of the altitude of aircraft. Successful communications below 5° elevation from mobiles of other services are regularly reported. Figure 9.3 (a) (Richharia, 1994) shows 5° elevation coverage from Inmarsat's operational satellite and Figure 9.3(b) (Richharia, 1994) shows spot beam coverage of each Inmarsat-3 satellite.

Transmission band Feeder links use a 6 GHz band in the Earth–space direction and 3–4 GHz band in the space–Earth direction. As the feeder link frequencies depend on the translation frequency of satellites, LES frequency tuning range has to comply with each satellite so as to enable transfer of operation from one satellite to another. The polarization of LESs must comply with the satellite C-band antenna polarization.

The mobile links use 1.6 GHz in the mobile–space direction and 1.5 GHz in the space–mobile direction, and only one type of polarization (RHCP) is used in the mobile link.

The transmission bandwidth has been progressively increased in each satellite generation, as a consequence of an increase in traffic demand and additions to frequency allocations by the ITU. However, spectrum demand by other MSS operators is also increasing, and hence Inmarsat is taking steps, such as the use of a significantly higher number of spot beams, to make the most efficient use of spectrum.

Figure 9.3
(a) 5° elevation coverage
of Inmarsat satellites
(Courtesy: Inmarsat Ltd.)

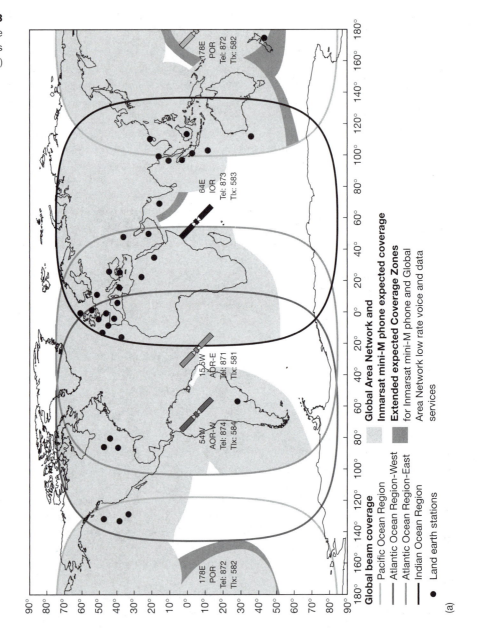

Figure 9.3
(a) 5° elevation coverage of Inmarsat satellites (Courtesy: Inmarsat Ltd.)

Table 9.4 lists the frequency ranges used by each satellite generation.

Demand assigned operation

Inmarsat uses various types of demand assigned (DA) satellite accessing schemes for communications. These schemes provide efficient utilization of satellite

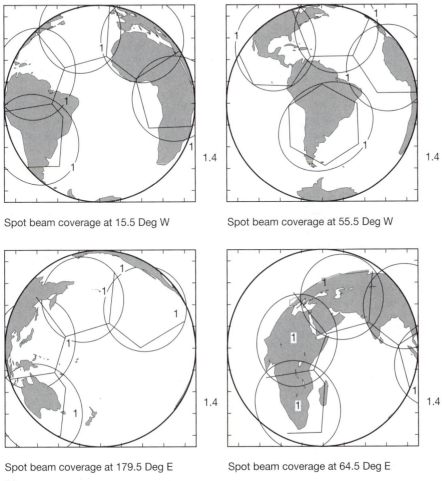

Spot beam coverage at 15.5 Deg W

Spot beam coverage at 55.5 Deg W

Spot beam coverage at 179.5 Deg E

Spot beam coverage at 64.5 Deg E

(b)

Figure 9.3

(b) Approximate spot beam coverage of Inmarsat satellites (courtesy Inmarsat Ltd.)

power and bandwidth. Inmarsat-A, B, C and M use a centrally controlled DA scheme. A frequency (or time slot) pool is maintained at an NCS specific to each service and assigned to users on demand. Inmarsat-B and M NCS share a number of functions, hence are co-located. To minimize network failure caused by NCS outage, each NCS is backed up by a stand-by NCS.

At present the Inmarsat-Aeronautical service uses a distributed resource management architecture, where each GES manages its individual frequency pool independent of other GES. This type of distributed approach is more resilient to a single point (i.e. NCS) failure, but less efficient in terms of spectrum utilization.

Types of channel

In addition to the DA traffic channels, a number of fixed assigned channels are used for signalling between the NCS, LES and MES components of each service. Such channels generally do not earn revenue but are essential for network operation.

In a single channel per carrier system, it is necessary to transmit a reference pilot signal via each satellite to enable each LES to correct for frequency uncertainties caused by satellite, LESs local oscillator and the Doppler effect due to satellite motion. Uncorrected transmissions/reception from/at LESs can cause interference to/from adjacent channels because carriers in the network are closely spaced, the occupied bandwidth ranging from 2.5 to 100 kHz. Because of the location dependence of the Doppler effect, pilots are transmitted for the northern hemisphere, equatorial region and southern hemisphere to provide the most accurate reference for each region. To improve the reliability of the network, there are provisions for stand-by AFC transmissions, should AFC transmissions from the designated LES fail.

There are other types of carriers for specific applications, such as special services, orderwire, tests, etc.

Satellite transmission monitoring

A computer-controlled spectrum analyser monitoring system known as the Satellite Spectrum Monitoring System (SSMS) is used for maintaining network RF integrity. The SSMS consists of a number of remote Earth stations linked to the NOC via land or satellite links.

To monitor traffic flow in real time and maintain a record of traffic flow through the network, data contained in the signalling channels of each service are extracted and processed. In addition to assistance in real-time fault detection and diagnosis, the processed data are used for billing and to provide useful statistics such as call holding time, satellite usage each day, type of calls, grade of service etc. The statistics have a number of operational uses, such as congestion management, monitoring usage of satellite EIRP, planning of space segment resources, traffic forecast, etc.

American Mobile Satellite Corporation System

The American Mobile Satellite Corporation (AMSC), then a consortium of eight shareholders, was granted a licence to provide a mobile satellite service for voice services in the USA in the L band. AMSC joined with Telesat Mobile Inc. (TMI) of Canada in the satellite procurement process, with a satellite sharing arrangement (Johanson *et al.*, 1993). There have been recent changes to the organization, which we will not address here. The coverage includes the continental United States, Canada, Alaska, Hawaii, the Caribbean Basin and offshore territorial waters to at least 200 n miles. The services are offered to vehicular, maritime and aeronautical terminals. The services include:

- voice service to private or public networks, including advanced calling features;
- circuit switched asynchronous transmissions at 1.2, 2.4 and 4.8 kbps;

- CCITT Group 3 facsimile service;
- packet-switched data service at 2.25–5.0 kbps.

AMSC satellites provide a six-beam L-band coverage for the service link and a single global beam at the K_u band for the feeder link. Antenna isolation provides reuse between the east and the west beam. Up to 1,800 6 kHz full duplex channels can be supported.

The main functional entities of the communication ground segment (CGS) and their relationships are shown in Figures 9.4 and 9.5.

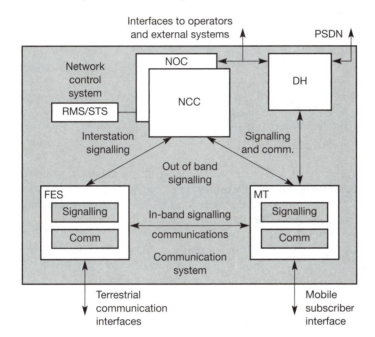

Figure 9.4

Main functional entities of the AMSC communication ground segment (CGS) (Johanson *et al.*, 1993)

Functionally, the ground segment can be broadly divided into two parts – the network control system (NCS) and the communications system (CS). These functions may reside together within a single fixed Earth station.

The NCS comprises the network operations centre (NOC) and network control centre (NCC). They are responsible for system management control, which includes commissioning, user authentication, paging, call set-up/clearing, channel assignment, congestion control, etc. Control and signalling are done in a combination of random access, TDM and TDMA schemes, which have an overhead of less than 5% of total bandwidth. Network management is done by the network management system (NMS), the network operations centre (NOC) and the data network management system (DNMS).

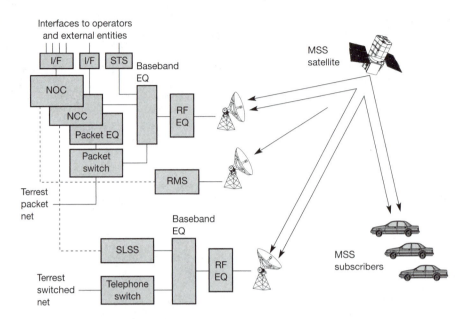

Figure 9.5
American Mobile
Satellite Corporation
System (Johanson *et al.*,
1993)

The NOC provides interfaces to internal entities such as data hub (DH) and a number of external entities, which include providers of aeronautical mobile satellite (route) services, independent users of satellites, other NOCs including TMIs, customer management information systems and the satellite operations centre. Most of these functions are non real time.

The NOC provides real-time control for the circuit switched network by assigning DA channels from a pool on a call-by-call basis. The system is designed to provide a connection within about 3 seconds and support a maximum call establishment rate of 70 call attempts per second. The NOC also has the capability to perform verification tests of mobile terminals.

The communication system, responsible for the main communications, comprises fixed Earth stations (FES) and mobile terminal(s) (MT), connected with each other per call through communication and in-band signalling links. The FESs can deploy up to 1,500 frequency agile channel units connected to K_u-band Earth station equipment. The interface to PSTN is through a mobile telephone switch, which manages calls and supports multimode terminal roaming between CGS and the cellular system. Each FES is capable of supporting up to 20 call attempts per second. FESs can be operated remotely.

Mobile terminals are small and low cost; they can be mounted easily on vehicles and may be adapted for maritime and aeronautical applications. The antenna gain, G/T and EIRP of the mobiles are respectively 8 dB, –16 dB/K and 12.5 dBW.

Access to packet switched services is provided to each mobile by a DH through packet data channels in each spot beam. The supported services include interactive data sessions, query-response sessions and data broadcasts. MT can interface with the packet services via the X.25 or an asynchronous protocol. The maximum packet throughput is 6,000 32-byte packets per second and the call placement rate is 130 calls/second. An MT has the capability to interrupt a packet data session to receive or place circuit switched calls.

Monitoring stations in each spot beam provide information on L-band spectrum usage with quality and performance of signalling channels. Testing of FES channel unit performance is available through a system test station located within the NCS.

Each MT is assigned a set of control channels at the time of commissioning, which essentially are the forward link primary system broadcast channel and the return link signalling channel. The forward channels are TDM while the return channels are TDMA and Slotted Aloha channels. All idle mobile terminals monitor the TDM broadcast channel, the bulletin board, which carries commands, pages, responses to signals from mobiles and other system information. Mobiles use the inbound random access channel for initiating functions such as request for a call. Subsequent MT signalling is carried on one of the TDMA channels. Signalling channels use a variety of error detection and retransmission schemes for reducing errors.

Circuit switched call set-up follows the typical set-up procedures as given in section 8.4. Packet switched calls follow a similar procedure. For a mobile-originated call, the user initiates a call request using a PC or an interface unit on the random access channel with the address of the called party. When the receiving host at the PSDN end responds with a call accept message, this information is transmitted on the forward broadcast channel. After a connection is established and the mobile is ready to send data, it sends a capacity request message on the random access channel to the data hub. The data hub assigns the capacity on the return TDMA channel. The capacity can be extended by the mobile by sending an extension request on the TDMA channel piggy-backed on data packets. The host-to-mobile packets are carried on the forward TDM channel. Packet sizes of up to 64 bytes are transmitted under an ARQ protocol. A similar procedure is used when calls are sent from the PSDN to the mobile.

The NMS functionally comprises system engineering (SE), network engineering (NE) and a customer management information system (CMIS). The SE function establishes long-range planning; which is used to predict when the network should be expanded or reconfigured to comply with forecasts or changes to system utilization. The NE function carries out the decisions made by the systems engineering entity; it formulates tactical plans such as the definition of circuit pools, satellite resource planning and network configuration and sends them electronically to the NOC and data network management system for implementation. The CMIS is an interface for mobile users through which new users are introduced to the network or user profiles altered; it also provides subscriber billing data.

The NOC and DMNS constitute the core of the circuit switched and packet switched services, forming centralized network management for satellite resource management, configuration, accounting, security, faults and performance.

Network entities involved in processing a call, collect billing and performance-related information pass to the NOC or DMNS as applicable. The information is held and passed on to the CMIS when requested by it. The performance data is transferred by the NOC and DMNS daily for NE and SE functions.

Non co-located network elements are interconnected through an inter-network employing various communications technologies, from dial-up modems to high-speed dedicated links, depending on data rate and frequency of data exchange.

Asia Cellular Satellite System (ACes)

The Asia Cellular Satellite System (ACes) is a regional system, originally planned to be operational in 1999, intended to provide digital voice, facsimile, data and paging services to hand-held, mobile and fixed terminals in regions encompassing China and Japan to the north, Indonesia to the south, the Philippines to the east and India and Pakistan to the west (Nguyen *et al.*, 1997; see Appendix for coverage). The network is designed on the GSM model to ensure integration with the terrestrial standard and to maximize reuse of proven terrestrial technology. Integration with the cellular networks will allow seamless roaming with partner terrestrial GSM systems. The first satellite, known as Garuda, was launched recently.

Figure 9.6 shows ACes system architecture, which comprises Garuda satellites controlled by the satellite control facility (SCF) located in Batam Island, Indonesia, the network control centre, gateways and customer management information system.

The salient features of the satellite are a dynamic digital channelizer, a beam-forming network, and multi-port power amplifiers with the capability to switch power and beamwidth to spot beams based on traffic requirements. The digital channelizer routes frequency sub-bands between C- and L-band beams. The routing table used by the channelizer is fully configurable through ground commands from the network control centre (NCC) via the SCF. Figure 9.7 shows an artist's impression of a Garuda satellite. The satellite operates in the extended C-band frequency range of 6,425–6,725/3,400–3,700 MHz and 1,626.5–1,660.5/1,525–1,559 MHz, deploying 140 spot beams in the service link formed by a separate 12 m deployable reflector antenna system to avoid the passive inter-modulation problem and based on flight-proven deployable mesh reflector technology. Using a seven-cell pattern, a frequency reuse factor of 20 is achievable. With a G/T of 15 dB/K and an L-band EIRP of 73 dBW, the satellite can provide up to 11,000 simultaneous voice links to hand-held terminals with a margin of 10 dB, and up to 28,000 voice links at a lower margin.

The ground segment comprises the network control centre, co-located with SCF, which controls and manages the ACes system; the customer management information system manages customer billing, etc.; gateways located at a number of places interface with the public switched telephone network (PSTN), public land mobile network (PLMN) and private network (PN), and include a gateway customer management information system. The main functions of the NCC are monitoring/configuring the channelizer routing table, control of the frequency reuse scheme, network resource allocation to gateways, monitoring the network load, system broadcast, call set-up, and performance monitoring. The ACes customer management information system gathers resource usage data, and performs customer and gateway accounting functions.

Figure 9.8 represents the main block schematic of a gateway. In addition to connecting the ACes network to the terrestrial network, the gateway manages frequency sub-bands and TDMA time slots allocated to it by the NCC; performs call set-up in conjunction with the NCC; controls frequency and timing offset, mobility management, and functions such as user authentication. The RF sub-system is used for radio link connectivity with the satellite and users, through a 13 m tracking antenna. The traffic channel equipment provides modulation, demodulation, baseband processing and RF power control. The gateway station controller performs call set-up and release in conjunction with the NCC. The mobile switching centre (MSC) interfaces the ACes network to the terrestrial network, handling the SS7 signalling of the fixed network. The home location register (HLR), visitor location register (VLR) and equipment identity register (EIR) are used for mobility management and authentication. The network synchronization subsystem (NSS) is slaved to the master NSS resident in the NCC, which comprises ranging equipment, a timing pulse generator and GPS equipment. The inter-station subsystem consists of equipment for communications between the NCC and the gateways through the $C \times C$ link. Finally, the gateway customer management information system (GCMIS) is a customer management information system for maintaining the subscriber and terminal database, in addition to subscribers' accounts.

There are three types of user terminal. Cellular-sized hand-held terminals are satellite/terrestrial dual mode with automatic or manual switching capability, while mobile and fixed terminals use larger, directive antennas. In addition to the standard services mentioned above, the system offers GSM features such as call transfer, call waiting, call holding, conference calls, etc. A high-penetration alerting facility informs the user that a call is not able to connect due to blockage, so that the user may move to a more favourable location. Dual-mode terminals can roam the terrestrial network on the same number. Optional services include voice mail, store and forward facsimile and high power paging.

The radio link, operating in the C band for the feeder link and the L band for the service link, uses a propriety TDMA/FDMA accessing scheme, based on GSM signalling. Mobile-to-mobile connectivity can be achieved by $L \times L$ cross-strapping in the satellite's digital channelizer.

Figure 9.6
ACes system
architecture (Nguyen
et al., 1997)

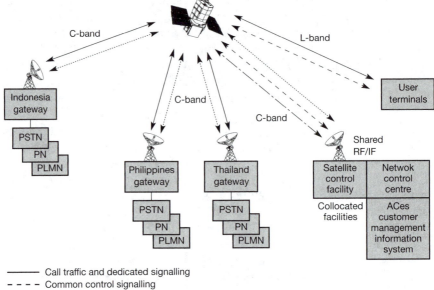

C-band

L-band

Indonesia gateway

PSTN
PN
PLMN

C-band

C-band

User terminals

Shared RF/IF

Philippines gateway

PSTN
PN
PLMN

Thailand gateway

PSTN
PN
PLMN

Satellite control facility	Netwok control centre
Collocated facilities	ACes customer management information system

———— Call traffic and dedicated signalling
- - - - Common control signalling
·············· Interstation communications
—·—· Spacecraft TT&C

Figure 9.7
Garuda-1 satellite view
(Nguyen *et al.*, 1997)

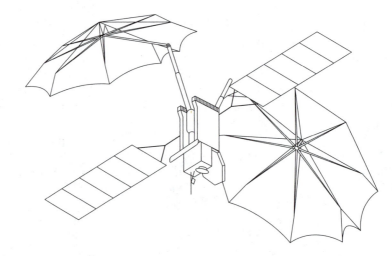

EUTELTRACS

EUTELTRACS is Europe's first commercial mobile satellite service and is operated by the European Telecommunications Satellite Organization (EUTELSAT), supporting an integrated message exchange and position reporting service within the footprint of EUTELSAT satellites, shown on Figure 9.9 (Colcy and Steinhäuser, 1993). The commercial arrangement consists of national or regional service providers; each interfaces with the hub station located just outside Paris, operated by EUTELSAT who have the overall responsibility for the service. The responsibility includes avail-

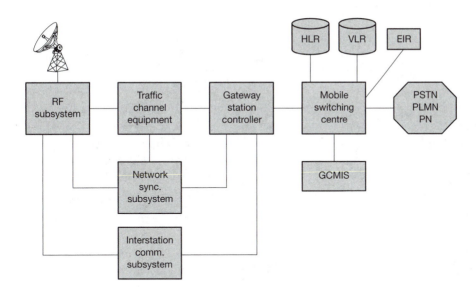

Figure 9.8
ACes gateway functional diagram (Nguyen *et al.*, 1997)

ability of space segment capacity. Aspects related to system hardware, software, mobile terminals and marketing are managed by Alcatel Qualcomm. Regional service providers operate their individual service network management centres and deal with the end users, invoicing them for transactions, equipment, software, training, etc. The system offers a fully integrated message exchange and position reporting system; it has a low-entry-cost architecture capable of exploiting capacity on non-dedicated satellites using well-proven technology; it is a secure system with a single point of contact for all services and billing. Applications include mobile data broadcasting, supervisory control and data acquisition (SCADA), and monitoring fishing activities, as well as low bit rate regional aeronautical communications.

The system architecture is based on the OmniTracs system operated by Qualcomm in the USA since 1988 (Jacobs, 1989). The main components, shown in Figure 9.10, comprise:

- Customer's terminal and dispatch centre.
- Service provider's network management centre (SNMC) used for sending/receiving messages to/from customers and to access position information for the customer's fleet.
- Hub station which consists of two antennas and associated RF equipment, and a hub terminal facility whose function is to process, control and monitor traffic flow and control satellite access.
- EUTELSAT satellites, comprising a 'data' satellite, for the main communications and a 'ranging' satellite to assist localization functions. A ranging signal is transmitted by the secondary hub antenna solely for localization and timing information.
- A mobile communication terminal, mounted on a vehicle, provides communication and measurements for the position reporting service.

Figure 9.9
Footprint of EUTELSAT
satellites (Colcy and
Steinhäuser, 1993)

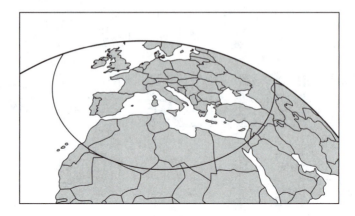

Figure 9.9
Footprint of EUTELSAT
satellites (Colcy and
Steinhäuser, 1993)

Figure 9.10
Main components
of EUTELTRACS
(Colcy and Steinhäuser,
1993)

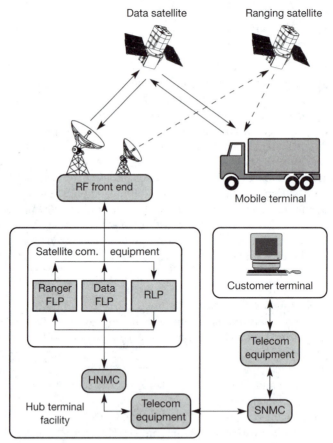

FLP = forward link processor
RLP = return link processor
HNMC = hub network management centre
SNMC = service provider's network management centre

The messaging service operates through two transponders on orthogonal polarization. The forward link comprises a high-power TDM stream spread with a chirp signal to minimize the effects of multipath and interference, resulting in a 2 MHz bandwidth signal. The data rate is 4.96 kbps, (BPSK rate $\frac{1}{2}$ Golay coded – 1× data rate) or 14.88 kbps (QPSK rate $\frac{3}{4}$ coded 3× data rate). The result is always 9.92 kbps PSK in 9.92 kHz bandwidth.

The return link uses the full 36 MHz of the transponder to support a maximum of 45,000 users transmitting at low levels. The low bit rate link uses rate $\frac{1}{3}$ convolution code of constraint length 9 with Viterbi decoding in conjunction with a powerful interleaving scheme to mitigate likely interference in its operational band. A 32-ary FSK scheme is combined with direct spread sequence at 1 MHz rate in an MSK modulator. The spread signal randomly frequency-hops over the full return band to reduce the effects of interference. The hopped sequence is synchronized at the user terminal and the hub to permit demodulation of data. During return link transmissions, the transmitter is enabled for 50% of the time or 15.12 ms, during which 55 bits of 1× data rate or 165 bit/s of 3× data rate are transmitted. During the period when transmission is switched off, the tracking system points to the satellite using the forward link transmissions, and at the same time the receiver performs frequency and time tracking. The selection of the data rate is dynamic, depending on the transmit environment, and is performed under the control of the hub network management computer. Before setting up a transmission, the mobile terminal must acquire the forward link broadcast and request for resource allocation. Calls are set up through the forward signalling channel using special system packets.

Data integrity is obtained by an acknowledged store and forward protocol. Each packet is acknowledged after an error-free reconstruction. If no acknowledgement is received from the mobile, the packet is retransmitted by the hub up to 12 times in an hour before declaring it unacknowledged. No new messages are transmitted to a mobile until the current message has been delivered successfully. In the return direction, the hub network management centre (HNMC) acknowledges error-free receipt. The mobile retransmits a message up to 50 times before discarding the packet(s).

The performance of a store and forward system can be estimated from the message delivery time, which includes queueing, acknowledgement and other processing delays. EUTELTRACS quantifies the performance in terms of four ranges of delay time distribution, defined as the number of attempts to establish a full transaction.

Table 9.7
Delay distribution
statistics in
EUTELTRACS

Number of tries	Delay time range	% of messages delivered successfully in each environment		
		Clear	**Nominal**	**Marginal**
1	20–30 sec	99.2	85.6	73.3
2 or 3	2–3 min	0.6	9.9	14.9
up to 8	12–20 min	0.2	4.5	8.1
up to 12	< 1 hour	0	0	2.9
Never delivered	–	0	0	0.8

Delay distributions have been obtained for three types of mobile environment:

1. Clear line of sight. Example: fixed sites.
2. Nominal land mobile. Example: land mobile travelling in flat, hilly country, city suburbs, etc.
3. Marginal. Example: low elevation angle, mountainous area, large city, etc.

Table 9.7 summarizes the distribution together with the results of a measurement campaign (Colcy and Dutronc, 1990). Note that in the clear and nominal conditions, all the messages are delivered within an hour. In the worst case considered, i.e. marginal, 97.1% of the messages are delivered in an hour and, in all, 99.2% of the messages are delivered successfully.

The EUTELTRACS system also offers position reporting. The technique involves the use of two separate geostationary-orbit satellites for timing measurement at the mobile. Measurements include a precise estimate of round trip delay and the time difference between the two waveforms transmitted by the hub. A multi-lateration technique which provides consistent, reliable, economic and accurate results was chosen for obtaining the position estimates. The time difference is measured at the mobile alternately for both satellites. Round trip delay is measured for all message packets as a part of normal demodulation, which provides the data of the main satellite. The secondary hub uplinks a low power signal, which is structurally identical to the forward message but without carrying the message packets. The period of the spreading function is long enough so that position ambiguities do not arise when the mobile antenna alternates between the two satellites. The mobile acquires the ranging signal transmitted by the ranging satellite, derives the timing information and returns to the primary satellite. The derived information is transmitted to the hub with the next return message or with acknowledgement packets.

The system acquires satellite position independently of the satellite control centre by multi-latering from a number of geographically spread fixed Earth stations. The mobile altitude necessary for accurate position estimation is obtained from a database at the hub terminal facility, given as the distance from the centre of the Earth based on WGS84 ellipsoid. The altitudes above this reference are taken from the United States Geological Survey world database which is updated with recent satellite survey data. The grid has a precision of 10 arc minutes and height precision is 100 feet. A linear interpolation is used within this grid, giving a worst-case error of half the peak-to-peak variations in rough terrain.

The error in the position reporting system has been analyzed in terms of two components – a bias error, mainly caused by altitude model inaccuracy manifesting as north/south bias in the position solution, and a random error caused by errors in timing measurements, satellite position and influenced by satellite separation. Figure 9.11 shows the random accuracy versus satellite separation; a maximum of 15° on the graph shows the current maximum separation possible from EUTELSAT satellites (Colcy and Steinhäuser, 1993). Table 9.8 lists bias errors, while Table 9.9 gives a summary of random accuracy versus satellite angular separation, measured in a trial.

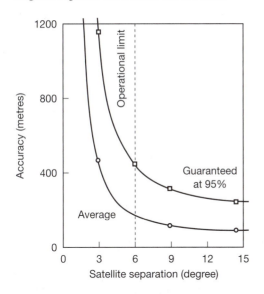

Figure 9.11
Random accuracy versus satellite separation; a maximum of 15° on the graph shows the current maximum separation possible from EUTELSAT satellites (Colcy and Steinhäuser, 1993)

	Flat terrain	Hilly terrain	Mountainous
Average accuracy (m)	10	40	200
Guaranteed accuracy (m)	20	120	420

Table 9.8
Accuracy due to bias errors versus environment (Colcy and Steinhäuser 1993)

Satellite Separation (Deg)	3	6	9	14.5
Average Accuracy (m)	460	170	120	80
Guaranteed accuracy (m)	1150	440	320	240

Table 9.9
Accuracy due to random errors versus environment (Colcy and Steinhäuser, 1993)

9.2.2 Big-LEO systems

Iridium

The announcement of the Iridium satellite communications system in late 1989 by Motorola of the USA began one of the most innovative phases in the satellite communications industry. Maintaining its lead, Iridium became the first operational system in November 1998. Unfortunately, the company failed to meet its business objectives, resulting in bankruptcy shortly afterwards and subsequent takeover by another group. Despite its commercial failure, the system remains an example of innovative engineering.

The system is named after the element Iridium which has 77 electrons in its atom, corresponding to the 77-satellite constellation originally proposed by Motorola engineers. The size of the constellation was later scaled down to 66, but the original name was retained for its appeal. The system offers voice communication to pocket-size telephones, and other services such as facsimile, two-way messaging, duplex 2.4 kbps data bearers, position determination and paging. Handsets could select a local terrestrial cellular system, such as GSM, CDMA, PDC, D-AMPS, when a user was within coverage of a partner terrestrial cellular system, communicating through the space segment when outside the cellular coverage; the system also offered a world roaming service, permitting users to roam across a number of wireless protocols with the same number and a single bill. Iridium also provides aeronautical services, such as voice, facsimile, and data services, to travellers on commercial, business and general-aviation aircraft. Pocket-sized pagers provide world-wide reception of alphanumeric messages in a number of international character sets and use an off-the-shelf disposable battery, which has an average lifetime of one month. At the start of

Figure 9.12

An overview of the Iridium system components (Hutcheson and Laurin, 1995)

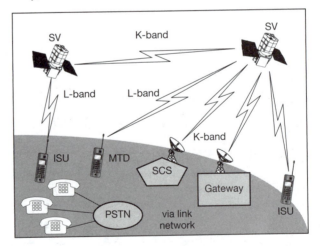

LEGEND:
ISU – IRIDIUM® Subscriber Unit
PSTN – Public Switched Telephone Network
SCS – System Control Segment
MTD – Message Termination Device
SV – Space Vehicle

operation, the communicators cost from \$300 for a paging or position determination receiver to ~ \$3,000 for a voice communicator, and voice call charges were set at \$3 per minute to anywhere in the world.

An overview of the Iridium system components is portrayed in Figure 9.12 (Hutcheson and Laurin, 1995).

(a)

Figure 9.13
(a) Iridium's constellation pictorial (Hutcheson and Laurin, 1995.
(b) Close-up of satellites within constellation (Hutcheson and Laurin, 1995)

(b)

Figure 9.13
(c) Satellite footprint near the equator (Hutcheson and Laurin, 1995)

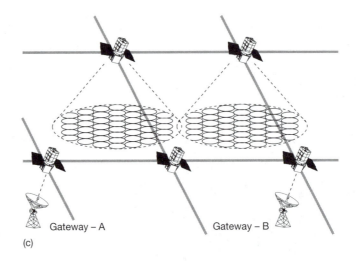

Gateway – A   Gateway – B

(c)

The space segment consists of a LEO constellation comprising regenerative satellites interconnected via intersatellite links (ISL). The fixed ground segment comprises a few strategically located gateways connected to the terrestrial fixed and mobile networks (see Appendix for their location). The service link supports a number of services through a variety of terminals, as mentioned above.

The constellation consists of six, 86° inclined orbital planes at an altitude of 780 km, with 11 equi-spaced satellites in each plane, providing a seamless world-wide coverage at a minimum elevation angle of 8°. Figure 9.13(a) and (b) illustrates a pictorial view of the constellation and a close-up portraying the intersatellite link, respectively.

Each Iridium satellite deploys 48 circular spot beams, with a footprint of ~4,800 km diameter, per satellite in the service link. Spot beams have a frequency reuse scheme similar to the terrestrial cellular system. Figure 9.13 (c) shows the antenna pattern of a single satellite near the equator. However, as the spot beams move with the satellites, the reuse scheme must take into consideration the dynamics of the constellation (see section 8.5.1). In all, there are 2,150 active beams on the surface of the Earth which give 180 times frequency reuse; within the USA alone, the system achieved a reuse of five times. Three phased array antennas, located on the side panels of the satellite, could each form 16 cellular beams. Each beam could support an average of 236 channels. A spot beam could be used by more than one gateway, and spot beams, as such, were not 'owned' by any Earth station. Iridium constellation visibility analysis and a comparison with other constellations are demonstrated in section 2.3.8.

Due to the limited visibility of satellites in a LEO constellation, a large number of gateways are required for international call connectivity when only terrestrial routing is used. In the Iridium system, the problem was solved by intersatellite links. Calls were routed to the gateway nearest to the destination

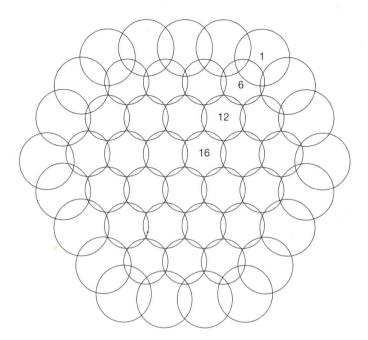

Figure 9.14
Iridium's spot beam
coverage pattern
near the equator
(Freeman, 1996).
Reprinted by permission
of John Wiley & Sons Ltd

without the need for intermediate terrestrial routing. Each satellite could be linked via a fixed wave guide slot antenna to the forward and rear satellite of the same plane and a mechanically steered wave guide slot array for each adjacent orbit. Antennas provided a 5° elevation beamwidth so that steering was required only in the azimuth plane. The gain of each array was 36 dBi. Each ISL used a QPSK modulation scheme and operated at 25 Mbps, thus the channels were spaced at 25 MHz intervals and eight frequencies were required to support all intersatellite communication of the network. RF signals were demodulated, decoded, switched and again coded/modulated before transmission. A BER of 1×10^{-7} was supported with $\frac{1}{2}$ rate FEC coding. Under the most severe loading, up to 1,100 Iridium users could access the satellite and up to 5,000 calls could be transiting through the satellite, and thus the worst-case throughput of the onboard processors was estimated as 100 Mbps. In a commercial system, regenerative transponders with this scale of satellite processing were introduced for the first time. Link budgets for the intersatellite links are given in section 3.4.

Service link spot beams were arranged in a regular pattern, as seen in Figure 9.14, and satellite-fixed, i.e. moved with the satellite. Being a polar constellation, coverage tended to concentrate as the latitude increased and therefore spot beams were switched off progressively as satellites moved to higher latitudes, thus conserving satellite power and avoiding unwarranted coverage redundancy and interference. The constellation did not use satellite diversity and therefore a high margin of ~16 dB was built into the RF link.

Table 9.10 shows the frequency bands used in the Iridium system. The system used 1,616–1,626.5 MHz in both the up and down service link, using the time division duplex accessing (TDD) scheme. The feeder link used 27.5–30.0 GHz in the uplink and 18.8–20.2 GHz in the downlink, and the intersatellite links operated in the 22.55–23.55 GHz band. The Iridium system employed a combination of frequency and time division multiple access (FDMA/TDMA) signal multiplexing to make efficient use of limited spectrum.

Table 9.10
Iridium frequency bands

Link	Frequency (GHz)
Service up and down link	1,616–1,626.5
Feeder link (uplink/downlink)	27.5–30.0/18.8–20.2
Intersatellite link	22.55–23.55

Figures 9.15 and 9.16 respectively represent the service link frequency plan and the TDD multiplexing format; each frequency channel could support four full-duplex voice channels.

Figure 9.15
(a) Service link frequency plan (Freeman, 1996) © John Wiley & Sons

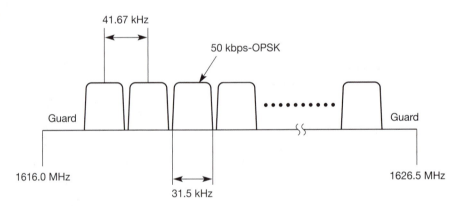

Figure 9.16
Time division duplex (TDD) frame format (Freeman, 1996) © John Wiley & Sons

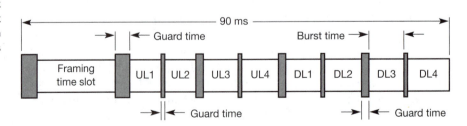

Each transmission is Doppler corrected so as to arrive at the correct frequency. Mobiles use up to four 8.28 ms TDD bursts in 90 ms frames, the peak transmit power being 3.7 W at an average of 0.34 W. Transmissions are QPSK modulated at 50 kbps, with carrier spacing of 41.67 kHz. Iridium units use a propriety voice compression technique called AMBE, a trademark of Digital Voice Systems Inc., providing voice at 2.4 kbps which is then convolutionally coded at $\frac{3}{4}$ rate. The bit error rate for voice transmission for the worst case is specified as 2×10^{-2} – typical BER ranged 1×10^{-3} – 1×10^{-4}. For data transmission, lower BER of the order of 1×10^{-5} are obtained by using more powerful coding and protocols, installed in the subscriber's unit. The subscriber units are similar in size and shape to a cellular phone and use a quadrifilar helix antenna of 1 dBi gain, 8.5 dBW peak EIRP and –23 dB/K G/T.

Iridium gateways operate in the K_a band in the feeder link, using $\frac{1}{2}$ FEC and transmission rate of 12.5 Mbps. Six channels are assigned at a spacing of 15 MHz. The supported BER is better than 1×10^{-7}. Two gateway antennas placed 34 nautical miles apart achieve spatial diversity gains, mitigating the effects of rain, and remove outage due to Sun interference; the availability on these links is of the order of 99.8%. Gateway positions are given in Appendix A to give an insight into gateway selection for such an architecture. The end-to-end network connectivity of the Iridium system is shown in Figure 9.16 (Freeman, 1996) and connectivity arrangements with regard to business management are discussed in Chapter 7.

To reduce risk in launch delays, Iridium satellites were launched through three providers – Boeing, China Great Wall and the Khrunichev State Research and Production Centre of the Russian Federation. These providers support launch services respectively through Delta 2 with six satellites per launch; Long March 2C/SD with two satellites per launch; and Proton 2K with seven satellites per launch respectively. There were nine Delta launches, three Proton launches and the remainder were on Long March. The first satellite was launched on 5 May, 1997.

The constellation is managed by the system control segment, comprising a ground station which performs TT&C functions together with the task of computing and loading the satellites with routing data and frequency plans.

Gateways comprise K-band tracking Earth stations which connect the space segment to the public networks via an international switching centre. The connection used PCM transmission and SS7 multi-frequency compelled response (MCFCR2) signalling.

The underlying architecture of the system is modelled after the well-established terrestrial standard, GSM. It was anticipated that this would allow easy integration with terrestrial systems and permit the Iridium system to benefit from advances in the terrestrial standard. Thus each gateway incorporated GSM functions for call processing, and additionally, incorporated Iridium-specific functionality to manage communications with the satellite constellation, as depicted in Figure 9.17.

Figure 9.17

Main elements of an
Iridium gateway
(Hutcheson and Laurin,
1995)

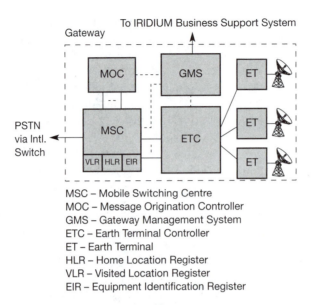

Figure 9.17 Main elements of an Iridium gateway (Hutcheson and Laurin, 1995)

MSC – Mobile Switching Centre
MOC – Message Origination Controller
GMS – Gateway Management System
ETC – Earth Terminal Controller
ET – Earth Terminal
HLR – Home Location Register
VLR – Visited Location Register
EIR – Equipment Identification Register

A mobile switching centre (MSC) constitutes the main switching element connecting the gateway's Earth terminal controller (ETC) to the public network via an international switching centre. The ETC, akin to the GSM base site subsystem, is connected to the MSC by an 'A' interface. On the space segment end, it controls three K-band Earth station terminals, one carrying traffic, the second providing redundancy and rain diversity where separation is adequate, and the third prepared to acquire a new satellite to take over connection from the current satellite. The visitor location register (VLR) and home location register (HLR) maintain location information about subscribers in much the same way as in GSM. The equipment identification register (EIR) is a database for keeping the identity of physical subscriber equipment. The message origination controller (MOC) supports Iridium's paging services. Operations, administration and maintenance support is provided by a gateway management system (GMS), which is also connected to Iridium's business support system (BSS). BSS has the function of managing usage and making settlement statements between Iridium Inc. and gateway operators, for which it received call detail records from each gateway's GMS over an Ethernet link.

Each Iridium subscriber has a home gateway which maintains a record of the subscriber in an HLR. An Iridium subscriber can be identified by several types of numbers: mobile subscriber ISDN number (MSISDN), which is used by a land calling party; a temporary mobile subscriber identification (TMSI), sent over the radio link while establishing connection, which is changed periodically to   protect user identity; and an Iridium network subscriber identity (INSI), a number stored in the user's phone, sent on the radio path when a valid TMSI is unavailable. Table 9.11 shows the structure of a user's MSISDN number (Hutcheson and Laurin, 1995).

Table 9.11

Number structure of
Iridium's MSISDN

Iridium country code	Iridium significant number		
	Geo-political entity	Service provider	Subscriber number
3 digits (assumed)	3 digits 200–799	2 digits 0–99	7 digits

The 'geo-political entity' and 'service provider' fields together are used to identify the home gateway of a subscriber. When a call is made to or initiated from a subscriber, the gateway locates the subscriber and sends the information to the home gateway using SS7 capability. The gateway evaluates the location information and grants the subscriber access if the call is permissible at the location to ensure compliance with restrictions in a territory where access to the system is not permitted. The home gateway at the same time uses the location information to determine the subscriber's location area code (LAC). The LAC is used to determine the visitor's gateway through a LAC map, which defines the service area of each gateway on a world map in terms of LAC. Each gateway has the LAC association of every gateway; 2,047 LACs are available to each gateway. The visited gateway keeps a record of the subscriber in its VLR until the subscriber moves away to another gateway's jurisdiction. All call management is done by the visited gateway as long as it is within the LAC served by it. When a call is made from the fixed network through a gateway where the subscriber is not registered, the gateway can determine the location of the subscriber through the home gateway and set up the call via intersatellite links. Similarly, a mobile-originated call can call a PSTN number outside the jurisdiction of the visited gateway. The location of the connecting gateway can be determined through the dialled PSTN number. The call can be established using a transit connection between the visited gateway, PSTN gateway and ISLs. Figures 9.18 (a) and (b) show the concept. In each case, the dotted line shows the signalling path, the top connection to each block shows the K-band link and the bottom one shows the land connection.

In Figure 9.18 (a), when a mobile is to be routed through a PSTN served by a different gateway, the voice call is looped back through the serving gateway for call completion. In Figure 9.18 (b), the voice call is established directly to the destination gateway through an ISL; however, the signalling path through the serving gateway is maintained, giving it full control of the call. Thus, if necessary, the call can be terminated. This feature of the system, known as 'cut-through', can also be used for establishing a mobile–mobile call to minimize delay. Cut-through provides more efficient delivery, lower path delay and reduced loading on the K-band feeder link.

Intersatellite links, GSM call processing architecture and geographically controlled system access feature permits flexibility in adding gateways when necessary for business or to provide redundancy arrangements between gate-

Figure 9.18
(a) Mobile is routed
through a PSTN served
by a different
gateway (Hutcheson
and Laurin, 1995)
(b) Voice call is
established directly
to the destination
gateway through
an ISL (Hutcheson and
Laurin, 1995)

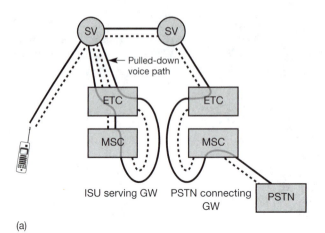

(a)

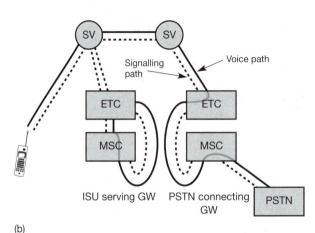

(b)

ways. Because the home gateway is determined by MSISDN and the visited gateway by the geographical map, it is possible to assign service areas to gateways as desired. For example, an operator with multiple gateways can partition subscribers in terms of best business, subscriber mailing address, service centre, etc. If a new gateway is added, new or existing subscriber blocks can be added to the gateway's home register and its LAC map would be distributed to all gateways. Geographic regions can even be shared between visited gateways by providing LAC maps with common territory.

Globalstar

The Globalstar system was propounded by Loral Aerospace Corporation with Qualcomm Inc., the former being responsible for the space segment and the latter taking responsibility for most of the ground and user segments. The system is based on a combination of recent technological developments – cellular CDMA technology (US EIA/71A IS-95 standard) and LEO satellite system technology. The CDMA technology is based on the company's experience with the OmniTracs radio determination and messaging service and development of the US's CDMA terrestrial standard IS95.

The goal was to keep the system simple, affordable, and low risk by employing well-proven technologies which also offered high spectral efficiency. Despite a launch failure resulting in the loss of 11 satellites, the system began operation in 1999.

Enabling technologies used in the Globalstar system include (Schindall, 1995):

- spread spectrum/CDMA with efficient power control;
- high efficiency vocoder with voice activation;
- diversity, including soft handover technique;
- spot beams with weighted antenna gains.

The system offers voice and a 7.4 kbps data bearer service. The service is targeted at rural and thin route requirements of the public, government and private networks – essentially attempting to meet telecommunication needs where terrestrial services are expensive or non-existent.

The main system components are shown in Figure 9.19.

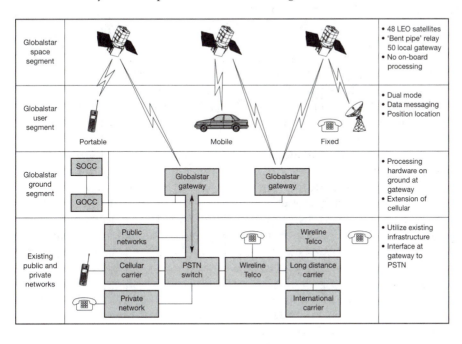

Figure 9.19

Main components of Globalstar's MSS (Dietrich, 1997)

The Globalstar space segment consists of a 48-satellite Walker (48/8/1) constellation. The 390 kg satellites are placed in eight 52° inclined circular orbits at an altitude of 1,414 km, which has an orbital period of 114 minutes. Full coverage with dual satellite diversity is available up to ± 70° latitude, with intermittent coverage up to ± 80°. The constellation is optimized for the 30–60 latitude region, where the highest market was perceived. Figure 2.29 (Chapter 2) represents satellite visibility statistics as a function of latitude. The diameter of each satellite's footprint is about 7,600 km. Low-risk spacecraft technologies include a Sun tracking solar cell array and Nickel-Hydrogen battery cells. Onboard GPS receivers are used for position and attitude control to simplify orbital maintenance and provide an onboard timing and frequency reference. The constellation does not use intersatellite links and satellites use transparent transponders.

Each three-axis stabilized satellite consists of 16 spot beams comprising a central beam and two rings of outer beams. The beams are generated with phased array technology using 91 elements which have individual high power amplifiers (or low noise amplifiers for the receive pattern). The spacecraft power sytem consists of solar arrays which have a beginning-of-life power of 1.9 kW and use a Nickel-Hydazine battery.

The system utilizes frequency division spread spectrum/CDMA (FD/SS/CDMA) for satellite access at the L band (1.62–1.63 GHz) in uplink and S band (2.48–2.5 GHz) in downlink. The maximum end-to-end delay is less than 100 ms, of which only 18 ms are incurred in the satellite links. A FD/SS/CDMA scheme allows path diversity by combining signals from multiple satellites and multipath to mitigate fading effects. The technique simplifies handover between satellites. Globalstar originators' experience with the terrestrial CDMA system enabled reuse of terrestrial technology in a satellite environment. A CDMA scheme can provide frequency reuse with spot beam gain isolation as low as 2–3 dB, whereas in FDMA or TDMA typical isolation for frequency reuse is of the order of 15–20 dB.

The system provides around 11 dB fade margin – a 1 dB margin for Ricean fading is provided in the link design, and power control allows increments of power by up to 10 dB in steps of 0.5 dB. When the 11 dB margin has been exhausted, the system can route the signal through another satellite by utilizing path diversity. A CDMA scheme permits graceful degradation of channel quality, unlike TDMA and FDMA which have a hard limit. At the limit of capacity for a specific quality, the operator may either permit degradation of channel quality or redistribute traffic between satellites in view – thus giving the potential to exceed the capacity under peak loading conditions. Radio resource management is relatively easy, as all frequencies may be reused in all spot beams. Signals transmitted from users are picked up by all satellites in view of the user. The gateway passing the call manages power control as well as path diversity. It receives signals from all satellites in view and combines signals at an appropriate time. Each path is monitored for level and stability and the user is diverted to the best path at a pre-set threshold. A variable rate voice codec operating between 1.2 and 9.6 kbps, averaging 2.4 kbps, is used with $\frac{1}{3}$ FEC in the uplink and $\frac{1}{2}$ in the downlink. Vocoder rate is changed every 20 ms depending on channel conditions. In the absence of voice, the bit rate drops to 1.2 kbps, which reduces the transmitted power, thereby increasing system capacity (see section 4.2.2 and 4.4.1). The CDMA is a direct sequence spread spectrum scheme spread in a 1.25 MHz band with QPSK modulation. The transmission band is spread over thirteen 1.25 MHz bands. The voice-activated carriers operate at mobile powers of up to 2 W.

There are three types of user terminals – *fixed* for residential or rural locations, *portable* for mobile access to satellite components and *portable dual-mode* units for accessing satellite and terrestrial systems. The terminals use rake receivers which allow reception of many signal paths simultaneously, irrespective of their being direct or reflected, and combine them coherently to give a high signal level to the decoder. Fixed terminals use a small antenna and a 3 W power amplifier and, in addition to the rake receiver, incorporate software to manage calls, etc. Portable units are slightly bigger than conventional cellular telephones with a specially designed antenna for providing near omni-directional coverage of 1–1.5 dB gain. The antenna protrudes slightly above the head to minimize blockage from the user's head. The output power averages to around –10 dBW, with a peak of about –4 dBW. Battery life is over 2 hours for talk and 15 hours in stand-by mode. Globalstar-specific functions include call management, local Earth model for position location, etc. A car version of the phone uses an outdoor antenna with a power amplifier. The dual-mode terminal is essentially integration of the Globalstar portable unit with a cellular phone, reusing components wherever possible to preserve as much as possible the features of cellular systems. The phone design includes a subscriber identification module (SIM), allowing the user to change handset at will.

The gateways are operated by local service operators and each gateway has an independent control, giving the operators full control for call routing. Each gateway uses up to four tracking antennas, which ensures that all satellites in view – normally three – are tracked. If a lower number of satellites are visible, such as at high latitudes, gateway antennas may be reduced accordingly. Figure 9.20 shows a block schematic of the Globalstar gateway. Further details of network connectivity of the Globalstar system are discussed in section 7.3.

9.2.3 Little-LEO system
ORBCOM

ORBCOM, a subsidiary company of Orbital Sciences Corporation (OSC), in cooperation with OSC, has developed and deployed a LEO mobile satellite system offering world-wide low-cost, low bit rate two-way communication services. OSC is a developer of small satellite technology and the low-cost air launch system, Pegasus. ORBCOM was fully funded by 1993 when it entered a strategic financing and marketing alliance with Teleglobe, Canada (Schoen and Locke, 1995).

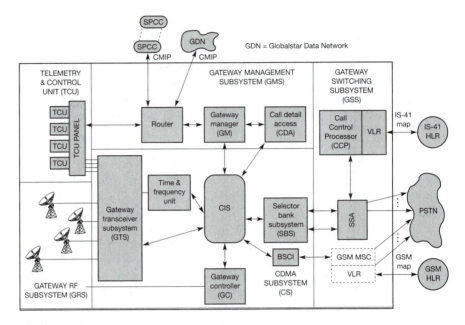

Figure 9.20
Block diagram of a
gateway in the Globalstar
system (Dietrich, 1997)

The ORBCOM system provides low-cost world-wide two-way low bit rate communications in the very high frequency (VHF) band for remote users, via a constellation of 26 LEO satellites. The VHF band permits very low cost technology to be used, with unit costs being notably lower than the terminals of big-LEO systems. The market is directed at applications requiring 100–200-byte short message transmissions from remote areas, such as supervisory control and data acquisition (SCADA). The service charge includes a fixed monthly charge plus a cost levied per 100 bytes of message. ORBCOM services include personal and data messaging, such as e-mail and position determination. The system also provides a geo-location service to an accuracy of 500 m, by combining Doppler frequency shift measurement with the satellite's ephemeris data available from the satellite's GPS receiver.

Some of the key features of the system are low cost for space segment deployment, low transmission delay, low-cost, proven VHF electronics, and simple low-cost spacecraft. Clearly, ORBCOM's strengths lie in its choice of low-risk, low-cost technology and niche market. The LEO system was the first to enter the market.

Applications include tracking of assets during shipment, remote interrogation of electronic tags, collection of environmental data, monitoring of remote sensors from field command posts, military applications, etc. In a typical application, a user terminal interfaces with a monitoring device, such as an electronic tag, through an RS-232 link. An application program is used to activate the monitoring at the appropriate time and data is sent via the ORBCOM space segment to the monitoring end.

The constellation comprises 36 satellites (note: other numbers have also been cited), optimized such that users in temperate zones are in view of a satellite at 5° elevation for 98% of the time. The target for 98% outage caused by satellites not being visible is less than 2 minutes and for 99% outage, the target is < 5 minutes. The orbital altitude of the constellation is 775 km (see Appendix A). The low constellation altitude in conjunction with tiny, 75 lb, satellites enables lower launch costs, of the order of $15 million for each Pegasus launch placing eight satellites per launch. In the case of satellite failure, the remaining satellites are repositioned within their plane to minimize the impact of the lost satellite. The system can absorb a loss of up to four satellites without significant impact on system capacity.

A single satellite control centre co-located with a network control centre (NCC) in Dulles, Virginia, provides satellite TT&C and other functions:

- message handling, which includes management of delivery of inter- and intra-system data messages;
- network management, which includes statistics, diagnostics and configuration control;
- message transfer;
- supporting value-added services;
- message gateway, wherein messages are reformatted to match source and destination;
- customer services and billing;
- gateway Earth station (GES) control and monitoring.

The NCC is essentially an X.400 backbone messaging system interfacing the ORBCOM network to national public and private networks comprising X.25 and X.400 gateways. It is the centre of a star network of GESs, connected to each by leased lines and the public data network, with plans to have VSAT as backup. NCC in different countries are interconnected through public networks. A message transmitted by a user is relayed by the satellite to a regional GES, which relays the information to the NCC through an optical fibre link. In areas where an appropriate GES (or user) is not in a satellite footprint, the satellite can store and forward a message when a suitable GES (or the subscriber) comes within the view of the satellite. The NCC routes the signal to the addressee, who may be either a terrestrial party or another ORBCOM user. It has the capability to translate messages from a number of e-mail systems, such as the internet, to an X.400 protocol for communicating with users of the ORBCOM network.

A GES comprises 14-dBi tracking antennas with associated RF and baseband equipment to transmit packets between satellites and the NCC. Equipment, including the antenna, has redundancy and is designed for unattended operation. GESs uplink in the 148.0–150.05 MHz band at 56.7 kbps transmitted at 250 W and receive in the 137.0–138 MHz band. The main functions of a GES are:

- acquisition and tracking of satellites using orbital parameters supplied by the NCC;
- to transmit/receive from satellites and the NCC;
- to monitor status of its own hardware/software;
- to monitor system-level performance of the satellites to which it is connected.

There are four unmanned GESs within the USA to support operation there. It is anticipated that licensees in each country will own their NCC and GESs.

At the start of service, user terminals weigh about $\frac{1}{2}$ lb (16–20 ounces), with transmitter power of 5 W, and use either an internal or an external power supply. Manufacturers produce terminals under a licensing agreement with ORBCOM. Under this agreement, ORBCOM provides the communications software which allows terminals to communicate with the space segment. Terminals include keypads, RS-232 interface port, integrated GPS receivers, encryption capability using Digital Encryption Standard chips, and an LCD screen. The terminals are type certified by ORBCOM to ensure that the design complies with ORBCOM's operating licence. Each terminal has a unique ID given by the manufacturer and a unique X.400 address at the time of its activation, which are verified by the NCC before initiating a call.

Satellites are three-axis gravity gradient stabilized, transmit 70 W average power and weigh 95 lb, which permits up to eight satellites to be launched at a time. It is anticipated that next generation satellites will be tailored according to market requirements. Satellites comprise message routing and queueing computers accessible by VHF radio links through eight receivers and three transmitters. Satellites include GPS receivers for attitude determination and provide orbital location and velocity information to ground receivers which support position determination capability. Satellite downlinks operate in the 137–138 MHz little-LEO band. The band was chosen because it is shared with the space operations service, which is diminishing in this band as the band is relatively narrow for modern spacecraft operations. The 148.0–149.9 MHz band used in the uplink was selected because of its proximity to the downlink, allowing the same antenna to be used, and it would be easier to coordinate the frequencies. This band is shared by a number of terrestrial mobile systems. To minimize interference, the satellites use a channel management technique known as the dynamic channel activity assignment system (DCAAS). Each satellite monitors power levels on 2.5 kHz segments over the entire uplink band and channels are ranked 1 to N, from best to worst according to noise power; additionally, the spacecraft also keeps a record of packet error in each active channel. The spacecraft transmits this list to receive terminals on a broadcast channel for establishing a session. Table 9.12 summarizes the main features of the satellites (Schoen and Locke, 1995).

Mass	95 lb
Solar array power (orbital average)	160 W
Transmitters	VHF (user link): 2
	VHF (feeder link): 1
	UHF (beacon): 1
Receivers	VHF (user links): 7
	VHF (feeder links): 2
Attitude control	Autonomous/GPS
Design life	4 years
Cost	< $2 million

Table 9.12

Main characteristics of ORBCOM satellites

9.2.4 MEO systems

ICO system

A MEO system called the Intermediate Circular Orbit system or simply ICO is being developed by ICO Global Communications Ltd (now New ICO, after a recent acquisition), an affiliate company of Inmarsat created in 1995 to implement the system developed by Inmarsat under its project-21 programme (Inmarsat, 1992). The system, as originally conceived, was meant to provide world-wide voice telephony at 4.8 kbps and data services at 2.4 kbps, and a suite of messaging services to single- or dual-mode hand-held terminals. The system has since been modified to increase data rates, following a change of ownership. Here we will consider the baseline system (Bains, 1999). A feature of the service is a high penetration notification function with acknowledgement, used to alert a user to an incoming call when a normal communication link cannot be established due to excessive signal shadowing. Dual-mode terminals have the capability to select the ICO network or a terrestrial system. The constellation consists of 10 satellites in two 45° inclined orbits at an altitude of 10,355 km. Figure 9.21 represents a pictorial view of the constellation. The system architecture is shown in Figure 9.22(a).

The service links operate in the personal communications band at 1.98–2.01 GHz in the uplink and 2.17–2.2 GHz in the downlink. TDMA is used for accessing the satellite, with QPSK as the modulation scheme. Satellites support at least 4,500 TDMA telephone channels.

The feeder link operates in the 5 and 7 GHz bands, which are the WRC-95 allocation for feeder links of non-geostationary satellite systems.

The ground network comprises 12 ground stations known as satellite access nodes (SAN), interconnected with a terrestrial network known as ICONET as shown in Figure 9.22(b). Each SAN consists of multiple antennas, RF and switching equipment, together with associated databases. The Appendix lists the location of various SANs with their previous affiliation.

Figure 9.21

ICO constellation
geometry (Bains, 1999)

A MEO constellation was chosen as it could offer 'the best overall service quality for the desired market' by offering the following benefits:

- high elevation angle coverage;
- good satellite path diversity;
- slow satellite movement, of the order of 1 degree/minute, as viewed from the Earth;
- reasonable implementation and schedule. The size of a LEO constellation coupled with the short lifetime of satellites in such an orbit would impose logistical/manufacturing difficulties in constellation maintenance; satellites in a GEO constellation would become too complex due to the need for a large number of spot beams, and transmission delays of satellites in such an orbit would be relatively long.

The constellation provides world-wide coverage at an average elevation angle of about 40–50°. Figure A.1 (Appendix A) shows a snapshot of the constellation coverage footprint. Satellite footprints from such an altitude cover about 30% of the Earth's surface at any given time. Figure 2.29 shows visibility statistics of satellites as a function of latitude (also see Chapter 2). Two to four satellites are generally visible from a SAN and a user at any time.

Each satellite deploys a transparent transponder, which permits flexibility in transmission format. A high digital technology content provides flexibility in satellite reconfiguration, and also offers advantages in manufacturing and mass production when compared to analogue systems. Functions where digital

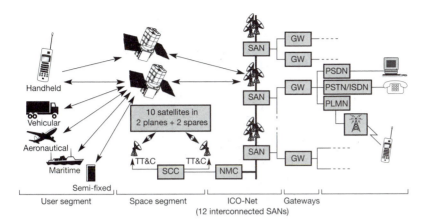

(a)

Figure 9.22
(a) ICO system overview
geometry (Bains, 1999).
(b) ICO-NET architecture

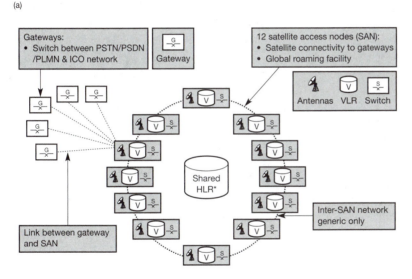

The ICO-NET consists of the satellite access nodes, the links between them, and the mobility databases.
* *VLR, HLR: mobility databases (using GSM terminology)*

(b)

subsystems have replaced analogue solutions include channelization and beam generation. Transmit and receive antennas in the service link are separate in order to achieve better inter-modulation protection and ease in manufacturing. Each satellite transmits 163 transmit/receive spot beams in the service link at an EIRP adequate to provide an 8 dB link margin. Figure 9.23 represents spot beam coverage from a single satellite.

Solar cells use Gallium Arsenide technology to give 8,700 W of power. Each satellite, weighing about 2,600 kg, can be injected directly into the MEO, thus

Figure 9.23
Spot beam coverage
from a single satellite
(Bains, 1999)

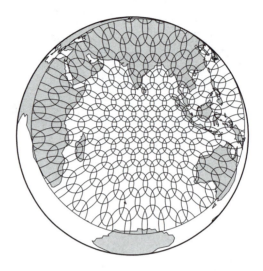

Figure 9.24
ICO satellite view
(Bains, 1999)

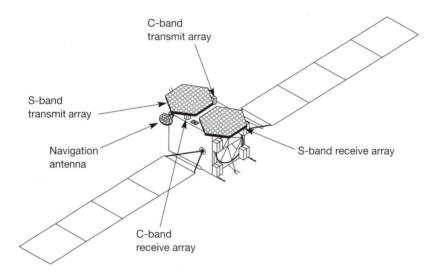

dispensing with the need for apogee-kick motors and permitting multiple launch. The design lifetime of each satellite is estimated at about 12 years. Figure 9.24 shows an artist's representation of an ICO satellite.

The function of the network's satellite control centre (SCC) is to manage the constellation by regularly adjusting the orbits of each satellite under normal conditions or when a satellite in the constellation fails. The control centre also monitors satellite health and supports the launch and deployment phase. An important operational task of the SCC is to configure feeder–service link mapping so as to distribute channels within spot beams according to traffic distribution.

Figure 9.25
ICO phone (Bains, 1999)

The dual-mode hand-held sets are to be competitively priced. They will include features such as external data ports/internal buffer memory to provide data communications, messaging and use of SIM. To comply with radiation safety standards, the average power will be around 0.25 W, around the same magnitude as cellular phones (~0.25–0.6 W). Other types of terminals include vehicular, aeronautical and rural communication terminals. Figure 9.25 portrays the initial proposal for the ICO phone.

A SAN interfaces between the space segment and the terrestrial network. Each SAN comprises five antennas and the associated RF; a switch for routing traffic within ICONET and to the public networks; and databases to support mobility management, i.e. the HLR and VLR. Each SAN tracks satellites within its view and routes traffic to a user through the most robust link. Mobility management follows the GSM standards. Information between SANs for the purpose are exchanged over the ICONET.

9.2.5 Hybrid orbit systems

ELLIPSO system

The ELLIPSO system is one of the five big-LEO systems granted licences in the USA for commercial MSS operation (Castiel and Draim, 1995; Draim and Davidson, 1999). The system is expected to provide public mobile communication services through cellular phone-sized sets, fixed and semi-fixed installations in Canada and most of the world around 2003. ELLIPSO intends to provide services to users whose communication needs are not served by existing mobile or fixed telephone systems. The target market will therefore be residents of remote areas, remote installations such as offshore mining and exploration sites, tourists, mountaineers, emergency services, security services, cross-country transportation fleets, aviation, etc.

Mobile services will be provided by a number of types of terminals. Cellular phone-sized hand-held phones for personal use resemble the terrestrial phone in appearance as well as radiated power, weight, size and operation. Other models envisaged include vehicle-mounted, ruggedized versions for maritime and military applications. Semi-fixed/fixed installations include coin/card-operated solar payphones, private residential phones consisting of rooftop antennas wired to building phone jacks where regular phones may be plugged in, and a ruggedized transportable version for moving sites. Other terminal designs include data-only terminals, and paging and polling terminals.

There are plans to sell satellite time wholesale. Services on offer include telephone services, data transfer, facsimile, paging, voice mail, messaging and geo-positioning services. Figure 9.26 shows the main components of the system.

Due to the rather tentative nature of the target market, the system has been designed for profit with a conservatively sized market. There are elements in ELLIPSO's system design which allow capacity to be tailored to suit geographical variations in demand. An initial market of fewer than 1 million subscribers is needed to obtain profitability, of which 400,000 users are targeted in the USA alone and the remainder in the rest of the world.

The system has been designed to deploy a unique patented constellation design, which judiciously combines highly elliptical and circular orbits to favour coverage over desired regions of the Earth during daytime hours when higher capacity is required. This approach allows an efficient matching of demand and capacity, with a lower number of satellites than in an unbiased constellation.

The Earth's distribution of land masses and population forms the basis of the constellation design. The northern hemisphere above 40° N contains several times more land mass than the southern hemisphere south of 40° S – the former includes Europe, about half of the United States and Canada, the CIS and parts of Japan. In the southern hemisphere, most of the land lies within about 52° S and most of the world's population is confined to areas above the 40° S latitude.

Inclined elliptical orbits are used to provide coverage to high latitudes in the northern hemisphere and a circular equatorial orbit provides coverage in the remaining portions. The orbital design comprises two 116.5° inclination, elliptical Sun-synchronous orbits, named BOREALIS[T], each having a perigee altitude of 520 km and an apogee of 7,846 km over high northern longitudes, such that they provide high elevation angle coverage over Canada and North America. The reason for the choice of Sun-synchronous orbits is that such orbits exhibit the same relationship with respect to the Sun and hence it is possible to maintain the same time of day relationship throughout the year (see Chapter 2). This enables a satellite to cover the northern hemisphere during daylight hours. In

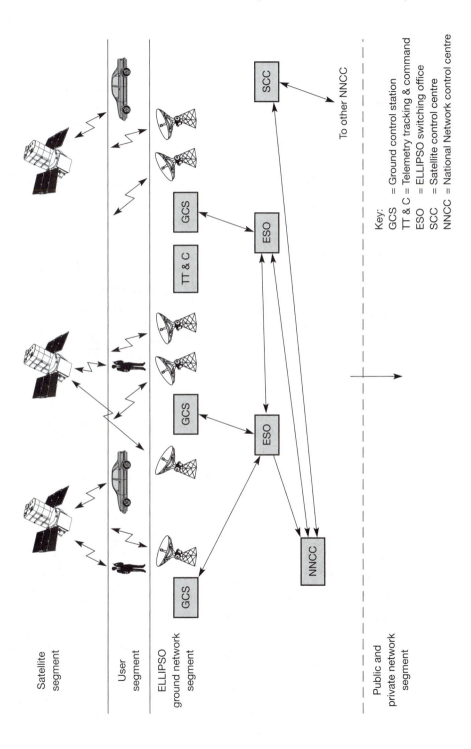

Satellite segment

User segment

ELLIPSO ground network segment

Public and private network segment

SCC

To other NNCC

GCS

TT & C

ESO

ESO

NNCC

GCS

GCS

Key:
GCS = Ground control station
TT & C = Telemetry tracking & command
ESO = ELLIPSO switching office
SCC = Satellite control centre
NNCC = National Network control centre

Figure 9.26
Main components of
ELLIPSO satellite
system (Castiel and
Draim, 1995)

practice, this property can be achieved by selecting the orientation of the plane with respect to the Earth and tilting the perigee–apogee line so that the apogee is over the desired northern region at the desired time of day. Launching elliptical orbit satellites is less expensive than a circular orbit satellite with similar coverage. The satellites remain quiescent below the equator, which reduces power consumption while the batteries are charged. ELLIPSO system designers estimate that four satellites in a Sun-synchronous orbit provide service equivalent to six circular orbit satellites. There are five satellites in each orbital plane, with a period of approximately 3 hours. The orbital parameters are chosen such that the satellites remain in the northern hemisphere during daylight hours for an average $\frac{2}{3}$ of the orbital period.

The lower northern latitudes, tropical regions and populated regions of the southern hemisphere are served by satellites in an equatorial circular orbit called CONCORDIA. The plan is to deploy six satellites at an altitude of 8,040 km (4.8-hour period) so that continuous coverage is available up to the 55° S latitude which contains the most populous regions. The Tropic of Cancer approximately divides the service area of the two constellations. There is a wide band of coverage overlap. Constellation deployment can be adjusted to meet market demands. The constellation is designed to provide dual coverage to achieve diversity gains above the 40° S latitude and single coverage up to a latitude of 55° S.

Figure 9.27 shows a representation of the ELLIPSO constellation. Figure 2.29 (d) presents coverage statistics of both the orbits taken over a two-week period.

Figure 9.27
Representation of
ELLIPSO constellation
(Castiel and Draim,
1995)

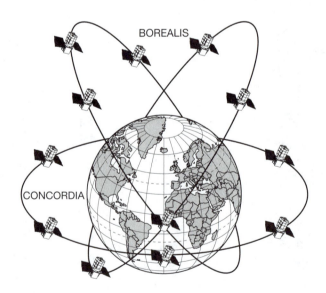

All satellites of the constellation use an identical design to minimize design risk and ease manufacturing. Each satellite, weighing 1,350 kg, is three-axis stabilized using two large solar arrays of about 40 sq.m. The transponders are transparent, with well-proven designs. The antenna uses 61 spot beams in the service link, arranged in a circular symmetry. The accessing scheme is CDMA, allowing the same 12 MHz band to be used in each spot beam. The service link operates in 1,610–1,621 MHz in the uplink and 2,480-2,500 MHz in the downlink. Diversity is used for improving link reliability. A single beam operating in the K_u-band uplink and C-band downlink is used in the feeder link. Satellites use reaction wheels and thrusters for attitude control and station keeping, with enough battery reserve to allow full eclipse operation. The mean lifetime of each satellite is five years. Up to six satellites may be launched at a time.

The ELLIPSO voice codec uses a technique known as the code excited linear predictive (CELP) algorithm which can provide high quality voice at 4.15 kbps. In addition, a number of digital services in the range 300–9,600 bps will be supported. These include Hayes modem data, facsimile, message forwarding, paging and geo-location.

Signals are convolutionally coded and interleaved, and the pseudo-random code spreads the signal by a factor of 1,000. Signals may be transmitted over a 3 MHz wide band or a 7.5 MHz band. Generally, the wideband transmissions are assigned for mobile and hand-held applications.

User terminals lock to a broadcast channel when idle. The broadcast channel announces system status and incoming calls; in addition, the channel is used by terminals for maintaining synchronization. Calls are established through exchange of a sequence of packets over the broadcast channel in the forward direction, and response and request channels originating from a mobile. After a call has been successfully established, the ground control station assigns a spreading code key to the user, which is used during the call. In-band signalling, as, for example, necessary for power control, is multiplexed with data signals. Some short messages or functions, such as paging and position location, are carried over the signalling channels. The system is capable of handling advanced signalling and call services offered by the PSTN.

ELLIPSO, like Globalstar, has chosen CDMA as the multiple access scheme, for similar reasons. In particular, CDMA offers frequency reuse in each beam and satellite, the capability to combine signals when using path diversity, the capability to provide soft handover, less difficulty in spectrum coordination, the capacity enhancement through voice activation, resistance to fading, the potential to make precise time measurements for geo-location and the potential for privacy.

ELLIPSO's ground network comprises the ground control stations (GCS), national network control centres (NNCC), ELLIPSO switching offices (ESO), system coordination centre (SCC) and tracking, telemetry and command centres (TTCC). Some of their main functions are:

- call connection;
- power control for optimum utilization of capacity;
- selection and maintenance of optimum transmission path;
- interface to public network;
- subscriber record keeping, validation, and transaction accounting;
- dissemination of network information to all entities, including user location information maintenance, dissemination and geo-location;
- system-related functions – planning, system health monitoring/maintenance, resource allocation, creation of optimum network configuration.

The ground network shown in Figure 9.28 has been modelled on the GSM architecture, equipment and standards to facilitate integration with terrestrial networks. The network utilizes modern switching and trunking facilities, and CCITT and ANSI signalling systems. Features include CCITT Signalling System 7 attributes such as call forwarding and call identification, domestic and international subscriber roaming, etc.

A ground control station (GCS) has three antennas, two of which are used for communication, while the third is used to acquire a new satellite. The GCS interfaces with the system's switching office for communications and signalling and with the area's NNCC for system status and control information. The GCS also determines the user's location for establishing connectivity within the network. Functionally, a GCS is similar to a terrestrial base site controller and emulates it to benefit from terrestrial technology and encourage integration.

The switching office (SO) interfaces with the PSTN and includes X.25 and SS7 network interfaces. The SO is also connected to NNCCs for control and status purposes, transfer of call records and subscriber information. The SO

Figure 9.28
ELLIPSO's ground network architecture (adapted from Castiel and Draim, 1995)

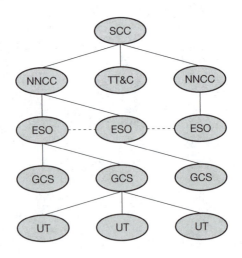

contains a fault-tolerant switch, data processing system, and subscriber and network databases for managing subscriber profiles and call connectivity. It can be located at any convenient position – at a GCS, NNCC, a terrestrial MSC – and may serve one or more GCSs.

The NNCC controls the ELLIPSO market in a region and is the central facility for network planning, management and accounting. Its functions are handling of subscriber affiliation, record keeping, accounting, roamer call routing, GCS management, satellite and network resource management system. The NNCC interfaces with the SO, GCS and SCC to carry out these functions. The interface with the SCC is used for transactions reporting global roaming management and system resource management.

The SCC manages the overall ELLIPSO system functions, interfacing with each NNCC of the network and with the TT&C. Its functions, in conjunction with these interfaces, include world-wide call transactions, maintenance of the subscriber database together with their current status/location, coordination of routing for international roaming, and management of network and spacecraft health and status. Global system planning and resource allocation is also a part of the SCC's function. It is envisaged that separate SCCs may be established for BOREALIS and CONCORDIA constellations.

ELLIPSO's TT&C will be situated in two locations so that together they can see each satellite in every orbital revolution. The function of the TT&C is to monitor, control the function and integrity of the spacecraft, maintain orbit, determine ephemeris, and assist the NNCC in managing satellite allocation and handoffs.

The location of each network component will be decided on the basis of market needs, which are believed by ELLIPSO to be influenced by population distribution, demand for ELLIPSO services, the status of the prevailing telecommunications infrastructure, etc. Technically, it is possible to provide a world-wide service with a minimum of 14 ground control stations strategically located. However, factors such as trunking costs and territorial interests will influence their number and location. For example, it is envisaged that markets such as the USA, spanning large geographical areas, will require more than one GCS to provide acceptable GCS–subscriber connectivity statistics, bearing in mind that common satellite visibility between the two is essential for making a connection. GCSs are likely to be located at the extremes of sub-satellite travel over the market area; intermediate stations may be added if better services can be offered through trunking economy.

Further reading

Bains, N. (1999) 'The ICO system for personal communications by satellite', *Sixth International Mobile Satellite Conference,* Ottawa, June, 88–93.

Castiel, D. and Draim, J.E. (1995) 'The Ellipso™ mobile satellite system', *International Mobile Satellite Conference,* Ottawa, IMSC 1995, The Fourth International Mobile Satellite Conference, Ottawa, co-sponsored by Communications Research Centre/Industry Canada and Jet Propulsion Laboratory/NASA, 409–18.

Colcy, J.N. and Dutronc, J. (1990) 'Land mobile communications in K_u band: result of a test campaign on EUTELSAT IF1', *International Journal of Satellite Communications,* 8.

Colcy, J.N.; Dutronc, J.; Ames, W.G. (1990) 'The EUTELTRACS position reporting system characteristics and performance', *NAV 90* – The 1990 International Conference of the Royal Institutes of Navigation, Warwick, UK.

Colcy, J.N. and Steinhäuser, R. (1993) 'EUTELTRACS: The European experience on mobile satellite service', *IMSC* 1993, 16–18 June, Pasadena, California, JPL Publication 93-009, 261–6.

Dietrich, F.J. (1997) 'The Globalstar satellite communication system design and status', *International Mobile Satellite Conference,* 1997.

Draim, J.E. and Davidson, C.S. (1999) 'The ELLIPSO ™ system – An optimal solution for the Canadian mobile satellite communications market', *IMSC '99, Sixth International Mobile Satellite Conference,* Ottawa, co-sponsored by Communications Research Centre and the Jet Propulsion Laboratory June, 76–80.

Freeman, R.L. (1996) *Reference Manual for Telecommunications Engineering*, John Wiley & Sons, second edition.

Hutcheson, J. and Laurin, M. (1995) 'Network flexibility of the Iridium^R global mobile satellite system', *International Mobile Satellite Conference,* Ottawa, IMSC 1995, The Fourth International Mobile Satellite Conference, Ottawa, co-sponsored by Communications Research Centre/Industry Canada and Jet Propulsion Laboratory/NASA, 503–7.

Inmarsat (1992) 'Project 21: The development of personal mobile satellite communication', June.

Inmarsat website **http://www.inmarsat.com**

Jacobs, I.M. (1989) 'An overview of OmniTRACS: the first operational two way mobile K_u band satellite communication system', *Space Communications*, 7 (1), December.

Johanson, G.A.; Davies, N.G.; Tisdale, W.R.H. (1993) 'The American mobile satellite system: implementation of a system to provide mobile satellite services in North America', *Space Communications*, 11, 121–8.

Nguyen, N.P.; Buhion, P.A.; Adiwoso, A.R. (1997) 'The Asia Cellular Satellite System', *IMSC 1997, Proceedings of the Fifth International Mobile Satellite Conference,* Pasadena, California, 16–18 June, co-sponsored by NASA/JPL and DOC/CRC, JPL Publication 97–11, Jet Propulsion Laboratory, Pasadena, California, 16 June, 145–52.

Richharia, M. (1994) 'Inmarsat space segment', *Infocom '94*, Bombay, December.

Schindall, J. (1995) 'Concept and implementation of the GLOBALSTAR mobile satellite system', *IMSC '95, Fourth International Mobile Satellite Conference*, Ottawa, co-sponsored by Communications Research Centre/Industry Canada and Jet Propulsion Laboratory/NASA, A-11 – A-16.

Schoen, D.C. and Locke, P.A. (1990) 'The ORBCOM data communication systems', *IMSC 1990, The Second Mobile Satellite Conference*, Ottawa, co-sponsored by NASA/JPL and Communications Canada, 267–72.

Schoen, D.C. and Locke, P.A. (1995) 'ORBCOM – initial operations', *IMSC 1995, The Fourth International Mobile Satellite Conference*, Ottawa, co-sponsored by Communications Research Centre/Industry Canada and Jet Propulsion Laboratory/NASA, 397–400.

Wiedeman, R.A. and Viterbi, A.J. (1993) 'The GLOBALSTAR mobile satellite system for worldwide personal communications', *IMSC 1993*, 16–18 June, Pasadena, California, JPL Publication 93-009, 291–6.

Related satellite systems   10

10.1 Introduction

Satellite-aided navigation, satellite-delivered radio broadcasts to individuals, distress and safety-related services, and fixed satellite service VSAT systems are similar to MSS services – all very useful satellite-delivered services for business, groups or individuals delivered through small, low-cost terminals. But formally they do not belong to the MSS due to differences in service characteristics, tradition and radio regulations. While each is a subject in its own right, it was felt that a chapter summarizing the main characteristics of such systems would be an appropriate and useful addition to the book, as they have certain technical similarities, appear similar to MSS products to users, and synergistic applications continue to emerge.

The chapter addresses distress and safety-related services, navigation, and direct radio broadcast to individuals in more detail. A brief section introduces the interesting topic of direct television to mobiles. In the introductory chapter, we mentioned similarities and competition between VSAT and MSS services at the higher bit rate applications end. Here we will introduce salient technical features of VSAT. In previous chapters, we mentioned terrestrial mobile system architecture in a number of instances because of an increasing convergence between these environments. Therefore a section outlining the basic concepts of a cellular system has been included for a better appreciation of terrestrial technology.

10.2 Distress and safety systems

Radio systems have been used for distress applications since the turn of the twentieth century. Until recently, the system used in ships comprised Morse telegraphy on 500 kHz, and installation of the system was mandatory in ships of 1,600 gross registered tonnage (GRT) and above, and in all passenger ships. The primary objective was for ship–ship communication within a maximum range of about 250 nautical miles. The system had a number of disadvantages, such as manual alerting and a need for ships to maintain a constant watch.

There were instances of ships being lost at sea without being able to contact anyone for help.

To address the problem, the International Maritime Organization (IMO) began to investigate ways to improve the system in the early 1960s by use of modern terrestrial and satellite technology (O'Neil, 1992). As a first step, the IMO established the International Maritime Satellite Organization (Inmarsat; now the International Mobile Satellite Organization), which gave a distress communication capability in all maritime regions below about ±75° latitude. The terrestrial system introduced included digital selective calling (DSC) and direct printing telegraphy, i.e. telex by radio, satellite and terrestrial systems. The Cospas-Sarsat system added further capability, enhancing satellite coverage up to the poles. These systems laid the foundation of an operational concept of a fully automatic, global distress communication system known as the Global Maritime Distress and Safety System (GMDSS). The ITU extended its cooperation by assigning specific frequencies for search and rescue (SAR).

The IMO introduced the GMDSS system from 1 February 1992. The system is quicker, simpler, more efficient and automatic, with the following features and goals:

- fully automatic, eliminating the need for manual watch keeping;
- applicable to general and distress communications;
- all distress messages should be heard and addressed on shore;
- equipment should be easy to operate.

The system is meant to provide a rapid alert to ships in the vicinity and to rescue authorities on shore in case of distress, enabling them to assist in a coordinated search for the affected party. Furthermore, it provides urgent and safety communications, and dissemination of maritime safety information such as navigation and meteorological warnings.

There are nine essential communications to be performed by ships falling into the category using equipment that would meet the requirement in each region specified by the IMO. The communication functions are: distress alerting (ship–shore, shore–ship and ship–ship), search and rescue coordination communications, on-scene communications, signal location, transmission/reception of maritime safety information and general radio communication; the sea areas are categorized as A1 to A4 on the basis of availability of communication facility. Sea area A3 refers to the area outside areas 2 and 3 within coverage of Inmarsat geostationary satellites which provide a continuous alerting facility. The equipment, depending on sea areas, comprises 9 GHz radar transponders, a NAVTEX receiver, equipment to receive Inmarsat's Enhanced Group Call system (an international SafetyNet Service), a satellite emergency position indicating radio beacon (EPIRB) which is free floating with automatic activation capability, and equipment for continuously monitoring DSC on channel 70. Equipment must comply with the performance standard specified by the IMO. Additionally, there are equipment availability specifications as well as a requirement for every ship to include personnel suitably qualified for distress/safety radio communication.

Cospas-Sarsat search and rescue system

Cospas-Sarsat is a global search and rescue system for sea, air or land, in use since 1982, and is credited with saving thousands of lives. Figure 10.1 illustrates the number of SAR events in which the Cospas-Sarsat system participated in the period 1989–97.

The system consists of small radio transmitters which transmit distress signals in the case of a distress situation; the signals are received by a constellation of LEO satellites and retransmitted to ground stations known as local user terminals (LUT), which send the location information to a mission control centre (MCC). The MCC initiates a search and rescue operation through an appropriate rescue co-ordination centre (RCC) or search and rescue points of contact (SPOC).

The concept of a satellite-aided search and rescue system, using LEO satellites, was proposed in the 1970s and proven in the USA, Canada and France, which helped the setting up of a joint experiment for Search and Rescue Satellite-Aided Tracking (Sarsat) by NASA (USA), the Communications Research Centre/Department of National Defence (Canada) and the French Space Agency. Later, the former USSR joined the experiment and agreed to develop the COSPAS (Cosmicheskaya Sistyema Poiska Avarinyich Sudov – translated as 'space system for search of vessels in distress') component, which would be a compatible search and rescue system. The combined system became known as the Cospas-Sarsat system (King, 1999). During the demonstration phase beginning in 1982, the benefits of the system were realized when many lives were saved in various countries. Due to the success of the experiment, the system, instead of being switched off in the mid-1980s at the end of the demonstration phase, was declared a world-wide operational system in 1985 by the four founding countries. In 1987, the Cospas-Sarsat secretariat was established at Inmarsat

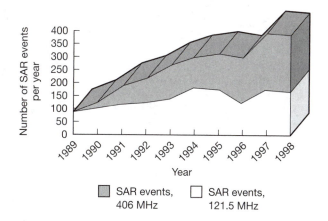

Figure 10.1

Number of search and rescue events in which the Cospas-Sarsat system participated in the period January 1989–97 (Courtesy: Cospas-Sarsat)

HQ in London; in 1988 a formal inter-governmental agreement was signed between the four founding countries, and associating the IMO, the International Civil Aviation Organization (ICAO) and the ITU, thus ensuring long-term availability of the system to all states on a non-discriminatory basis. In 1997, there were 30 member countries.

The implementation and operating cost is shared by member countries. The space segment is provided by the four founder countries. Users pay for purchasing the distress beacon transmitters but not for the use of the system. The programme is managed by a council consisting of representatives of member countries and supported by a secretariat located in London.

It is anticipated that the system will continue to provide SAR services, despite the proliferation of mobile satellite systems with similar facilities. Most of these systems are commercial, levying call charges for any service, except for Inmarsat's distress and safety related wing which has specific commitments to international search and rescue as part of the GMDSS.

The main system components are illustrated in Figure 10.2.

The space segment of the system comprises a constellation of four LEO satellites in a near polar orbit, called the Low Earth Orbit Search and Rescue (LEOSAR) system, and a geostationary system overlay, called the Geostationary Orbit Satellite Search and Rescue (GEOSAR) system, introduced recently. The GEOSAR system, however, operates only with 406 MHz beacons through a transparent repeater. The constellations complement each other and are illustrated in Figure 10.3. The GEOSAR system has the advantage of providing near-instantaneous alerting within ~ ±75° latitude and LEOSAR provides true world-wide coverage with its own position estimation capability. Furthermore, LEOSAR has an inherent capability to reduce the probability of total signal blockage from an obstacle due to the changing transmission path profile.

Figure 10.2
Main components of
Cospas-Sarsat system
(Courtesy: Cospas-
Sarsat)

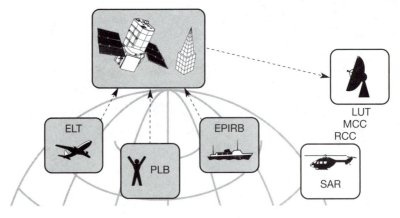

ELT   = Emergency Locator Transmitter
PLB   = Personal Locator Beacon
EPIRB = Emergency Position Indicating Radio Beacon
SAR   = Search and Rescue

LUT = Local User Terminal
MCC = Mission Control Centre
RCC = Rescue Coordination Centre

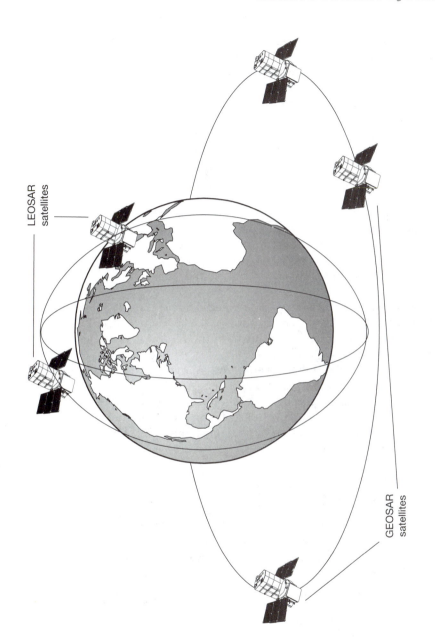

Figure 10.3
Cospas-Sarsat
constellation (Courtesy:
Cospas-Sarsat)

LEOSAR
satellites

GEOSAR
satellites

LEO was selected to maximize the Doppler change used for position determination, and to be able to receive low power transmissions from small transmitters. Each satellite in a polar orbit can cover the entire Earth; however, due to the limited number of satellites used in the system, the coverage at any location is intermittent. Typical waiting time at mid-latitude is less than 90 minutes. When the constellation is replenished, more than four satellites become available, until the oldest satellite(s) become unusable. Two Cospas satellites are supplied by Russia and the remaining two by the USA, piggy-backed on NOAA meteorological satellites. The Sarsat payload of 121.5 MHz and 406 MHz is supplied by Canada and France. Some satellites also have a 243 MHz payload. Cospas satellites are placed at an altitude of 800 km and inclination of 83° and NOAA satellites are 850 km-altitude Sun-synchronous satellites at an inclination of 98°. At these altitudes, the orbital period is about 100 minutes and satellites cover a 4,000-km-wide swathe as they move. The entire Earth is covered within 12 hours. Thus with four satellites distributed in various orbital planes, a waiting time of less than one hour is typical at mid-latitudes. Visibility time from the ground is of the order of 10–15 minutes on each pass.

The geostationary overlay uses three geostationary satellites, two of which – GOES East and GOES West – are supplied by the USA and the third, INSAT-2A, is provided by India. Figure 10.4 illustrates the footprint of the constellation (Cospas-Sarsat).

Due to negligible Doppler variations from geostationary satellites, the Doppler position determination technique is inaccurate. Thus the position information is obtained through another navigation source. The position can also be derived from the LEOSAR constellation. The signals transmitted by the distress transmitter are received by ground stations called geostationary LUT (local user terminal) or GEOLUT.

Figure 10.4
Geostationary footprint of Cospas-Sarsat (Courtesy: Cospas-Sarsat)

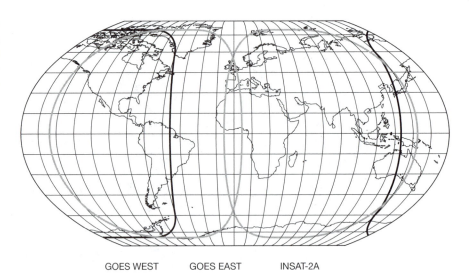

GOES WEST   GOES EAST   INSAT-2A

Three types of position location transmitters are used by the Cospas-Sarsat system – emergency locator transmitters (ELT) used in aviation, emergency position indicating radio beacons (EPIRB) used in maritime environments, and personal locator beacons (PLB) carried by individuals. ELTs can be activated automatically in the case of an aircraft crash; similarly, an EPIRB can float and transmit radio beacons automatically in the case of a sinking ship. The radio beacons, transmitted at a frequency of 121.5 or 406 MHz, are transponded back to a large ground station known as a LEO local user terminal (LEOLUT); as the LEO satellite in view of the transmitter moves rapidly with respect to the transmitter, the received signal is Doppler shifted. The position of the observer can be estimated at the LUT or the satellite (for 406 MHz transponders), by measuring the Doppler frequency and knowledge of the satellite's orbital location. Usually, two passes are needed to resolve the position ambiguity when using the older 121.5 MHz system; however, the introduction of a more stable 406 MHz beacon enables ambiguity to be removed in a single pass. To ensure reliability, consistency and compatibility, each LEO has to comply with Cospas-Sarsat specifications and procedures.

There can be up to four channels supported by a LEOLUT – these are the 406 MHz Search and Rescue Processor (SARP) channel; 406 MHz Search and Rescue Repeater (SARR); and 121.5 and 243 MHz SARR. The SARP satellite channel transmits partially processed data, which provide identification, transmit time and received frequency for each distress burst of 2,400 bps, as well as time-tagged Doppler measurements. The SARP channel provides intermittent global coverage. The SARR channel operates as a repeater, hence offers a local mode of operation and further processing is necessary at the LEOLUT. The 121.5 MHz/243 MHz SARR channels are similar to the 406 MHz SARR channel, except that these beacons do not contain identification information.

The 121.5 MHz payload retransmits to the LUT in real time and therefore the signal is received only if there is an LUT within the visibility of the satellite. This is known as the local mode of transmission. The 406 MHz beacon has a wider capability. In addition to transponding the beacon, the signal is partially processed, stored and retransmitted in real time. The stored data is transmitted continuously for several orbits, so that all LUTs would have received the location information, and hence this mode of operation is known as global mode. The message is received by a large number of LUTs, thereby building high redundancy in the system. Most LUTs are fully automated; many are unmanned and can be operated remotely from a mission control centre (MCC). The ground stations deploy a tracking antenna which generally uses program tracking. The transmission formats of all satellites are compatible with LUT capabilities, which may differ to some extent. Each LUT has to comply with system specifications. There are about 38 LUTs, dispersed around the world.

As mentioned, a distress message received at an LUT is passed to the MCC, which transfers the information to an appropriate rescue coordination centre (RCC) or rescue points of contact (RPOC) or other MCC where the data is processed, including removal of repetitions and false alerts. The MCC also provides system information to assist operations, including satellite ephemeris and time calibration used for beacon location, coordination of messages and status of the space and ground segments. To exchange a free flow of information, the MCCs are interconnected. MCCs have to comply with Cospas-Sarsat's performance specifications to meet the organization's reliability and system integrity standards. Regular exercises are performed to ensure correct functioning and compliance of various elements of the overall system.

The distress transmitters operate at 121.5 MHz or 405 MHz. The 121.5 MHz terminals were in use prior to the introduction of satellites and hence were not designed for satellite systems. Most of the 600,000 units of this type in use in 1998 were on aircraft. The amplitude modulated transmitters operate at 0.05–0.1 W but there are limitations in this technique – poor frequency stability limits the number of simultaneous transmissions to 10; location accuracy is limited to ~20 km; and little information about the user's identity is provided by such transmitters. Nevertheless, the system is credited with saving many lives.

The 406 MHz beacon transmitters were designed for satellite systems and therefore include improved features, such as high peak power output (5 W)/low duty cycle ($\frac{1}{2}$ s burst every 50 s), higher frequency stability, unique identification code for each transmitter, digital transmissions permitting storage on satellite and the use of ITU's distress band. Consequently, 406 MHz has a larger system capacity, with a capacity of 90 simultaneous transmissions, better accuracy (~2 km), user identification and negligible interference from aircraft emissions. The low bit rate message, containing information about country of origin, originator's identification and position data (optional), phase modulates the carrier. There were 156,000 406 MHz beacon transmitters in 1998.

10.3 Navigation systems

10.3.1 Background

Prior to the advent of radio systems, sailors used celestial bodies, known landmarks, lighthouses etc. as their navigation aids. Navigation received an impetus after the invention of radio by extending the navigation range far beyond the traditional visual, and overcoming the limitation of visual navigation during adverse weather conditions. The first radio systems, providing continuous reference signals, were installed in 1921 for terrestrial/maritime usage, at about the same period when airborne navigation aids were installed. Navigation systems have continually improved since.

The *Loran (long-range navigation) system* was developed for marine and airborne applications. The system is based on the time difference navigation principle, wherein the time difference in signals arriving from a reference station and a slave station are measured at a receiver. Points where signals are received with equal time difference lie on a hyperbola, and if two stations are used, a pair of parabolas are created. The receiver then lies at the intersection of these parabolas. The modern version, called Loran-C, operates at 100 kHz with a range of over 3,330 km from the master station.

The *Decca system* is named after the British company which introduced it in 1946. The system uses the same principle as the Loran system. However, here the reference and slave stations transmit continuous signals instead of the bursts transmitted in the Loran system, and phase difference is measured instead of time difference. The system is widely used in ships, providing highly accurate fixes in the 160–480 km range. It is used less for aeronautical systems.

The *Omega system* is also based on the same hyperbolic principle as Decca. This long-range system operates at very long wavelengths and provides world-wide coverage through eight synchronized sites, transmitting encoded signals continuously.

A terrestrial system known as very high frequency omni-directional radio range (VOR), which has evolved since 1930, continues to be used. The system consists of a very high frequency (VHF) omni-directional transmitter, transmitting two signals simultaneously. Aircraft measure phase differences which are calibrated and displayed to the pilot as a bearing. VHF transmissions minimize propagation-related effects and interference. The VOR, used in conjunction with distance measuring equipment (DME), provides a basic navigation system to aeroplanes. An internationally approved DME standard consists of an aircraft radio transmitter which transmits on one of the 126 approved frequencies when requiring a distance measurement. The signal is received at a ground station and transponded back after a 50 µs delay. The time difference between the transmitted and received signals is estimated at the aircraft and used to calculate the range.

The era of satellite navigation dawned in the early 1960s when a system named Transit was launched by the USA. At present, the GPS system is a dominant world-wide navigation system, with a myriad civilian applications entwining a multi-billion-dollar industry. The GPS system is addressed in more detail later, including current thinking on the possible evolution of the industry. Navigation systems have now become an integral part of mobile satellite systems, with applications ranging from vehicle fleet management and terminal position reporting for mobility management to spacecraft attitude and orbit control systems. But before addressing specific navigation systems, let us briefly review satellite navigation principles.

10.3.2 Satellite navigation principles

Navigation systems may be categorized as *active* or *passive* depending on the extent of user participation; and as *single* or *multi-satellite* contingent on the preferred navigation technique. In active systems, users transmit a signal which is processed at a hub for estimating user location. In passive systems, users receive signals and process them to estimate their location. For either mode, single or multiple (1–4) satellites can be used.

In single satellite systems, location is fixed by measuring the Doppler shift and estimating the user position using the satellite ephemeris, usually broadcast by each satellite. The solution for intersection of the Doppler 'sphere' with the Earth provides two possible locations – the ambiguity is resolved by considering the deviation caused by rotation of the Earth. Accuracy can be improved by transmission at two frequencies, which allows application of group delay correction. The principle is used in the Argos (French) system, which is an active system, and in the Transit (USA) and Cicada (former Soviet Union) systems, which are both passive systems.

Position can also be calculated by estimating the range from two satellites. The user position lies at the intersection of three spheres – those around each satellite and the Earth. This technique is used in the OmniTracs (USA) and EUTELTRACS (Europe) systems, which are identical active systems.

In a three-satellite system, range is estimated through three satellites; the user is at the intersection of three spheres centred at each satellite and the Earth. A fourth satellite becomes necessary to resolve uncertainties in range estimation (see next section). The Navigation System with Time and Ranging Global Positioning or Navstar Global Positioning System (GPS) (USA), more commonly called GPS, and the Global Orbiting Navigation Satellite System or GLONASS (Commonwealth of Independent States), both passive systems, use the principle. Technical details of GPS systems are presented in the next section. Included is the concept of differential GPS, where the accuracy of the fix provided by GPS is augmented by using correction data obtained at a reference site transmitted to users over a separate radio channel. Due to spatial and temporal correlation in error, the same corrections may be applied by users, giving a significant improvement provided they are not very far from the reference site.

10.3.3 Navigation system examples

A variety of navigation systems have developed and evolved over the years. Their applicability depends on the desired accuracy, environment, cost, and to some extent historic association. Table 10.1 compares the accuracy of various types of navigation systems (source: Schänzer, 1995).

The Transit system, which uses the Doppler signature technique, comprises a constellation in a 1,100 km circular polar orbit. Each satellite transmits departure of its nominal frequency and orbital location continuously at about 150 and 400 MHz. The signals are used to estimate the Doppler shift and rate of Doppler change by comparing the measured and expected frequency. The information can be processed to estimate an observer's position. The system provides an accuracy of about 160 m, which is adequate for ship navigation but not for aircraft navigation, due to uncertainties in position fixes introduced by Doppler components caused by aircraft motion. For example, a Doppler uncertainty of 1 m/s introduces an error of 1 km in a location fix. Moreover, the system is not available continuously – users having to wait up to 2 hours for a satellite pass. Cicada, operated by the former Soviet Union, was based on the same principle.

These limitations were removed in the Navstar GPS and GLONASS navigation systems (see the next section).

The French Argos system is a single satellite active navigation system. It uses navigation transponders on NOAA satellites which operate in ~805 km polar orbit. When requiring a fix, users transmit a signal which is relayed by the satellite

Table 10.1
Accuracy of various types
of navigation systems
(Schänzer, 1995)

System name	Accuracy (2 standard deviation) m	Comments
Differential GPS	0.3–6	Accuracy depends on distance to reference station
Instrument landing system (ILS)	5–10	Used for aircraft landing
Microwave landing system (MLS)	5–10	Developed to replace ILS but never took off for cost and other reasons
GPS (P-code)	20	Military applications
GLONASS	20	Military applications
VOR and DME	60–180	Used for aircraft navigation; see text
GPS (C/A code)	100	Civilian version of GPS
Transit/Cicada	450	Early satellite systems for military/civilian applications; see text
Omega	3,600–7,200	
Aircraft inertial navigation	300–20 km	Best accuracy near take-off/touch-down

to hub stations where the Doppler curve is extracted from the user signal. Intersection of the Doppler 'sphere' with the Earth provides the location of the user. East–West ambiguity is resolved by taking Earth rotation into consideration.

10.3.3.1 Global Positioning System (GPS)

The Global Positioning System (GPS) was developed by the US government for the Department of Defense. It is essentially a US military system, but it offers navigation services to civilians; however, at present, there is no law which mandates the service to be made available for commercial application (Chien, 1991). Position fix is obtained through passive receivers by the triangulation method, wherein estimated ranges from four satellites are used to derive the position and altitude of a point. Ranges from three satellites can provide the latitude and longitude of a point on the Earth; the addition of a fourth satellite can provide a user's altitude and correct receiver clock error. It is possible to derive the velocity of the user and precise time information originating from onboard atomic clocks, which have a drift rate of 1 s per 70,000 years. There are two rubidium and two caesium clocks aboard each satellite.

Two types of pseudo-random codes are transmitted on two frequency bands, L_1 (1,575.42 MHz) and L_2 (1,227.6 MHz) – the encrypted Precise-Code (P-Code) meant for US military operation is available on both frequencies, and the unencrypted coarse acquisition code (C/A-code), available without restriction, is transmitted only in the L_1 band, where it is combined with P-Code in phase quad-rature. Codes have low cross-correlation, allowing transmissions from each satellite on the same frequency. Spread spectrum modulation also provides some resistance against multipath and immunity to interference. The C/A code, operat-ing at 1.023 Mbps, is a 1023-bit pseudo-random code repeating each millisecond, and the P-code, operating at 10.023 Mbps, has a cycle of 267 days but it is reset every seven days. Each code is combined with a navigation message comprising the status of the satellite, time synchronization information for transferring from coarse to fine code, clock correction, satellite ephemeris, propagation delay correc-tions and approximate ephemeris and status of the constellation used for signal acquisition. Next generation (Block IIF) satellites are expected to transmit on up to five frequencies, providing unencrypted transmissions for civilian use on three frequencies. We will see later that transmissions on multiple frequencies are used for improving the accuracy of navigation fixes. The P-Code can provide accuracy of less than 8 m, whereas the C/A code, after selective availability (S/A), has a nominal accuracy of around 35 m. To avoid the system being used against itself, the military can degrade the accuracy by using S/A, which increases the error probability by satellite clock dithering and corruption of navigation message data. The S/A function is switched on or off as required by events. According to a recent report, the US government has decided to switch off the S/A function, giving improved accuracy for civilian use. Nevertheless, several techniques exist for improving the accuracy of C/A code fixes.

The GPS constellation comprises 21 satellites and three in-orbit spares. Satellites are in a circular orbit at an altitude of about 20,200 km (orbital period of about 12 hours), controlled by a master control station in Colorado Springs, with five monitoring stations and three ground stations dispersed around the world. The master control station estimates the orbital parameters of each satellite from ranging data collected by the monitoring stations, formats them and transmits them to each satellite. Table 10.2 (Daly, 1993) summarizes the main orbital para-meters of GPS, comparing it with the GLONASS system (see next section).

To estimate range precisely, each user clock must be synchronized to the satellite, which makes the receiver complex and conflicts with the need for a simple receiver. In the GPS system, the problem is resolved by estimating the range from a fourth satellite, which then allows resolution of user clock uncer-tainty. Range is measured by measuring the time shift between identical codes generated at the satellite and the receiver. The code generated at the receiver is time shifted until a maximum correlation is achieved between the transmitted and receiver codes; the time-shift provides an approximate range or 'pseudo' range comprising a number of errors, listed below. Note that user must have knowledge of code to be able to use the system; this feature permits the military to use the higher accuracy P-Code for its purpose. The received signal is around 22 dB below the receiver noise, and therefore it is necessary for the code to be acquired and signals despread prior to clock recovery.

Real range is estimated from the pseudo-range by solving a set of four simultaneous equations. Solution of equations from three measurements pro-

Table 10.2
Some Navstar and
GLONASS orbital
parameters (Daly, 1993)

Orbital parameter	Navstar	GLONASS
Type of orbit	Near circular	Near circular
Semi-major axis (km)	26,560	25,510
Period (minutes)	717.97	675.73
Inclination (degree)	63.0 – block I	
	55.0 – block II	64.8
Orbital separation	120	120
Ground track repeat (days)	1	8
Drift per day, minutes	−4.06	−4.07

vides two results; one of the two solutions can be discarded as it gives unrealistic results. The fourth equation is used to estimate the uncertainties of the receiver's clock. The speed with which position is calculated can be improved by using only three measurements, and may be traded off against accuracy and receiver complexity. GPS transmissions can also be used for determining user velocity by extracting the Doppler information.

There are a number of sources of error in range estimation:

- **Satellite clock offset relative to GPS system time and ephemeris errors:** GPS system time is maintained by the GPS master control station (MCS) through a set of highly accurate Caesium clocks; the clock offset of satellites is measured daily and transmitted to each satellite by MCS for retransmission to receivers which apply the necessary correction algorithmically; satellites themselves incorporate highly stable atomic clocks onboard.

  There is always a certain degree of inaccuracy in estimating satellite ephemeris. Certain components in ephemeris errors cannot be isolated from satellite offset errors and are therefore combined with the satellite clock error in the error budget (see Table 10.3).

- **User clock offset from GPS system time:** As mentioned in the preceding text, user offset can be removed by solving the range equation.

- **Error due to propagation delays:** Errors in range are caused by delays introduced by ray bending and velocity reduction while traversing the ionosphere. The delay is approximately inversely proportional to the square of the frequency and the correction can be derived by comparison at two frequencies, 1,227.6 and 1,575.42 MHz. The correction is derived at the MCS and downloaded to users via the satellites. Tropospheric delays are independent of frequency and can be readily estimated at the receiver by applying an elevation-angle-dependent correction.

- **Group delay** errors are caused by processing delay on satellites; their value is estimated in ground tests and transmitted to users along with other corrections.

- **Multipath** errors are caused by signals arriving at the receiver from different paths.

● **Receiver noise and resolution degradation** are caused by hardware and software limitations in the receiver. Receiver motion can introduce additional errors. Use of well-designed receivers and filtering algorithms such as the Kalman filter reduces the impact of the error.

Each source of error is corrected to the extent feasible, using the techniques mentioned above. Table 10.3 summarizes the extent of residual errors after corrections have been applied (Milliken and Zoller, 1978).

Table 10.3 Estimated 1 standard deviation range errors

Error source	Estimated 1 standard deviation residual range error (m)
Satellite clock and ephemeris	1.5
Atmospheric delays	2.4–5.2
Satellite equipment group delay	1.0
Multipath	1.2–2.7
Receiver noise and resolution	1.5
Root sum square	3.6–6.3

Figure 10.5 represents the concept of a two-dimensional navigation solution including range errors caused by receiver clock errors (derived from Milliken and Zoller, 1978). Note that position is estimated within a zone of uncertainty marked by hatched portions. When uncertainty is removed, the solution converges to a single point, assuming that all other sources of errors are negligible.

The overall accuracy of a fix is determined by a combination of ranging error and the geometry of the satellites used in range estimation; accuracy improves when satellites are wide apart. The effect of geometry is measured in terms of geometric dilution of precision (GDOP); a lower GDOP represents a better geometry. For example, the error is multiplied by 1.5 when satellites are spaced far apart, whereas the multiplier could be five or more for closely spaced satellites. Terminals use a simple algorithm based on the volume of the tetrahedron formed by the vector joining the user to the satellites for selecting the best satellite combination.

Figure 10.6 shows the overall architecture of the GPS system. We have already described the space segment. The control segment comprises a number of monitoring stations around the world, which receive GPS satellite transmissions and forward this data to a master control station located at Schriever Air Force Base, Colorado, USA, where this data is analyzed, GPS and universal standard times compared and corrections for satellite ephemeris and clock data prepared and then relayed to the satellites for user assimilation.

There are a number of improvements being considered for the next generation of satellites. These include:

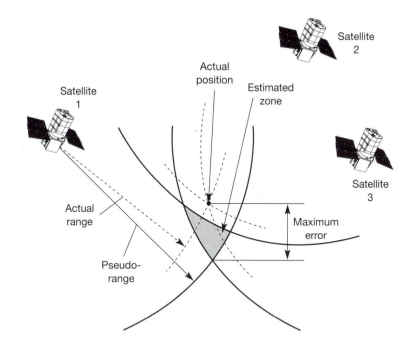

Figure 10.5
Concept of position fix and impact of errors in two dimensions (adapted from Milliken and Zoller, 1978)

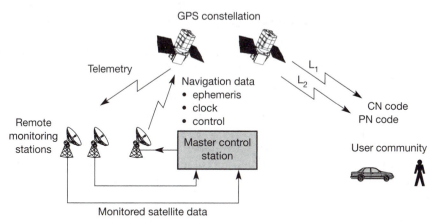

Figure 10.6
The GPS system architecture

- **Increase in satellite EIRP:** With 50 W satellite transmissions, the level received at the receiver is of the order of –160 dBW for C/A code and –166 dBW for the P-Code, making the system susceptible to extraneous noise. Higher power and frequency diversity (see next paragraph) are being considered to improve robustness.
- **Provision of two additional frequencies:** Next generation satellites, called Block IIF, are planned additionally to transmit C/A code in band L_2 and, as yet, undecided frequency, perhaps in the L_5 band (960—1,215 MHz)

currently used by the Aeronautical Radio Navigation Service. Use of frequency diversity improves transmission reliability and accuracy.

- **Auto-navigation feature on GPS satellites:** Satellites will cross-link in space to generate their own ephemeris, enabling them to provide ephemeris data for up to 180 days without ground uploads. At present, the data uploaded from the ground can be saved from 14 to 180 days (depending on satellite generation), but in the absence of regular refresh from the ground the accuracy deteriorates with time.

- **Mitigation of multipath by use of directional antennas:** Multipath noise degrades the accuracy of fixes, which may not be acceptable when very high accuracy is needed (see Table 10.3). Directional antennas tend to reduce the multipath and therefore their applicability is under investigation.

As mentioned, the accuracy of fixes depends on a number of factors, such as the prevailing atmospheric and ionospheric conditions which are corrected to the extent possible by signal processing and downloaded correction data. It has been observed that ephemeris and measurement errors are correlated in time and spatially. Spatial correlation decreases with distance but is quite insensitive to variation in distance – correlation is close even for a distance of a few tens of kilometres and temporal correlation decreases rather slowly with time, for propagation errors and S/A errors are correlated in a 5–10 s time span. An accurate estimate of errors can be obtained when the position of a location is known; if these corrections are transferred to users in the vicinity of the measurement site, the accuracy of the fix is improved considerably. Application of this technique is known as differential GPS (DGPS), where errors in the navigation solution are derived at a reference site and transmitted over a radio link to the receivers. Reference sites can be located from less than a kilometre to over 1,000 km, and as mentioned above, the accuracy of the fix improves with a reduction in distance to the reference station. Measurements and simulation demonstrate errors ranging from tens of centimetres when the reference site is a few kilometres, to 5 m for a distance of 1,000 km. A number of commercial and other bodies responsible for safety operate DGPS systems in various parts of the world. Service is offered through a radio carrier relayed through satellite or terrestrial transmitters. Commercial users include offshore oil platform operators, fleet managers in the trucking industry, etc. Safety-related services are offered, for example, by the US Coastguard, which provides differential corrections at 285–325 kHz in coastal areas free of charge. There are plans to extend the service throughout the USA. Potential beneficiaries include agriculture, forestry, the emergency services, etc.

The accuracy of fix estimates is reduced in the case of failure or unavailability of satellite(s) in the constellation, possibly without the user being aware. An external system which provides such an integrity monitor can be very useful. Geostationary satellites transmitting signals identical to GPS and GLONASS, containing the integrity of satellites and error estimates, can be used to improve the accuracy. Inmarsat-3 geostationary satellites include a navigation transponder and are used for providing this type of overlay in many regions.

This type of overlay approach is being developed in the USA, Europe, Japan and the Federal Aviation Administration (FAA) for application to civil aviation. In fact, such a system is expected to revolutionize air traffic control in the next few years. At present, aircraft navigation relies on ground-based transmitters or inertial navigation systems when aircraft are over oceans, which can cause them to drift several kilometres off course. Prior to landing, VOR and DME systems (see section 10.3.1) provide non-precision navigation aid to pilots, which is particularly useful in adverse weather conditions; pilots then use either visual information or an instrument landing system (ILS) for landing. Clearly, GPS or an equivalent satellite system offers promising solutions for in-flight and the approach part of landing, in particular when the cost of VOR systems is taken into consideration. Another limitation of the existing navigation aid is that it requires planes to fly wide apart – separation of 5–8 km requires use of radars, which are not available for transoceanic flights, causing inefficient use of the air corridor; VOR-based systems require separation of up to 8 miles (13 km). The solution being developed is called Automatic Dependent Surveillance-B (ADS-B), where an augmented GPS receiver on aircraft broadcasts its own position to the ground receiver, thereby removing the need for radar, permitting separation to be reduced considerably. The technique is also being developed for the next generation traffic-alert and collision-avoidance system (TCAS); the existing systems are based on radar monitoring. A GPS-based navigation solution, planned for implementation shortly by the FAA and known as the Wide Area Augmentation System (WAAS), is represented in Figure 10.7. WAAS is planned to provide GPS-based information for en route, departure and approaches under conditions where ceiling and visibility are 200 feet and $\frac{1}{2}$ mile respectively (Bretz, 2000). Systems compatible with WAAS are being developed in Europe and Japan. The FAA also intends to deploy the Local Area Augmentation System (LAAS), which will provide GPS navigation down to the surface, allowing near-blind landing, and airport surface navigation using ground receivers placed in areas within 30–45 km, giving an accuracy of 1 m.

GPS is used widely in mobile communications for applications where it is necessary for users to provide position updates, such as for tracking the positions of each mobile in a fleet, reporting the position for network mobility management, spot beam identification, etc. It is also used on yachts and ships and by explorers, mountaineers, scientists, surveyors, etc; specific examples include delivery or pick-up points for an accurate delivery/pick-up service; real-time response to marine hazards such as oil spills; attitude and orbit control (AOC) of LEO/MEO satellites, etc. Novel applications are continually being developed around the system. An insurance company monitors the usage of cars for insurance charges; a journey at midnight is more expensive than during the day, as people are more alert during the day! Low power GPS receivers can now be installed in watches and personal digital assistants; GPS time is in use for synchronizing CDMA

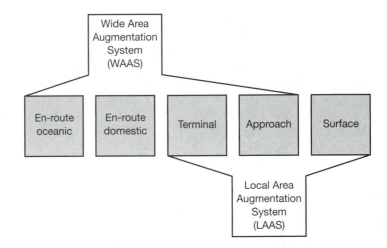

cellular base stations, essential for soft handover; medical services use it for responding to life-threatening emergencies; etc.

10.3.3.2 GLONASS

The GLONASS system, developed by the former Soviet Union for military application, was inaugurated in 1982. There are a number of similarities in these systems. Table 10.2 compares their main orbital parameters. The full constellation comprises 21 satellites with three in-orbit spares at an altitude of 25,510 km inclined at 64.8°. Satellites transmit on two frequencies, at about the same level as GPS satellites, but use the frequency-hopped spread spectrum technique. For civil applications, a single 511-bit code, repeating every 1 ms, is used with differential coding in an RZ format at 50 bps (Daly, 1993; Dale *et al.*, 1989); the military code operates at 10 times the civilian code rate. Frequencies lie in 1,240–1,260 and 1,597–1,617 MHz, the channel spacing for these two bands is respectively 0.4375 and 0.5625 MHz, and the number of channels is 24 in each case. A BPSK modulation scheme is used with a bandwidth of 1 MHz for civilian applications and 10 MHz for military applications. Navigation data and the principle of position-fixing are similar to GPS. However, there are a number of differences which cause difficulty in integrating them, but nevertheless integration is technically feasible. Some notable comparative system features include:

- The GPS system is well maintained with reliable satellites; the GLONASS constellation is not well maintained at present due to lack of financial resources; there were only nine satellites functioning in the first quarter of 1999, degrading system performance, and therefore not many receivers are being manufactured and sold.
- There is a large degree of compatibility in orbital geometry.
- While GPS satellites can be monitored continuously through a network of ground stations, GLONASS satellites can be monitored only within the former Soviet Union territory.

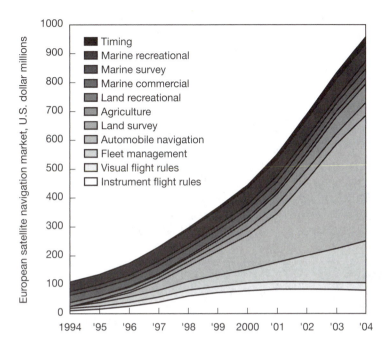

Figure 10.8
Anticipated European
market (Bretz, 2000)
© 2000 IEEE

- The time references of these two systems are different.
- Both systems can benefit from geostationary overlay available from Inmarsat-3 satellites.

10.3.3.3 Galileo

Independently, the European Union is defining a system known as Galileo in order to reduce dependency on the GPS system and foster the European market, estimated to run into hundreds of millions of dollars (see Figure 10.8). The EU is considering a number of alternatives, including the possibility of using the GLONASS constellation. Present plans are to use a 21-satellite constellation placed at an altitude of 24,000 km, with a possible overlay of geostationary satellites. The target is to start operation by 2005 and to become fully operational by 2008, though the project had yet to be agreed by all signatories. One scenario is to have a GPS-compatible system which would provide two complementary systems to users, with obvious benefits.

10.4 Direct satellite sound broadcast

Broadcast Satellite Service (Sound), also known as Direct Broadcast Service – Radio, or Digital Audio Broadcast (or DAB – its digital version), is a service conceived for broadcasting sound directly to mobiles, outdoor and indoor portable sets and receivers using fixed outdoor antennas. DAB service applies to terrestrial or satellite only, or to hybrid systems comprising a synergistic combination of both media. Satellite transmissions cover a large geographical expanse, while terrestrial services cover areas where satellite services are unreliable, such as in dense urban environments. Through some pioneering work in Europe, the USA and by the CCIR, sufficient interest was aroused in the service by the late 1980s, leading to spectrum allocation in WARC-92. A frequency range between 500 MHz and 3 GHz was considered the most appropriate for the service due to spacecraft EIRP constraints, the relatively benign propagation environment and the developed technology status of these bands. Below 500 MHz, spacecraft antenna size becomes prohibitive, and above ~3 GHz the satellite EIRP per channel gets too high for the desired flux density on the ground. The spectrum awarded to DBS (sound) in WARC-92 is shown in Table 10.4 (see the Radio Regulations for the most recent situation and full details). Some nations have applied certain restrictions in the use of this band until 2007.

Table 10.4
Spectrum awarded to
DBS (sound) in WARC-92

Frequency range	Bandwidth (MHz)	Countries
1,452–1,492 MHz	40	All countries except those listed below
2,310–2,360 MHz	50	India and USA
2,535–2,655 MHz	120	Bangladesh, Belarus, China, Japan, Pakistan, republic of Korea, Russian Federation, Singapore, Sri Lanka, Ukraine

From a technical standpoint, there are a number of similarities in direct sound broadcasts and MSS technologies. Their frequency bands are in close proximity, which allows similar RF receiver technologies; their operational environment and propagation characteristics are similar; and finally, there are similarities in user expectations in terms of service provision and receiver design.

Recent developments in digital systems in general, and computing/multimedia in particular, have opened interesting opportunities in the broadcast field. For example, it is possible to multiplex useful data with audio programmes for downloading to home PCs. Real-time interactivity, such as offered by direct television broadcasts, can be introduced through terrestrial feedback to the source, or feedback through low-cost leased MSS channels. In view of such trends, it was concluded that future sound broadcast systems would be digital. The ITU has consequently recommended a number of generic requirements for DAB services (ITU, 1995). A DAB system is expected to:

● provide a range of receiver signal qualities up to CD quality; and stereophonic, two- or multi-channel sound;

- use state-of-the-art source/channel coding, modulation and digital signal processing techniques;
- be more spectrally and power-efficient than conventional analogue schemes in the presence of multipath and shadowing;
- operate with terrestrial systems synergistically through hybrid/mixed satellite-terrestrial systems and maximize commonality in dual-mode receivers used in such systems to benefit from economies of scale;
- provide programme-related facilities such as service identification, programme labelling, programme delivery control, copyright control, etc.;
- provide value-added services such as business data, paging, graphics, etc. using a variety of data rates;
- provide coverage for national, regional or international services.

Two systems have received particular attention by the ITU in formulating recommendations. These are the European Eureka 147 system, called Digital System A, and another system called Digital System B, sponsored by the Voice of America and the Jet Propulsion Laboratory of the USA – both of which are under evaluation. Eureka 147 is a European consortium comprising the CETT laboratory of France, the IRT laboratory of Germany, manufacturers, the European Broadcasting Union and several European governments. The European approach, based on COFDM (coherent orthogonal frequency division multiplexed; see Chapter 4) over a 1.5 MHz band, is based on the hypothesis that frequency diversity of this order is necessary to combat frequency selective fading and multipath observed in land mobile systems; whereas the American developers intend to use techniques such as adaptive equalization to transmit within 50 kHz (monophonic FM quality) to 200 kHz (CD quality). Another approach, pursued under Project Acorn, conducted by USA Digital, intends to demonstrate simulcast FM and digital signals using COFDM on the same transmitter, with digital signals 30 dB lower than the FM signals (Messer, 1993).

Subsequently, a number of companies, notably in the USA, proposed satellite direct sound broadcast with CD quality for both national and international markets.

10.4.1 Technical considerations

Direct broadcast from satellites to fixed terminals at home has been used for several years through direct broadcast satellites, such as ASTRA, INSAT and JCSAT-2. The terminals for such broadcasts typically use 50–90 cm fixed antennas with high G/T, and therefore reception of high quality signals does not pose much of a problem; neither is bandwidth a major problem, as satellite operators can offer narrowband sound broadcasts relatively easily from their large bandwidth pool. Such broadcasts, however, cannot provide services to portable terminals and hence a different system approach is necessary.

A widely used system design approach during the early studies was to use an analogue FM system with transmission characteristics similar to terrestrial FM broadcasts, which would allow reception by conventional receivers through a front-end converter, as is common for reception of satellite television broadcasts. This approach was overtaken by the advent of digital systems.

Table 10.5 (a) and (b) (CCIR Report 955-1) illustrates radio link characteristics of an FM broadcast in various types of environment as a baseline system. Table 10.5 (b) demonstrates that a satellite antenna size of 20 m is very demanding and satellite power requirements of ~ 3 and ~18.5 kW for environments C and D are prohibitive. By extrapolation, the same conclusion can be applied to digital transmissions.

A summary of propagation conditions is fundamental to the technical viability of the service. It has been established that, in general, shadowing loss

Table 10.5 (a)
System parameters of
the baseline FM
broadcast system

System parameter	Characteristic
Modulation	FM
Carrier deviation (kHz)	±75
Noise bandwidth (kHz)	250
Carrier to noise ratio	10
Coupling loss (dB)	1
Receive antenna gain (dBi)	3
Receive system noise temperature (K)	2,000
Carrier frequency (GHz)	1.0

Table 10.5 (b)
Satellite requirements
for various environments

System parameter	Characteristics for complying to requirements in Table 10.5 (a)			
	Environment A (note 1)	Environment B (note 1)	Environment C (note 1)	Environment D (note 1)
Link margin	6	15	25	33
Satellite EIRP (beam centre)	59.6	68.6	78.6	86.6
Satellite antenna gain; diameter = 20 m; beamwidth = 1°	43.9	43.9	43.9	43.9
Input power to satellite antenna (W)	37	295	2,951	18,621

Note 1: Environment a: Rural; elevation angle > 70°; receiver threshold exceeded for >90% of area; Environment b: Urban; elevation angle = 20°; receiver threshold exceeded for >90% of area; Environment c: Urban; elevation angle = low; receiver threshold exceeded for 90% of time for >90% of area; Environment d: Urban; elevation angle = low; receiver threshold exceeded for 95% of time for >90 % of area.

increases with frequency; this happens because the diffraction advantage reduces as the frequency is increased; this has also been established through multiband experiments conducted at the L, S and K_u bands in the UK. However, the difference in attenuation loss between the L and S bands is quite marginal (see Chapter 3), and therefore propagation models in the L band apply, with minor modifications. The reader may refer to section 3.3.2 for further details. A notable difference is that while most MSS studies have concentrated on narrowband channel characterization, DAB systems prefer wideband transmissions to maximize frequency diversity advantage (see section 3.3.2, Wideband). Yet another consideration, particularly applicable to DAB systems, is the need to estimate building penetration loss, as some of the portables could well be used within buildings. Data collected through ATS-6 transmissions at 860, 1,550 and 2,569 MHz within various types of house demonstrated that penetration loss within buildings depends on frequency, receive antenna polarization, the material used in house construction (e.g. wood siding or brick veneer), thermal insulation within the house, and proximity of rooms to an outside wall. It was noted that penetration loss is independent of the elevation angle down to 5° when there is a clear line of sight to the satellite, and furthermore that the loss was normally distributed with a standard deviation of 3 dB within each type of house (Wells, 1977; Miller, 1985).

As listed in Table 10.5(b), high-power transmissions are necessary to mitigate the effects of shadowing, which increases the cost of the system, and therefore transmission techniques for power reduction are essential. FEC coding and diversity methods have been investigated in detail. For example, it has been shown that $\frac{1}{2}$ rate FEC convolution code of constraint length 7 with a Viterbi maximum likelihood soft decision decoding algorithm offers advantage of the order of 3–6 dB over uncoded QPSK with coherent detection in a memoryless Rayleigh fading channel for a BER of 10^{-5}. However, the coding gain cannot be realized when the channel has memory – the case when the fading rate is much less than the symbol rate. When a receiver is in deep fade for several seconds, coding gain is no longer realizable, as the signal will be irretrievable. The duration of signal loss caused by deep fades can be reduced by frequency, spatial or time diversity.

In *frequency diversity*, the same information is transmitted on a number of carriers, separated by more than the coherence bandwidth of the channel (see section 3.3.2, Wideband). Thus it is possible to obtain the information from the carriers which are unfaded at a given time. Measurements performed on simulated satellite paths demonstrate that the coherence bandwidth is narrower in urban environments than in rural environments and therefore carrier spacing has to be larger in rural environments. Frequency diversity can be achieved by a modulation scheme, called the COFDM scheme, which was discussed in section 4.2.1.

Spatial diversity can be used for analogue or digital modulation by deploying multiple antennas separated far enough from each other so that the signal fading received in each is not correlated to each other. Signals from each antenna can be combined to form a composite signal which has a lower fade than individual components. Measurements have shown that diversity gain is possible when the separation is of the order of $\frac{1}{2}$ wavelength or greater (Hess, 1980). Maximum ratio combining is one possible method for signal combination. It can be implemented by introducing phase lock loop to each path to obtain phase coherence before amplitude weighting and combining. Evaluation for $\frac{1}{2}$ rate convolution code of constraint length 7 and a fourth order diversity using maximal-ratio combining in a Rayleigh fading channel demonstrates energy per bit by noise density ratio (E_b/N_o) of 7.0 dB for BER of 10^{-5}. Similarly, an evaluation of a fourth order diversity applied to an FM system illustrates an improvement of 26 dB at 0.001 probability.

Time diversity is possible with digital systems by interleaving bits at the transmitter and de-interleaving at the receiver (see section 4.3). The interleaving depth can be adjusted such that adjacent symbols are separated far enough so that fade events remain uncorrelated. Note, however, that interleaving depth is limited and hence this scheme is better suited for rapid fades such as are caused when receiving signals on a moving mobile. Hence this scheme can best be utilized in conjunction with other fade mitigation methods. It is also possible to achieve time diversity by transmitting the same signal block twice, one delayed with respect to the other. The receiver should then either combine the two or extract the better of the two.

Table 10.6(a) (Miller, 1988) demonstrates a substantial reduction in satellite power vis-à-vis the baseline system (Table 10.5 (a)), through a combination of convolution coding and spatial diversity.

Table 10.6
(a) System parameters of a digital sound broadcast system using $\frac{1}{2}$ rate FEC convolution coding, Viterbi decoding and quad-spatial diversity on a vehicle

System parameter	Characteristic
Modulation	QPSK
Broadcast mode	Stereophonic
Audio bandwidth (kHz)	15
Bit rate (kbps) (Adaptive Delta Modulation; 220 kbps/channel)	440
Code rate	$\frac{1}{2}$
RF bandwidth (kHz)	880
E_b/N_o vehicle (house)	7.0 (3.8)
Implementation margin	1.0
BER	10^{-5}
Subjective quality [Q = 4 perceptible but not annoying degradation; Q = 5 imperceptible degradation (CCIR Rec 562-2)]	Q = 4.5
Receiver antenna gain	5
Receive system noise temperature (K)	600
Carrier frequency (GHz)	1.0

Table 10.6
(b) Satellite
requirements for various
environments

System parameter	Characteristics for complying with requirements in Table 10.6 (a)		
	Vehicle: light foliage	**Vehicle: heavy foliage**	**Inside house**
Link margin	0	10.0	12.0
Satellite EIRP (beam centre)	47.0	57.0	55.8
Satellite antenna gain; diameter = 20 m; beamwidth = 1°	43.9	43.9	43.9
Input power to satellite antenna (W)	2.0	20.4	15.5

Because of increasing spectrum demands, considerable development effort is being directed towards a reduction in the bit rate of sound coders. A high quality stereophonic broadcast would require a bit rate of about 2×800 kbps, with a conventional waveform coding technique. A number of efficient source coding algorithms have been developed with sub-band coding or transform coding which can compress the bit rate by a factor of 4–12. Table 10.7 summarizes some recent developments.

10.4.2 Transmission standards

Standardization is essential for public broadcast services, as it permits equipment manufacturers to develop equipment which can interwork, opening a wider market with consequent economies of scale. A number of transmission schemes have been studied and experimented with for standardization. ITU-R has converged on two transmission schemes – a scheme called Digital System A, developed by the Eureka 147 DAB project team and supported by the European Broadcasting Union, and another scheme called Digital System B. The former has now been chosen as a European Broadcast Satellite Service and Broadcast Service (sound) standard for sound broadcast service to vehicles, and to portable and fixed receivers.

Figure 10.9 shows the transmission section of Digital System A (ITU, 1995).

The audio coder uses an MPEG-Audio layer II. The sub-band coding section of the coder uses the MUSICAM system, giving bit rates of 32, 48, 56, 64, 80, 96, 112, 128, 160 and 192 kbps monophonic channels and 2 kbps of programme-related data. General and auxiliary data services can be combined with the audio. The data are multiplexed into a frame, which consists of a synchronization channel for synchronizing, a fast information channel which has multiplex-related information, service information and other general/auxiliary data, and a number of

Table 10.7

A summary of some
recent developments

Type	Comment	Compression	Developer
Sub-band coding – ADPCM in low frequency bands	32 sub-bands	$\frac{1}{6} - \frac{1}{8}$	NHK (Japan)
Masking pattern adapted universal sub-band integrated coding and multiplexing (MUSICAM)	32 sub-bands	$\frac{1}{4} - \frac{1}{12}$	IRT (Germany): original developer); technique improved by CCETT (France); IRT (Germany); Philips (Netherlands); Matsushita (Japan)
Precision adaptive sub-band coding (PASC)	32 sub-bands	$\frac{1}{4}$	Philips (Netherlands)
SB-ADPCM	4 sub-bands	$\frac{1}{4}$	PTT (Switzerland)
apt-X100	4 sub-bands, ADPCM	$\frac{1}{4}$	APT (UK)
Adaptive transform acoustic coding (ATRAC)	Sub-bands + transform coding	Not applicable (N/A)	Sony (Japan)
AC-2	Modified discrete cosine transform + adaptive transform coder	N/A	Dolby (USA)
Spectrum efficient digital audio technology (SEDAT)	Discrete cosine transform + adaptive transform coder	N/A	Scientific Atlanta (USA)

programme service channels. The frame duration is 24 ms for mode 3 transmission, which is best suited for satellite and complementary terrestrial transmissions up to 3 GHz. There are, in all, four modes with different frame duration, each suited to a specific propagation environment. The channels are convolution coded with a constraint length of 7; the audio channels have a variable code rate at an average coding rate between $\frac{1}{3}$ and $\frac{3}{4}$; and the fast information channel is encoded at a constant rate of $\frac{1}{3}$. The convolution encoded data is further interleaved to a depth of 16. The coded data is then transmitted using a COFDM technique (see Chapter 4). The total occupied bandwidth of the channel is 1.5 MHz. Each carrier of the COFDM signal is modulated by a differentially encoded PSK. The receiver of Digital System A consists of a demodulator, de-interleaver, convolution decoder, de-multiplexer and voice/data decoder.

Figure 10.10 (ITU 1995) illustrates the functional blocks of Digital System B.

Each audio programme is encoded and multiplexed with other digital data. The data are multiplexed encoded in a Reed-Solomon encoder having a rate of 140/160, convolutionally encoded at a rate of $\frac{1}{2}$ with a constraint length of 7 followed by a block time interleaver and frame sync insertion. This signal is then

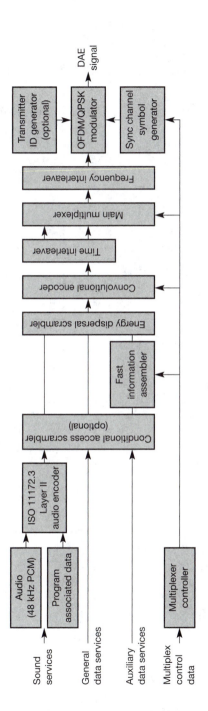

Figure 10.9
Transmission section of
ITU-R Digital System A,
developed by Eureka 147
Digital Audio Broadcast
consortium (Rec ITU–R
BO. 1130–1, ITU, 1995)
© 1995 ITU

Figure 10.10
The main blocks of
Digital System B
(Rec ITU-R BO. 789–2,
ITU, 1995) © 1995 ITU

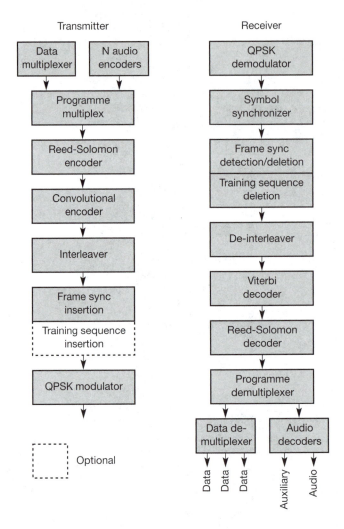

Figure 10.10 The main blocks of Digital System B (Rec ITU-R BO. 789–2, ITU, 1995) © 1995 ITU

QPSK modulated, up-converted and transmitted. It is also possible to include a training sequence if equalization is used, and time diversity by repetition of signals. The frequency division multiplexed signal is received, demodulated and synchronized; frame synch and training sequence data are removed; the signal is de-interleaved and decoded in a Viterbi decoder, followed by a Reed-Solomon decoder. The composite signal is de-multiplexed and individual channels separated. If time or signal diversity is used, appropriate combination is performed at the de-multiplexer stage to give a more accurate received signal.

10.4.3 DAB system configuration

Satellite direct broadcast systems can operate as standalone systems which are suitable for remote areas with very little existing transmissions or for a niche market (e.g. hi-fidelity music channels for motorists). Alternatively, they can be overlaid on an existing terrestrial broadcast system to fill in coverage gaps – as in the Eureka 147 system. Conversely, terrestrial systems can be used to augment service to areas where satellite propagation conditions are difficult. Satellite transmissions are retransmitted terrestrially at low level in these areas. Satellite and terrestrial signals are combined at a receiver to improve signal quality. Such a scheme is anticipated to reduce the required satellite link margin by up to 5 dB. Yet another variant is transmission of satellite and terrestrial signals in the same format and frequency band but at different frequencies. Both terrestrial and satellite transmissions are received and the user may select the more appropriate one, or the intelligence can be built into the receiver.

10.4.4 Space segment

One of the major considerations in selection of the space segment is the necessary propagation margin. From Chapter 3, we know that the fading loss is inversely proportion to the elevation angle. Generally, geostationary satellite systems are suited to equatorial – mid-latitude – regions. At mid-latitudes and above, the propagation margin begins to increase.

One way to mitigate the problem of fading at mid/high latitudes such as Europe is to deploy satellite constellation in a highly elliptical Molniya or Tundra orbit (see section 2.2). This concept has been studied extensively in the European Space Agency's Archimedes project. The Tundra orbit is considered a possible candidate for European DAB (Galligan, 1989), as satellites in this type of orbit appear at elevation angles of ~55 to 90° offering reduced link margin and simple upward pointing antennas on vehicles. An added advantage is that the satellite eclipse occurs near the perigee when satellites are not operating. However, two satellites phased 180° apart are necessary to provide continuous coverage (>52°). This requires satellite handover and thereby complicates the system architecture. The Molniya orbit was discarded in this study as radiation doses in such orbits are higher; altitude excursion during operations is large; launch from the European launcher (Ariane) is difficult; and three (rather than two) satellites are necessary. Assuming a single beam antenna and an omni-directional receiving antenna, compared to the geostationary system, the Archimedes system requires 30 times lower power per channel of digital telephony (Giovagnoli, 1989) – the same order of advantage would apply to sound broadcasts.

10.4.5 Status of DAB systems

The first step towards the commercial introduction of such systems was taken in 1992 after allocation of frequencies for DAB. Frequencies have been allocated for satellite and terrestrial digital broadcasts, as it is recognized that both the components will coexist synergistically. Standardization activities have further assisted in the process. A number of terrestrial systems are beginning to be introduced, and at least one US satellite system has started deploying satellites for world-wide coverage. Despite the high investment costs of satellite-delivered radio systems at the outset, a number of commercial systems are emerging, most of them originating in the USA.

In the USA, a number of companies have been granted licences and spectrum for providing DAB. These include CD Radio, the American Mobile Radio Corporation and Worldspace. We will consider one representative example here. Worldspace Inc. has begun deploying satellites for world-wide DAB coverage using a format similar to Digital System B (Campanella, 1996). The three satellites, called AfriStar (21° E), CaribStar (95° E) and AsiaStar (105° E) are being deployed to provide sound broadcasts for small portable radio sets in Africa, Asia and Central and South America, using the 1,467–1,492 MHz band. Bit rates of 16, 32, 64 and 128 kbps will be available using the MPEG 2 compression algorithm. The system can also provide data commensurate with multimedia applications for downloading to personal computers. Satellites deploy a global beam in the uplink and three spot beams in the downlink, using FDMA in the uplink with a QPSK modulation scheme and a TDM in the broadcast link. The transponders are regenerative, converting the received FDMA signals to a TDM stream at a rate of ~1.767 megasymbols/sec. Up to 288 prime rate carriers can be uplinked, and each downlink can transmit up to 96 such carriers. Antennas under consideration for the receivers are a patch antenna of 4–6 dBi gain, a phased array antenna of 10–12 dBi gain and a rod-shaped helical antenna. Multipath can be reduced by using spatial diversity by mounting antennas at various points on a vehicle. Programmes can be uplinked through a 2.4 m X-band antenna with the capability to transmit 128 kbps using a 25 W power amplifier.

Terrestrial DAB are already in use in Canada and satellite DAB are expected in the 2003–5 time frame (Paiement *et al.*, 1995). A geostationary orbit has been favoured. To minimize the propagation link margin, particularly significant for Canada, the approach has been to deploy a hybrid architecture where areas requiring a high link margin, such as dense urban areas, are served by a complementary terrestrial service operating in the same band. This way, a link margin of around 4 dB is adequate to serve vast geographical areas. A COFDM modulation scheme with time interleaving provides resistance to frequency selective fading. An EIRP of 62.5 dBW is required per programme channel, with an eight spot beam system which can be constructed for ~10 antenna diameter. Two programme channels are planned per spot beam. The MPEG-1 source coder gives five CD quality stereo programmes at a bit rate of 1,152 kbps. The COFDM scheme uses 1.536 MHz bandwidth. Hybrid receivers are able to operate within the 40 MHz band using terrestrial or satellite reception, as necessary.

Considerable strides have been taken in Europe in both terrestrial and satellite DAB systems. A number of countries have begun operating terrestrial DAB following the Eureka 147 system, and in others pilot programmes are under way. Some broadcasters are considering providing multimedia download through DAB systems. Another European satellite DAB initiative is through the Archimedes system.

In Japan, a digital PCM broadcasting service is already available at the K_u band from JCSAT-2 satellites to fixed receivers. The WARC-92 compatible satellite DAB service is expected to be tried out through an ETS-VIII satellite, which has an S-band mobile system as well. The sound broadcast payload uses a large antenna system with onboard processing and shares a number of onboard systems with the mobile payload. The feeder link frequency is at 30 GHz and the service link operates at 2,535–2,540 MHz within the WARC-92 band. The COFDM/QPSK modulation scheme is used in a satellite–terrestrial hybrid network. Six channels at a data rate of 256 kbps are available. User receiver antennas have a gain of 7 dBi and receiver G/T is –18.8 dB/k. The satellite is planned to be located at 135° E and has a lifetime of three years (Ohmori *et al.*, 1998).

Other countries where satellite DAB has been tried out or is being planned are Australia, where DAB trials were conducted using the Optus B3 satellite, India, where a DAB service is planned for 2003 following on from the introduction of terrestrial DAB, and Mexico, where L-band satellite DAB tests were conducted through the Solidaridad-2 satellite.

10.5 Direct television broadcast system

Since the advent of direct broadcast satellite (DBS) systems for homes, there has been an interest in providing a similar service to mobiles. The service is generally provided in the K_u band through small 50–100 cm directive antennas and high-power wideband channels using analogue (older systems) or digital modulation schemes broadcast from geostationary satellites. The main difference between a mobile terminal and its fixed version in this respect is the need for a mobile tracking antenna and the ability of the receiver to absorb Doppler shift of the order of 100 Hz (40 m/s) for a slow-moving vehicle, such as a ship, to 16 kHz for a fast-moving aircraft (400 m/S) at 12 GHz, which amounts to frequency uncertainty of 1.2×10^{-8} to 1.3×10^{-3}. Ship-borne DBS terminals have been available for several years; the design specifications of these terminals are relatively relaxed when compared to aircraft systems, where the making of high gain tracking antennas is technically quite challenging. Phased array designs offer advantages in terms of drag, weight, maintenance and reliability, and are less obtrusive when compared to mechanical or hybrid mechanical–phased array designs, but are more expensive. An example of a development in this area is Boeing's Satellite Television Airplane Receiving System (STARS), which is being developed for commercial operation with a regular DBS system in the 12.2–12.7 GHz band in the USA. The system comprises a phased array system with a low noise block assembly with a G/T of 8.2 dB/K and a television receiver offering the required signal quality with a satellite EIRP of 50.4 dBW (Vertatschitsch and Fitzsimmons, 1995). Television broadcasts during off-peak telecommunication traffic hours, for transmitting important events to US ships, have been made regularly via Inmarsat satellites within a bandwidth of 400 kHz using a QPSK modulation scheme for a number of years. The company has plans to provide regular television broadcasts to aircraft which will be able to receive them via existing Inmarsat communication antennas, thereby reducing the cost and complexity of the receiver.

10.6 Very small aperture terminal systems

Very small aperture terminals or VSAT systems have been widely dealt with in the literature. Keeping within the scope of the book, here we will only summarize features of VSAT vis-à-vis MSS. VSAT systems are suited for high volume users operating from fixed sites. As such users usually operate on dedicated channels, high channel usage offers savings. However, if satellite usage is low and uncertain, a demand assigned (DA) mobile satellite service is the preferred choice. At present, VSAT terminals are bulkier and not as well suited for transportation as MSS terminals. Installing VSAT is somewhat complex, requiring several hours of specialists' time, while MSS terminals can be installed in a few minutes with minimal training. VSAT systems are generally more sensitive to antenna depointing because their beam width is narrower compared to MSS terminals, which operate with wider beamwidth antennas, and VSATs are not suitable for communication during transportation. Obtaining an operating licence is not straightforward in all countries. This applies both to transmission and reception sites. Trans-border communication is not permitted in large parts of the world. Furthermore, a VSAT may not be erected anywhere at will, due to the possibility of local interference because VSATs operate in the FSS band. Trans-border communication is a significantly lower problem for MSS.

VSATs are generally used by closed user groups without connection to the public networks for regulatory reasons, whereas MSS systems are connected to the public networks of the world. However, certain types of users prefer to use a closed environment for various reasons. Interworking of equipment is no problem for an MSS, as manufacturers follow the same transmission standard recommended by the MSS operator. VSATs do not as yet have a single standard and therefore interoperability of equipment is not always possible, because each VSAT operator optimizes their system for their specific use. It is anticipated that next generation VSATs operating in the K_a band will be significantly smaller in size, suitable for individuals, and offering megabits of throughput far higher than envisaged for next generation MSS. An interesting synergy appears to be on the horizon as MSS operators position themselves to accept the reality of MSS terminal-sized VSAT units.

10.7 Terrestrial cellular system

In this section, we focus specifically on the technical principles of cellular mobile systems due to their relevance to the mobile satellite communication systems and the interesting architecture deployed in such systems. In Chapter 9, we noted that several MSS system architectures have been based on the cellular model in order to facilitate their integration with terrestrial cellular counterparts. Cellular nomenclature is often used in such systems.

The basic objectives for a large-scale mobile telephone service may be summarized as:

1   large subscriber capacity;
2   efficient use of spectrum;
3   national/international compatibility;
4   widespread availability;
5   adaptability to traffic density (such as urban, rural, etc.);
6   service to vehicle and portables;
7   preferably 'add on' services – in addition to the telephone;
8   high quality (as good as the fixed network telephones);
9   affordability.

The need to increase capacity almost indefinitely, without requiring additional spectrum, led to the development of the cellular concept. The cellular
architecture is based on the principle of frequency reuse and cell splitting.

Frequency reuse

Radio channels are reused at a distance large enough to reduce co-channel interference to acceptable limits. The coverage area is partitioned into small units, or
cells, of a few hundred metres to tens of kilometres, depending on the traffic.
Each cell is served by a base station, which establishes a connection on a frequency selected from a demand assigned pool. The frequency set is reused in
other cells in accordance with a predefined reuse pattern. By reducing the transmitted power on each channel, cell size and hence reuse distance can be
reduced arbitrarily, thereby increasing system capacity. Thus we have the concept of cell splitting and macro (several kilometres), micro (hundreds of metres
to kilometres) and pico cells (tens of metres). Figure 10.11 illustrates the cellular
concept. In practice, cell shapes are not necessarily hexagonal as shown, but the
hexagonal shape is used in planning because the shape is close to reality and is
easy to analyze geometrically.

Cell splitting

Assume that T traffic channels are divided equally between M base stations.
Then each base station has a pool of T/M channels. Eventually, the requirements of certain base stations will reach their maximum capacity. Further
growth can then be accommodated by a process called cell splitting. The process
involves reducing the transmitted power of base stations, which has the effect
of reducing the cell size and increasing the number of channels serving the
region without the need to increase spectrum. Cell splitting matches the spatial
density of available spectrum to the spatial density of traffic demand, since a
lower demand area may be served by larger cells and a higher demand area by
smaller cells. Figure 10.12 shows the cell-splitting process in the earlier stages
and advanced stages of deployment.

In practice, cell splitting is complicated by logistics and other practical
considerations.

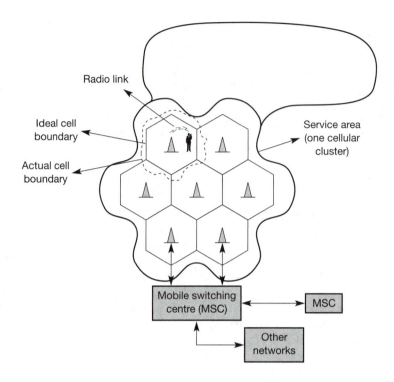

Figure 10.11
The cellular concept

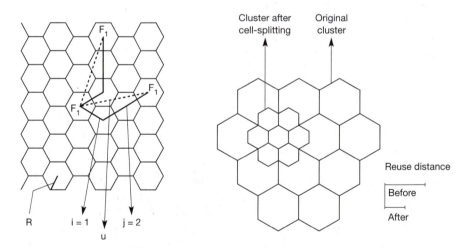

Figure 10.12
The cell-splitting process
(cluster size = $(itj)^2 - ij$;
reuse ratio = $\frac{u}{R}$; F_1 =
cochannel frequency)

Cellular system architecture

The main elements of any cellular mobile system are (see Figure 10.11):

1. mobiles;
2. a network of cells each served by a base station;

3.   one or more mobile switching centres (MSC), depending on the size of coverage area, interfaced with all base stations at one end and the public network at the other.

The radio link between mobile units and a base station consists of two types of channels – a control channel and traffic channels. The control channel transfers system messages and traffic channels carry traffic and supervisory signals during a call. A base station is connected, usually by land lines, to an MSC with a group of voice trunk and data links for exchange of information to process calls. The MSCs are connected to the public networks.

Typically, each mobile telephone is assigned a 'home area'. The home MSC maintains a location database in which it keeps the most recent position of mobiles registered with the MSC. This information is used for routing calls to the appropriate section of the network. Each mobile automatically registers its location with the visiting MSC whenever the mobile migrates outside its home MSC. The visited MSC transfers the information to the mobile's home MSC.

Whenever a mobile telephone is switched on, it scans all system control channels and locks to the strongest channel. This operation is continuous, ensuring that the mobile always operates with the strongest signal. The control channel is used for two types of messages: general system information, which contains network identity, channels available, area code, and other facilities/requirements; and mobile control information, which consists of paging messages to notify a mobile and channel assignment messages used to set up calls. If the mobile signal quality degrades during a call, the call is handed over to the cell which can provide a better link quality. The handover involves signalling between base stations and the MSC and is transparent to a user. There are a large number of systems in use, which differ in detail. To harmonize growth throughout the world, therefore, third generation systems have been developed using common standards (see Chapter 11).

Further reading

Bretz, E.A. (2000) 'X marks the spot, maybe', *IEEE Spectrum*, April, 26–36.

Butt, G.; Evans, B.G.; Richharia, M. (1992) 'Multiband propagation experiment for narrowband characterisation of high elevation angle land mobile-satellite channels', *Electronics Letters*, **28** (15), 16th July, 1449–50.

Campanella, S.J. (1996) 'The worldspace satellite-to-radio multimedia broadcast system: a technical overview', *Proceedings 16th Int Communications Satellite Systems Conference AIAA*, Washington DC, 25–29 February, 826–31.

Chien, P. (1991) 'GPS commercial applications', *Via Satellite*, October, 37–41.

Cospas-Sarsat http://www.cospas-sarsat.org/overview/overview.htm

Dale, S.A.; Daly, P.; Kitching, I.D. (1989) 'Understanding signals from GLONASS navigation satellite', 7, 11–22.

Daly, P. (1993) 'Navstar GPS and GLONASS: global satellite navigation systems', *Electronics and Communication Engineering Journal*, December, 349–57.

Galligan, K.P. (1989) 'Sound broadcasting applications of highly elliptical orbits', *IEE Colloquium on Highly Elliptical Orbit Satellite Systems* (1989/86), 24 May, Digest, 7/1–7/4.

Giovagnoli, F. (1989) 'Land mobile communications from non-geostationary orbits (Review of Archimedese ESA study)', *IEE Colloquium on Highly Elliptical Orbit Satellite Systems*, (1989/86), 24 May, Digest, pp. 3/1–3/5.

Hess, G.C. (1980) 'Land mobile satellite excess path loss measurements', *IEEE Transactions on Vehicular Technology*, **VT-29** (2), May, 290–7.

ITU (1986) 'Satellite sound broadcasting with portable receivers and receivers in automobiles', *CCIR report 955-1*, **X and XI**, Part 2, XVI Plenary Assembly, Dubrovnik, 107–28.

ITU-R (1995) 'Recommendation ITU-R BO.1130-1', 'Systems for digital sound broadcasting to vehicular, portable and fixed receivers for broadcasting-satellite service (sound) in the frequency range 1400-2700 MHz', *BO series Fascile, Broadcasting-Satellite Service (Sound and Television)*, Geneva, Switzerland.

ITU-R (1995) 'Recommendation ITU-R BO.789-2', 'Services for digital sound broadcasting to vehicular, portable and fixed receivers for broadcasting-satellite service (sound) in the frequency range 1400-2700 MHz', *BO series Fascile, Broadcasting-Satellite Service (Sound and Television)*, Geneva, Switzerland.

King, J.V. (1999) 'Overview of the Cospas-Sarsat Satellite System for Search and Rescue', *IMSC '99, Sixth International Mobile Satellite Conference*, Ottawa, co-sponsored by Communications Research Centre and the Jet Propulsion Laboratory, 31–6.

Kozammernik, F. (1995) 'Digital audio broadcasting – radio now and for the future', *EBU Technical Review*, Autumn, 2–27.

Levey, R.J. (ed.) (1988) *Advanced Digital Techniques for UHF Satellite Sound Broadcasting*, EBU Technical Centre, Brussells, Belgium, August.

Messer, D. (1993) 'World-wide survey of direct-to-listener digital audio delivery systems development since WARC-92', *IMSC 1993, Proceedings of the Third International Mobile Satellite Conference*, Pasadena, California, JPL Publication 93-009, 16–18 June, 3–8.

Miller, J.E. (1985) 'Satellite sound broadcasting around 1 GHz', *IEEE Journal on Selected Area in Communications*, **SAC-3** (1), January, 204–10.

Miller, J.E. (1988) 'Application of coding and diversity to UHF satellite sound broadcasting systems', *IEEE Transactions on Broadcasting*, **34** (4), 465–75.

Milliken, R.J. and Zoller, C.J. (1978) 'Principle of operation of NAVSTAR and system characteristics', *Navigation Journal*, The Institute of Navigation, vol. 25, no. 2, summer, pp. 95–106.

O'Neil, W.A. (1992) 'GMDSS', *Ocean Voice*, Inmarsat.

Ohmori S.; Wakana, H.; Kawase, S. (1998) *Mobile Satellite Communications*, Artech House Publishers, Norwood, MA.

Paiement, R.V.; Voyer, R.; Prendergast, D. (1995) 'Digital radio broadcasting using mixed satellite/terrestrial approach: an application study', *IMSC 1995*,

The Fourth International Mobile Satellite Conference, Ottawa, co-sponsored by Communications Research Centre/Industry Canada and Jet Propulsion Laboratory/NASA, 6–8 June, 439–44.

Rosetti, C. and Fromm, H.H. (1977) 'Possibilities for national coverage by FM sound broadcasting from satellites in certain parts of the frequency spectrum', *EBU Review – Technical Part*, April.

Schänzer, G. (1995) 'Satellite navigation for precision approach: technological and political benefits and risks', *Space Communications*, 13, 97–108.

Vertatschitsch, E.J. and Fitzsimmons, G.W. (1995) 'Boeing Satellite Television Airplane Receiving System (STARS) performance', *IMSC 1995, The Fourth International Mobile Satellite Conference*, Ottawa, co-sponsored by Communications Research Centre/Industry Canada and Jet Propulsion Laboratory/NASA, 301–5.

Wells, P.I. (1977) 'The attenuation of UHF radio signals by houses', *IEEE Transactions on Vehicular Technology*, **VT-26** (6), November, 358–62.

Future trends **11**

11.1 Introduction

In this chapter we attempt to trace the evolution and trends of mobile satellite services. As an MSS is part of a much larger telecommunications network, its evolution will depend strongly on general telecommunications development trends. In the past few years, we have observed an escalating integration between broadcasting, computing and telecommunications, and their evolution into a single so-called information society seems imminent. Figure 11.1 (Rudge, 1993) shows the concept of such an evolutionary path, although projected dates are not precise.

Evolution will depend on the market growth dictated by the almighty user, who is likely to demand better quality and ease of use. An all-encompassing network will provide the user with the flexibility to include a satellite service when necessary, and thus provide satellite operators with a better business opportunity with the advantage of maximizing network hardware commonalities and economies of scale. To enable integration on this scale, efforts to standardize network interfaces are being made.

In the information society, individuals and businesses will exchange vast amounts of data as images, video, etc., demanding an ever-increasing network capacity, as already evident – compare the throughput of second and third generation terrestrial mobile systems, as an example. Initiatives exploring the best architecture and technology to meet such demands are well under way. Several wideband mobile and fixed satellite systems have been proposed to meet the anticipated demand. Due to limited spectrum in the traditional MSS bands, higher frequency bands are being investigated. The wideband personal communication markets are being pursued by both fixed and mobile satellite service communities. A number of promising technologies are on the horizon.

11.2 Trends

Specific areas of recent growth include CD-ROM technology, digital television, the internet, intranets, IP-based services, terrestrial and satellite mobile communications. ITU projections indicate that world-wide mobile subscribers will overtake fixed subscribers around 2010, totalling over 1.5 billion, as illustrated in Figure 11.2 (ITU website).

Figure 11.1

Evolution to information society (Rudge, 1993)

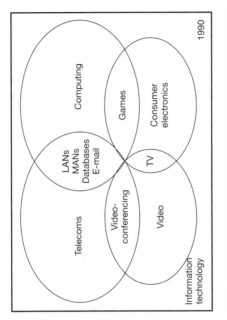

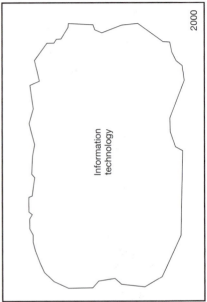

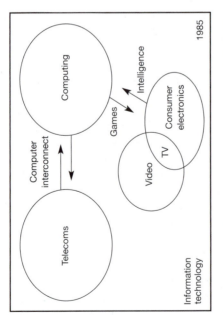

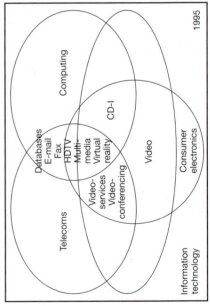

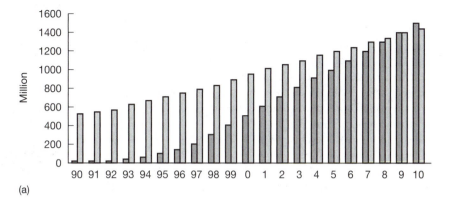

(a)

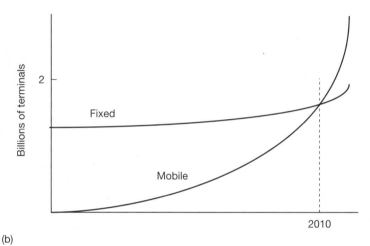

(b)

Typical multimedia applications and their transmission characteristics are
listed in Table 11.1.

Application	BER	Allowable end-to-end delay (ms)	Data rate (kb/s)	
			Forward link	Return link
PC networking	10^{-6}	200	64	64
World Wide Web access	10^{-6}	500	1–5	64
Video conference	10^{-6}	200	64–2,000	64–2,000
Tele-education	10^{-3}–10^{-4} (Voice)	Voice: 200	4–8	4–8
	10^{-6} (Data)	Data: 1 sec	1,000	64

These advances are likely to have a profound influence on society in the next decade, and indicate opportunities in various fields, including in the area of mobile communications, as users expect extension of the services of the fixed network to a go-anywhere mobile environment in the information society of tomorrow. To this end, mobile systems will assist in a variety of ways, such as providing access to public and company databases and business and government services to skilled workers, allowing flexibility in working practices and thereby increased productivity.

This opportunity has been well recognized by various international consortia and regulatory bodies. The future of mobile communications is being shaped in international forums such as the International Telecommunications Union and the European Union in initiatives such as the IMT-2000, UMTS, RACE mobile and fourth generation (4G) mobile system definition programmes. The common vision of such initiatives is provision of a full range of universal wideband communication services, including voice, graphics and video, directly to users irrespective of the user's location, network or terminal. Such personal communication services will offer portability between fixed, terrestrial mobile and satellite mobile segments as a result of a synergistic convergence between the fixed and mobile networks and the evolution of an intelligent network (UMTS Forum Report # 8). These services must improve quality and security and must be developed so as to offer flexibility in creating and introducing future services. Examples of such services are passive audio-visual services (e.g. video-on-demand, direct to home television); passive audio services (e.g. audio-on-demand as an alternative to CDs); information, education and entertainment services (e.g. interactive education, booking, home shopping, online media); personal communication services (e.g. video-telephony, video-conferencing, telemedicine); corporate communication services (e.g. LAN–LAN interconnection, remote collaborative working). It is anticipated that most of the fixed multimedia services will be using packet mode, such as IP through an ATM switch, and may be asymmetric.

11.3   Integrated mobile communications networks

At present, there is a plethora of mobile communications products and standards, resulting in difficult and inefficient spectrum management, limitations to user roaming, fragmented markets and limited economies of scale at global level. Therefore the international bodies are developing global standards, which would permit users to access a range of mobile telecommunication services through a single low-cost communicator which would be transparent to the transmission medium. The international effort is being pursued through an ITU body called International Mobile Telecommunication – 2000 (IMT-2000; Working Party 8F, formed recently, is a follow-on working group after IMT-2000 completed its tasks). In Europe, the effort is called the Universal Mobile Telecommunications System or UMTS, which complements the IMT-2000 but has a special focus on European needs. Similarly, Japanese effort in this direction is called Core-A. Within the IMT-200 requirement a satellite component is embedded, as it is recognized that satellites are essential for providing universal mobile communications.

Integration between the terrestrial and mobile networks has been the subject of study since the 1980s. Figure 11.3 shows the concept of an integrated system, where satellite cells are overlaid on cellular cells. Cellular cells abound in areas of high traffic density, tapering off as traffic density reduces until they become uneconomic, at which point satellite cells take over the service. There is no distinct boundary between the satellite and cellular cells, and therefore, if necessary, a user may communicate via the satellite even within the cellular coverage area.

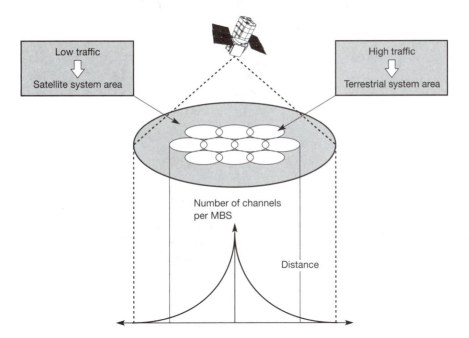

Figure 11.3
The concept of an integrated system (ITU–R report 1177) © ITU

Integration between satellite and terrestrial mobile systems is progressing in a number of phases, as identified below:

1 Terrestrial and satellite systems are separate entities. Integration is by way of dual-mode phones, which package terrestrial and satellite phones within a single unit, offering the user the option to select the desired network. Here the networks are separate and the terrestrial system may consist of a single or multiple standards. This is essentially a handset packaging technology, where common hardware is shared as much as possible. The terminals are bulkier than a single standard terminal. The user selects the desired network manually.

2   In the next level of integration, a limited network integration is built so that the dual-mode phones may automatically select the best-suited system, e.g. in terms of call cost or geographical region, perhaps selecting the satellite system when the (cheaper) terrestrial system is unavailable. The network may similarly direct the call to the mobile by the most cost-effective route. The user has a single number; there is no handover between systems during a call; the physical layers of both systems are optimized independently and therefore are not necessarily compatible with each other. The choice of the complementary terrestrial network depends on the geographical region and the satellite operator's preference of local terrestrial operator. A suitable operational arrangement for commissioning, billing, etc. is necessary between the satellite and terrestrial operators. The concept is presented in Figure 11.4. Note the independent architecture of each system. The communication in this particular scheme occurs between mobile switching centres.

Figure 11.4

A network integration concept. In this scenario, the mobile calling party selects the routing and the network is transparent to the fixed party (ITU-R report 1177; slightly adapted) partial network integration © ITU

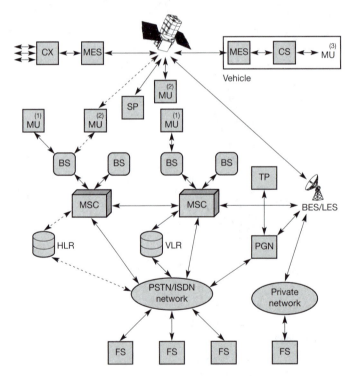

(1) Terrestrial terminal (Mobile Station, MS or Personal Station, PS)
(2) Satellite-type terminal
(3) Dual-mode terminal
       MSC interconnected with
       a satellite network

Key
CS = personal base station (cell site for personal station)
SP = satellite pager; MES = mobile Earth station; FS = fixed subscriber; MU = mobile user;
TP = terrestrial pager; BS = base station; CX = small rural exchange, etc.; BES = base Earth station; MSC = mobile services switching centre; HLR = home location register; VLR = visited location register; PGN = terrestrial paging network; LES = land Earth station

3   There is full integration of terrestrial and satellite systems at network level
    and the physical layer maximizes commonalities. Handover of calls between
    the two segments is possible, operating frequencies are made as close as per-
    missible, and accessing schemes and modulation/coding schemes are
    harmonized as much as possible. This approach allows maximum benefit to
    the user as well as the operator by allowing common equipment. In such an
    integrated network, the user terminal accesses any network anywhere
    through an integrated multimode single terminal. The user is able to choose
    from a range of services. Such an approach is impossible if operators con-
    tinue to develop systems independently, and hence, as mentioned above,
    the concept of such a system is being standardized by international bodies in
    forums such as IMT-2000 (formerly known as Future Public Land Mobile
    Telecommunications Systems or FPLMTS) and UMTS. One approach being
    favoured is to define a number of standard interfaces, which will allow the
    systems of different operators to be developed independently but with suit-
    able provision for exchanging relevant information.

Table 11.2 summarizes the current status of integration.

System	Terminal type	Extent of integration	Comments
Inmarsat-phone	Satellite phone	Some models with dual-mode capability may emerge depending on market demands, but generally little integration	Forerunner of hand-held telephones
Iridium	Dual mode	Various terrestrial standards possible; operating arrangements with terrestrial operators	
Globalstar	Dual mode	Can integrate with various terrestrial standards, depending on operating arrangements with terrestrial operators	
ICO	Dual mode	Can integrate with various terrestrial standards, depending on operating arrangements with terrestrial operators	

Table 11.2
Current status of integration between terrestrial and satellite systems

11.3.1 Satellite-International Mobile Telecommunication 2000

A global system cannot be conceived without a satellite component, and therefore a satellite component was included in the IMT-2000 initiative. ITU studies
are being conducted within the radio communication sector (ITU-R) study

group (group 8) responsible for all mobile services (Task Group 8/1). The study has also been supported by several study groups of ITU's telecommunications standardization sector (ITU-R). One of the first tasks was to identify a suitable frequency band for global operation of such a service. After much deliberation and a series of World Radio Conferences, parts of the 2 GHz band were allocated to the service due to its attractiveness in terms of implementation. Preferred bands and the amount of spectrum were identified in the World Administrative Radio Conference for Mobile Service (WARC MOB-87). Initial allocations were made in World Administrative Radio Conference-92 (WARC-92); minor modifications were made in the World Radio Conference (WRC) of 1995. The Satellite-IMT (S-IMT) allocations in uplink/downlink lie in the range 1,980–2,010 and 2,170–2,200 MHz in regions 1 and 3; and 1,990–2,025 and 2,160–2,200 MHz in region 2. Regulatory procedures for these bands are specified in Resolution-46 of the regulations.

Appealing as the idea may appear, despite continuing efforts, agreement on one or even multiple S-IMT standards has not been possible so far (2000). Six proposals for radio access were received by the ITU, as listed in Table 11.3. All have been included in the ITU recommendation.

Table 11.3
Six proposals for radio were received by the ITU

System	Brief description	Access scheme	Originator
SAT-CDMA	49-satellite/ 7-plane constellation at an altitude of 2,000 km	CDMA	S Korea TTA
SW-CDMA	Satellite wideband CDMA	CDMA	European Space Agency
SW-CTDMA	Satellite wideband hybrid CD/TDMA	TD/CDMA	European Space Agency
ICO	MEO system [10-satellite/2-plane constellation at an altitude of 10,390 km]	TDMA	ICO Global Communications
HORIZON	Geostationary wideband system	TDMA	Inmarsat
IRIDIUM	66-satellite LEO system	CDMA/TDMA	Motorola

Four rather competing standardization approaches have been proposed (Barani et al., 1999).

- **No a-priori standard:** This approach is favoured by the US participants, as studies indicate that different technologies are necessary for differing services.

- **Family of standards**: The standards have maximum commonality with terrestrial standards. In this approach, MSS operators are competing for service rather than on technology. This approach is favoured by the European Space Agency.
- **Hybrid approach**: where only the upper layers of the service are standardized, leaving the operators the flexibility to choose their own physical/MAC layers. This approach has been developed within the European Satellite Integration into Network for UMTS Services (SINUS) project.
- **Medium-term convergence using software radio**: This approach permits independent standards to develop, while integration is achieved through a software-reconfigurable terminal. Software radio or flexible architecture terminals are a focus of considerable world-wide industrial activity. Figure 11.5 shows a possible architecture for an ideal software radio terminal. The ideal circulator is introduced to eliminate the need for a diplexer, which restricts the tuneability of a terminal due to its 'fixed' frequency. The analogue to digital (A/D) converter converts the signal to a digital form for the DSP, where all functions such as modulation/demodulation, coding/decoding, protocol management, etc. are performed. The linear amplifier ensures ideal transfer of RF to minimize distortion and adjacent channel interference.

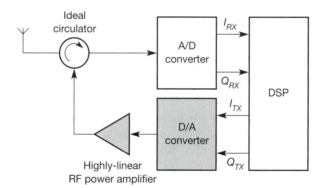

Figure 11.5
A possible architecture for an ideal software radio terminal
(Kenington, 1999)

The interested reader may refer to the abundant S-IMT 2000 literature (see further reading section).

Some of the issues under investigation are:

- integration with terrestrial networks for seamless service provision, in particular, hand-over issues;
- support of a service while traversing networks with different capability (e.g. when migrating from terrestrial to satellite services);
- seamless roaming across a range of networks;
- service adaptation to different types of terminal;
- terminal features such as human-machine interface, including machine-independent programming;
- service portability between operators;

- traffic sharing in a hybrid network;
- possibility of integration of GSM with a satellite system.

In the following text, we focus on the European initiative, Universal Mobile Telecommunication System or UMTS, which should give an insight into the underlying issues. UMTS is a definition of a third generation broadband mobile communication system. The UMTS terrestrial standard was submitted to the ITU in the context of the IMT-2000 family of standards. The purpose of the European initiative is to assist development of the IMT-2000 standards and provide recommendations for European systems in greater detail.

The UMTS systems provide:

- broadband capability with flexible bandwidth allocation to both mobile and fixed users transparently. Data rates up to 2.048 Mbps will be supported, giving the capability of multimedia and other broadband usage such as high-speed image transfer, fast internet access, low-medium quality video, etc.;
- integrated service provision, i.e. cordless, paging, private mobile radio, broadband mobile services all integrated within a single system offering services in business, domestic, vehicular, public and multi-user environments;
- global roaming and service provision;
- mass market to be supported in an efficient and cost-effective manner;
- backward compatibility;
- support of personal mobility through a personal number (such as a subscriber identity module or SIM card).

Due to the need for UMTS to communicate in a variety of environments, four types of cells will be used. Satellite cells are expected to serve areas of thin population through spot beams of hundreds of km; macro cells will comprise cells of up to 35 km to cover areas of low population and back up smaller cells; micro cells of less than 1 km will serve densely populated areas; and pico cells will be used for coverage of confined areas, such as indoors.

Keeping in view some of the unique advantages – wide area coverage, quick implementation of services, etc. – a satellite component is being standardized within the UMTS framework (S-UMTS). The general objectives of the satellite component have been identified in an ETSI document (ETSI/ETR SMG-51201) and a technical report (ETSI, 1994). The system is expected to provide seamless coverage, support small pocket-size telephones and larger terminals, provide paging services where reception is poor, etc. The following roles have been identified for the satellite component:

- **coverage filler and extension**: the satellite component fills terrestrial coverage gaps and extends service to areas where terrestrial systems have not been or cannot be deployed, e.g. maritime or air corridors.
- **backup**: it backs up terrestrial systems in instances of terrestrial system unavailability, such as caused by natural disaster or war.

- **offload terrestrial traffic load**: it offloads heavily congested terrestrial routes.
- **global roaming**: the system provides wide area roaming.

The system is expected to offer cost-effective services with efficient use of RF spectrum; equitable access to users; equitable opportunities for equipment manufacturers and terrestrial network operators if they wish to offer a satellite extension.

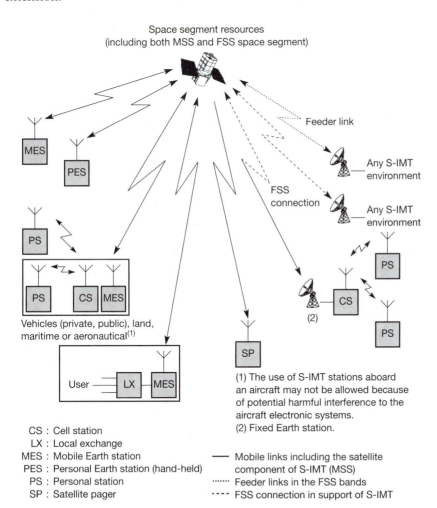

Figure 11.6
Satellite system as envisaged by ITU's IMT-2000 (Donald, 1995 – adapted) © 1995 IEEE

Space segment resources
(including both MSS and FSS space segment)

Feeder link

Any S-IMT environment

FSS connection

Any S-IMT environment

MES

PES

PS

PS   CS   MES

PS

CS

PS

Vehicles (private, public), land, maritime or aeronautical[1]

SP

(2)

User — LX — MES

(1) The use of S-IMT stations aboard an aircraft may not be allowed because of potential harmful interference to the aircraft electronic systems.
(2) Fixed Earth station.

CS : Cell station
LX : Local exchange
MES : Mobile Earth station
PES : Personal Earth station (hand-held)
PS : Personal station
SP : Satellite pager

— Mobile links including the satellite component of S-IMT (MSS)
······ Feeder links in the FSS bands
---- FSS connection in support of S-IMT

S-UMTS is expected to support five categories of user terminals: hand-held, vehicular, transportable, fixed and paging. A number of general requirements have been recommended for these terminals, such as compliance with radiation safety standards, low terminal and usage charge, compactness, ease of use, etc.

Due to the evolving state of satellite systems in the 1990s, the standardization had to keep in view the rapid changes in technical characteristics of satellite systems and it was recognized that the experience of emerging systems will influence the final outcome. It is generally believed that S-UMTS standardization will follow ITU-R's recommendation on the integration of satellite components into future public mobile communications networks, as it would then be possible for a UMTS terminal to be useable globally. Figure 11.6 illustrates such a system was being considered by IMT-2000 (Rec. ITU-R M.818-1).

Whether a terrestrial or satellite UMTS component is used by a terminal is distinguished only by the service link. Satellite links which serve other parts of the network, such as connecting base stations, are not considered in the standardization of S-UMTS. Furthermore, a satellite link falls within standardization only if the service link operates within the IMT-2000 personal communications band, which lies in the range 1,980–2,010 MHz in the uplink and 2,170–2,200 MHz in the downlink. This does not, however, preclude the use of satellite links directly to users in other bands; such an operation is referred to as 'satellite' operation within the UMTS forum, following the ITU IMT-2000 guidelines (ITU-R recommendation M.818-1). Interfaces are standardized so that satellite systems may be developed independently.

Particular satellite system designs have not been standardized but their characteristics and limitations, such as propagation delay and Doppler shift, have been considered in standardization, ensuring that the satellite component is not a burden to other components. Thus a satellite system should apply solutions within itself where possible; for example, Doppler compensation is a part of the satellite network and hence should be applied within the satellite systems. Furthermore, inherent satellite limitations, such as propagation delay, do not exclude satellite systems. For example, a switching centre protocol must exhibit an adequate time window to permit call establishment with a mobile terminal if a satellite component is included in the network.

Due to a variety of possible architectures and orbits, standardization of the radio interface is complex. Possible solutions vis-à-vis orbits include (Dondl, 1995):

- standardize on the worst parameter across orbits, e.g. propagation delay of GEO; Doppler of LEO.
- Standardize individual orbits, so as to extract maximum advantage from each.

The latter would appear to provide a better option, as it creates the opportunity for various types of systems to become a part of UMTS.

It was concluded that standardization would be possible at network or system level but not for lower levels (such as geographical or service-specific level). It was noted that satellite components may exist either as extensions of terrestrial land mobile systems, wherein both satellite and terrestrial cells are served, or as exclusive satellite components serving satellite cells only. UMTS terminals may access the satellite component either through a dual-mode terminal or through a satellite cell station. It is envisaged that it will be possible to hand over calls between satellite and terrestrial systems. Figure 11.7 represents a UMTS satellite system integrated as an extension to a terrestrial mobile system.

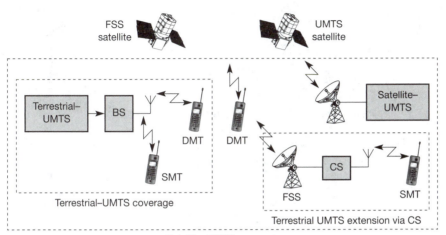

Figure 11.7
UMTS terminals may access the satellite component either through a dual-mode terminal or through a satellite cell station

BS = Base station
CS = Cell station
DMT = Dual-mode terminal

SMT = Single-mode terminal
FSS = Fixed satellite service

A number of means for network integration are possible. An approach considered to be quite promising is to let existing systems operate together with the future access system and the core network. This scheme, portrayed in Figure 11.8, has been investigated in Europe under the ACTS project (discussed later) and the European Telecommunications Standards Institute/Special Mobile Group 3 (ETSI/SMG 3). The S-UMTS component interfaces with the UMTS transport network, which may be ATM/IP based, through an Inter Working Unit (IWU). The IWU manages network-based functions such as mobility management and handover between segments. A number of handover methods have been proposed and have been investigated under the ACTS projects. These include network-controlled or assisted handover and mobile-controlled or assisted handover.

The requirements for S-UMTS, like its terrestrial counterpart, is being defined generically in terms of transport quality, transport quantity and maximum transport delivery delay. A set of satellite-specific services will be defined for S-UMTS, which will permit existing systems to integrate through appropriate adaptation to interface, terminal type etc. Table 11.4 gives an example of a proposal for the 'bearer' services (i.e. a telecommunications transport service), which could be used for supporting satellite-specific applications as well as services offered by terrestrial UMTS, such as voice, facsimile, multimedia, Internet access, LAN connections, etc.

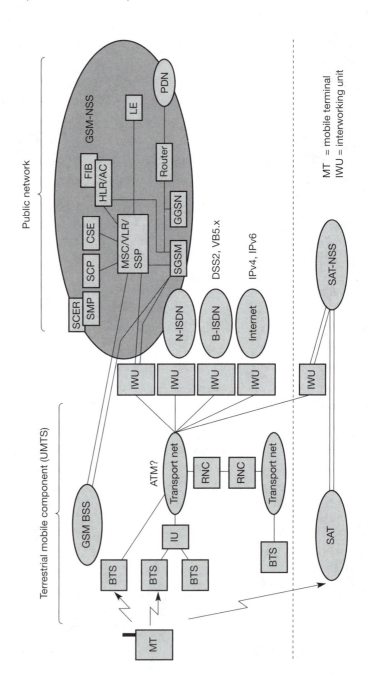

Figure 11.8

Existing systems operate together with the future access system and the core network – European Telecommunications Standards Institute/Special Mobile Group 3 scenario (Guntsch *et al.*, 1998) © 1998 IEEE

Bearer service	Transport rate (kbps)		
	Hand-held (near term)	Hand-held (long term)	Portable
Low rate	0.3–4.8	0.3–64	1.2–64
Medium rate	Not available	64–144	N × 64 (N = 1, 6) Symmetric and asymmetric transport
High rate	Not available	144–432	N × 64 (N = 6 to 32) Symmetric and asymmetric transport

Table 11.4
A proposal for S-UMTS transport services in the short and long term (adapted from Guntsch *et al.*, 1998)

A number of research projects addressing the needs of S-UMTS have been covered in the Advanced Communications Technologies and Services (ACTS) programme of the European Union's fourth research and development framework (Guntsch *et al.*, 1998, ACTS web page). Specific S-UMTS-related issues being addressed include integration into terrestrial networks, the air interface, the user terminals, services and applications, new technologies, e.g. neural networks and EHF band technologies. The ACTS projects are developing system demonstrators for field trial to illustrate proof of concept and further technology.

The aim of the Integrated Satellite UMTS Real Environment Demonstrator (INSURED) project is to define an S-UMTS reference model by simulations and demonstrations using modified GSM and Iridium networks. The trial demonstrates intersystem handover using a network-assisted scheme and roaming functions using a dual-mode terminal. An L/S channel model using Iridium transmissions was planned.

A project named Neural Network Signal Processing Schemes for Wireless Terrestrial and Satellite Transmissions (NEWTEST) is investigating new neural-network-based methods for adaptive equalization of S-UMTS channels at UMTS terminal level and carrying out a feasibility study of application-specific integrated circuit (ASIC) based subsystems.

The Satellite EHF Communications for Multimedia Mobile Services (SECOMS) project is targeting to demonstrate 4 kbps – 2 Mbps transmissions through portable and mobile terminals at both the K_a and EHF bands (see section 3.3). A sub-project, ABATE, is aiming for development of a mobile multimedia service specifically through aeronautical terminals using the K_a band.

The SINUS project (Satellite Integration into Network for UMTS Services) characterized and proposed solutions for air-interface-related issues, including propagation characterization, access and handover techniques (in particular, mobile-assisted handover), etc.; network procedures such as inter- segment procedures; interworking with terrestrial UMTS and B-ISDN; and possible S-UMTS environments to assess the feasibility of S-UMTS service provision in each.

Yet another project, known as Inter-Trial Testbed of Mobile Applications for Satellite Communications (TOMAS), demonstrated and evaluated high bit rate applications and components of S-UMTS. ISDN and ATM access field trials at bit rates up to 2 Mbps using portable terminals to operate through Inmarsat's and the European Space Agency's geostationary satellites have been conducted within its framework.

11.3.2 Beyond UMTS

Effort is under way in Europe, under a project called RACE II, to develop concepts and technologies for a mobile broadband system (MBS) (Chelouche and Plattner, 1993; Fernandes, 1995), which should allow universal mobility at data rates commensurate with broadband ISDN. Though the focus seems to be on terrestrial systems, it is likely that a satellite component will become a part of the activity, similar to the satellite component of the UMTS. This poses further challenges, far beyond those already being resolved for the UMTS, such as operating at millimetre wave frequencies and more stringent network performance requirements due to the types of user applications envisaged. The relationship of the MBS with second/third generation mobile systems is illustrated in Figure 11.9. The MBS service is meant to be an enhancement to the UMTS service rather than a replacement, and will be made available wherever there is a need for it.

Figure 11.9
Relationship of the MBS with respect to second/third generation mobile systems (Chelouche and Plattner, 1993)

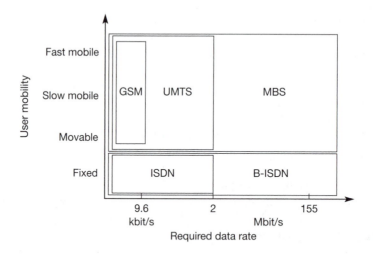

Possible applications, together with the mobility and bit rate expectations, are illustrated in Figure 11.10. They may be broadly classified as mobile-extensions to B-ISDN or mobile specific applications. Example of the former are video-telephony, business communications such as banking/insurance, multimedia library access, and teleworking; examples of the latter are emergency services support, repair assistance, city guidance, transport and travel information, wireless LAN and TV broadcasts.

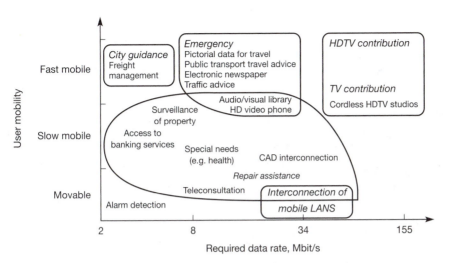

Figure 11.10
Applications with mobility
and bit rate expectations
(Chelouche and Plattner,
1993)

The main objectives of the terrestrial component of the project are to develop a system architecture, functionality and interface to B-ISDN; characterization of mobile communications channels in the 60 GHz band; development of low-cost antennas and other RF technology; methods and techniques for cellular coverage planning; and demonstration of a terrestrial mobile link operating at a channel bit rate of 32 Mbps (user bit rate = 16 Mbps).

It is anticipated that satellite systems can play a role in the future broadband applications extending beyond the capability of the present or near-future mobile satellite systems, which are at best expected to be operating up to a few hundred kbps. Clearly, a significant gap exists currently in terms of MSS capability in this respect.

11.4 Broadband systems

Development of MSS architecture beyond the IMT-2000 era will be influenced by trends in telecommunications growth, technology, spectrum demands and regulatory issues. Recent trends in the mobile industry illustrate that developments in the mobile environment closely follow those in the fixed network, as users expect fixed service facilities to be available in the mobile environment. The most notable technical development influencing telecommunications has been in the field of information technology, and in particular the internet, resulting in an exponentially increasing demand for large throughputs as individuals access remote databases and download large files. Furthermore, a variety of multimedia services, such as video retrieval, image transfer, etc., are on offer through the fixed network and asymmetric direct broadcast satellite system, which indicates the potential of personal wideband systems. Fixed satellite systems are already catering to the internet demand at the ISP level, whereas DBS systems offer the facility from the user's satellite dish. A large number of broadband FSS systems are under development or consideration, targeting the personal and small enterprises market (e.g. Spaceway, Teledesic). Wideband MSS have been experimented with under the USA's ACTS programme.

Table 11.5 Features of next generation broadband satellite systems

System name	Planned year of operation	Number of satellites; type of transponder	Orbital altitude (km)	Frequency band	Multiple access	Network protocol	Typical services offered
Cyberstar	2001	3	36,000 (GEO)	K_a	FDMA TDMA	IP/ATM Frame relay	Internet access, broadband services
iSky (KaStar)	2001	2	36,000 (GEO)	K_a	Not available	IP/ATM	Internet access, direct broadcast service, and personal communication system
SkyBridge	2001	80 LEO Walker constellation; transparent transponder	1,469	K_u	CDMA TDMA FDMA	IP/ATM	High bit rate internet access, interactive multimedia service
Spaceway	2002	16 GEO	36,000	K_a	FDMA	IP/ATM	High-speed internet, bandwidth on demand, multimedia
		20 MEO: Transparent transponder	10,352		TDMA	ISDN, Frame Relay	
Teledesic	2002	288; regenerative transponder	1,375	K_a; 60 GHZ inter-satellite link	Multi-frequency TDMA asynchronous TDMA	IP/ATM ISDN	Internet, high quality voice, data, video
Astrolink	2003	9; regenerative transponder	36,000 (GEO)	K_a	FDMA TDMA	IP/ATM ISDN	High-speed multimedia

Assuming that users will demand in the mobile environment the services on offer in the fixed band, it can be anticipated that there will be a demand for broadband services in the mobile environment. Thus it is possible to postulate a basic set of requirements of a future mobile personal communication service:

- wideband;
- bandwidth on demand;
- protocols to support a mix of services;
- circuit-and packet-mode services;
- hand-held or small portable sets;
- propagation delay compliant with multimedia requirements.

A major problem in the provision of wideband service in the L and S band is spectrum scarcity. Studies conducted as part of S-UMTS development demonstrate a shortfall of up to 17 and 61 MHz in global hot spots for the years 2005 and 2010, respectively (Barani *et al.*, 1999). Throughput can be increased by improving the modulation efficiency, restricting mobility, employing more efficient accessing techniques, narrowing the size of spot beams, and migrating to less crowded parts of the spectrum. While all techniques are being refined and improved, migration of services to less problematic bands appears promising in a longer perspective, and therefore there is considerable interest in utilizing bands above 20 GHz for wideband MSS.

Most proposed commercial wideband systems are in FSS parts of the K_a band and therefore do not formally belong to a mobile service, but the terminals are small enough to be portable and thus there is a fuzziness between applications of these services. A variety of orbits and architectures have been proposed, including low, geostationary orbits or hybrid orbits, intersatellite links, onboard signal processing, etc. In the USA alone, the FCC has awarded licences to over a dozen applicants for K_a-band FSS systems for a variety of broadband solutions, such as interactive multimedia personal communications. By 1997, the ITU had received filing for over a thousand K_a-band satellites. The main features of a sample of US systems are summarized in Table 11.5 (Farserotu and Prasad, 2000). The list is representative but not exhaustive and is included to demonstrate trends; note that a number of broadband systems have also been proposed in Europe and elsewhere.

Consider, as an example, the K_a-band broadband technology demonstrator system proposed under the SECOMS/ABATE (Satellite EHF Communications for Multimedia Mobile Services) project under the European Union's Fourth Framework Advanced Communications Technologies and Services (ACTS) programme, which aims to develop a broadband K_a (20/30 GHz) and EHF (40/50 GHz) system to cover Europe and the North Atlantic with data rates in the range 4 kbps to 2 Mbps.

The proposed system is a multi-spot, K_a/EHF band, multi-satellite, geostationary orbit, private satellite network to provide 4 kbps – 2 Mbps mobile multimedia service for Europe and the North Atlantic (Losquadro *et al.*, 1998). Implementation is envisaged in two phases; in phase one, a K_a-band payload will be introduced, followed by the introduction of an EHF payload in phase two; both systems will be interconnected through intersatellite links to ensure compatibility.

The network architecture is illustrated in Figure 11.11 (Losquadro *et al.*, 1998).

Figure 11.11

Network architecture of the SECOMS/ABATE system (Losquadro *et al.*, 1998) © 1998 IEEE

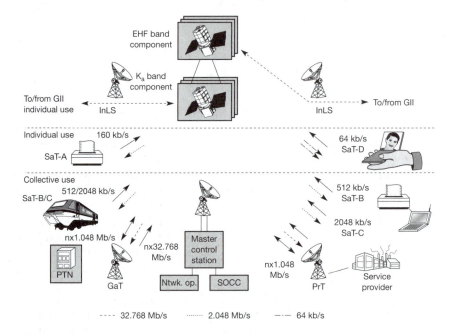

The main components are: four types of user terminals; K_a and EHF-band regenerative transponders; gateway and service provider Earth stations; and a master control station interfaced to a network operator and a satellite operation control centre. The SECOM network is considered as a private network which interfaces with terrestrial fixed networks through a user–network interface at the SECAM gateway stations. There are three rates of user terminals for the K_a band – SaT-A, SaT-B and SaT-C – capable of delivering data rates of 160, 512 and 2048 kbps, and a single terminal type, SaT-D, at 64 kbps for the EHF band. The main categories of users are individuals and groups, with a further differentiation on mode of operation, i.e. portable, fixed and mobile. Provided that data rates are compatible, users can be interconnected. Gateways, service provider centres and the MCS operate up-link and downlink transmissions at 32 Mbps in the K_a band sub-network, and at 1,024 Mbps with the capability to operate more than one link in the EHF sub-net. Protocol adapters at fixed and mobile stations allow interworking with a host of terrestrial systems, such as ISDN, B-ISDN, IP, etc. The satellite operation control centre is used for satellite operations, and the network operator manages administration and billing of external public terrestrial networks and user equipment. The satellite payload concept is illustrated in Figure 11.12(a) and its traffic routing processor in Figure 11.12(b) (Losquadro *et al.*, 1998).

Data, encapsulated in cells, received during a single frame interval, are stored and on the following frame interval routed through the memory to the output buffer and on to the multi-carrier modulator. The traffic resource manager generates input and output traffic rate variation matrices (as beam status matrices) which are used by the terminals to filter the uplink traffic statistics in accordance with the payload resource utilization status. The feedback resource management scheme maximizes satellite resources for circuit and packet switched traffic.

To provide services consisting of a mix of data types and rates, circuit and packet modes, a variety of protocols and an ATM-type transport mechanism have been developed. The network is designed to manage granularity of 16 kbps for circuit mode and 64 bytes for packet networks, by encapsulating the information data in a form similar to that used in ATM. The resulting cell structure is able to support an ATM cell without affecting it. The 11-bit header comprises space segment related information – in-band request, source identifier, destination identifier, protocol identifier and sequence number for cell numbering, as shown in Figure 11.13. Cell coding and multiple access parameters are illustrated in Table 11.6.

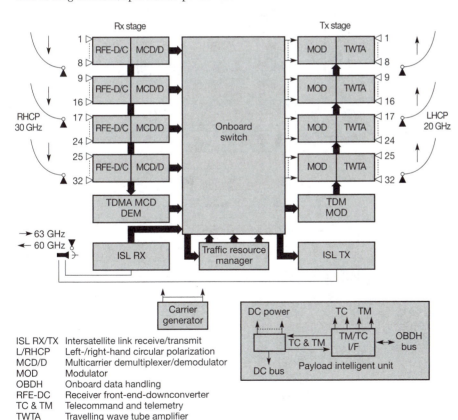

ISL RX/TX Intersatellite link receive/transmit
L/RHCP   Left-/right-hand circular polarization
MCD/D   Multicarrier demultiplexer/demodulator
MOD   Modulator
OBDH   Onboard data handling
RFE-DC   Receiver front-end-downconverter
TC & TM   Telecommand and telemetry
TWTA   Travelling wave tube amplifier

(a)

Figure 11.12

(a) Satellite payload concept (Losquadro *et al.*, 1998)
© 1998 IEEE

Figure 11.12
(b) Traffic routing
processor (Losquadro
et al., 1998)
© 1998 IEEE

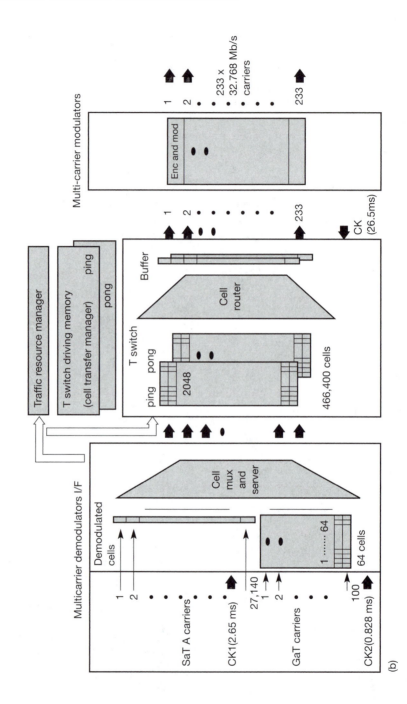

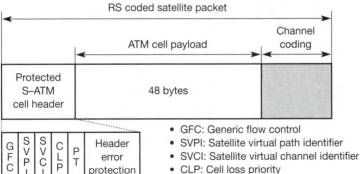

RS coded satellite packet

ATM cell payload

Channel coding

Protected
S–ATM
cell header

48 bytes

| G F C | S V P I | S V C I | C L P | P T | Header error protection |

- GFC: Generic flow control
- SVPI: Satellite virtual path identifier
- SVCI: Satellite virtual channel identifier
- CLP: Cell loss priority
- PT: Payload type

Figure 11.13
Encapsulation of an ATM cell (Mertzanis, I. *et al.*, 1999, International Journal of Satellite Communications). Reproduced by permission of John Wiley & Sons Ltd.

Table 11.6
Cell coding and multiple access parameters

Uplink

Cell encoding
A shortened version of Reed-Solomon code: RS(80, 64, 16)

Cell transmission duration
ms @ 160 kbps
0.828 ms @ 512 kbps
ms @ 618.264 kbps
0.01325 ms @ 32768 kbps
Cells/frame K_a band satellite: 72,462 cells (4.6 Mbyte)

Downlink

Cell encoding
Concatenated code: RS (208, 192, 16) + Convolution: $R = \frac{3}{4}$

Figure 11.14 depicts the end-user terminal's protocol adaptation and network interface scheme.

Various types of applications and protocols are passed to a terminal's protocol adapter/network interface unit, where the protocol is adapted for the SECAM system. A number of dedicated signalling protocols are used within the network for managing transmissions; these are protocols for synchronization of terminals and intersatellite links, traffic assignment/monitoring, and terminal mobility/ubiquity.

In addition to the K_a band, for which there is already an interest, attention is now extending to the EHF band, such as 40–50 GHz and 94 GHz (see Chapter 3 for propagation consideration), where a several-fold increase in bandwidth is possible. Few commercial MSS systems are being considered in the EHF band; however, a number of FSS personal communication systems have been proposed.

It appears that MSS and FSS market offerings will converge for broadband applications, especially from the users' and service providers' perspective. However, regulatory barriers may prevail and cause artificial barriers.

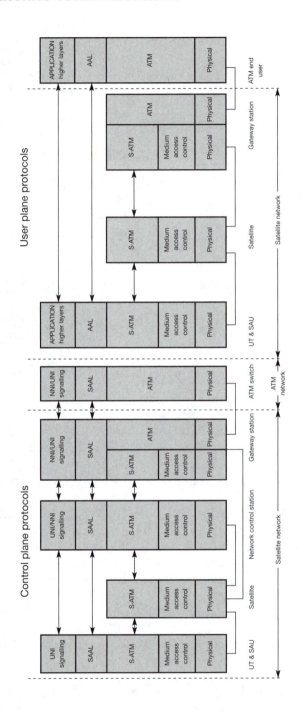

Figure 11.14
End-user terminal's
protocol adaptation and
network interface
scheme
(Losquadro et al., 1998)
Reproduced by
permission of John Wiley
& Sons Ltd

K-band technology

There are a number of definitions of the K-band, which designate it as 10.9–36 GHz (IEEE RDE, General) or 18–27 GHz (Standard 521 Radar) or 12.4–40 GHz (Old English), etc. The K band is further subdivided into the K_u and K_a bands, which are again further subdivided. Here we refer to the K band as the IEEE band 10.9–36 GHz and the technology being considered here refers to the K band above 20 GHz.

There have been a number of experiments conducted in Japan, Europe and the USA since the 1970s. The K band has been, in fact, in regular use in Japan since 1970. The experiments include Japanese, ETS-II (34 GHz propagation experiment), CS-2 and CS-3 (K_a band communications experiment; Italian SIRIO (18 GHz) and ITALSAT (K_a band technology demonstrator); ESA's Olympus (K_a band technology demonstrator and 20–40–60 GHz propagation experiments); the USA's ATS series of satellites (K_a band propagation experiment) and ACTS experiment (technology demonstrator, including mobile communications). These experiments have collectively demonstrated new K_a band technologies and concepts such as K_a band regenerative transponders, multiple hopping spot beams, K_a band channel characterization and adaptive rain fade control systems, forming the basis of many K_a band proposals. These technologies have formed the basis of several proposed constellations to provide world-wide multimedia, bandwidth-on-demand type services envisaged in the next decade (see previous section).

EHF band status

The EHF band is defined by the ITU as the frequency range 30–300 GHz. The frequency bands above the K band are known as the Q band (36–46 GHz), V band (46–56 GHz) and W band (56–100 GHz); the designations are used in radars.

A number of Q-, V- and W-band experiments have been proposed for future personal communication applications. Table 11.7 summarizes some recent initiatives for experimentation in this band.

Notable amongst these is the Data Audio and Video Interactive Distribution (DAVID) experiment proposed by the Italian space agency, planned for 2002 (Ruggieri *et al.*, 1998). The mission plans use of the 94 GHz band, which lies above the Oxygen absorption band (50–70 GHz), for the service link to a LEO satellite. The LEO satellite is connected via an ISL operating in the 23–27 GHz band to ESA's ARTEMIS GEO satellite and then on to a gateway and backbone packet switching network, as illustrated in Figure 11.15.

The DAVID satellite is planned to deploy a regenerative transponder with data rate adaptation for matching the service data rate to the intersatellite data rate. The mission will provide data for channel characterization, new service trials and is expected to support an Italian Antarctic mission. Its application is envisaged for the *next generation* internet service; the system is planned to comprise a hybrid constellation. Some of the salient features of such a network for the internet services are stated as follows:

Frequency band (GHz)	Goal	Spacecraft	Comments	Country/year	Reference
38–43	New frequency bands; inter-satellite link (ISL) development; encourage millimetre wave device development; experiment personal communication system	ETS-VI	Frequency selected as a trade-off between achievable technology level and atmospheric attenuation; elliptical orbit used for experiments due to launch failure; ground terminal used for simulating ISL	Japan /1994	Gargione *et al.* (1999)
44–47 & 21–31	High bit rate personal communication experiment	COMETS	Elliptical orbit (apogee 17, 700 km, Perigee 500 km) used due to launch failure; 21–30 GHz and 44–47 GHz regenerative transponders interconnected at IF and baseband	Japan/1998	(Wakana *et al.*, 1998)
40–50 & 20–30	Channel characterization; land and aeronautical field trials	ITALSAT		Italy	
94	Personal wideband communication (120 Mbps user data rate) experiments	Data Audio and Video Interactive Distribution (DAVID) mission; LEO satellite with ARTEMIS GEO satellite	LEO–GEO system with ISL	Italy/ESA (planned 2002)	

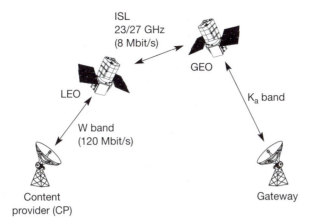

Figure 11.15
DAVID network
architecture (Gargione
et al., 1999)
© 1999 IEEE

- LEO satellites may be deployed gradually, as required, avoiding the need for a fully interconnected constellation from the outset.
- GEO K_a band satellites already deployed by then may be used as an overlay.
- 94 GHz band and a LEO constellation can provide enormous space segment capacity.
- Relatively lower availability caused by propagation outages is expected to be acceptable for internet services.
- Such a system is robust to space segment and network failures due to inherent redundancy in the architecture.

Propagation loss at EHF bands is high, which must be countered by high satellite EIRP. For example, 6 mm/h rain rate fade at 40 and 50 GHz are of the order of 11 and 16.5 dB and the corresponding atmospheric loss around 2.1 and 0.5 dB respectively. As EHF power amplifiers have a limited capability, satellites must use very narrow spot beams and include a capability to distribute power to each spot beam on demand to maximize the use of the satellite power resource.

11.5 Little-LEO systems

While broadband systems are suited for a variety of applications envisaged in the next decade, there is an array of applications which can be served by low bit rate/low cost systems. In this respect little-LEO systems could well seize a substantial market. Little-LEO satellite systems operate in a different band, use low-cost simple satellites and deliver small amounts of data to small low-cost user terminals. The markets targeted include pipeline monitoring, reading meters from remote areas, fleet management, hand-held messaging, etc. The market for such systems is expected to be driven by the success of a number of other complementary markets, such as hand-held computers, GPS receivers, cellular voice and data, internet/electronic mail, etc. One of the key features of such systems is the low-cost space segment which comprises simple small satellites.

As geostationary satellites become larger and more complex, little-LEO satellites are expected to get smaller, smarter and cheaper. On the extreme end of this shift, researchers are pursuing radically different technologies which could enable satellites to shrink to palm-size, weighing around half an ounce and using control systems based on the principles of animal neurones (Page, 1998). These satellites do not use microprocessors for control or fixed algorithms for operation. The so-called nervous net technology 'works like the neurones in animal nervous systems which put out spiked pulses that hold information in the timing between the pulses'. The gradients of the Earth's magnetic fields are used for stabilization and the light gradient from photosensors for orienting the satellite with respect to sun. Short-range antennas can be used for short intersatellite links and long-range antennas will establish more distant communications. The purpose of this particular development is for scientific studies, but suffice to say that such concepts, if applied to communication satellites, can bring a significant reduction in the cost of LEO systems in the long term. Low-cost LEO satellite technologies are being investigated in a number of companies, institutions and universities, such as the University of Surrey in the UK.

11.6 ATM over satellite

Future telecommunication needs are expected to comprise a mix of services requiring throughputs of a few kilobites per second to hundreds of megabits per second. Traditional transport mechanisms – circuit and packet modes – cannot provide the flexibility required for supporting users with variable bandwidth/throughput requirements. A flexible transport mechanism, which may adapt itself to user demands and provide the best use of network resources, is necessary. A flexible transport mechanism, known as the Asynchronous Transport Mechanism (ATM), was developed for the purpose, with the aim of implementing the Broadband-Integrated Switched Data Network (B-ISDN), which is anticipated to offer services ranging from voice to multimedia. The vision was that, in future, most communication may use ATM as a transmission standard (However, see page 529.)

Recognizing the need for the extension of ATM technology for satellite and terrestrial wireless systems, a committee was formed by the ATM forum to address wireless-related issues. The question was not whether ATM should be extended to satellite environment, but how best to achieve it.

ATM technology

Asynchronous transfer mode or ATM technology has been developed for carrying broadband services. The technology has been standardized for broadband ISDN (Jeffery, 1994; De Prycker, 1991; Cuthbert and Sapanel, 1993). The primary rate of transmission supported is 155 Mbps; in the short term lower rates will be used, and in the long term, higher rates such as 622 Mbps may be used.

Incoming information is segmented in small fixed-length packets or cells and transported individually across a network at very high speeds. Incoming data streams are sliced into ATM cells and transmitted so as to meet the desired quality of service. Thus circuit-mode transmission is supported by transmitting continuous streams of cells, whereas transmission of data streams can be discontinuous. Each cell comprises 53 octets, eight of which comprise the header, the remainder carrying information. The size has been selected as a compromise between opposing stipulations of the data-services and telephone services communities. Larger packet lengths achieve higher link efficiency for data transmissions, whereas short packets minimize network transmission delay, a desirable property for telephony.

The header part of an ATM cell contains a number of fields, which contain network information associated with the user data. The header for the user–network interface (UNI) is slightly different from the network–node interface, due to the need for the UNI to arbitrate usage between sharing terminals at the access point. The header provides a number of messages for controlling the flow of information in the network (see Table 11.8).

Header name	Function
Generic flow control (GFC)	Arbitration of messages for controlling usage between several terminals (only in UNI)
Virtual path identifier (VPI)	Coarse routing
Virtual channel identifier (VCI)	Fine routing
Payload type (PT)	Type of data used for congestion control, maintenance cell identification, etc.
Cell loss priority (CLP)	Used by the network to decide whether to discard a cell or not when a cell must be dropped from the network
Header error check (HEC)	Can detect and sometimes correct errors

Table 11.8
The header provides a number of messages for controlling the flow of information in the network

ATM cells are transported within the network through ATM switch nodes. Each switch has a translation table, which contains an entry for VPI/VCI supported within it. ATM calls are initiated on a shared virtual signalling channel – the metasignalling channel. In response to a request on the metasignalling channel, a virtual signalling channel is assigned, which is used by the terminal for further signalling. The ATM exchange then determines a route to the destination by seeking information from the ATM layer. Once the route has been found, the exchange generates a message to the next node using a CCITT signalling No. 7 protocol. The process is repeated at subsequent nodes until the destination is reached. A new route is determined at each node and a new value of VPI VCI is assigned. When the called party has answered a call, a signalling message is sent back along the line of exchanges, causing each to switch the connection. The call can then progress along the virtual route until completion, when entries on all exchanges are cleared.

Four classes of service have been identified. Service classes are identified by their bit rate behaviour, i.e. constant bit rate, or variable bit rate, and by the manner in which connections are established, i.e. connection oriented or connectionless (see Table 11.9).

Table 11.9
ATM service classes

Service class	Description
A	Constant bit rate over established connection, e.g. telephony
B	Variable bit rate with burst traffic, e.g. compressed video
C	Variable bit rate data over established connections, e.g. Frame relay
D	Connectionless data, e.g. LAN such as Ethernet

A service is specified by the traffic descriptor and peak cell rate (PCR). The variable bit rate (VBR) service permits a guaranteed rate but allows sharing of network resources, thereby improving resource utilization through statistical multiplexing. The traffic descriptors associated with VBR are PCR, sustainable cell rate (SCR) and maximum burst size (MBS). These parameters can be monitored using an algorithm called the Generic Cell Rate Algorithm (GCRA) or 'leaky bucket'. The available bit rate (ABR) service has associated with it the service descriptors PCR and minimum cell rate (MCR). The ABR service is meant to support non-real-time service, and hence there are no specifications on delay requirements. Finally, unspecified bit rate (UBR) is a best-effort service, with no guarantees regarding its quality of service or associated traffic descriptor.

To transfer the cells, each service has to add some underlying transport functions. The ATM adaptation layer (AAL) is used for the purpose. The AAL information is embedded in the user information field and is therefore carried transparently across the network, which leaves the choice of the preferred AAL to the communicating terminals. Various types of AAL have been standardized or are being studied. For example, type 1 AAL is intended to carry Class A services, which are essentially constant bit rate data streams over established connections (e.g. telephony).

An ATM network requires extensive network management and control. Intelligent bandwidth management is essential to support variable rate services, as otherwise a bursty message may suffer loss when a high burst of activity occurs. Quality of service must be ensured for each service. Delay and cell loss are important parameters for wideband transmissions. In particular, queueing delay becomes important for interactive services and due care must be exercised in allocating bandwidth to a session. Delay variations caused by variations in arrival rate, due to queueing in the path, can cause annoyance. As an example, in a video transmission, audio and video should be synchronized – jitter can cause loss of synchronization between sound and video. When a queue is too large, incoming cells are likely to be lost. Cells are therefore marked as 'low priority' in the priority bit, giving unmarked cells a higher probability of arrival. The ATM network also requires monitoring to ensure that users are not violating their requested allocation by using more network resources than requested. Monitoring also provides a means of determining faulty terminals, etc.

Call charging in an ATM network can be complicated, as ATM calls have variables which have not been used traditionally – these include requested bandwidth, requested quality of service, cell count, and quantity of successful delivery.

ATM can be carried over a number of physical layers, as it does not depend on any technology. When transmitting ATM over satellite systems and, in particular, mobile satellite systems, a number of considerations apply. These are discussed in the main text. The interested reader is encouraged to refer to the literature, as this is a vast and rapidly evolving discipline.

ATM and IP

The recent popularity of the internet and intranets has created a tremendous interest in their underlying packet routing protocol, known as Transmission Control Protocol/Internet Protocol (TCP/IP). Hence there is great industry interest in the evolution of ATM vis-à-vis IP (or vice versa), and IP, it seems, is indispensable for some applications (Williamson, 1999; Bray, 1998). IP is a layer-3 protocol using packet routing, whereas ATM is based on layer-2 cell switching. As already mentioned, ATM is a multi-service transport platform, whereas IP has been developed for data and is currently not suitable for voice, as it carries considerable overheads, although substantial progress has been made in the past few years to enable transport of voice over IP. Furthermore, ATM has a standardized benchmark for all telecommunication attributes, such as quality of service, whereas IP at present is a best-effort technology. However, IP has a powerful addressing capability. It is not as if IP cannot be transported over ATM. But IP cannot take full advantage of the ATM's underlying intelligence and also, whereas ATM is connection oriented, IP is connectionless. Therefore considerable activity is in progress to combine IP routing with the ATM switching in the ATM forum, Internet Engineering Task Force (IETF) and the research community (e.g. Barnett, 1997). Solutions have been proposed, but they currently have limi-

Figure 11.16
ATM and IP usage
prediction
(Williamson, 1999)

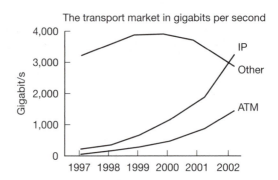

tations for some applications. Synergistic solutions offer IP the low delay and high quality of service possible from ATM. However, if IP-focused companies prefer to develop their individual solutions to achieve these metrics, as is clearly the trend, then the possibility of an independent IP network is a distinct possibility (see Figure 11.16).

ICP/IP is used widely in terrestrial systems. Throughput for IP-over-satellite suffers due to bit error rate, congestion and queues. Satellite delay in geostationary systems is perceived as congestion by the protocol, causing TCP/IP to slow down, thus reducing the throughput. The delay also causes a break in packet flow between connection sites (Brown, 1998). Nevertheless, there is at present considerable interest in using the transport means over satellite by internet service providers (ISP). Solutions being considered include 'spoofing' TCP, wherein premature acknowledgements are sent to simulate acknowledgement and thereby continue transmissions, and selective repeat acknowledgement, which improves protocol efficiency and congestion control over links exhibiting long delays.

As far as satellites are concerned, both technologies may be supported over satellite channels but neither has been optimized for satellite delivery. Nevertheless, ATM seems set to be an influence on next generation transport, evidenced by adoption of the technology by a large number of service providers throughout the world. Table 11.5 demonstrates that most of the planned broadband systems have adopted IP/ATM as the network transport technology. It is anticipated that IP, ATM or a combination of the two will form the backbone of next generation broadband services, as the trend is towards flexible packet-oriented services with the assurance of quality of service (QOS) and bandwidth-on-demand giving efficient resource utilization. Although, at present, IP offers a best-effort quality of service, future versions of IP are expected to offer multiple QOS classes. IP routing also presents problems over satellite networks, particularly for LEO networks due to their dynamics. Solutions being considered include tunnelling, network address translation, Border Gateway Protocol, IP/ATM with multi-protocol label switching and other proprietary solutions.

ATM standards assume an almost perfect channel, hence ATM protocols have very few overheads. One of the main concerns is that satellite channels, and in particular mobile satellite channels, are noise and/or interference limited and hence do not guarantee such low bit error rates or cell drop-out. The problem may worsen in mobiles operating via LEO or MEO constellations due to variability in channel performance. Measurements over satellite ATM channels indicate that cell loss ratio may be the most sensitive, as the ATM header cannot tolerate more than 1 bit error. The ATM layer protects only the header; payload errors remain undetected at this layer. Some AAL layers, however, have error detection features. Other types of impairments include increased jitter, reduction in the efficiency of protocols due to latency, and timeouts. Hence satellite ATM systems require the following considerations:

- To comply with CLR and BER requirements and possible degradation to ATM protocols, link design must be improved by incorporating counter-measures such as ATM-link enhancement techniques, interleaving of appropriate parts of cells, powerful channel coding, etc.
- Latency in satellite communication will require revision of availability conditions.
- Management of bandwidth-on-demand.
- Signalling overheads will be necessary to manage satellite resources and mobility, which should include terminal and satellite movements and various types of handover.
- As MSS ATM network throughput is limited, it will be necessary to take suitable measures to avoid congestion when demands from terrestrial systems exceed satellite throughput capability.

Several concepts have been proposed for meeting the requirements of wireless extension, both in the ATM forum and by researchers. Two groups of ATM network architectures are defined – one for networks using transparent transponders and the other for regenerative satellites. To minimize the impact of propagation delay, it is advantageous to put all delay-sensitive functions, such as switching, queueing, flow control and scheduling, on board the satellite. Connection control and resource allocation may be put on board or on the ground, depending on the complexity of the payload and the effect of the resulting delay on these particular functions. Other delay-insensitive functions such as billing, network monitoring and maintenance can be kept at the ground network control centre.

Definition and standardization of wireless ATM protocol remain one of the issues under intense investigation. Generally, the ATM cell is considered as the unit for protocol processing and switching for the wireless medium. Each ATM cell is then encapsulated within outer layers, which cater for wireless and mobility management functions. The satellite-specific protocols terminate at the terrestrial interface within the gateway. This approach requires no modification to the basic ATM layer and hence is quite straightforward from an implementation viewpoint. This scheme has been proposed for the European SECOMS project discussed elsewhere in this chapter. To remove certain inefficiencies of the encapsulation protocol approach, a satellite-specific ATM (S-ATM) protocol has been proposed (Mertzanis *et al.*, 1999). An S-ATM layer, which includes all satellite network functionality, replaces the ATM layer at the gateway.

As a satellite is a shared node, a medium access control (MAC) layer is necessary, plus a data link layer for error control on the RF link. The MAC layer is particularly challenging for ATM applications due to the need to guarantee the desired quality of service and maintain efficient statistical multiplexing (Bostic, 1999).

Figure 11.17(a) depicts an example satellite ATM architecture and Figure 11.17(b) the underlying protocol stack (Akyildiz and Jeong, 1997).

Figure 11.17
(a) An example satellite ATM architecture (Akyildiz and Jeong, 1997)
(b) The underlying protocol stack (Akyildiz and Jeong, 1997)
© 1997 IEEE

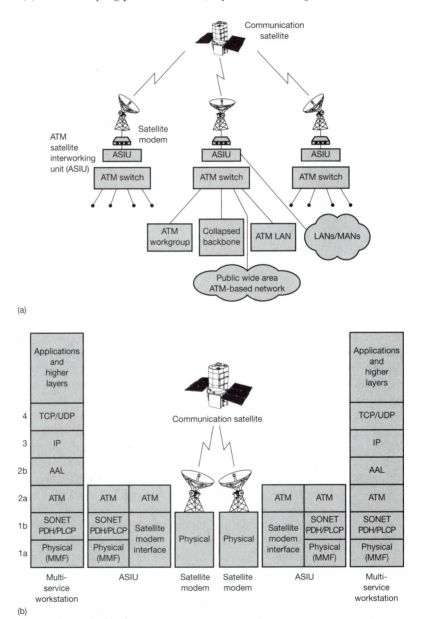

The terrestrial ATM is assumed to be transported on currently used terrestrial transport technologies, such as Synchronous Optical Network/Synchronous Digital Hierarchy (SONET/SDH), Plesiochronous Digital Hierarchy (PDH) and Physical Layer Convergence Protocol (PLCP). The ATM network interfaces with the satellite network through the ATM satellite interworking unit (ASIU). The ASIU manages and controls key system resources, such as bandwidth, network access, timing, call monitoring, traffic control, etc. Figure 11.18 shows the functional block of the ASIU.

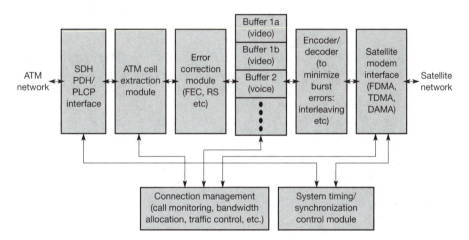

Figure 11.18
Functional block of ATM satellite interworking unit (Akyildiz and Jeong, © IEEE)

The terrestrial frames containing ATM cells are received at the front end, ATM cells are extracted, suitable error correcting schemes are applied, and cells are classified according to traffic class, prioritized and placed in a buffer. Further error protection by way of interleaving is applied for improved CLR performance and the signal transmitted using the chosen multiple access scheme.

One approach for better protection of ATM cells is to increase coding as link performance degrades; other approaches include compressing the header and the payload, and reformatting both into a new frame which has a suitable Reed-Solomon coding and permits synchronization at the receiver. The approach is bandwidth efficient, and capable of delivering error-free transmission.

Bandwidth-on-demand satellite services require satellite resource to be assigned to users on a request basis. Requests must therefore be arbitrated by the network in order to manage satellite capacity and meet user demands. TDMA/FDMA schemes are one way of ensuring flexible capacity utilization.

Figure 11.19 shows an architecture of a multi-service satellite network which can support five different traffic sources – ATM, Frame Relay, ISDN, SS7 and IP – which could serve a variety of users (Evans, 2000). The transmissions can be dynamically changed on a burst-by-burst basis, and TDMA frame length can be selected on a network basis; each burst can include multiple channels, and its structure would depend on the type of traffic channel and transport mechanism

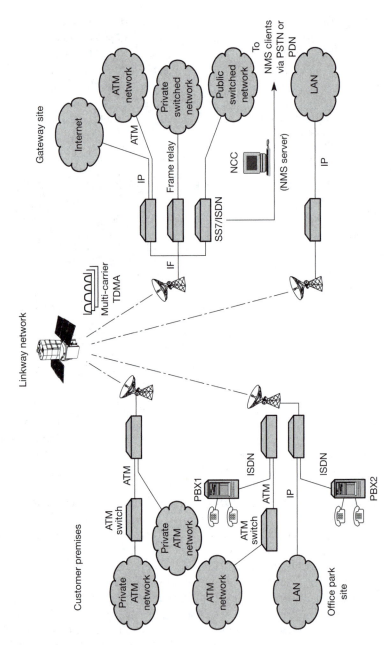

Figure 11.19

A multi-service satellite
network offering
bandwidth-on-demand
(Evans, 2000)

– packet, circuit, ATM, IP, etc. Coding can be altered, depending on the type of
channel and signal quality. Demand assignment for both circuit- and packet-
mode transmissions on a committed and incremental basis, coupled with
frequency hopping on a burst-by-burst basis, improves bandwidth efficiency.

A web network management system (NMS), based on Java client–server technology, provides network management, including control and monitoring. Simulations show that a network of this type can achieve higher efficiency when the number of nodes exceeds approximately five, depending on the traffic load and mix. Networks of this type are currently of great interest, as they are particularly attractive for regions with less developed infrastructure.

11.7 Role of satellites in future networks

A vision of telecommunications in this century is the Global Information Infrastructure (GII) – a high capacity backbone network. The network will comprise a plethora of wired and wireless systems, which would together meet service needs economically and with the desired flexibility and mobility (Pelton, 1994). Within such a meshed broadband network comprising optical fibres and terrestrial wireless transmission systems, satellite systems are considered as one of the weakest links in terms of bandwidth, throughput, quality of service and latency. For example, a 25 Gigabit link operating at a bit error rate of 10^{-11} and transmission delay of <100 ms is clearly not a norm at the dawn of the new millennium in the fixed satellite system, let alone the MSS.

Is this a fundamental limitation due to some law of physics? Take the Teledesic FSS LEO (Tuck *et al.*, 1994) satellite system as an example; the system is designed to provide fibre-optic-like performance through a huge constellation of LEO satellites. Technology deployed in the network has been stretched as far as possible at this point in time. It is difficult to believe that such a technology is the end of the road for satellite communications, bearing in mind that in the past 20 years satellites have become 1,000 times more powerful and about the same order of magnitude more cost-effective, and last 10–15 times longer. However, there are major regulatory and technical hurdles which have to be surmounted before satellite technology can provide digital rates of hundreds of GBps through portable and mobile terminals.

Consider the main limitations of the current MSS systems:

- excessive transmission delay for GEO and MEO systems when compared to optical fibre systems;
- severe shortage of spectrum;
- inability to provide consistently high signal quality;
- susceptibility to rain fades at mm waves, scintillation and shadowing.

Here are some interesting possibilities – the material has been extracted from the literature (e.g. Pelton, 1994; Purchase, 1995) and includes the author's views.

The problem of latency can be solved by using low Earth orbits, stratospheric platforms or their combination. In such a system, stratospheric platforms will cover the densely populated areas, while thinly populated areas will be served by a LEO system. For applications where latency is not critical, an umbrella of GEO satellite systems is a possible solution. It is not necessary for the entire system to be owned by a single entity or organization, if interfaces are standardized so that the system can be owned and operated by different organizations and systems added at will, emulating the internet model; it may be possible to plug in terrestrial wireless systems as necessary. Figure 11.20 portrays an impression of a system consisting of such a conglomerate.

Figure 11.20
A hybrid mobile communication system using terrestrial cellular systems, stratospheric platforms and LEO and GEO satellite constellations to support future broadband transmissions.
(a) coverage (b) entities

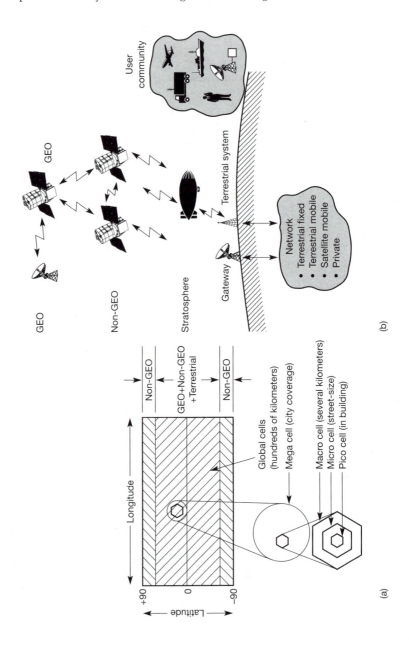

To offer Giga-bytes of information rate, several Giga-Hertz of spectrum will be necessary, which can be available above 35 GHz, and this is where regulatory decisions are necessary, coupled with large investments to develop the necessary hardware; large segments of continuous spectrum could provide optical-fibre-like throughputs. Wideband CDMA could permit large numbers of simultaneous accesses. Coupling this with the advantages of onboard processing satellites, advanced spot beam technology and improved coding techniques, it is estimated that up to a hundred-fold frequency reuse is possible, giving several hundred GHz of spectrum in the 35–50 GHz band. The problems of severe rain fades and scintillation compound with usage of high efficiency modulation schemes which are highly susceptible to flat and frequency selective fading, and thus powerful coding and fade resistant modulation schemes become desirable. The technology for generating thousands of spot beams is already being developed in some laboratories, and regenerative onboard capabilities have made vast improvements recently. Dynamic power control can alleviate the problem of rain fades and shadowing by diverting large amounts of power to fading channels. With more powerful transmitters and antennas deploying thousands of micro spot beams, link margins of 15–20 dB are conceivable, giving extremely high and stable BER. The distinction between MSS and FSS may have to be compromised for personal communications, as we can even now observe VSAT sizes gradually shrinking and approaching mobile portable size. An FSS-MSS integrated architecture is conceivable in the foreseeable future, allowing a user to roam between a home 'Giga-byte' dish and a personal communicator while away on a beach holiday. Clearly there does not appear to be an inherent limitation which precludes satellites performing as well as their terrestrial counterpart. Admittedly, it may not happen in this decade, but who is willing to bet that it won't happen in two decades?

Further reading

ACTS **http://www.infowin.org/ACTS**

Akyildiz, I.F. and Jeong, S.–H. (1997) 'Satellite ATM networks: A survey', *IEEE Communications Magazine*, July, pp. 30–43.

Arcand, S; Murthy, K.M.S.; Hafez, R. (1995) 'Traffic sharing algorithm for hybrid mobile networks', *IMSC 1995, The Fourth International Mobile Satellite Conference*, Ottawa, co-sponsored by Communications Research Centre/Industry Canada and Jet Propulsion Laboratory/NASA, 42–7.

Astrolink Lockheed-Martin ***http://www.astrolink.com***

Ayyagari, D. and Ephremides, A. (1998) 'A satellite-augmented cellular network concept', *Wireless Networks*, 4, 189–98.

Barani, B; Schwarz da Silva, J; Pereira, J; Arroyo-Fernandez, B.; Ikonomou, D. (1999) 'S-UMTS in the wireless information society: The challenges ahead',

IMSC '99, Sixth International Mobile Satellite Conference, Ottawa, co-sponsored by Communications Research Centre and the Jet Propulsion Laboratory, 454–468.

Barnett, R. (1997) 'Connectionless ATM', *Electronics and Communications Engineering Journal*, October, 221–30.

Bostic, J. (1999) 'MAC protocol issues for multimedia satellite systems', *IMSC '99, Sixth International Mobile Satellite Conference*, Ottawa, co-sponsored by Communications Research Centre and the Jet Propulsion Laboratory, 303–9.

Bray, A. (1998) 'IP over ATM: A switch for the better?', *Telecommunications*, October, 67–71.

Brown, P.J. (1998) 'IP-over-satellite – A global solution now', *Via Satellite*, October, 16–28.

Caini, C; Corazza, G.E.; Falciasecca, G.; Ruggieri, M.; Vatalaro, F. (1992) 'A spectrum and power efficient EHF mobile satellite system to be integrated with terrestrial cellular systems', *IEEE Journal on Selected Areas in Communications*, **10** (8), October, 1315–25.

Chelouche, M. and Plattner, A. (1993) 'Mobile broadband system (MBS): Trends and impact on 60 GHz band MIC development', *Electronics and Communication Engineering Journal*, June, 187–97.

Chitre, D.M.; Gokhale, D.S.; Henderson, T.; Lunsford, J.L.; Mathews, N. (1994) 'Asynchronous Transfer Mode (ATM) operation via satellite: issues, challenges and resolutions', *International Journal of Satellite Communications*, 12, 211–22.

Cuthbert, L.G. and Sapanel, J.C. (1993) 'ATM: the broadband telecommunications solution', *IEE*.

Cyberstar Loral *http://www.cyberstar.com*

De Prycker, M. (1991) *Asynchronous Transfer Mode: solution for broadband ISDN*, Ellis Horwood.

Del Re, E. (1988) 'Satellite system integrated with terrestrial cellular network for mobile communications', *ESA/ESTEL Memorandum*, XR/1988/ML/mt.

Donald, P. (1995) 'Standardisation of the satellite component of the UMTS', *IEEE Personal Communications*, October, 68–74.

ETSI (1994) 'Special Mobile Group (SMG); Framework for satellite integration within the Universal Mobile Telecommunication System', *Draft ETR*, (12-01) V.0.6.1

Evans, B.G. and Tafazolli, R. (1996) 'Future multimedia communications via satellite', *International Journal of Satellite Communications*, October, 467–74.

Evans, J.V. (2000) 'Network interoperability meets multimedia', *Satellite Communications*, February, 30–6.

Farserotu, J. and Prasad, R. (2000) 'A survey of future broadband multimedia satellite systems, issues and trends', *IEEE Communications Magazine*, June, 38 (6), 128–33.

Fernandes, L. (1995) 'Developing a system concept and technologies for mobile broadband communications', *IEEE Personal Communications*, February, 54–9.

Gargione, F.; Iida, T.; Valdoni, F.; Vatalaro, F. (1999) 'Services, technologies, and systems at K_a band and beyond – A survey', *International Journal on Satellite Communications*, **17** (2), February, 133–44.

Guntsch, A.; Ibnkahla, M.; Losquadro, G.; Mazzella, M.; Roviras, D.; Timm, A. (1998) 'EU's activities on third-generation mobile satellite systems (S-UMTS)', *IEEE Communications Magazine*, February, 104–10.

Hughes Spaceway *http://www.hns.com/spaceway*

Hung, A.; Montpetit, M.; Kesidis, G. (1998) 'ATM via satellite: A framework and implementation', *Wireless Networks*, 4, 141–53.

Ingley, C. (1999) 'Little LEOs: Riding the wave of market forces', *Satellite Communications*, December, 28–32.

iSky *http://www.ka-star.com ; http://www.isky.net*

ITU IMT-2000 website **http://www.itu.int/imt/l-info/mkt-growth/index.html**

ITU (1990) 'Integration of terrestrial and satellite land mobile systems', ITU–R Report 1177.

Jeffery, M. (1994) 'Asynchronous transfer mode: the ultimate broadband solution?', *Electronics and Communications Engineering Journal*, June: 143–51.

Kenington, P.B. (1999) 'Emerging technologies for software radio', *Electronics and Communication Engineering Journal*, April, 69–83.

Leite, F.; Engelman, R.; Kodama, S; Mennenga, H.; Towaij, S. (1997) 'Regulatory considerations relating to IMT-2000', *IEEE Personal Communications*, August, 14–19.

Losquadro, G.; Aerospazio, A.; Sherif, R.E. (1998) 'Requirements of multi-regional mobile broadband satellite networks', *IEEE Personal Communications*, April, 26–30.

Mertzanis, I.; Sfikas, G.; Tafazolli, R.; Evans, B.G. (1999) 'Satellite-ATM networking and call performance evaluation for multimedia broadband services', *International Journal of Satellite Communications*, 12, 107–27.

Mohorcic, M. and Kandus, G. (1996) 'Performance study of an integrated satellite/terrestrial mobile communication system', **14** , 413–25.

Morgan, W.L. and Gordon, G.D. (1989) *Communications Satellite Handbook*, New York, John Wiley & Sons.

Page, D. (1998) 'Paradigm shifting without a clutch', *Satellite Communications*, September, 70–2.

Pelton, J.N. (1994) 'Low Earth Orbit Satellites: best hope for advanced satellite communication and the global information infrastructure', *Space Communications*, **12**, 233–247.

Priscoli, F.D. and Muratore, F. (1996) 'Radio and network comparisons for MSBN and GSM systems in a geostationary satellite environment', *International Journal of Satellite Communications*, **14**, 439–54.

Purchase, J.F. (1995) 'Establishing the satellite-based Infobahn in a multicultural world', *Via Satellite*, **10** (7), 18–22.

Re, D.E. and Iannucci, P. (1995) 'The GSM procedures in an integrated cellular/satellite system', *IEEE Journal on Selected Areas in Communications*, **13** (2), February, 421–30.

Richharia, M. and Evans, B.G. (1988) 'Synergy between land mobile satellite and terrestrial systems – possibilities in the European region', *Fourth International Conference on Satellite Systems for Mobile Communications and Navigation*, London, 17–19 October, 111–17.

Rudge, A.W. (1993) 'I'll be seeing you: Multimedia communications in the 21st century', *Electronics & Communication Engineering Journal*, October, 293–302.

Ruggieri, M.; Vatalaro, F.; Paraboni, A.; Bonifazi, C. (1998) 'DAVID: A small satellite mission for data distribution', *Proceedings CNES-98*, Antibes, Juan les Pins, France, September.

SkyBridge *http://www.skybridgesatellite.com*

Tuck, E.F.; Patterson, D.P.; Stuart, J.R.; Lawrence, M.H. (1994) 'The Calling Network: A global wireless communication system', *International Journal of Satellite Communications*, 12, 45–61.

Teledesic: *http://www.teledesic.com*

TIA/EIA (1991) 'Satellite ATM: Architectures and Guidelines', *Telecommunications Systems Bulletin* (TSB-91), May.

UMTS Forum (1999) 'The future mobile market – global trends and developments with a focus on Western Europe', Report no. 8, March.

Wakana, H.; Tanaka, M.; Kozono, S.; Li, H.; Takahashi, M. (1998) 'The COMETS experiments for advanced mobile satellite communications and advanced satellite broadcasting', *Proc Int Symposium Space Technology and Science*, 98-h-05, Omiya, Japan, May.

Williamson, J. (1999) 'Hitting the IP jackpot', *Global Telephony*, March, 23–30.

Zhao, W.; Tafazolli, R.; Evans, B.G. (1995) 'Handover performance in an integrated GSM and satellite mobile communication systems', *IMSC 1995, The Fourth International Mobile Satellite Conference*, Ottawa, co-sponsored by Communications Research Centre/Industry Canada and Jet Propulsion Laboratory/NASA, 447–52.

APPENDIX 1

1

A.1 Orbital characteristics of proposed and operational systems

System	Number of satellites	Planes	Satellites /plane	Altitude (km)	Inclination (deg)	Features
ICO Global Communications (Renamed: New ICO)	10	2	5	10,350	45	Service: voice – 4.8 kbps, data – 2.4 kbps, and higher bit rate; operation anticipated in 2003; financial difficulties led to system being taken over by private investors; system cost estimate: around $4 bn; 163 spot beams/satellite; approximate spot beam area – 950,000 km^2; channels/beam 28; service link – 1.98 to 2.01 GHz (downlink) and 2.17 to 2.2 GHz (uplink) ; feeder link – 6.5 GHz band (uplink)/ 3.6 GHz band (downlink); multiple access: TDMA; see Chapter 9; see Figure A.1 for snapshot of constellation coverage.
Ellipsat	Borealis orbit: 16 (highly elliptical orbit)	2	8	520/ 7,800	116.6	Voice, data and facsimile service; service link: 1.61 to 1.6165 GHz (uplink) and 2.4835 to 2.5 GHz (downlink); system cost[1]: $700 m; cost/satellite[1]:
	Concordia orbit: 8 (Equatorial orbit)	1	8	7,800	0	sponsored by Mobile Holdings Inc., Washington; see Chapter 9; see Figure A.2 for snapshot of constellation coverage.

Main characteristics of non-geostationary satellite systems proposal (Crossman, 1999; Hernandez *et al.*, 1996; Finean, 1996; Logsdon, 1995; Evans, 1998; and other miscellaneous sources)

System	Number of satellites	Planes	Satellites /plane	Altitude (km)	Inclination (deg)	Features
Archimedes	4	1	4 (90° separation between satellites)	(Period = 12 hours)	63.4	Sponsors: European Space Agency; voice, data, facsimile and broadcast services; 3-axis stabilized satellites; six spot beams/satellite; service link 1.5 GHz band.
Iridium	66	6	11	780	86	Service: voice – 4.8 kbps, data – 2.4 kbps; operation ceased in 2000 after financial difficulties leading to bankruptcy; subsequenlty bought by another investor; system cost: around $4.7 bn; 48 spot beams/satellite; Approximate spot beam area – 350,000 km^2; channels/beam 23; minimum elevation angle: 8°; service link 1.62 to 1.63 GHz; feeder link: 29.1–29.3 GHz (uplink)/ 19.4–19.6 GHz (downlink); inter-satellite link 23.2–23.4 GHz; multiple access: TDD; see Figure A.3 for snapshot of constellation coverage.
Globalstar	48	8	6	1,414	52	Service: voice – 4.8 kbps, data – 7.2 kbps; operation started in 1999; early financial difficulties; initial system cost estimate: around $2.6 bn; 16 spot beams/ satellite; approximate spot beam area – 2,900,000 km^2; channels/beam 175; Service link – 1.61 to 1.63 GHz (downlink) and 2.48–2.5 GHz (uplink); feeder link – 6.7 to 7.08 GHz (uplink)/ 5.09 to 5.25 GHz (downlink); multiple access: CDMA
ECCO	46	2	11 in 0° inclination	2,000	0	See Figure A.4 for coverage (note the equatorial-only coverage).
			35 in 62° inclination		62	
ORBCOM	36	4	8	775	45	System cost[1]: $350m; cost/satellite[1]:13; operational since 1998; first in market; near real-time service; service link: 148–149 MHz (uplink), 137–138 MHz (downlink); spacecraft mass: 40 kg.
		2	2	775	70	
Starsys	24	6	4	1,000	53	Messaging and positioning; 148–149 MHz (uplink), 137–138 MHz (downlink); spacecraft mass: 150 kg; global coverage.

System	Number of satellites	Planes	Satellites /plane	Altitude (km)	Inclination (deg)	Features
Vita	3	N/A	N/A	670	N/A	Filed in USA; store and forward messaging service; non-profit organization; almost fully financed; system cost[1]: $15 m; cost/satellite[1]: $8 m; partially financed; planned in-service year: 2000; 148–149 MHz (uplink), 137–138 MHz (downlink).
CTA (Gemnet)	38	5	4 × 6 and 2 × 1	1,000	N/A	Filed in USA; messaging service
E-Star	6	N/A	N/A	1,261	N/A	Filed in USA; store and forward messaging service; system cost[1]: $90 m; cost/satellite[1]: $15 m; partially financed; partly backed by Echostar; planned in-service year: 2001.
Faisat	26	5	4 × 6 + 1 × 2	1,000	N/A	Filed in USA; near real-time and store/forward messaging service; some satellites launched; system cost[1]: $350 m; cost/ satellite[1]: $11 m; partially financed; planned in-service year: 2001.
GE Americon	24	N/A	N/A	800	N/A	Filed in USA; messaging service.
LEO ONE	48	N/A	N/A	950	N/A	Filed in USA; near real-time messaging service; system cost[1]: $600 m; cost/ satellite[1]: $13 m; partially financed; planned in-service year: 2000.
Gonetz	36	6	6	1,300– 1,500	83°	Filed in Russia; near real-time messaging service; some satellites launched; planned in-service year: 2000.
Koskon	32	N/A	N/A	1,200	74°	Filed in Russia; messaging service; some satellites launched.
Signal	48	12	4			Filed in Russia; messaging service.
Globalstar	N/A	N/A	N/A	N/A	N/A	Filed in Russia – SPS Ural, Pallada; messaging service.
Ariadne	N/A	N/A	N/A	N/A	N/A	Filed in Ukraine; messaging service.
Safir	2	N/A	N/A	N/A	N/A	Partially financed; originating countries: Germany and Italy; store and forward messaging

System	Number of satellites	Planes	Satellites /plane	Altitude (km)	Inclination (deg)	Features
						service; Europe primary market; planned in-service year: 1999.
IRIS	6	N/A	N/A	N/A	N/A	Store and forward messaging service; originating countries: Germany, Belgium; system cost[1]: $50 m; cost/satellite[1]: $8 m; partially financed.
Teledesic (original proposal)	840 (original)	21	40	700	98.2	FSS service but provision for mobile service; Sun-synchronous orbit; K_a-band service link – 28.6 to 29.1 GHz in the uplink and 18.8 to 19.3 GHz in the downlink; intersatellite link – 60 GHz; 16 kbps to 2.048 Mbps services, including video services; 795 kg satellites; capacity of 2 million subscribers; Earth-fixed cells; system cost[1]: $9 bn; user terminal cost: $2,000.
Teledesic (final proposal)	288 (Scaled down)	12	24	700		Ibid

Note 1: Published initial cost estimates

Figure A.1

Snapshot of ICO
Coverage (Bains, 1999)

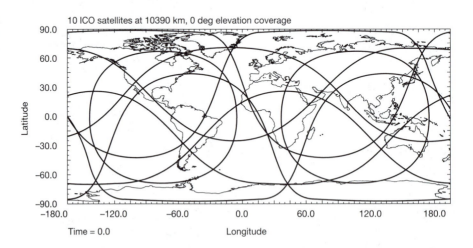

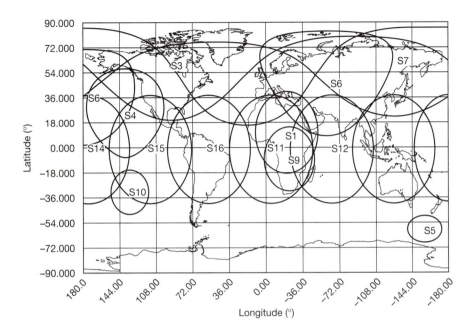

Figure A.2
Snapshot of Ellipsat coverage (Krewel and Maral, 1998) Reproduced by permission of John Wiley & Sons Ltd

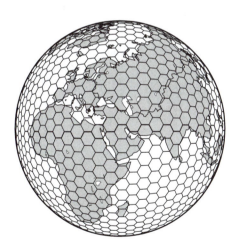

Figure A.3
Snapshot of Iridium coverage (Wu *et al.*, 1994) © 1994 IEEE

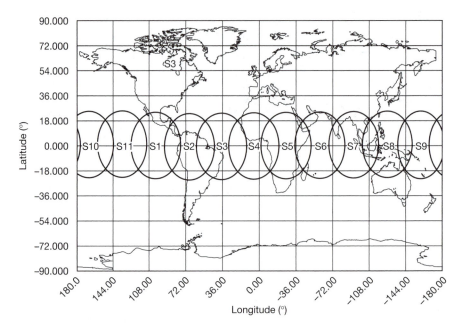

Figure A.4
Snapshot of ECCO
coverage (Krewel and
Maral, 1998)
Reproduced by
permission of John Wiley
& Sons Ltd

A.2 GEO constellations

System name	Coverage	Service	Frequency band (service link)	Comments
Inmarsat	World-wide	Voice/high rate data	L	Operational since early 1980; see Chapter 9.
AMSC/TMI	American subcontinent	Voice/medium rate data	L	Operational since 1995; see Figure A.5 for coverage pattern; see section A.4 for link characteristics; see Chapter 9.
OPTUS	Australia	4.8 kbps voice/ 2.4 kbps data/ 4.8 kbps facsimile	L	Commercial service started 1994; satellites located at 156° E and 160° E, with an EIRP of 46 dBW and service link G/T of –2 dB/K, provide a single beam coverage (Harrison, 1995).

System name	Coverage	Service	Frequency band (service link)	Comments
N-Star	Japanese Island and surrounding sea area	5.6 kbps voice, 4.8 kbps data and facsimile	S	Operated by Japanese NTT Mobile Communications Network Inc; service started 1995; service link: 2.6/2.5 GHz; feeder link: 6/4 GHz; satellite-only and dual-mode user terminals; four spot beams, each of 1,200 km diameter; two satellites (N-Star-a at 132° E and N-Star-b at 136° E); four spot beams are used to cover areas up to 370 km into sea around Japan.
Solidaridad	Regional/Mexico	Voice/medium rate data	L	Satellite launched in 1994; expected to provide same type of service as AMSC/TMI.

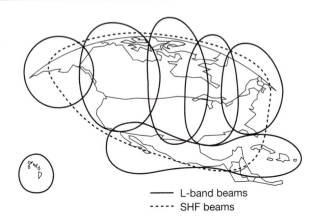

—— L-band beams
- - - - SHF beams

Figure A.5
AMSC/TMI coverage
(Evans, 1998)
© 1998 IEEE

A.3 Proposed regional super-geostationary systems

System name	Coverage area	Planned launch year (dates are revised regularly)	Comments
Agrani	Turkey to Singapore; Sri Lanka to Russia	1998	2 satellites, manufactured by Hughes.
Aces	Southern China, Thailand, Indonesia (see Figure A.6)	1999	2 satellites, Lockheed Martin; see Chapter 9.
Thuraya	Middle-East and North Africa	2000	Probably 2 satellites, manufactured by Hughes
EAST	Europe	2000	1 satellite, manufactured by Matra Marconi (now called Astrium).
Asia Pacific Mobile Telecommunications (APMT)	Far East	1998	Backed by China/Singapore; manufactured by Hughes.

A.4 Typical link parameters of an operational first generation regional GEO system

Main link parameter	Forward link	Return link
Uplink		
Frequency (GHz)	13.2	1.65
EIRP (voice activated) (dBW)	47.2; antenna diameter = 4.7m	14.5
Satellite G/T (dB/K)	−3	2.8
C/NO (dB-Hz)	66	57.2
Downlink		
Frequency (GHz)	1.545	11.3
EIRP (dBW)	30	8.6
Receiver G/T (dB/K)	−15	28.2; antenna diameter = 4.7m
C/No (dB-Hz)	55.4	59.6

A.5 Gateways of non-geostationary satellite systems

A.5.1 Iridium
Note: Iridium operation in its original form has now terminated

Gateway	Operating company
Beijing, China	Iridium China (HK) Ltd
Seoul, Korea	SK Telecom (Iridium Korea)
Nagano, Japan	Nippon Iridium Corporation
Taipei, Taiwan	Pacific Iridium Telecom Corporation
Bangkok, Thailand	Thai Satellite Telecom Iridium South Pacific
Tempe, Arizona, USA	Iridium North America Iridium Central America and Mexico
Hawaii, USA	US Government
Rio de Janeiro, Brazil	Iridium S America Corporation
Rome, Italy	Iridium Italia Iridium Communications Germany
Moscow, Russia	Iridium Eurasia
Jeddah, Saudi Arabia	Iridium Middle East Corporation
Mumbai, India	Iridium India Telecom Ltd

A.5.2 ICO

Note: The company has been renamed New ICO after being taken over by new investors; the list below refers to its predecessor company, ICO.

Country	Jurisdiction (previous affiliations)
Australia	Telstra
Chile	ENTEL SA
Germany	T-Mobil
India	VSNL
Indonesia	PT INDOSAT
Korea	Korea Telecom
South Africa	Telecom SA
United States	COMSAT
UAE	ETISALAT
Brazil	UGB Telecomunicacoes
Mexico	Telecomm Mexico
China	Being negotiated

A.6 Satellite launchers

Country	Launcher	Manufacturer	Performance in a LEO circular orbit	
Russia/Ukraine	START 1	KOMPLEX	480 kg, 300 km, 76°	Under development
	START	KOMPLEX	750 kg, 300 km, 76°	
	ROKOT	KHRUNICHEV	1,600 kg, 300 km, 65°	
	COSMOS	POLIOT	1,400 kg, 200 km, 51°	
	TSYKLON	NPO YOUJNOE	4,000 kg, 200 km, 51°	
	PROTON	KHRUNICHEV	20,000 kg, 200 km, 51°	
	ZENIT	NPO YOUJNOE	14,000 kg, 200 km, 51°	
	MOLNYA	SP KOROLEV	7,000 kg, 200 km, 51°	
	SOYUZ	SP KOROLEV	7,500 kg, 200 km, 51°	
India	PSLV	ISRO/ANTRIX	3,200 kg, 200 km, 43°	
China	CZ3	CGWIC	5,000 kg, 185–200 km, 28.5–31°	
	CZ4	CGWIC	4,000 kg, 200 km, 97.8°	
	CZ 2E	CGWIC	8,800 kg, 185–200 km, 28.5–31°	
ISRAEL	NEXT	IAI	500 kg, 185–200 km, 28.5–31°	Under development
Japan	MU-3	ISAS/Nissan Motor Co	780 kg, 185–200 km, 28.5–31°	
	H2	NASDA/ Mitsubishi	10,500 kg, 185–200 km, 28.5–31°	
USA	Pegasus (standard version)	OSC	360 kg	
	Pegasus XL	OSC	450 kg	Under development
	Taurus	OSC	1,250–1,400 kg	
	Taurus 2	OSC	2,300–4,600 kg	Under development
	Conestoga	E.E.R. System	350–1,700 kg	Under development
	LLV (1 to 3)	Lockheed	800–3,650 kg	Under development
	Aquila	American Rocket	1,500 kg	Under development
	Delta (6920 & 7920)	McDonnell Douglas	4,000 & 5,000 kg	
	Atlas (II, IIA, II AS)	Martin Marietta	6,580, 7,280 and 8,640 kg	
	MSLS	Martin Marietta	160–550 kg	Under development
	LCLS-A	Hercules	2,300 kg	
Europe	Ariane 4	Arianespace	4,600–9,400 kg	
	Ariane 5	Arianespace	18,000 – 550 km, 28.5°	Under development
	ESL			Under development

Source: Hernandez *et al.*, 1996

Further reading

Bains, N. (1999) 'The ICO system for personal communications by satellite', *IMSC 99, The Sixth International Mobile Satellite Conference*, Ottawa, 1999, co-sponsored by the Communications Research Centre and the Jet Propulsion Laboratory, 16–18 June, 88–93.

Crossman, M. (1999) 'The Little LEO scorecard', *Via Satellite*, January, 70.

Evans, J.V. (1998) 'Satellite systems for personal communications', *Proceedings of IEEE*, 86 (7), July, 1325–41.

Finean, R.J. (1996) 'Mobile satellite communications proposals', *BT Technology Journal*, **14** (3), July, 74–80.

Harrison, S. (1995) 'Mobilesat® – the world's first domestic land satellite system', *Space Communications*, 13 (3), 249–56.

Hernandez, D.; Cussac, T.; Ecoffet, R.; Foliard, J.; Thoby, M. (1996) 'Constellation for mobile satellite services, an overview', *Space Communications*, **14**, 101–10.

Ingley, C. (1999) 'Global vision: How big LEO and little LEO service providers will interconnect with regional gateways', *Satellite Communications*, 23 (2), 38–45.

Krewel, W. and Maral, G. (1998) 'Single and multiple satellite visibility statistics of first-generation non-geo constellations for personal communications', *International Journal of Satellite Communications*, 16, 105–25.

Logsdon, T. (1995) *Mobile Communication Satellites*, McGraw-Hill Inc International Edition, Singapore.

Nguyen, N.P.; Buhion, P.A.; Adiwoso, A.R. (1997) 'The Asia cellular satellite system', *IMSC 1997, Proceedings of the Fifth International Mobile Satellite Conference*, Pasadena, California, 16–18 June, co-sponsored by NASA/JPL and DOC/CRC, JPL, Publication 97–11, Jet Propulsion Laboratory, Pasadena, California, 16 June, 145–52.

Wu, W.W.; Miller, E.F.; Pritchard, W.L.; Pickholtz, R.L. (1994) 'Mobile satellite communications', *Proceedings of the IEEE*, 82 (9), September, 1431–49.

Index

accessing schemes, 196–8, 207, 214
Adams, W.S., 67
administrative management, 373
Advanced Communications Technologies
    and Services (ACTS) programme,
    190, 511, 513, 517
Advanced Communications Technology
    Satellite (ACTS), 276–9, 296, 523
Advanced Mobile Phone System (AMPS), 5
Advanced Relay and Technology Mission
    (ARTEMIS), 282, 294, 523
aeronautical applications, 26, 140–3, 219,
    225, 349, 370, 374, 387, 412, 432
air interfaces, 338, 513
air traffic control, 412, 477
aircraft separation, 477
Aloha, 196
altitude, orbital, 42–3, 55, 67, 286–8, 296,
    300, 343, 347, 349–53
American Mobile Satellite Corporation
    (AMSC), 8, 420–4
amplitude shift keying, 164
antennas, 54, 282–5, 288, 294
    types, 217–20
apogee, 49–50, 62
application specific integrated circuits
    (ASICs), 295
applications of MSS *see* aeronautical
    applications; maritime applications
Archimedes system, 50, 489–90
architecture of satellite systems, 13–22, 33,
    271–4, 341, 345, 354–65, 383
Argos navigation system, 470–1
Arianespace, 323
Asia Cellular Satellite (ACes) programme,
    284, 348, 405, 424–7
Astro Aerospace Corporation, 285
asynchronous transfer mode (ATM), 21, 31,
    199–200, 295, 519–21, 526–33
    satellite-specific, 526, 531
AT&T (company), 4–5
atmospheric drag, 53, 61
attenuation, 96–9, 114, 140–1, 221, 292–3

attenuation frequency scaling model,
    118–19
Aulin, T., 120
Automatic Dependent Surveillance-B, 477
automatic repeat request (ARQ) schemes,
    183, 192
available bit rate (ABR) service, 528
azimuth angle, 39, 45–6, 75–6

Bahl, L.R., 187
Ballard, A.H., 70, 73
bandwidth-on-demand, 533–4
baseband processing, 269–75, 282
battery technology, 286
Bell Laboratories, 4
Berlekamp-Massey decoding, 192
Bessel functions, 89
Beste, D.C., 67–9
billing, 321–2
binary phase shift keying (BPSK), 164, 170,
    208
bit error rate (BER), 127, 165–7, 171, 175,
    183, 266
block codes, 183–4, 188, 190–1
Boeing (company), 491
Brewster angle, 117, 137
broadband services, 34, 73, 327–8, 508,
    514–27
    comparison of systems, 191
broadcasting *see* Digital Audio Broadcast;
    digital broadcasting; radio broad-
    casting; television broadcasting
buffers, 17, 19, 362, 533
Burton, G.C., 61
business plans, 22, 303–7, 312, 344, 402

call barring, 325, 344
call charges, 26, 32, 146, 312, 316, 318,
    335, 529
carrier recovery, 162
carrier switching, 269–70
carrier-to-noise ratio, 266–7
celestial horizon coordinate system, 39

553